DEDICATION

To Nick Schmader, thank you for a decade of music and friendship.

CONTENTS

CHAPTER 2
Introduction to Systems Architecture 21

CHAPTER 7
Input/Output Technology   **259**

PREFACE

INTENDED AUDIENCE

This text is intended for undergraduate students majoring or concentrating in information systems (IS) and as a reference for IS professionals. The text provides a technical foundation for systems design, systems implementation, hardware and software procurement, and computing resource management. Computer hardware and system software topics that are most useful to IS students and professionals are described at an appropriate level of detail. For some topics, the reader will gain sufficient knowledge to solve technical problems. For other topics, the reader will gain sufficient knowledge to effectively communicate with technical specialists.

Computer science students are exposed to computer hardware and system software technology in many undergraduate computer science courses. Many texts designed for such computer science courses focus specifically on a subset of the topics in this book. However, coverage of computer hardware and system software in an IS curriculum is usually limited. Most IS curricula emphasize applications development, systems analysis and design, and computer resource management. A brief overview of computer hardware and system software may be provided in an introductory IS course and some specific technical topics may be covered in other courses, but there is at most one course exclusively devoted to computer hardware and system software.

The topical coverage in this text is a superset of the requirements for IS 2002.4—Information Technology Hardware and System Software in the most recent recommended IS curriculum.[1] This text covers all of the IS 2002.4 topics and includes several related topics including networks, application development software, and system administration.

The text can also serve as a supplement to other texts in system design and computer resource management courses. With respect to system design, the text covers many technical topics that must be addressed when selecting and configuring computer hardware and system software. With respect to computer resource management, the text provides the broad technical foundation needed to effectively manage computer resources.

---

[1] *IS 2002 Model Curriculum and Guidelines for Undergraduate Degree Programs in Information Systems*, published by the Association for Computing Machinery, Association for Information Systems, and Association of Information Technology Professionals, 2002, http://www.aisnet.org/Curriculum/IS2002-12-31.pdf.

READER BACKGROUND KNOWLEDGE

Because IS 2002.4 is placed early in the recommended IS curriculum, this text makes few assumptions about reader background knowledge. Unlike many computer science texts, the reader is not assumed to have an extensive background in mathematics, physics, or engineering. Where necessary, background information in these areas is presented in appropriate depth.

The reader is not assumed to know any particular programming language. However, classroom or practical experience with at least one programming language is helpful to fully comprehend the coverage of CPU instruction sets, operating systems, and application development software. Programming examples are provided in several programming languages and in pseudocode.

The reader is not assumed to have detailed knowledge of any particular operating system. However, as with programming experience, practical experience with at least one operating system is helpful. Lengthy examples from specific operating systems are purposely avoided but there are some short examples from MS-DOS, UNIX, and recent Windows versions.

The reader is not assumed to have any knowledge of low-level machine instructions or assembly language programming. Assembly instructions are described in several chapters, but a generic assembly language is used and no detailed coverage of assembly language program organization is provided.

CHANGES IN THIS EDITION

Although there are no major organizational changes in this edition, there are some significant changes to individual chapters, including:

- ► Chapter 1 – Introduced the Unified Process systems development model referred to throughout the text.
- ► Chapter 2—Updated typical computer specifications, revised definitions of computer classes, added coverage of clusters, grids, and blade servers, updated the quantum computing Technology Focus.
- ► Chapter 4—Updated discussion of RISC and the Pentium Technology Focus, added new material on benchmarks, pipelined processing, and multiprocessing.
- ► Chapter 5—Updated and expanded discussion of nonvolatile memory, new Technology Focus and Business Focus sections.
- ► Chapter 6—Added section on parallel processing covering multicore CPUs, multi-CPU architecture, and clustering.
- ► Chapter 8—Updated Technology and Business Focus sections and added a new Technology Focus covering differences between parallel and serial versions of ATA (IDE) and SCSI.

- ▶ Chapter 9—Updated Business Focus and Ethernet Technology Focus, scaled back discussion of ring topology and token passing, and added a Voice over IP (VoIP) Technology focus.
- ▶ Chapter 12—Updated RAID Technology Focus and added material on storage consolidation including network-attached storage and storage area networks.
- ▶ Chapter 13—Added J2EE Technology Focus.

In addition, there are many minor updates throughout the book to modernize the technology coverage, increase readability, and improve and update end of chapter materials.

RESOURCES FOR INSTRUCTORS

Systems Architecture, Fifth Edition includes the following resources to support professors in the classroom. These tools can be found in the Instructor's Resource Kit, which is available from the Course Technology Web site (www.course.com) and on CD-ROM.

- ▶ **Instructor's Manual** The Instructor's Manual provides materials to help professors make their classes informative and interesting. It includes teaching tips, discussion topics, and solutions to end-of-chapter materials.
- ▶ **Classroom Presentations** Microsoft PowerPoint presentations are available for each chapter of this book to assist instructors in classroom lectures or to make available to students.
- ▶ **ExamView®** ExamView® is a powerful testing software package that allows instructors to create and administer printed, computer (LAN-based), and Internet exams. ExamView® includes hundreds of questions that correspond to the topics covered in this text, enabling students to generate detailed study guides that include page references for further review. The computer-based and Internet testing components allow students to take exams at their computers, and also save the instructor time by grading each exam automatically.
- ▶ **Distance Learning** Course Technology is proud to present online content in WEBCT and Blackboard to provide the most complete and dynamic learning experience possible. For more information on how to bring distance learning to your course, contact your local Course Technology sales representative.

WORLD WIDE WEB SITES

Two support sites for this book (instructor and student) are located at http://averia.mgt.unm.edu. The sites provide:

- ▶ The Instructor's Manual
- ▶ Figure files
- ▶ Interactive end-of-chapter questions and answers
- ▶ Web resource links for most text topics and research problems
- ▶ Text updates and errata
- ▶ Hyperlinked glossary

TEXT ORGANIZATION

This text is organized into four chapter groups. The first group contains two chapters that provide overviews of computer hardware, software, and networks and describe sources of technology information. The second group contains five chapters covering hardware technology. The third group contains two chapters that cover data communications and computer networks. The fourth group contains five chapters covering software technology and system administration.

The chapters are intended for sequential coverage, though other orderings are possible. The prerequisites for each chapter are described below. Alternate chapter orders can be constructed based upon these prerequisite relationships. Chapter 2 should always be covered before other chapters. Some chapters and sections may be skipped without loss of continuity.

There should be time to fully cover between 9 and 12 chapters in a three-credit-hour undergraduate course. The text contains 14 chapters to provide flexibility in course content. Topics in some chapters may be covered in other courses in a specific curriculum. For example, Chapters 8 and 9 are often covered in a separate networking course and Chapter 14 is often covered in a separate systems administration course. The instructor can choose specific chapters to best match the overall curriculum design and teaching preferences.

CHAPTER DESCRIPTIONS

Chapter 1: Computer Technology: Your Need to Know. This chapter briefly describes how knowledge of computer technology is used in the systems development life cycle. The chapter also describes sources for computer hardware and system software information and provides a list of recommended periodicals and Web sites. The chapter can be skipped entirely or assigned only as background reading.

Chapter 2: Introduction to Systems Architecture. This chapter provides brief overviews of computer hardware, system and application software, and computer networks. The chapter describes primary classes of hardware components and computer systems and describes the differences between application and system software. The chapter introduces many key terms and concepts that are used throughout the text.

Chapter 3: Data Representation. This chapter describes primitive CPU data types and common coding methods for each type. Binary, octal, and hexadecimal numbering systems and a limited set of data structures are also described. Chapter 2 is a recommended prerequisite.

Chapter 4: Processor Technology and Architecture. This chapter describes CPU architecture and operation, including instruction sets and assembly language programming. The chapter covers traditional von Neumann architectural features, including instruction and execution cycles, word size, clock rate, registers, and instruction formats. Semiconductor and microprocessor fabrication technology are also described. Chapters 2 and 3 are necessary prerequisites.

Chapter 5: Data Storage Technology. This chapter describes primary and secondary storage implementation with semiconductor, magnetic, and optical technologies. Principles of each storage technology are described first, followed by details of storage hardware based on the technology. Chapters 3 and 4 are necessary prerequisites. Chapter 2 is a recommended prerequisite.

Chapter 6: System Integration and Performance. This chapter describes communication among computer system components and various performance enhancement methods. The chapter begins with a discussion of the system bus and bus protocols followed by a description of device controllers, mainframe channels, and interrupt processing. Performance enhancement method descriptions include buffering, caching, multiprocessing, and compression. Chapters 4 and 5 are required prerequisites, and Chapters 2 and 3 are recommended prerequisites.

Chapter 7: Input/Output Technology. This chapter describes I/O devices including keyboards, pointing devices, printers and plotters, video display terminals, video controllers and monitors, optical input devices, and audio I/O devices. The chapter also covers fonts, image representation, color representation, and image description languages. Chapter 3 is a necessary prerequisite. Chapters 2, 5, and 6 are recommended prerequisites.

Chapter 8: Data and Network Communication Technology. This chapter describes data communication technology beginning with analog and digital signals, transmission media, and bit encoding methods. Communication and protocol topics include serial and parallel transmission, synchronous and asynchronous transmission, channel sharing methods, and error detection and correction. Chapters 2 and 3 are recommended prerequisites.

Chapter 9: Computer Networks. This chapter describes computer network architecture and hardware. The chapter begins with network topology, addressing, and routing. The chapter then describes media access control and network hardware devices such as routers and switches. The chapter concludes with a discussion of IEEE and OSI networking standards and an in-depth look at TCP/IP. Chapters 3, 4, and 8 are necessary prerequisites. Chapter 2 is a recommended prerequisite.

Chapter 10: Application Development. This chapter begins with a brief overview of the software development process and software development tools. The chapter then describes programming languages, compilation, link editing, interpretation, and symbolic debugging. The final section describes application development tools including CASE tools, code generators, and integrated development environments. Chapters 2, 3, and 4 are necessary prerequisites.

Chapter 11: Operating Systems. This chapter begins with an overview of operating system architecture and resource allocation. The chapter then describes how the operating system manages the CPU, processes, threads, and memory. Chapters 2, 4, and 5 are necessary prerequisites. Chapter 10 is a recommended prerequisite.

Chapter 12: File and Secondary Storage Management. This chapter begins with an overview of file management functions, including the differences between logical and physical secondary storage access. The chapter then describes file content and structure and directories. Next, the chapter describes storage allocation, file manipulation, and access controls. The chapter concludes with file migration, backup, recovery, and networked storage. Chapters 5 and 11 are necessary prerequisites. Chapter 10 is a recommended prerequisite.

Chapter 13: Internet and Distributed Application Services. This chapter begins by discussing distributed computing and network resource access. The chapter then describes networking software starting with the lowest OSI layers. The chapter also describes services for tracking distributed resources and applications. Chapters 2, 8, and 9 are necessary prerequisites. Chapters 4, 11, and 12 are recommended prerequisites.

Chapter 14: System Administration. This chapter begins with an overview of system administration and the strategic role of hardware and software resources in an organization. The chapter then describes the hardware and software acquisition process. Next, the chapter describes requirements determination and performance monitoring methods. The next section describes various aspects of system security including access controls, auditing, virus protection, software updates, and firewalls. The last section discusses physical aspects of computer installation and operation. Chapters 2, 4, 8, 11, and 12 are recommended prerequisites.

Appendix: Measurement Units. The appendix briefly describes common measurement units, abbreviations, and usage conventions for data storage capacity and data transfer.

ACKNOWLEDGMENTS

The first edition of this text was a revision of another text entitled *Systems Architecture: Software and Hardware Concepts*, by Leigh and Ali. Some of their original work has endured through all five text editions. I am indebted to Leigh, Ali, and Course Technology for providing the starting point for all editions of this text.

I thank everyone who contributed to this edition and helped to make previous editions a success. Jim Edwards took a chance on me as an untested author for the first edition. Kristen Duerr and Jennifer Locke shepherded the text through the second, third, and fourth editions. Mac Mendelsohn and Maureen Martin oversaw the fifth edition and Eunice Yeates-Fogle, BobbiJo Frasca, Kelly Murphy, and Deb Kaufmann kept the project running smoothly. Thanks to all past and present development and production team members. And a special thanks to Deb Kaufmann for helping me improve content, clarity, and organization in the third and fifth editions, and for continually providing me with a timely mix of helpful advice, gentle prodding, and cheerful praise. Thanks also to the peer reviewers of the fifth edition: Daniel Ziesmer, San Juan College; Peter Strunk, University of Cincinnati; Tate Redding, University of Maryland; and Andrew Tarnowski, Haworth College of Business, Western Michigan University.

I thank students in the undergraduate MIS concentration at the Anderson Schools of Management, University of New Mexico, who have used manuscripts and editions of the text over the last two decades. Student comments have contributed significantly to improving the text. I also thank my faculty colleagues, Ranjit Bose, Bill Bullers, Nick Flor, David Harris, Laurie Schatzberg, and Alex Seazzu for their continued support and encouragement of my textbook-writing activities.

Finally, I'd like to thank Dee, my wife, and Alex and Amelia, my children. Developing this textbook through multiple editions has been a time-consuming process that often impinged on family time and activities. My success as an author would not be possible without my family's love and support.

Chapter 1

Computer Technology: Your Need to Know

Chapter Goals

▶ Describe the activities of information systems professionals

▶ Describe the technical knowledge of computer hardware and system software needed to develop and manage information systems

▶ Identify additional sources of information for continuing education in computer hardware and system software

The words you are reading now result, in part, from computer-based information systems: the author, editors, production team, and distribution team all relied on computer systems to organize, produce, and then convey this information. Although many different kinds of computer information systems were involved, they all share similar technology: each consists of computer hardware, system and application software, data, and communications capabilities. In this chapter you learn why you need to study hardware and software technology if you work, or plan to work, in information systems. You also learn about additional sources of information that can help to expand and keep current your knowledge of hardware and software.

TECHNOLOGY AND KNOWLEDGE

The world is filled with complex technical devices that ordinary people use every day, such as automobiles and televisions. Fortunately, we do not need detailed understanding of how these devices work in order to use them. Imagine what it would be like if three months of training were required to use a refrigerator, or if detailed understanding of mechanics and electronics were required to drive an automobile. It took years of training to use the earliest computers in the 1940s, but today, even though computers are increasingly complex and powerful, they are also easier to use. As a result, computers have proliferated beyond their original scientific applications into businesses, classrooms, and homes. If computers have become so easy to use, then why do you need to know anything about their inner technology?

ACQUIRING AND CONFIGURING TECHNOLOGICAL DEVICES

The knowledge required to purchase and configure technically complex devices is far greater than the knowledge required to use them effectively. Many people can use complex devices such as automobiles, home theatre systems, and computers, but few people feel comfortable purchasing or configuring them. Why is this so?

When you walk into a store or visit a Web site to purchase a computer, you are immediately confronted with a wide range of choices, including processor type and speed, hard disk speed and capacity, amount of memory, and operating system. To make an informed choice, you must know your preferences and requirements (for example, the application software that you plan to use and whether the computer will be upgraded or discarded in a year or two). In order to evaluate the alternatives to determine their compatibility with your preferences and requirements, you must be able to comprehend technical terms (for example, megahertz, gigabyte, IDE, and PCI), technical documentation, product and technology reviews, and the advice of friends, experts, and salespeople.

An information systems (IS) professional faces computer acquisition, upgrade, and configuration choices that are significantly more complex. Larger computer systems employ more complex technology than smaller ones. There are many more components and, therefore, more complex configuration, compatibility, and administrative issues. And, of course, the stakes are higher. Employers and users rely on the expertise of IS professionals and invest substantial sums of money based on their recommendations. Are you (or will you be) able to meet the challenge?

INFORMATION SYSTEM DEVELOPMENT

When developing an information system, IS professionals follow a series of steps called a **systems development life cycle (SDLC)**. Figure 1-1 shows a modern SDLC called the **Unified Process (UP)**. Under the UP, an information system is built in a series of 4-6 week repeated steps called **iterations** (the vertical columns separated by dashed outlines). Although Figure 1-1 shows six iterations, the number of iterations is tailored to the specifics of each development project. Typically, the first iteration or two produces documentation and a prototype system that is refined and expanded in subsequent iterations until it becomes the final system.

Each iteration includes whatever activities are needed to produce testable models or working software. Related activities are grouped into UP **disciplines**. For example, the testing discipline includes activities such as creating test data, conducting unit tests, conducting integration tests, and evaluating test results. Activities and effort within each discipline vary across iterations, as shown by the shaded curves in Figure 1-1. For example, the figure shows activities in iteration #1 drawn primarily from the business modeling, requirements, design, and deployment disciplines while activities in iteration #6 are drawn primarily from the implementation, testing, and deployment disciplines. As with the number of project iterations, the activities of each iteration are tailored to the specifics of each development project. Thus, effort in each discipline will not always be distributed across iterations exactly as depicted in Figure 1-1.

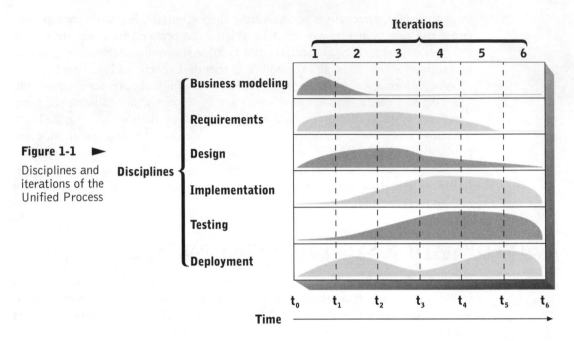

Figure 1-1 ►

Disciplines and
iterations of the
Unified Process

The following sections explore the disciplines in more detail and describe the
knowledge of computer hardware and system software required in each.

Business Modeling and Requirements Disciplines

Activities of the **business modeling discipline** and the **requirements discipline** are
primarily concerned with building models of the organization that will own and
operate the system, models of the system's environment, and models of system
requirements. The models constructed could include narratives, organization
charts, workflow diagrams, network diagrams, class diagrams, and interaction
diagrams. The purpose of building business and requirements models is to
understand the environment in which the system will function and the tasks that
the system must perform or assist users to perform.

While building business and requirements models, developers ask many
questions about the needs of the organization, users, and other constituents, and
the extent to which those needs are (or aren't) currently met and how they'll be
addressed by a new system. Technical knowledge of computer hardware and system software is required to assess the degree to which user needs are being met,
and to estimate the resources required to address unmet needs.

For example, an analyst surveying a point-of-sale system in a retail store
may pose questions about the current system, such as:

► How much time is required to process a sale?

► Is the system easy to use?

▶ Is sufficient information being gathered (for example, for marketing purposes)?

▶ Can the hardware and network handle peak sales volumes (for example, holidays)?

▶ Can the system capacity be expanded?

▶ Can hardware and application software be supported by a different operating system?

▶ Are there cheaper hardware alternatives?

Processing time and ease of use depend on the capabilities of the hardware and software and the structure of the user interface. Determining whether a system can respond to peak demand requires detailed knowledge of processing and storage capabilities, operating systems, networks, and application software. Potential for expansion depends on the unused hardware and network capacity and the limitations of both hardware and software. Determining whether cheaper alternatives exist requires technical knowledge of a wide range of hardware and software options.

Design Discipline

The **design discipline** is the set of activities that determine the structure of a specific information system that fulfills the system requirements. The first set of design activities, collectively called **architectural design**, select and describe the exact configuration of all hardware, network, system software, and application development tools that will support system development and operations (see Figure 1-2). These selections affect all other design decisions and provide a blueprint for later systems implementation.

Specific systems design tasks include selecting:

▶ Computer hardware (processing, storage, I/O, and network components)

▶ Network hardware (transmission lines, routers, firewalls)

▶ System software (operating system, database management system, network services, network protocols, security protocols and software)

▶ Application program development tools (programming languages, component libraries, integrated development environments)

Figure 1-2 ►

Design activities
of the Unified
Process

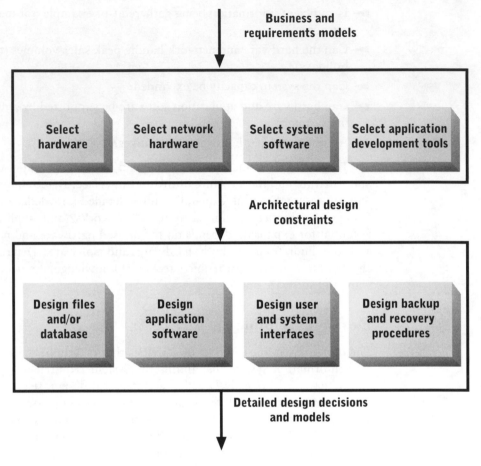

Business and
requirements models

Select
hardware

Select network
hardware

Select system
software

Select application
development tools

Architectural design
constraints

Design files
and/or
database

Design
application
software

Design user
and system
interfaces

Design backup
and recovery
procedures

Detailed design decisions
and models

Collectively, these choices define an **information architecture**—a set of requirements and constraints that define important characteristics of information processing resources and how those resources interact with one another. When actual hardware, network, and system software components are acquired and installed, they comprise an information technology infrastructure for one or more information systems. Operating and maintaining the infrastructure is a complex and costly endeavor in most organizations.

The remaining design activities, collectively called **detailed design**, are narrower in scope and are constrained by the information architecture. Detailed design activities include:

► Files or database design (such as grouping data elements into records and files, indexing, and sorting)
► Application software design

▶ User and external system interface design (input screen formats, report formats, and protocols to interact with external services and systems)

▶ Design of system backup and recovery mechanisms

Technical knowledge of computer hardware and system software is most important for completing architectural design activities. Selecting hardware and network components requires detailed knowledge of their capabilities and limitations. When multiple hardware and network components are integrated into a single system, the designer must evaluate their compatibility. The hardware, network, and overall performance requirements of the system affect the choice of system software. The designer must also consider the compatibility of new hardware, network components, and system software with the organization's existing information systems and computing infrastructure.

Selecting appropriate program development tools requires knowledge of information system requirements and the capabilities of the hardware, networks, and operating system. Development tools (and the software components built with them) vary widely in their efficiency, power, and compatibility. Tool selection will also affect future system development projects.

Implementation and Testing Disciplines

The **implementation discipline** of the UP includes all of the activities that build, acquire, and integrate application software components. The **testing discipline** includes activities that verify correct functioning of infrastructure and application software components and ensure that they satisfy system requirements. Implementation and especially testing activities require relatively specific knowledge of hardware, networks, and system software.

For example, developing an application software component that interacts with an external Web service to schedule a shipment requires specific knowledge of the network protocols used to find and interact with the service. Diagnosing an error that occurs when the software executes requires detailed knowledge of the operating system, network services, and network protocols by which the Web service request is created and transmitted and the response is transmitted and received.

Deployment Discipline

The **deployment discipline** is the set of activities that install and configure infrastructure and application software components and bring them into operation. Technical knowledge of computer hardware and system software is needed to complete many deployment tasks. Installing and configuring hardware, networks, and system software is a specialized task that requires a great deal of understanding of the components being installed and the purposes for which they will be used. Tasks such as formatting storage devices, setting up system security, installing and configuring network services, and establishing accounting and auditing controls require considerable technical expertise.

Systems Evaluation and Maintenance

Although not a formal UP discipline, systems evaluation and maintenance is nonetheless an important group of activities that accounts for much of the long range system cost. Over time, various problems with the system can and do arise. Errors that escaped detection during testing and deployment might appear. For example, the system may become overloaded because of inadequate estimates of processing volume. Information needs can change, necessitating additional data collection, storage, and/or processing.

Relatively minor system changes (such as correcting application software errors or minor processing adjustments) are normally handled as maintenance changes. Maintenance changes may require extensive technical knowledge, and some technical knowledge may be required to properly classify a proposed change as major or minor. Will new processing requirements be the "straw that breaks the camel's back" in terms of hardware, network, or software capacity? Do proposed changes require application software development tools that are incompatible with the current systems design or configuration? The answers to these questions determine whether the existing system will be modified or replaced by an entirely new system.

If the existing system is to be modified, the application software components and files to be changed are identified, modified, tested, and deployed. The specific technical knowledge requirements depend heavily on the specific hardware, network, and software components affected by the change. If an entirely new system is required, a new development life cycle will be initiated.

MANAGING COMPUTER RESOURCES

Thus far, the need for technological knowledge has been discussed in the context of developing a single information system. But consider the complexities and knowledge required to manage the hundreds or even thousands of computer resources within a large organization, in which many new development projects or system upgrades can be in progress at any one time.

Such an environment requires increased attention to two very important technological issues—compatibility and future trends. Both issues have become increasingly important as the number of computer systems and applications has increased and as information systems have become more tightly integrated. For example, accounts payable and accounts receivable programs usually share a common hardware platform and operating system. Data from both systems is often input to a financial reporting system on an entirely different computer. Also, data from many sources within the organization is often stored in a common database and accessed via a communications network.

The manager of an integrated collection of information systems and supporting infrastructure must contend with a great deal of technical complexity. He or she must ensure that each new system not only operates correctly by itself, but also operates smoothly with all of the other systems in the organization. The manager also must ensure that hardware and software acquisitions provide a good foundation for both current and future systems.

Given the rapid pace of change in computer technology, a manager must have a broad understanding of current technology and future technology trends. Will the computer purchased today be compatible with the hardware available three years from now? Can the organization's communications network be easily expanded to meet future needs? Should the organization invest only in "tried and true" technologies, or should it acquire cutting-edge technologies in hopes of greater performance or competitive advantage?

The answers to these questions require a great depth of technical knowledge—far more knowledge than any one person possesses. Typically, a manager confronted by these questions relies on the advice of experts and other sources of information. Even so, he or she must have a sufficient base of technical knowledge to understand this information and advice.

ROLES AND JOB TITLES

Many people in many different jobs are called computer professionals, and an even larger number use computers in the workplace, in school, or at home. There is a bewildering array of job titles, specializations, and professional certifications to accompany this wide range of roles. The following sections attempt to classify computer professionals into groups, explain some of their common characteristics, and describe the computer hardware and system software knowledge needed by members of each group.

Application Developers

Many computer professionals create application software for specific processing needs. These professionals have many different job titles, including programmer, systems analyst, and systems designer. Each role contributes to a different part of the systems development life cycle. A **systems analyst** performs activities of the business modeling and requirements disciplines. A **systems designer** performs activities of the design discipline and, sometimes, the deployment discipline. An **application programmer** builds and tests software.

Many computer professionals have responsibilities that do not fit neatly into these job descriptions. For example, a systems analyst is often responsible for business modeling, requirements, design, and management of a development project. Programmers often perform some requirements and design discipline tasks in addition to building and testing software.

Further confusion about job titles, activities, and responsibilities arises from differences in the types of software applications that are developed and the training required for each type. Applications can be classified loosely into three types—information processing, scientific, and technical. Information processing applications typically process business transactions or provide information to managers. Developers of these applications usually have college or technical degrees in management or business with a specialization in information processing.[1]

Scientific applications meet the data processing and numerical modeling needs of researchers in areas such as astronomy, meteorology, and physics. Technical applications usually control or directly interact with hardware devices. Examples of technical applications include robotics, flight navigation, and scientific instrumentation. Developers of scientific and technical applications typically have degrees in **computer science** or some branch of engineering.

Application developers need technical knowledge of computer hardware and system software as described earlier in the Information System Development section. Developers of technical applications typically need in-depth hardware knowledge due to the hardware control characteristics of many technical applications. Scientific application developers must also have in-depth hardware knowledge if the applications they develop push the boundaries of hardware capabilities, for example, simulating weather patterns on a global scale with cutting-edge supercomputers.

Systems Programmers

Another large class of computer professionals, called **systems programmers**, develops system software such as operating systems, compilers, database management systems, and network security monitors. Systems programmers typically have degrees in computer science or computer engineering. Organizations with large amounts of computer equipment and software employ systems programmers to perform tasks such as hardware troubleshooting and software installation and configuration. Many systems programmers are employed by organizations that develop and market systems and application development software, such as Microsoft, ORACLE, and Borland.

Systems programmers must have in-depth knowledge of system software as well as computer hardware and networks, because many types of system software, such as operating systems, directly control computer hardware or interact

---

[1] Other names for the field include management information systems, data processing, and business computer systems.

with networks. For this reason, computer science programs usually cover the subjects within this textbook in several different courses.

Hardware Personnel

Computer hardware vendors employ a variety of people for design, installation, and maintenance. Lower-level personnel usually have technical degrees and/or vendor-specific training, while higher-level personnel usually have degrees in computer science or computer engineering. Employees require extensive knowledge of computer hardware including processing, data storage, input/output, and networking devices. Hardware designers require the greatest depth of knowledge, far exceeding the scope of this text.

Systems Managers

The proliferation of computer hardware, networks, and software applications in modern organizations has created a need for large numbers of computer-related managers and administrators. The job descriptions vary widely from one organization to another due to differences in organizational structure and the nature of the organization's information systems and infrastructures. Common job titles include computer operations manager, network administrator, database administrator, and chief information officer.

A **computer operations manager** oversees the operation of a large information-processing facility. Such facilities typically have one or more large computer systems and all related peripheral equipment housed in a central location. They also usually have large databases, thousands of application programs, dozens to hundreds of employees, and significant amounts of batch processing. Organizations requiring this type of computing facility include large banks, credit reporting bureaus, the Social Security Administration, and the Internal Revenue Service. The management of day-to-day operations is extremely complex. Scheduling, staffing, security, system backups, maintenance, and upgrades are some of the more important responsibilities of a computer operations manager.

A computer operations manager can have many technical specialists on staff. Staff members usually have considerable technical knowledge within narrow specialties such as data storage hardware, data communications, mainframe operating systems, and performance tuning. A computer operations manager requires a broad base of technical knowledge to understand the organization's information systems and infrastructure. The manager also must be capable of understanding the advice of technical staff.

The title of **network administrator** is typically applied to one of two roles. The first is responsibility for the network infrastructure of an organization such as an Internet service provider or a large multinational corporation. The design, operation, and maintenance of a large network require substantial technical expertise in computer hardware, telecommunications, and system software. The role of network administrator in this environment is an important and high-level position. Technical knowledge requirements are similar to those of a computer operations manager, though the emphasis is on network and data communication technology.

In a smaller organization, the title of network administrator is applied to the manager of a local area network. Such networks connect anywhere from a half dozen to a few hundred computers (mostly microcomputers) and provide access to one or more shared databases. The network administrator can be responsible for many tasks other than operating and maintaining the network itself. These tasks might include installation and maintenance of end-user software, hardware installation and configuration, training users, and assisting management in selecting and acquiring software and hardware. This position is one of the fastest growing areas of employment for IS professionals. It also is one of the most demanding positions in terms of breadth and depth of required skills and technical knowledge.

The technology by which large collections of data, called databases, are managed and accessed is specialized and highly complex. This complexity, combined with managerial recognition of the importance of data resources, has resulted in the creation of many positions with the title of **database administrator**. This role requires both technical expertise and the ability to help the organization exploit its data resources.

A **chief information officer** (CIO) is a high-level manager of a large organization with a substantial investment in computer, network, and software technology. Many of the previously defined positions (database administrator, network administrator, and computer operations manager) report to the CIO. The CIO is responsible for the organization's computers, networks, software, and data, as well as strategic planning and the effective use of information and computing technology.

A CIO cannot possibly be an expert in every aspect of computer technology relevant to his or her organization. But the CIO must possess a broad enough base of technical knowledge to interact effectively with all the technical specialists employed by the organization. The CIO must have a vision of how technology is changing and how best to respond to those changes to support the organization's mission and objectives.

COMPUTER TECHNOLOGY INFORMATION SOURCES

This book provides a foundation of technical knowledge for a career in information system development or management. Unfortunately, that foundation will erode quickly because computer and information technology rapidly changes. How will you keep up with the changes?

There are many resources you can use to keep your knowledge current. Periodicals and Web sites can provide a wealth of information regarding current trends and technologies. Training courses offered by hardware and software vendors can teach you the specifics of current products. Additional coursework and self-study can keep you up to date on technologies and trends not specifically geared toward particular products. By far, the most important of these activities is reading periodical literature.

Periodical Literature

The volume of literature on computer topics is huge. It often is difficult to determine which sources are most important and relevant to IS professionals. Because of differences in training and focus among computer professionals, sources of information that are oriented toward one specialty are difficult reading for others. For example, a detailed discussion of user interaction and validation of requirement models targeted for IS professionals may be beyond the training of a computer scientist or engineer. Similarly, detailed descriptions of optical telecommunications theory may be beyond the training of an IS professional. These differences pose a problem for IS professionals who need current information on hardware and system software technology. The periodicals listed below are good information sources for IS professionals:

► *ACM Computing Surveys*—an excellent source of information on the latest research trends in computer software and hardware. Contains in-depth summaries of technologies or trends geared toward a readership with a moderate to high level of familiarity with computer hardware and software.

► *Computerworld*—a weekly magazine primarily geared toward computer news items. Provides coverage of product releases, trade shows, and occasional coverage of technologies and trends.

► *Communications of the ACM*—a widely used source of information about research topics in computer science. Many of the articles are highly technical and specialized, but some are targeted to a less research-oriented audience.

► *Computer*—a widely used source of information on computer hardware and software. Many of the articles are research-oriented, but it also provides occasional coverage of technologies and trends for a less technical audience.

These periodicals are only a small sample of the available literature. A complete printed list of today's recommended periodicals become out of date due to rapid change in computer technology and the computer-related publishing industry. A current list of recommended periodicals can be found on the textbook Web site (http://averia.mgt.unm.edu).

Most periodical publishers have a Web site that provides content and services beyond those provided in printed periodicals, including:

► Content from back issues of the printed periodical

► Additional current content that doesn't appear in the printed periodical

► Search engine

Web-based periodicals often provide Web links to article references and other related material. For example, an article that refers to another article in a back issue might contain a hypertext link to the back issue. Reference lists and bibliographies can also be hyperlinked. The task of accessing background and reference material is reduced to the click of a mouse. Similarly, reviews of specific software or hardware might contain links to product information and specifications maintained on vendor or manufacturer Web sites. Figure 1-3 shows the Web site for *Computerworld* magazine.

Figure 1-3 ►

Computerworld home page

Technology-Oriented Web Sites

Computer technology is a big business; there are millions of computer professionals worldwide, and many Web sites that are devoted to serving their needs. Table 1-1 lists some of the most widely used sites. Check the textbook Web site for updates to this table.

Table 1-1 ►			
Technology-related Internet sites	C/net	www.cnet.com	Oriented to consumers though it does have some content of interest to IS professionals.
	Earthweb	www.earthweb.com	Contains a broad range of information for IS professionals.
	Gartner Group	www.gartnergroup.com	A consulting and research company specializing in services for CIOs and other executive-level decision makers.
	Internet.com	www.internet.com	Specializes in Internet technology and Internet-related business. It is primarily a news channel, though it does have other resources for IS professionals.
	ITworld	www.itworld.com	Contains a broad range of information for IS professionals.
	TechWeb	www.techweb.com	Contains a broad range of information for IS professionals.

Consolidation among periodical publishers has created large corporate families of technology-related publications. Many technology-oriented Web sites are owned by, or closely associated with, periodical publishers. For example, TechWeb is owned by CMP Media, which publishes *InformationWeek*, *Network Computing*, and many other print periodicals. Technology-oriented Web sites provide a common interface to these publication families. They also enable a publisher to provide content and services that transcend a single paper-based publication, such as cross-referencing among publications, online discussion groups, employment services, and Web-based interfaces to hardware, software, and technology service vendors.

Technology Web sites can make money in several ways. A few companies charge customers directly for Web-based information and services (for example, Gartner Group). But most companies earn revenue in other ways, including:

► Advertising

► Direct sales of goods and services

► Commissions on goods and services sold by advertisers and partners

Any site that generates revenue from advertising, referrals, commissions, or preferred partner arrangements may have biased content. Some examples of bias are:

► Ordering of content, links, or search results that favor organizations that have paid a fee to the Web site owner

► Omission of information from organizations that have not paid a fee to the search provider

► Omission of information that is against the interests of organizations that have paid a fee to the search provider

Note that these same biases may be present in general Internet search engines. It is not always obvious to the reader which, if any, of these biases are present in any particular Web site.

Unbiased information is present on the Web, though it is not always easy to find. The old saying that "you get what you pay for" applies. High-quality, unbiased information is the product of intensive research. Some computer professional societies and governmental organizations do this as a public service, but much of the information you will find is produced for a profit. Expect to pay for completely unbiased information. When dealing with publicly accessible information sources, be sure to use information from multiple unrelated sources in order to balance the biases of each individual source.

Vendor and Manufacturer Web Sites

Most vendors and manufacturers of computer hardware and software have an extensive Web presence (see Figure 1-4) because most purchasers of computer hardware and software use the Web. Vendor Web pages are oriented to sales, but they usually contain detailed information on specific products either directly or as links to manufacturer Web sites. Manufacturer Web sites contain detailed information on their products and also provide technical and customer support services.

Figure 1-4 ▶

A typical computer hardware manufacturer's Web page

Manufacturers' Web sites are primarily marketing and customer support tools. They provide technical product information that is far more detailed and current than was ever available via paper-based brochures and technical documents. The breadth and depth of this information potentially allow IS professionals to make faster, better, and more informed choices. The downside is that this information is often biased in favor of the vendors' products and the technologies on which they are based. As with any source of information, you must consider the motives and objectivity of the information provider. Hardware and software manufacturers are not in the business of providing unbiased information; they are in the business of selling what they produce. You should expect the content of a manufacturer or vendor Web site to be biased toward the products it produces or sells.

In the best case, biased content may manifest itself as marketing hype that a reader must wade through or filter out to get to the real information. In the worst case, the information may be biased purposely by content, omission, or both. Many sites contain technology overviews, often called white papers, and similar information that easily is confused with independent research. A few contain links to a biased subset of supporting reviews or research. It is the reader's responsibility to balance potentially biased vendor and manufacturer information with information from unbiased sources.

Professional Societies

There are several professional societies that are excellent sources of information about computer technology. These include:

- ► *AITP*—The membership of the **Association for Information Technology Professionals** consists primarily of IS managers and application developers. AITP has local chapters throughout the country and publishes several periodicals including *Information Executive*. Their Web site is www.aitp.org.

- ► *ACM*—The **Association for Computing Machinery** is a well-established organization with a primary emphasis on computer science. ACM has dozens of special interest groups. ACM sponsors hundreds of research conferences each year and publishes many periodicals and technical books. The membership represents a broad cross section of the computer community, including hardware and software manufacturers, educators, researchers, and students. Their Web site is www.acm.org.

- ► *IEEE Computer Society*—The **Institute for Electrical and Electronics Engineers Computer Society** is a subgroup of the IEEE that specializes in computer and data communication technology. The membership is largely composed of engineers with an interest in computer hardware, but there are many members with software interests and many with academic and other backgrounds. The IEEE Computer Society sponsors many conferences (many jointly with the ACM). Its publications include several periodicals, such as *IEEE Computer*, and a large collection of technical books and standards. Their Web site is www.computer.org.

SUMMARY

► Technical knowledge of computer hardware, networks, and system software is required to develop information systems. The type and depth of required knowledge differs among disciplines of the Unified Process (UP). A broad base of knowledge is required to complete business modeling and requirements discipline activities. More in-depth knowledge is required to complete activities of the implementation, testing, and deployment disciplines.

► Technical knowledge is also required to manage an organization's information systems and infrastructure. Particular attention must be given to compatibility and future trends. Compatibility is important because computer hardware and system software typically are shared among organizational units and subsystems. Future trends must be considered in acquisitions due to the long-term nature of hardware and software investments.

► Technical knowledge must be constantly updated due to changes in hardware and software technology. IS professionals must engage in continuing education and study to keep pace with these changes. You can obtain training through vendors, educational organizations, and self-study. Self-study relies heavily on periodical literature and Web resources. You must exercise care when selecting periodical literature due to the differences in intended audience and required background and training.

► Information about computer hardware and software is readily available on the World Wide Web. Sites of interest include those maintained by publishers, vendors, manufacturers, and professional organizations. The amount of available information is substantial, but you must be cautious because the available information may be biased and/or incomplete.

In this chapter you learned why you need to understand computer technology and how you can keep that knowledge current. In the next chapter you will take a broad look at hardware, software, and networking technology. You will examine concepts and terms that will be explored in detail in the rest of the text. Your journey through the inner workings of modern information systems is about to begin.

Key Terms

application programmer
architectural design
Association for Computing
    Machinery (ACM)
Association for Information
    Technology Professionals
    (AITP)
business modeling discipline
chief information officer (CIO)
computer operations manager
computer science

database administrator
deployment discipline
design discipline
detailed design
discipline
implementation discipline
infrastructure
Institute for Electrical and
    Electronics Engineers
    (IEEE) Computer Society
iteration

network administrator
requirements discipline
systems analyst
systems designer
systems development life cycle
    (SDLC)
systems programmer
systems survey
testing discipline
Unified Process (UP)

Vocabulary Exercises

1.   Students of _____ generally focus on system software, while students of information systems generally focus on application software.

2.   Configuration of hardware and system software is an activity of the UP _____ discipline.

3.   Professional organizations with which an information systems student or professional should be familiar include _____, _____, and _____.

4.   Selection of hardware, network components, and system software is an activity of the UP _____ discipline.

5.   A(n) _____ typically is responsible for a large computer center and all of the software executed therein.

6.   Computer specialties most concerned with hardware and the hardware/software interface are _____ and computer engineering.

7.   During the UP _____ disciplines, the business, its environment, and user processing requirements are defined and modeled.

8.   The job titles of persons directly responsible for developing application software include _____, _____, and _____.

Review Questions

1. In what way(s) is the knowledge needed to operate complex devices different from the knowledge needed to acquire and configure them?

2. What knowledge of computer hardware and system software is necessary to successfully complete activities of the UP business modeling and requirements disciplines?

3. What knowledge of computer hardware and system software is necessary to successfully complete activities of the UP design and deployment disciplines?

4. What additional technical issues must be addressed when managing a computer center or local area network, as compared to developing a single information systems application?

Research Problems

1. The U.S. Bureau of Labor Statistics (BLS) compiles employment statistics for a variety of job categories and industries. BLS also produces predictions of employment trends by job category and industry. Most of the information produced by the BLS is available via the World Wide Web. Access the BLS Web site (www.bls.gov) and investigate the current and expected employment prospects for IS professionals.

2. Assume that you are an IS professional and you have been asked by your boss to prepare a briefing for senior staff on the comparative advantages and disadvantages of three competing tape drive technologies—digital linear tape (DLT), Advanced Intelligent Tape (AIT), and Linear Tape Open (LTO). Search each of the technology Web sites in Table 1-1 for source material that may aid you in preparing the briefing. Which site(s) provide the most useful information? Which site(s) allow you to find useful information with relative ease?

3. Read several dozen job listings at www.itcareers.com or a similar site. Which jobs (and how many) require a broad knowledge of computer technology? At what level are those jobs? Try searching the listings based on a few specific keywords, such as programmer or network. Look closely at the hiring companies and where their job postings show up in the search results. Can you draw any conclusions about how the listings are ordered?

Chapter **2**

Introduction to Systems Architecture

Chapter Goals

- Discuss the development of automated computing

- Describe the general capabilities of a computer

- Describe computer system components and their functions

- List computer system classes and their distinguishing characteristics

- Define the role and function of application and system software

- Describe the economic role of system and application development software

- Describe the components and functions of computer networks

Computer systems are complex combinations of hardware, software, and network components. The term **systems architecture** describes the structure, interaction, and technology of computer system components. The term architecture is a misnomer, because it implies a concern with only static structural characteristics. In this text you also learn about the dynamic behavior of a computer system, that is, how its components interact as the computer operates.

This chapter lays the foundation for the text with a brief discussion of the major components and functions of hardware, software, and networks. Each component is described in detail in later chapters. Avoid the temptation to rush through this chapter, because it is as important to know how all the components of a computer system interrelate as it is to know their internal workings.

AUTOMATED COMPUTATION

A simple definition of a computer is any device that can:

► Accept numeric inputs
► Perform computational functions, such as addition and subtraction
► Communicate results

This definition captures the basic functions of a computer, but it also can apply to people and simple devices, such as calculators. These functions can be implemented by many methods and devices. For example, a modern computer may implement computation electronically (using transistors within a microprocessor), implement storage optically (using a laser and the reflective coating on an optical disc), and implement communication using a combination of electronics and mechanics, for example, the mechanical and electrical components of a printer. Some experimental computers have even used quantum physics to implement data storage and computation.

Mechanical Implementation

Early mechanical computation devices were built to perform repetitive mathematical calculations. The most famous of those machines was the difference engine, built by Charles Babbage in 1821 (see Figure 2-1), which computed logarithms by moving gears and other mechanical components. Many other mechanical computation machines were developed well into the twentieth century. Mechanical computers were used during World War II to compute firing solutions for naval guns and torpedoes. Mechanical adding machines were commonly used by bookkeepers and accountants as late as the 1970s.

Figure 2-1 ▶

Charles
Babbage's
difference engine

© 2004 IEEE

The common element in all such computation devices is a mechanical representation of a mathematical calculation. A mechanical clock driven by a spring and pendulum. Each swing of the pendulum allows a gear to move one step under pressure from the spring. As the pendulum swings, the gears advance the hands of the clock. The user inputs the current time by manually adjusting the hour and minute hands. Starting the pendulum in motion activates a calculation that repeatedly increments the current time and displays the result using the clock hands and numbers printed on the clock face.

Mechanical computation devices can also perform more complex calculations. Multiplication of whole numbers, for example, can be mechanically implemented as repeated addition. A machine capable of addition can perform multiplication by executing the addition function multiple times (6 times 3 can be calculated by adding 6 plus 6, temporarily storing the result, and then adding 6 a third time). Appropriate combinations of moving parts also can be used to implement complex functions such as logarithms and trigonometric functions.

The inherent limitations and shortcomings of mechanical computation include:

▶ Complex design and construction
▶ Wear, breakdown, and maintenance of mechanical parts
▶ Limits on operating speed

Automated computation with gears and other mechanical parts requires a complex set of components that must be designed, manufactured, and assembled to exacting specifications. As the complexity of the computational function increases, the complexity of the mechanical device that performs it also increases, exacerbating problems of design, construction, wear, and maintenance.

Electronic Implementation

Much as the era of mechanical clocks gave way to the era of electrical clocks, the era of mechanical computation eventually gave way to electronic computers. The biggest impetus for the change to electronic computing devices came during World War II. The military needed to solve many complex computational problems such as navigation and breaking enemy communication codes. The mechanical devices of the time were simply too slow and unreliable.

In an electronic computing device, the movement of electrons performs essentially the same functions as gears and wheels in mechanical computers. Numerical values are stored as magnetic charges or by positioning electrical switches, rather than by the position of gears and wheels. When necessary, electromechanical devices convert physical movement into electrical signals, or vice versa. For example, a keyboard converts the mechanical motion of keystrokes into electrical signals. The ink pumps in an ink-jet printer convert electrical signals into mechanical motion to force ink through a nozzle and onto paper.

Electronic computers addressed most of the shortcomings of mechanical computation. They were faster due to the high speed of moving electrons, and as electronic devices and fabrication technology improved, they became more reliable and easier to build than their mechanical counterparts. Electronic computers made it possible to perform complex calculations at speeds previously thought impossible. Larger and more complex problems could be addressed and simple problems could be solved much faster.

Optical Implementation

Light can also be used as a basis for computation. A particle of light, called a photon, moves at a high rate of speed. As with electrons, the energy of a moving photon can be harnessed to perform computational work. Light can be transmitted over appropriate conductors, such as laser light through a fiber-optic cable. Data can be represented as pulses of light and stored either directly (such as an image stored as a hologram) or indirectly by materials that reflect or don't reflect light (such as the surface of a CD or DVD).

Optical data communication is now common in computer networks that cover relatively large distances. Optical devices are gradually replacing electronic and magnetic storage devices. Some input/output devices (laser printers and optical scanners) are based on optical technologies and devices. Optical and hybrid electro-optical devices connect system components in some experimental and high-performance computers. Experimental optical computer processors have already been developed, and optical and hybrid electro-optical technologies will find much wider application in the computer hardware of the twenty-first century.

Technology
Focus

Quantum Computing

Current computer technology is based on principles of classical physics developed during the seventeenth through twentieth centuries including electronics, magnetism, and optics. These principles are based on mathematical rules that describe the behavior of matter at the level of atoms, molecules, and larger units. By manipulating matter at these levels, mechanical, electrical, and optical processors perform the mathematical functions that underlie classical physics.

Quantum physics describes the behavior of matter at a subatomic level, which is very different from the behavior described by classical physics. For example, in classical physics, an atom or molecule has a specific position in space at any point in time. In quantum physics, a subatomic particle such as a photon can be in multiple places at one time. Larger particles can also be in multiple states at once, though we can only observe one of those states at any given time.

As with classical physics, quantum physics describes subatomic behavior with mathematical rules. The rules differ from classical physics and are often much more complex, but they are still rules based on specific computations. In theory, a processor that manipulates matter at the quantum level can perform the calculations of quantum mathematics.

In a modern digital computer, data is represented by groups of bits, each having a clearly defined binary physical state. For example, a bit value might be represented by a switch that is open or closed or an atom that has positive or negative electrical charge. Each bit has two possible states, which represent the values zero and one. Each bit, obeying the rules of classical physics, must represent either zero or one at any specific time. A processor that manipulates the bit using principles of classical physics "sees" only one physical state and produces only one computational result per bit.

At the quantum level, matter can be in multiple states at the same time. Just as a photon can be in two places at once, an atom can be both positively and negatively charged at the same time. In effect, the atom stores both zero and one at the same time. Such an atom, or any other matter that stores data in multiple simultaneous quantum states, is called a **qubit**.

At first glance, a qubit may seem little more than a physicist's toy. But the ability to store two data values at once has significant advantages when tackling certain types of computational problems such as cryptography. One advantage arises from greater storage capacity. A group of three classical bits can store only one of 8 (2^3) possible values at a time. A group of three qubits can store all 8 possible values at once, an eightfold increase in storage capacity.

Another advantage arises from computational efficiency. A computer that can manipulate 3 qubits at the quantum level can perform a calculation on all eight values at the same time, producing eight different results at once. A conventional computer must store the eight values separately and perform eight different computations to produce eight results. As the number of qubits increases, so does the comparative efficiency of a quantum computer. For example, a 64-bit quantum computer is 2^{64} times more efficient than a 64-bit conventional computer.

A number of prototype components for quantum computing have already been built, though a fully functional quantum computer has yet to be publicly demonstrated. Figure 2-2 shows two prototype quantum cryptography systems deployed in 2004 to encrypt messages sent across the Internet.

Figure 2-2 ►

Prototype quantum cryptography systems

© 2004 IEEE

COMPUTER CAPABILITIES

All computers are automated computing devices, but not all automated computing devices are computers. The primary characteristics that distinguish a computer from other automated computation devices include:

► General-purpose processor capable of performing computation, data movement, comparison, and branching functions
► Storage capacity sufficient to hold large numbers of program instructions and data
► Flexible communication capability

Each of these characteristics is considered in the following sections.

Introduction to Systems Architecture

Processor

A **processor** is a device that performs data manipulation and transformation functions including:

► Computation (addition, subtraction, multiplication, and division)

► Comparison (less than, greater than, equal to, and not equal to)

► Data movement among memory, mass storage, and input/output devices

An **instruction** is a signal or command to a processor to perform one of its functions. When a processor performs a function in response to an instruction, it is said to be **executing** that instruction.

Each instruction directs the processor to perform one simple task (for example, add two numbers). The processor performs complex functions by executing a sequence of instructions. For example, to add a list of 10 numbers, a processor is instructed to add the first number to the second and to store the result temporarily. It is then instructed to add the stored result to the third number and to store that result temporarily.

More instructions are issued and executed until all 10 numbers have been added. Most useful computational tasks, such as recalculating a spreadsheet, are accomplished by executing a long sequence of instructions called a **program**. A program is a stored set of instructions that implements a specific task, such as calculating payroll and then generating paychecks or electronic fund transfers. Programs can be stored and reused over and over again.

A processor can be classified as either general-purpose or special-purpose. A **general-purpose processor** can execute many different instructions in many different sequences or combinations. A general-purpose processor can be instructed to do many different tasks such as payroll calculation, text processing, or scientific calculation simply by supplying it with an appropriate program.

A **special-purpose processor** is designed to perform only one specific task. In essence, it is a processor with a single internal program. Many commonly used devices such as microwave ovens, compact disc players, and computer printers contain special-purpose processors. Although such processors or the devices that contain them can be called computers, the term computer usually refers to a device containing a general-purpose processor that can execute a variety of programs.

Formulas and Algorithms   Some processing tasks require little more than a processor's computation instructions. For example, consider the following calculation:

GROSS_PROFIT = (QUANTITY_SOLD × SELLING_PRICE) – SELLING_EXPENSES

The only processor functions needed to compute GROSS_PROFIT are multiplication, subtraction, and storing and accessing intermediate results. In statement 30 (see Figure 2-3), QUANTITY_SOLD and SELLING_PRICE are multiplied and the result is stored temporarily as INTERMEDIATE_RESULT. In statement 40, SELLING_EXPENSES is subtracted from INTERMEDIATE_RESULT. The GROSS_PROFIT calculation is a **formula**. A processor executes a sequence of computation and data movement instructions to solve a formula.

Figure 2-3 ▶

A BASIC
program to
compute gross
profit

```
10      INPUT QUANTITY_SOLD

20      INPUT SELLING_PRICE

30      INTERMEDIATE_RESULT = QUANTITY_SOLD * SELLING_PRICE

40      GROSS_PROFIT = INTERMEDIATE_RESULT - SELLING_EXPENSES

50      OUTPUT GROSS_PROFIT

60      END
```

Computer processors also can perform a more complex class of processing tasks called algorithms. An **algorithm** is a program in which different sets of instructions are applied to different data input values. The program must make one or more decisions to determine what instructions to execute in order to produce correct data output values. Depending on the data input values, entirely different subsets of instructions may be executed. In contrast, all of the instructions that implement a formula are always executed in the same order regardless of the data input.

The procedure for computing United States income tax is an example of an algorithm. Figure 2-4 shows a set of income tax computation formulas. Note that different income values require different formulas in order to calculate the correct amount of tax. A program that computes tax based on this table must execute only the set of instructions that implements the appropriate formula for a particular income value.

Figure 2-4 ▶

A tax table

Single—Schedule X

If gross pay is:		The tax is:	of the amount over—
Over—	But not over—		
$0	$7,150	-------------- 10%	$0
7,150	29,050	$715.00 + 15%	7,150
29,050	70,350	4,000.00 + 25%	29,050
70,350	146,750	14,325.00 + 28%	70,350
146,750	319,100	35,717.00 + 33%	146,750
319,100	-----------	92,592.50 + 35%	319,100

Comparisons and Branching Decisions within a processing task are based on numerical comparisons. Each numerical comparison is called a **condition**, and the result of evaluating a condition is either true or false. In the tax example, the income value is compared to the valid income range for each formula. Each formula is implemented as a separate instruction or set of instructions within the program. When a comparison condition is true, the program **jumps** or **branches** to the first instruction that implements the corresponding formula.

Figure 2-5 illustrates a BASIC program that uses comparison and branching instructions to calculate income taxes. In statements 20, 50, 80, 110, and 140, INCOME is compared to the maximum income applicable to a particular tax calculation formula. The comparison result (either true or false) determines which instruction is executed next. If the comparison condition is false, then the next program instruction is executed. If the comparison condition is true, then the program jumps to a different point in the program by executing a GOTO, or branch, instruction.

Figure 2-5 ▶

A BASIC program to calculate income taxes

```
10  INPUT INCOME
20  IF INCOME > 7150 THEN GOTO 50
30  TAX = INCOME * 0.10
40  GOTO 180
50  IF INCOME > 29050 THEN GOTO 80
60  TAX = 715.00 + (INCOME - 7150) * 0.15)
70  GOTO 180
80  IF INCOME > 70350 THEN GOTO 110
90  TAX = 4000.00 + (INCOME - 29050) * 0.25)
100 GOTO 180
110 IF INCOME > 146750 THEN GOTO 140
120 TAX = 14325.00 + (INCOME - 70350) * 0.28)
130 GOTO 180
140 IF INCOME > 319100 THEN GOTO 170
150 TAX = 35717.00 + (INCOME - 146750) * 0.33)
160 GOTO 180
170 TAX = 92592.50 + (INCOME - 319100) * 0.35)
180 OUTPUT TAX
190 END
```

Comparison instructions are part of a group of **logic instructions**, implying a relationship to intelligent, decision-making behavior. A general-purpose processor within a computer is more restricted in its comparative abilities than a human being. A person can compare complex objects and phenomena and also handle uncertainty in the resulting conclusions, whereas a computer can perform only simple comparisons (equality, less than, and greater than) with numeric data, where the results are either completely true or completely false. Despite these limitations, comparison and branching instructions are the building blocks of all computer intelligence. They are also the capabilities that distinguish a computer processor from the processors of simpler, automated computation devices such as calculators.

Storage Capacity

A computer stores a variety of information, including:

► Intermediate processing results
► Data
► Programs

Because computers break complex processing tasks into small parts, a computer needs to store intermediate results (for example, the variable INTERMEDIATE_RESULT on line 30 in Figure 2-3). Programs that solve more complex, real-world problems may generate and access hundreds, thousands, or millions of intermediate results during execution.

Larger units of data such as customer records, transactions, and student transcripts must also be stored for present or future use. A user may need to store and access thousands or millions of data items, and a large organization may need to store and access trillions of data items. This data may be used by currently executing programs, held for future processing needs, or held as an historical record.

Programs must also be stored for present and future use. A simple program can contain thousands of instructions. Complex programs can contain millions or billions of instructions. A small computer may store thousands of programs, and a large computer may store millions of programs.

Storage devices vary widely in characteristics such as cost, access speed, and reliability. A computer uses a variety of storage devices because each device provides a set of storage characteristics best suited to a particular type of data. For example, program execution speed is increased when the processor can rapidly access intermediate results, current data inputs, and instructions. Thus, those data types are stored in devices with high access speed but relatively high cost. Programs and data held for future use can be stored on slower and less expensive storage devices. Data that must be physically transported is stored on removable media storage devices, such as CD-ROMs or USB memory keys, which are slower and less reliable than fixed media devices.

Input/Output Capability

Processing and storage capabilities are of little use if a computer cannot communicate with users or other computers. A computer's input/output devices must encompass a variety of communication modes—sound, text, and graphics for humans and electronic or optical communication for other computers. A typical small computer has up to a dozen input/output devices, including a video display, keyboard, mouse, printer, and modem or network interface. A large computer can have many more input/output devices (for example, multiple printers and network interfaces) and may use substantially more powerful and complex devices than a smaller computer.

Introduction to Systems Architecture

COMPUTER HARDWARE

Not surprisingly, computer hardware components parallel the processing, storage, and communication capabilities just outlined (see Figure 2-6). Computer hardware has four major functions, including:

► Processing—executing computation, comparison, and other instructions to transform data inputs into data outputs

► Storage—storing program instructions and data for temporary, short-term, and long-term use

► External communication—communicating with entities outside the computer system, including users, system administrators, and other computer systems

► Internal communication—transporting data and instructions among internal and peripheral hardware components such as processors, disk drives, video displays, and printers

Figure 2-6 ►

The primary functions of computer hardware

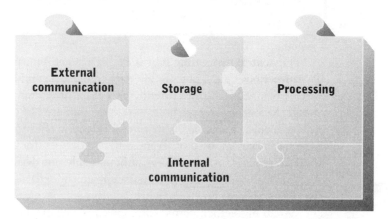

All computer systems include hardware to implement each function. However, each function is not necessarily implemented within a single device. For example, a computer processor implements processing and some storage functions, and external communication is implemented by many different hardware devices, such as keyboard, video display, modem, network interface device, sound generating device, and speakers.

Figure 2-7 shows the hardware components of a computer system. The number, implementation, complexity, and power of these components can vary substantially from one computer system to another, but the functions performed are similar.

Figure 2-7 ▶

The components
of a computer
system

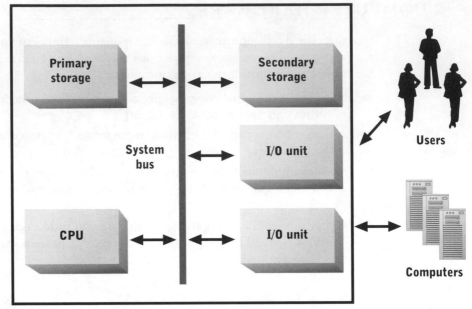

The **central processing unit** (CPU) is a general-purpose processor that executes all instructions and controls all data movement within the computer system. Instructions and data for currently executing programs flow to and from **primary storage**. **Secondary storage** holds programs that are not currently being executed, as well as groups of data items that are too large to fit in primary storage. Secondary storage can be composed of several different devices (for example, multiple hard disk drives, Zip drives, or tape drives), although only one device is shown in Figure 2-7.

Various **input output** (**I/O**) **units** implement external communication functions. Two I/O units are shown in Figure 2-7, though a typical computer has many more. The **system bus** is the internal communication channel that connects all other hardware devices.

Central Processing Unit

Figure 2-8 illustrates the components of the central processing unit, including the:

▶ Arithmetic logic unit
▶ Registers
▶ Control unit

Figure 2-8 ▶

Components of
the central
processing unit
(CPU)

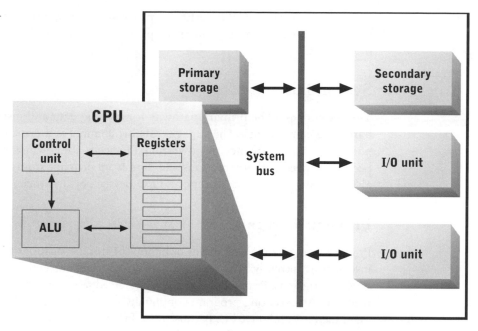

The **arithmetic logic unit** (**ALU**) contains electrical circuits that implement each instruction. A CPU can implement dozens or hundreds of different instructions. Simple arithmetic instructions include addition, subtraction, multiplication, and division. More advanced computation instructions such as exponentiation and logarithms also can be implemented. Logic instructions include various forms of comparison (equal to, greater than, less than) and other instructions that will be discussed in Chapter 4.

The CPU contains a small number of internal storage locations called **registers** that can each hold a single instruction or data item. Registers store data or instructions that are needed immediately or frequently. For example, two numbers that are about to be added together are each stored in a register. The ALU reads these numbers from the registers and stores the sum in another register. Because they are located within the CPU, the contents of registers can be accessed quickly by the other CPU components.

The **control unit** has two primary functions:

▶ Control the movement of data to and from CPU registers and other hardware components

▶ Access program instructions and issue appropriate commands to the ALU

As program instructions and data are needed, they are moved from primary storage to registers by the control unit. The control unit examines incoming instructions to determine how they should be processed. Computation and logic instructions are routed to the ALU for processing. The control unit executes data movement instructions to and from primary storage, secondary storage, and I/O devices.

System Bus

The system bus is the primary pathway for moving data and instructions among hardware components. The capacity of this channel is a critical factor in computer system performance. A powerful CPU needs a high-capacity system bus to keep it supplied with instructions and data from primary storage. Bus capacity is also critical to secondary storage and I/O device performance.

Primary Storage

Primary storage contains millions or billions of storage locations that hold program instructions currently being executed, as well as data currently being processed by those instructions. Primary storage is also referred to as **main memory,** or simply as **memory**. An executing program continually moves instructions and data between main memory and the CPU. Because the CPU is a relatively fast device, main memory devices must be capable of rapid access.

In current computer hardware, main memory is implemented with silicon-based semiconductor devices commonly called **random access memory (RAM)**. RAM provides the access speed required by the CPU and allows the CPU to read or write to specific memory locations. Unfortunately, RAM is relatively costly and this often limits the amount of main memory that can be included in a computer system.

Another problem with RAM is that it does not provide permanent storage. When the power is turned off, the contents of RAM are lost. This characteristic of RAM is called **volatility**. Any type of storage device that cannot retain data values indefinitely is said to be volatile. In contrast, storage devices that permanently retain data values are said to be nonvolatile. Due to both the volatility and limited capacity of primary storage, a computer system must have other devices to store data and programs over long periods of time.

Secondary Storage

Secondary storage is composed of high-capacity nonvolatile storage devices that hold:

► Programs not currently being executed

► Data not needed by currently executing programs

► Data needed by currently executing programs that does not fit within available primary storage

Within a typical information system, the number of programs and amount of data is quite large. A typical computer system must have a large amount of secondary storage capacity in comparison to its primary storage. For example, a microcomputer may have 1 billion primary storage locations and 100 billion secondary storage locations. Differences in the content and implementation of the various storage devices are summarized in Table 2-1.

Table 2-1 ►

Comparison of
storage types

Storage type	Implementation	Contents
CPU registers	Very-high-speed electronic devices within the CPU	Currently executing instruction; a few dozen items of related data
Primary storage	High-speed electronic devices (RAM) outside of the CPU	Programs currently being executed; hundreds to billions of data items
Secondary storage	Low-speed electromagnetic or optical devices (for example, magnetic and optical disc)	Programs not currently being executed; billions to trillions of data items

Relatively slow devices implement secondary storage, keeping the total cost within acceptable limits. The most common secondary storage devices are magnetic disks, optical discs, and magnetic tape. Magnetic disks provide relatively fast access, compared with optical discs and magnetic tapes. Optical discs provide greater storage capacity than magnetic disks at reduced cost. Magnetic tape provides the slowest but cheapest method of storage.

Input/Output Devices

The variety of input/output devices in modern computer systems addresses the many different forms of human-to-computer and computer-to-computer communication. From an architectural perspective, each I/O device is a separate hardware component attached to the system bus. I/O devices all interact similarly with other computer hardware devices, but their internal implementation varies depending on the exact form of communication they support.

I/O devices can be classified broadly into human-oriented and computer-oriented communication devices. Examples of input devices for human use include keyboards, pointing devices such as a mouse or trackball, and voice-recognition devices. The purpose of such devices is to accept input from a person (voice, touch, or physical movement) and convert that input into something the computer can understand (electrical signals). Output devices for human use include video displays, printers, and devices for speech and sound output. All of these devices convert electrical signals into a format that a person can understand, such as pictures, words, or sound. Computer-oriented input/output devices include modems and network interface units. These devices implement communication among computer systems, or between a computer and a distant I/O device.

COMPUTER SYSTEM CLASSES

Computer systems are available in many configurations that vary in CPU power, storage capacity, I/O capacity, number of simultaneous users, and intended application software. Computer systems can be loosely classified into the following classes:

► Microcomputer

► Midrange computer

► Mainframe

► Supercomputer

A **microcomputer** is a computer system designed to meet the information processing needs of a single user. A microcomputer also can be called a **personal computer (PC)** or **workstation**. Portable computers such as laptop, notebook, and handheld computers are also microcomputers. Examples of processing performed with microcomputers include word processing, computer games, and small- to medium-sized application programs, such as programs to compute an individual's income tax, perform budgeting for a home or small business, and compute payroll for a small business.

To some people, the terms personal computer and workstation are interchangeable. To others, the term workstation implies something more powerful than a typical microcomputer. This is especially true of microcomputers in scientific and engineering organizations. Processing tasks in such organizations require more computer hardware power than do typical business and home processing. Examples of hardware-intensive tasks include complex mathematical computation, computer-aided design (CAD), and manipulation of high-resolution video images. The power of a workstation often is similar to that of a midrange computer, but the overall design of a workstation is targeted toward a single-user operating environment.

A new type of microcomputer emerged in the late 1990s. A **network computer** is a microcomputer with minimal secondary storage capacity and little or no installed software. An ordinary microcomputer loads its operating system and application software from secondary storage devices directly attached to the computer. A network computer establishes a connection to a server when it is powered on. The server transfers operating system and application software via a computer network. Network computers are slightly less expensive than ordinary microcomputers. Server-based software installation and configuration make network computers easier to reconfigure than ordinary microcomputers.

A **midrange computer**, sometimes called a **minicomputer**, is designed to provide information processing for multiple users and to execute many application programs simultaneously. Supporting multiple users and programs requires fairly powerful processing, storage, and input/output subsystems. It also requires more sophisticated system software than is typically installed on microcomputers.

There are many ways of supporting multiple users. They can be connected to a computer system with relatively simple I/O devices such as video display terminals. In this case, all processing, data storage, and network communication is handled by the shared computer system. Multiple users can also share one or more resources (for example, printers, databases, and Web pages). A midrange computer can support several dozen people using video display terminals, or respond to a few hundred simultaneous requests for shared resources.

A **mainframe** computer system handles the information processing needs of a large number of users and applications. A mainframe can support hundreds of people using video display terminals and execute hundreds or thousands of programs at one time. A mainframe can also respond to thousands of simultaneous requests for shared resources. A typical use may involve 250 users entering customer orders, several programs generating periodic reports, dozens of users querying the contents of a large corporate database, and an operator making backup copies of disk files—all at the same time.

The feature that distinguishes mainframes from other computer classes is the ability to store large quantities of data and move it from one place to another quickly. Data can be moved among up to a dozen CPUs, several dozen secondary storage devices, and hundreds or thousands of users connected via a network. Fast CPUs and large amounts of primary and secondary storage also are required, but a mainframe is primarily optimized for rapid and efficient data movement.

A **supercomputer** is designed for one purpose—rapid mathematical computation (billions, trillions, or more computations per second). Supercomputers are used for computationally intensive applications such as simulation, 3-D modeling, weather prediction, computer animation, and real-time analysis of large databases. Such tasks require multiple CPUs with the highest possible computational speed. Storage and communications requirements also are extremely high but are secondary to computational speed. Supercomputers employ the very latest (and most expensive) computer technology.

The term **server** can describe computers as small as microcomputers and as large as supercomputers. The term does not imply any minimum set of hardware capabilities; rather, it implies a specific mode of use. A server is a computer system that manages one or more shared resources such as file systems, databases, Web sites, printers, and high-speed CPUs, and allows users to access those resources over a local- or wide-area network.

Server hardware capabilities depend on the resources being shared and the number of simultaneous users. For example, an ordinary microcomputer may be more than sufficient for sharing a file system and two printers among a dozen users on a local area network. A mainframe server may be required for sharing more demanding resources like large databases accessed by thousands of users. Supercomputer servers are sometimes used to augment the computational capabilities of workstations used for computer-aided design and animation.

Table 2-2 summarizes the configuration and capabilities of each class of computer system. Each class is represented by a typical model available in 2004. Specifications and performance are in a constant state of flux. Rapid advances in computer technology lead to rapid improvements in performance as well as a redefinition of both the classes themselves and the specifications of computers within each class. There is also a technology pecking order: that is, the newest and most expensive hardware and software technology usually appears first in supercomputers. As experience is gained and costs decrease, those technologies move downward to the less powerful and less costly computer classes.

Table 2-2 ▶

Representative products in various computer classes (2004)

Class	Typical Product	Typical Specifications	Approximate Cost	CPU(s)
Micro	Dell Dimension 4700	1 billion main memory cells 80 billion disk storage cells DVD ROM drive	$1,500	1
Workstation	Dell Precision 670	4 billion main memory cells 300 billion disk storage cells high speed storage subsystem rewritable DVD drive high-speed graphics processor	$8,250	2
Midrange	Dell PowerEdge 6600	8 billion main memory cells 900 billion disk storage cells high speed storage subsystem redundant power supply tape drive	$26,500	2
Mainframe	Hewlett-Packard AlphaServer GS1280 Model 8	32 billion main memory cells 2 trillion disk storage cells 2 high capacity I/O channels 2 high capacity network channels tape drive	$770,000	8
Super	Hewlett-Packard AlphaServer GS1280 Model 64	256 billion main memory cells 1 trillion disk storage cells	$5,600,000	32

Another factor that blurs the lines among various classes of computers is the ability to configure subsystems such as processors, primary and secondary storage, and input/output. For example, microcomputers can support some multiuser applications simply by upgrading the capacity of a few of the subsystems and utilizing more powerful system software.

Introduction to Systems Architecture

Technology
Focus

IBM POWER-Based Computer Systems

International Business Machines (IBM) first introduced the POWER CPU architecture in 1990. With this CPU, IBM broke away from its previous designs and laid the foundation for its future computer systems. Some microcomputers manufactured by Apple Computer, Incorporated include POWER CPUs. IBM has incorporated POWER CPUs in current systems ranging from workstations to supercomputers.

The POWER architecture has been through several generations and variations since 1990. The latest generation of the architecture is called the POWER5 and it incorporates IBM's latest microprocessor technologies, including:

▶ 0.13 micron fabrication technology

▶ Copper chip wiring

▶ Large on-chip and off-chip memory caches

▶ Simultaneous multithreading

Sample computer systems based on POWER CPUs are summarized in Table 2-3. The Intellistation p630 is a graphical workstation intended for 3-D modeling and visualization. Typical applications include computer-aided design, computer-aided animation, and data visualization. The system incorporates sophisticated graphical I/O hardware to speed image display and refresh.

Table 2-3 ▶

IBM system configurations (2004) for POWER CPUs

Model	System Type	Number of CPUs	CPU Type and Speed	Memory Capacity	Disk Capacity
Intellistation p630 6E4	Workstation	2	POWER4 1.2 GHz	2 billion bytes (GB)	72 GB
pSeries 670	Midrange	4	POWER4 1.5 GHz	4 GB	72 GB
pSeries 690	Mainframe	16	POWER5 1.7 GHz	128 GB	up to 4.5 TB
1600 Cluster	Supercomputer	2176	POWER5 1.9 GHz	2.75 trillion bytes (TB)	25 TB

The pSeries 670 is a midrange computer based on two or four POWER5 processors. A single POWER5 processor can operate as two virtual processors via a technique IBM calls simultaneous multithreading (a similar technique in Intel's advanced processors is called hyperthreading). Internal disk capacity is limited to 72 GB, but the system can be expanded well beyond those limits with external storage units. The pSeries 670 is primarily intended as a server for computationally intensive applications such as factory floor control and data analysis.

The pSeries 690 is a mainframe or supercomputer depending on its configuration. Table 2-3 shows a high-end mainframe configuration for computationally complex applications such as data mining and analysis. The specifications for the 1600 Cluster describe a computer system installed at the European Centre for Medium-Range Weather Forecasts. The cluster contains 68 pSeries 690 computers.

Multicomputer Configurations

A final factor that makes distinctions among computer classes is the trend toward multicomputer configurations. Simply put, a **multicomputer configuration** is any organization of multiple computers to support a specific set of services or applications. There are various common multicomputer configurations including clusters, blades, and grids.

A **cluster** is a group of similar or identical computers, connected by a high-speed network, that cooperate to provide services or execute a common application. A modern Web server farm is one example of a cluster. Incoming Web service requests are routed to one of a set of Web servers, each of which has access to the same content. Any server can respond to any request, so the load is distributed among the servers. In effect, the cluster acts as a single large Web server.

The advantages of a cluster include scalability and fault tolerance. Overall Web servicing capacity can be expanded or shrunk by adding or subtracting servers from the cluster. This is usually simpler than adding or subtracting capacity from a single larger computer system. Also, if any one server fails, the remaining servers continue to function, thus maintaining service availability at a reduced capacity. The primary disadvantages of clusters are complex configuration and administration.

A **blade** is a circuit board that contains most of a server computer. Typically, a blade has one or more CPUs, memory, and network interfaces. It lacks secondary storage, external I/O connections, and a power supply. Up to a dozen or so blades can be installed in a single cabinet, which also contains a shared power supply and external I/O connections. Secondary storage is typically provided by a storage server in close proximity to the blade cabinet.

In essence, a blade is a specialized cluster. It has the same advantages and disadvantages as a cluster though it usually simpler to modify a cluster of blades than a cluster of standalone computer systems. Blades also concentrate more computing power in less space than a typical cluster.

A **grid** is a group of dissimilar computer systems, connected by a high-speed network, that cooperate to provide services or execute a common application.

Besides dissimilarity of the component computers, grids have two other differences from clusters:

- Computers in a cluster are typically located close to one another (for example, in the same room or rack). Computers in a grid may be in separate rooms, buildings, or continents.
- Computers in a cluster work exclusively on the same set of services or applications. Computers in a grid work cooperatively at some times and independently at others.

Grids are typically implemented by installing software on each machine that accepts tasks from a central server and performs them when not busy doing other work. For example, all of the desktop computers in a building could form a grid that performs computationally intensive tasks at night or on weekends. Grids of mainframe or supercomputers are sometimes used to tackle computationally intensive research problems.

Bigger Isn't Always Better

In 1952, computer scientist H. A. Grosch asserted that computing power, as measured by millions of instructions per second (MIPS), is proportional to the square of the cost of hardware. According to Grosch, large and powerful computers will always be more cost-effective than smaller ones. The mathematical formula that describes this relationship is called **Grosch's Law**. For many years, computer system managers pointed to this law as justification for investments in ever larger computers.

Many modern trends in computer design and use have combined to rewrite Grosch's Law, including:

- Multiple classes of computers
- Expanded abilities to configure computers for specific purposes
- Increased software costs relative to hardware costs
- Large computer databases
- Widespread adoption of graphical user interfaces
- Multicomputer configurations

In Grosch's time there was basically only one class of computer—the mainframe. Purchasing larger machines within a single class offers more CPU power for the money, and thus Grosch's Law appears to hold. But if all classes of computers are considered as a group, Grosch's Law does not hold. *The cost of CPU power actually increases on a per-unit basis as computer class increases.*

Also, CPU power is an incomplete measure of real computer system power. For example, midrange computers and mainframes differ little in raw CPU power, but they differ substantially in their ability to store, retrieve, and move large amounts of data.

Cost-efficiency comparisons are even more difficult when networked computers are considered. In Grosch's time, computer networks that enable multicomputer configurations did not exist. But today, high-speed computer and storage networks have enabled many organizations to achieve mainframe or supercomputer capabilities with clusters or grids of smaller computers.

Within a given class or model of computer system, a deliberate choice can be made to tune the performance of the system to a particular application or class of applications cost-effectively. For example, in a transaction-processing application, the system may be configured with fast secondary storage and I/O subsystems while sacrificing CPU performance. The same system may be cost-effectively tailored to a simulation application by providing a large amount of high-speed main memory, multiple CPUs with a high computational capacity, and a high-capacity communication channel between memory and the CPUs. Secondary storage and I/O performance would be far less important, and less costly alternatives could be used.

Another difficulty in applying Grosch's Law today is the reality of costly software. Developing software requires a substantial commitment of labor and many other resources. The combination of rapidly falling hardware costs and rising software development costs has made the cost of computer hardware an ever smaller factor in the total cost of an information system. Although hardware costs should be managed effectively, an IS professional is far more likely to maximize the cost effectiveness of an information system by concentrating on its software components.

Business
Focus

Do We Need a Mainframe?

Bauer Industries (BI) is a manufacturer and wholesale distributor of jewelry-making machinery and components. The company has 200 employees and approximately 2,500 products. The current inventory of automated systems is small for a company of its size. Applications include order entry, inventory control, and general accounting functions like accounts payable, accounts receivable, payroll, and financial reporting. BI's applications currently run on an older IBM RS/6000 S70 midrange computer. Dedicated character-based video display terminals are used to interact with application programs.

BI has grown rapidly over the last decade. Increased transaction volume has strained existing computer hardware capability. Typical order-entry response

Introduction to Systems Architecture

time is now 30 to 60 seconds compared with 2 to 5 seconds a few years ago. The order-entry system fully utilizes memory and CPU capacity during daytime hours. All non-essential processing has been moved to nighttime hours to minimize order-entry response time. The RS/6000 currently is configured with 4 CPUs, 1 gigabyte of primary storage, and 500 gigabytes of disk storage. No further enhancements to the hardware are possible.

Bauer Industries management realizes the seriousness of the situation, but they are uncertain how to address it. There are additional factors complicating the decision, including:

▶ An expected four-fold increase in business over the next 10 years

▶ Relocation to new facilities within the next year

▶ Plans to move the Web-based catalog and online order system in house (they're currently contracted out).

▶ A desire on the part of some members of management to substantially increase and modernize automated support of operating and management functions

Management has identified three viable options to address the current problem:

1. Purchase one or more used RS/6000 systems (they are no longer manufactured), connect them with a high speed network, and partition the application software among the computer systems.

2. Accept an offer from IBM to upgrade to a pSeries 670 or 690 computer system.

3. Develop a highly scalable hardware platform consisting of IBM blade servers or a cluster of IBM midrange computer systems.

Option 1 is least disruptive to the status quo but has two obvious disadvantages. The first is the use of outdated hardware. Parts for RS/6000 computers are in short supply. The second problem is data overlap among application programs. The current operating and application software will not support access to data stored on one of the machines by applications executing on the other machine. Thus, data will have to be stored on both machines and periodically synchronized to reflect updates made by different application programs. Estimated two-year costs are $75,000 for hardware and $100,000 for labor.

Option 2 theoretically preserves the existing application software and allows some hardware, such as video display terminals, to be reused. It also provides sufficient capacity to operate current software at many times the current transaction volume. However, the new hardware requires a new version of the UNIX operating system. The new operating system version was designed to support applications executing under older versions on RS series hardware. But management has heard of many problems in migrating application software. Estimated two-year costs are $350,000 for hardware and $150,000 for labor.

Option 3 is the most expensive and the biggest departure from current operating methods. Migration problems are similar to Option 2. Existing video display terminals can be used initially and gradually replaced with workstations or network computers. Option 3 appears to hold the most promise for implementing a new generation of application software. But the company has little experience or expertise in operating a multicomputer configuration. Estimated (guesstimated?) two-year costs are $150,000 for hardware and $400,000 for labor (including hiring or contracting three new IS staff members for two years).

Questions

1. What problems would you anticipate within two to five years if Option 1 is implemented? Option 2? Option 3?

2. Management has historically opted for lowest-cost solutions. How can you justify to them the higher costs of options 2 and 3?

3. Which alternative would you recommend to management? Why?

THE ROLE OF SOFTWARE

The primary role of software is to translate user needs and requests into CPU instructions that, when executed, will produce a result that satisfies the need or request. People usually ask computers to perform relatively complex tasks such as "generate my company's income statement," "spell-check my term paper," or "print a weekly sales report" (see Figure 2-9).

Figure 2-9 ▶

The role of software as a translator between user requests and CPU instructions

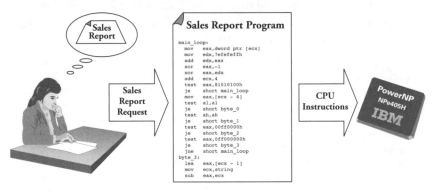

Software is complex because it performs a complex translation process that bridges two gaps:

▶ Human language to machine language

▶ High-level abstraction to low-level detail

People think, speak, and write in their native language, which has specific vocabulary, grammar, and semantics. Computers communicate in binary languages that consist of zeros and ones, which are grouped together by specific grammatical (syntax) rules to form meaningful statements or instructions.

The need or idea that motivates a request for computer processing is stated at a relatively high (abstract) level. The tasks that are performed (instructions that are executed) to satisfy a request are very specific and low-level. Because CPU instructions are very small units of work, the number of instructions that must be executed to satisfy a processing request is large.

Software Types

A program is a set of detailed instructions that directs a computer to perform a complex task. Much like a recipe, a program can be written down, stored, and retrieved when needed. Computer programs are usually stored in files on secondary storage devices. When needed, they are retrieved by name, copied into primary storage, and then executed by the CPU.

An **application program**, or **application software**, is a stored set of instructions for responding to a very specific request, much as you might access a specific recipe to prepare a particular dish. Examples of application programs in a system devoted to payroll processing include programs to print checks, enter new employee information, and produce annual tax reports. Application software is frequently purchased off-the-shelf; examples include accounting packages such as Quickbooks, project management software such as Microsoft Project, and graphics software such as PaintShop Pro.

Utility programs contain instructions for performing relatively basic tasks that are a necessary component of many application-oriented tasks, such as printing a text file, opening a new window on a video display, and copying a file from one disk to another. Although these tasks are relatively basic, they can still require thousands of CPU instructions to complete.

System software is a collection of programs that:

► Implement utility functions needed by many application programs

► Allocate computer resources to application programs

► Manage computer resources

The most significant distinction between application and system software is specificity of purpose and use. Application software is targeted to specific information-processing tasks like generating customer credit card bills. System software is targeted to general-purpose tasks that support many application programs and users.

Most application software is used directly by end users. In contrast, most system software does not interact with end users. Most utility programs used by application software operate invisibly in the background. Programs that allocate computer resources among application programs also operate in the background. System administrators may set policies that govern allocation rules and methods, but the allocation programs do not interact directly with users or system administrators during execution.

Utility programs that manage computer resources can be used directly by either end users or system administrators. Examples of user-oriented management programs include those that copy files and directories, install software on personal computers, and create and manage personal computer connections to Internet service providers. Examples of management programs intended for system administrators include those that create and modify user accounts, install hardware devices on servers, and implement computer security policies.

Figure 2-10 illustrates the interaction between the user, application software, utility programs within system software, and computer hardware. User input and output is limited to communication directly with the application program. Application programs, in turn, communicate with system software to request basic services (for example, opening a file or creating a new window). System software translates a service request into a sequence of machine instructions, passes those instructions to the hardware for execution, receives the results, and passes the results back to the application software.

Figure 2-10 ►

The interaction between the user, application software, system software, and computer hardware

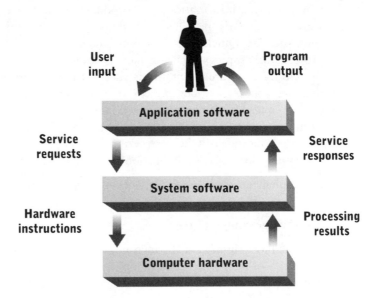

Figure 2-10 also illustrates what is commonly called a layered approach to software construction and operation. Application software is layered above system software, which is in turn layered above the hardware. A key advantage to this approach is that users and applications programmers do not need to know the technical details of computer hardware. Instead, they interact with hardware through a set of standardized service requests. Knowledge of the machine's physical details is embedded into the system software and hidden from the user and application programmers. This advantage is commonly referred to as **machine independence** or **hardware independence**.

System Software Layers

Figure 2-11 shows a more detailed view of software layers and their relationship to hardware functions. System software functions have been divided into four layers:

► System management—utility programs used by end users and system administrators to manage and control computer resources

► System services—utility programs used by system management and application programs to perform common functions

► Resource allocation—utility programs that allocate hardware and other resources among multiple users and programs

► Hardware interface—utility programs that control and interact with individual hardware devices

Figure 2-11 ►

Software layers and their relationship to hardware functions

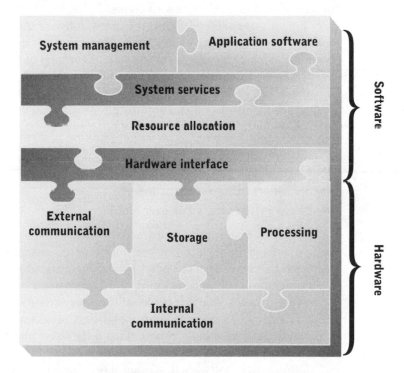

The principle of software layers applies not only to the relationship between application and system software but also to the relationship among various components of system software. Machine independence is achieved by placing all hardware interface functions within a single system software layer. In theory, system software can be modified to operate on new computer hardware by modifying only the hardware interface layer. In practice, machine independence is not always that simple, as you will learn in Chapter 11.

Resource allocation is an important, but relatively invisible, function of system software. A modern computer system can support multiple users and hundreds or thousands of programs executing simultaneously. Under such conditions, resources such as files and I/O devices cannot always be provided to a program or user immediately. System software balances the competing resource demands of users and processes. It does this in such a way that users and programs do not interfere with one another, while also ensuring that all users and programs receive the resources they need.

Operating Systems

An **operating system** is a collection of utility programs that supports users and application programs, allocates computer resources among multiple users and application programs, and controls access to computer hardware. Note that this definition is very similar to the earlier definition of system software. Also, note that the definition spans a very large range of functions. Operating systems are the most common, but not the only, type of system software. Examples of operating systems include Windows, UNIX, Linux, Mac OS X, and OpenVMS.

Operating systems include utility programs that meet the needs of most or all users, administrators, and application programs. Functions in most operating systems include:

► Program storage, loading, and execution
► File manipulation and access
► Secondary storage management
► Network and interactive user interfaces

Although operating systems are the most important and pervasive type of system software, they are not the only type of system software. Some system software functions aren't needed by the majority of computer users and thus aren't included as a standard part of all operating systems. For example, most operating systems don't include utility software for database manipulation because relatively few users and application programs need that capability. A separate type of system software called a database management system is used for that purpose.

Other system software functions may be provided by optional operating system components or by standalone systems software utilities. For example, Web server software is included as an optional component of some operating systems such as Windows Server, but it can also be obtained from other vendors including Macromedia and Apache Digital.

Windows operating systems tend toward an all-inclusive approach to system software, bundling most system software functions within the operating system. UNIX and Linux tend toward a less inclusive approach, with basic resource allocation and system services embedded in the operating system and third-party system software used for extended systems services such as Web services, distributed application support, and network security. Some users prefer the "bundled" approach of Windows though that approach has occasionally resulted in antitrust litigation and claims of unfair competition.

Application Development Software

Early programmers developed application programs without assistance from the computer itself. They wrote programs that consisted entirely of binary CPU instructions. There were no intermediaries such as compilers, interpreters, programming languages, or operating system services. As computers became more powerful, it became possible to write much larger application programs; but such programs were too complex to develop without automated assistance.

Application development software describes programs used to develop other programs. Though the term implies otherwise, most application development programs can also be used to develop system software or other application development programs. Examples of application development software include many compilers and interpreters for programming languages such as Java and Visual Basic and integrated software development packages such as Microsoft Visual Studio, IBM WebSphere, and Oracle JDeveloper.

Earlier in the chapter, software complexity was said to have resulted from the need to bridge the gaps between human-oriented and machine-oriented language, and between abstract, high-level statements of need and the low-level machine instructions that satisfy those needs. The same challenge of translating human intelligence into computer instructions applies to the process of developing programs.

A **program translator** is a program that translates instructions in a **programming language** into CPU instructions. Examples of programming languages include FORTRAN, Java, C++, and Visual Basic. A program translator doesn't just translate one language into another; it also translates high-level program instructions into detailed groups of CPU instructions. Modern programming languages allow programmers to express complex processing tasks in a single statement or instruction. For example, consider the BASIC program shown earlier in Figure 2-3. Hundreds or thousands of CPU instructions must be executed to perform that task. Programming languages free programmers from the need to specify every CPU instruction individually.

Program translators are not the only type of application development software. Other types include:

▶ **Program editors**—writing tools, similar to word processors, but customized for writing programs instead of documents
▶ **Debugging tools**—tools that simulate program execution and enable programmers to trace errors
▶ **System development tools**—tools that enable systems analysts and designers to develop models of entire information systems, models then used as the starting point for developing application programs

Like any tool, application development software increases the productivity of programmers and the quality of what they produce. The complexity of modern software demands application development tools of at least equal complexity.

ECONOMICS OF SYSTEM AND APPLICATION DEVELOPMENT SOFTWARE

The system and application development software we use today bears little resemblance to software used in the past. Programming languages, program development tools, and operating systems were not developed until the late 1950s, and were primitive in comparison to their modern equivalents.

Why is modern software so advanced? The answer derives from three economic facts of computer hardware and software:

▶ System software consumes hardware resources
▶ The cost per unit of computing power has rapidly decreased
▶ Software is more cost-effective when it is reused many times

These facts underlie a continually shifting tradeoff between the cost of developing and supporting application programs and the cost of computer hardware resources.

In the early years of computers, application development was a manual process. Programmers wrote binary CPU instructions with pencil and paper. They desk-checked their program and, when it looked correct, they used a plugboard, paper tape punch, or paper card punch machine to convert the written zeros and ones into something the computer could read. Then they tested the program to see if it worked as they expected. If too many rounds of testing and correction were required, the programmers were criticized for using too much valuable computer time. Computer hardware was simply too expensive to "waste" it debugging software.

As the cost of computer hardware decreased, the economic balance between programming labor and computer hardware resources shifted. Tedious programming using binary CPU instructions and extensive manual checks of code became more expensive than writing and executing programs with automated assistance. The trend toward greater automated support for application development and computer operation, as well as software reuse, had begun.

When a programmer uses a programming language, application development software, and a computer to develop a new program, the programmer is substituting software and hardware resources for the labor required to write binary CPU instructions. As programming languages and development tools become more powerful, application development software becomes more complex and consumes ever greater quantities of computer hardware resources. As hardware costs fall, the hardware resources consumed by application software become cheaper than the extra labor resources that would be needed if the application development tools were not employed. As you can see in Figure 2-12, the balance of hardware and software cost has shifted over time because of the declining cost of hardware and the emergence of system software. Hardware is typically the cheapest component of current information systems, and system and application software are nearly equal components in the total cost.

Figure 2-12 ▶

The change over time in the relative cost of hardware, application software, and system software in a typical information system

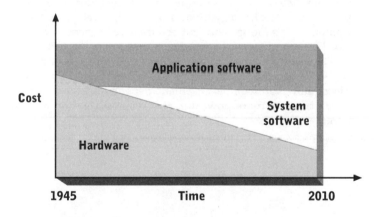

Both system and application development software also increase economic utility by promoting software reuse. Each system service is a utility program, and application software reuses those programs each time it makes a service request. Each time a program translator translates a generic instruction, such as Print Customer_Name into the equivalent set of CPU instructions, it is reusing the small utility program that implements a PRINT statement. Modern application development tools provide more overt examples of software reuse by providing libraries of utility programs and generating program skeletons from templates.

Technology
Focus

Intel CPUs and Microsoft Operating Systems

IBM introduced the first mass-market microcomputer in 1981. The computer contained an Intel 8088 CPU and ran the Microsoft MS-DOS 1.0 operating system. By the late 1980s, Intel and Microsoft had become the standard-setters for microcomputers, and they have continued in that role.

The original IBM PC platform, the Intel 8088, and MS-DOS 1.0 were primitive by modern standards. MS-DOS's primary functions were to load and execute application programs one at a time, provide minimal application support services such as file I/O, and provide device drivers for the small set of PC hardware resources such as the diskette drive, keyboard, and character-oriented video display. More sophisticated operating system capabilities were available on other more expensive hardware platforms using other CPUs. But MS-DOS's capabilities were about all that could be provided given the hardware capabilities of the 8088 and the original PC.

Figure 2-13 shows a time line for Intel CPUs and Microsoft operating system versions. Major advances in operating system function have typically lagged supporting CPU technology by approximately three years because software development is a long and labor-intensive process. Also, operating system development depends on hardware other than the CPU, such as video displays, network communication, primary storage, and secondary storage, that may be slower to develop. The evolution of Microsoft operating systems is a good example of how software development depends on hardware technology. As Intel CPUs became more powerful, the Windows operating system became more complex.

Introduction to Systems Architecture

Figure 2-13 ▶

The timing of Intel CPU and Microsoft operating system releases

Time	Intel	Microsoft
1980	8088	
1981-1985	8086 80286	MS-DOS 1.0
1986-1990	80386 80486	Windows 1.0 – 3.0
1991-1995	Pentium Pentium Pro	Windows NT 3.0 Windows 95
1996-2000	Pentium II Xeon Pentium III	Windows NT 4.0 Windows 98 Windows 2000 Windows 2000 Server
2001-2005	Itanium Pentium 4 Xeon MP Itanium 2	Windows Me Windows XP Windows Server 2003 (Datacenter Edition)

- ► **80386**—replaced earlier approaches to memory addressing that had limited application programs to no more than 640 kilobytes of memory, included hardware support for executing multiple programs simultaneously, simplified the partitioning of primary storage among programs, and provided mechanisms to protect programs from interfering with one another. It made it simpler for Microsoft developers to implement operating system functions such as task switching and virtual memory management. Windows 3.0 was the first Microsoft operating system to fully utilize the new capabilities of the 80386.

- ► **80486**—integrated memory caches and the CPU on a single chip, enhanced computational capabilities, and increased raw CPU speed. Windows 95 was developed to take better advantage of the chip's capabilities.

- ► **Pentium and Pentium Pro**—improved memory access and raw CPU speeds, and added features such as support for higher-speed system buses and multimedia processing instructions. Microsoft operating system development split into two distinct paths. Windows 95 evolved into Windows 98 and finally Windows Me. But increased speed and improved memory management enabled Microsoft to implement much more sophisticated memory and hardware management capabilities introduced in Windows NT 3.0.

- ► **Pentium II, III, and 4**—increased the speed at which the CPU could communicate with other hardware components, enlarged memory caches for improved performance, and supported simultaneous execution of multiple instructions. The improvements enabled Microsoft to continue building on capabilities introduced in Windows NT leading to Windows 2000, XP, and Server.

- ► **Xeon and Itanium**—the Xeon extended the Pentium architecture with larger memory caches and more efficient methods of splitting work across multiple CPUs. Latest generation Xeons can extend memory management beyond 4 gigabytes. The Itanium architecture abandoned many of the microcomputer-oriented characteristics of other Intel CPUs, enabling a design optimized for servers and high-performance workstations. Both CPUs enable Microsoft Server operating systems to compete against UNIX and other competitors for the medium- and large-scale server operating system market.

In the push-pull relationship of hardware power and software capability, hardware is usually the driving force. Software developers have a "wish list" of capabilities that hardware manufacturers try to satisfy to gain competitive advantage. Software development doesn't proceed in earnest until required hardware is available to support development and testing. However, the situation is occasionally reversed. Software makers may develop software that pushes past the limits of current hardware, confident that hardware manufacturers will soon provide more power. More demanding software in turn fuels users' drive to acquire new and more powerful machines, and the cycle continues. Competition among software developers guarantees continual and fairly rapid assimilation of new hardware technology.

COMPUTER NETWORKS

Computers once were processing islands; people physically carried data and programs to them. Computer networking and the Internet have forever changed the landscape of computing. Today, a computer isn't considered useful if it doesn't have the ability to interact with virtually every other computer on the planet.

A **computer network** is a set of hardware and software components that enables multiple users and computer systems to share information, software, and hardware resources. A computer network also enables many types of personal communication, such as e-mail, newsgroups, and team collaboration. The number and complexity of computer network functions and components has grown as network technology has matured and become more pervasive.

Figure 2-14 ▶

Computer network functions and their relationship to computer hardware and software

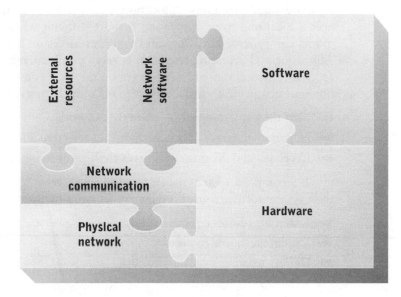

External Resources

Most network functions are extensions of the hardware and software functions discussed earlier in the chapter (see Figure 2-14) that allow a computer system to interface with a physical network and external resources. The complexity of modern networks arises from the huge quantity of distributed resources and the difficulties inherent in finding, accessing, and managing those resources. In the early days of networks, the only significant distributed resource type was raw data. As networks have matured and expanded, so have the types of resources that can be accessed. Data is now available in many different forms, including text files, sound and video, databases, and Web pages. A computer can ask another computer to execute a program and transmit the results, or it can retrieve a file containing the program and execute it locally.

Hardware devices of all types also can be shared across a network. One computer can use the CPU, storage, or I/O devices of another. Devices such as printers and secondary storage arrays can be directly attached to the network. The ability to share data, programs, and hardware resources among computers gives modern organizations considerable flexibility to deploy and redeploy computing and information resources to satisfy rapidly changing needs.

Network Software

Recall that a key function of system software is allocating resources among multiple users and programs. Allocation is simple when a single user and program access local resources. Allocation is more difficult when multiple users or programs compete for hardware and other resources on a single computer system. Allocating and accessing resources is very complex when a user or program can request resources that are not present on the local computer system and also are not managed by locally installed system software. In that case, system software must:

► Find the requested resources on the network
► Negotiate resource access with distant resource allocation software
► Receive and deliver the resources to the requesting user or program

If a computer system makes its local resources available to other computers, then system software must implement an additional set of functions to:

► Listen for resource requests
► Validate resource requests
► Deliver resources via the network

In essence, system software plays two roles in each network resource access: request and response. Many computer systems fill both roles and must implement both system software functions.

System software implements the intelligence needed to make and respond to external resource requests. Most operating systems include support for both functions. Specialized response functions are often provided as optional operating system components or as standalone system software packages. Examples include Web server software, e-mail distribution software, and Internet security software.

Network Communication and the Physical Network

A computer system requires at least one hardware device to implement its connection to a network. Network communication devices were once exotic and expensive but are now commonplace. Network communication devices differ from I/O devices in two important ways. First, they usually are simpler because they do not need to convert data represented electronically into another form. Second, they must support communication at relatively high speeds so that external resource access is not significantly slower than access to local resources.

Each computer system's network communication hardware is attached to a physical network. The physical network is a complex combination of communication protocols, methods of data transmission (cables and/or wireless), and network hardware devices (network interface cards, switches, routers, and other devices to make the physical data connections, as well as networked printers or other resources). The physical network can be implemented with a dizzying array of technologies and architectural approaches, which will be discussed in more detail later in the book.

SUMMARY

► A computer is an automated device for performing computational tasks. It accepts input data from the external world, performs one or more computations on the data, and then returns results to the external world. Early computers were mechanical devices with very limited capabilities. Modern computers have more extensive capabilities and are implemented with a combination of electronic and optical devices. The advantages of electronic and optical implementation include speed, accuracy, and reliability.

► Computer capabilities include processing, storage, and communication. A computer system contains a general-purpose processor that can perform computations, compare numeric data, and move data among storage locations and I/O devices. A command to the processor to perform a function is called an instruction. Complex tasks are implemented as a sequence of instructions called a program. Changing the program changes a computer system's behavior.

► A computer system consists of a central processing unit (CPU), primary storage, secondary storage, and input/output (I/O) devices. The CPU performs all computation and comparison functions and directs all data movement. Primary storage holds programs and data that are currently in use by the CPU. Primary storage is generally implemented by electronic devices called random access memory (RAM). Secondary storage consists of one or more devices with high capacity and relatively low cost and access speed. Secondary storage holds programs and data that are not currently in use by the CPU. I/O devices allow the CPU to communicate with the external world, including humans and other computers.

► A computer system can be classified as a microcomputer, midrange computer, mainframe, or supercomputer. Microcomputers are designed for use by a single user. Midrange computers and mainframes are designed to support many programs and users simultaneously. Mainframe computers are designed for large amounts of data storage and access. Supercomputers are designed to perform large amounts of numeric computation very quickly. Clusters and grids can mimic the performance of mainframes and supercomputers with groups of smaller computers.

► The role of software is to translate user requests into machine instructions. The two primary types of software are application software and system software. Application software consists of programs that satisfy specific user processing needs such as payroll calculation, accounts payable, and report generation. System software consists of utility programs that are designed to be general in nature and can be used many times by many different application programs or users.

► The operating system is the most important system software component in most computer systems. An operating system provides administrative utilities, utility services to application programs, resource allocation functions, and direct control over hardware.

► System and application development software reduce the cost of developing and deploying application programs, and allow organizations to substitute relatively cheap computer hardware for relatively expensive labor.

► A computer network is a set of hardware and software components that permits information, software, and hardware resources to be shared among multiple users and computer systems. Computer networks contain software and hardware components. Networking software allows a computer system to find and retrieve resources on other computer systems or to respond to requests from other computer systems for access to local resources. Network hardware implements direct communication among computer systems.

In this chapter you've learned about the basic elements of computer system architecture: hardware, software, and networks. You've been introduced to many new terms and concepts but haven't learned about them in detail. You'll get a chance to digest and expand your knowledge of these topics in later chapters.

Figure 2-15 shows key computer system functions that have been covered so far, and identifies the chapter(s) that explore each function in depth.

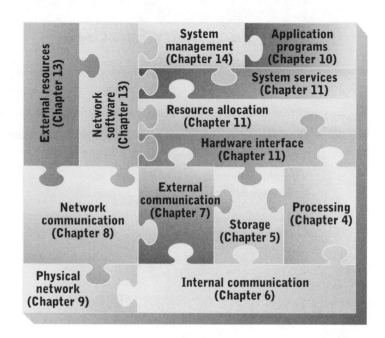

Figure 2-15 ►

Topics covered in the rest of the textbook

In the figure:
- External resources (Chapter 13)
- Network software (Chapter 13)
- System management (Chapter 14)
- Application programs (Chapter 10)
- System services (Chapter 11)
- Resource allocation (Chapter 11)
- Hardware interface (Chapter 11)
- Network communication (Chapter 8)
- External communication (Chapter 7)
- Processing (Chapter 4)
- Storage (Chapter 5)
- Physical network (Chapter 9)
- Internal communication (Chapter 6)

Key Terms

algorithm
application development
   software
application program
application software
arithmetic logic unit (ALU)
blade
branch
central processing unit (CPU)
cluster
computer network
condition
control unit
debugging tool
execute
formula
general-purpose processor
grid
Grosch's Law

hardware independence
input output (I/O) units
instruction
jump
logic instructions
machine independence
main memory
mainframe
memory
microcomputer
midrange computer
minicomputer
network computer
operating system
personal computer (PC)
primary storage
processor
program

program editor
program translator
programming language
quantum measurement problem
qubit
random access memory (RAM)
register
secondary storage
server
special-purpose processor
supercomputer
system bus
system development tool
system software
systems architecture
utility program
volatility
workstation

Vocabulary Exercises

1. A(n) _____ generally supports more simultaneous users than a(n) _____. Both are designed to support more than one user.

2. A(n) _____ is a storage location implemented within the CPU.

3. The term _____ refers to storage devices, not located within the CPU, that hold instructions and data of currently executing programs.

4. A problem-solving procedure that requires executing one or more comparison and branch instruction is called a(n) _____.

5. A(n) _____ is a command to the CPU to perform one processing function on one or two data inputs.

6. The term _____ describes the collection of storage devices that hold large quantities of data for long time periods.

7. A(n) _____ is a computer that manages shared resources and allows other computers to access them through a network.

8. A program that solves a(n) _____ requires no branching instructions.

9. The primary components of a CPU are the _____, _____, and a set of _____.

10. Primary storage also can be called _____ and is generally implemented using _____.

11. A set of instructions that is executed to solve a specific problem is called a(n) _____.

12. A(n) _____ is typically implemented with the latest and most expensive technology.

13. A(n) _____ is a group of similar or identical computers, connected by a high-speed network, that cooperate to provide services or execute a common application.

14. A(n) _____ is a group of dissimilar computer systems, connected by a high-speed network, that cooperate to provide services or execute a common application.

15. A(n) _____ is a hardware device that enables a computer system to communicate with humans or other computers.

16. A CPU is a(n) _____ processor capable of performing many different tasks by simply changing the program.

17. The _____ is the "plumbing" that connects all computer system components.

18. The CPU _____ program instructions one at a time.

19. The term _____ describes a computer system's components and the interaction among them.

20. Most programs are written in a _____ such as C or Java, which is then translated into an equivalent set of CPU instructions.

21. Resource allocation and direct hardware control is the responsibility of a(n) _____ .

22. _____ software is general purpose. _____ software is specialized to a specific user need.

23. A(n) _____ is a set of hardware and software components that enables information, software, and hardware resources to be shared among multiple users and computer systems.

Review Questions

1. What similarities exist among mechanical, electrical, and optical methods of computation?

2. What shortcomings of mechanical computation were addressed by the introduction of electronic computing devices?

3. What shortcomings of electrical computation will be addressed by the introduction of optical computing devices?

4. What is a CPU? What are its primary components?

5. What are registers? What is/are their function(s)?

6. What is main memory? In what way(s) does it differ from registers?

7. What are the differences between primary and secondary storage?

8. How does a workstation differ from a microcomputer?

9. How does a supercomputer differ from a mainframe computer?

10. Describe three types of multicomputer configurations. What are their comparative advantages and disadvantages?

11. What class(es) of computer system(s) normally are used to implement a server?

12. What is Grosch's Law? Does it hold today? Why or why not?

13. How can a computer system be tuned to a particular application?

14. What characteristics differentiate application software from system software?

15. In what ways does system software make the development of application software easier?

16. In what ways does system software make application software more portable?

17. What are the primary components of an operating system?

18. Why has the development of system software paralleled the development of computer hardware?

19. List at least five types of resources that can be shared among computers on a local-area or wide-area network.

20. Describe the dual roles played by most operating systems with respect to external resource access.

21. Describe the relationship between the resource allocation and management functions of system software and external resources accessible via a network. What system software functions must be provided to access external resources?

Research Problems

1. Obtain a catalog or access the Web site of a major distributor of microcomputer equipment such as Computer Discount Warehouse (www.cdw.com) or Dell (www.dell.com). Assume that you have a budget of $2,500 to purchase workstation hardware, exclusive of peripheral devices such as a scanner or printer. Select or configure a system that provides optimal performance for the following types of users:

   ► A home user who uses word-processing software such as Microsoft Word or Corel WordPerfect, a home accounting package such as Quicken and/or TurboTax, and children's games.

   ► An accountant who uses office software such as Microsoft Office, statistical software such as SYSTAT, and who regularly downloads large amounts of financial data from a corporate server for statistical and financial modeling.

   ► An architect who uses typical office software and computer-aided design software.

   Pay particular attention to the adequacy of CPU power, disk space, and I/O capabilities.

2. Assume that you have been asked to recommend a computer system that will be used as a server for a 250-person office. The server must provide shared access to a 500 GB file system, several printers, e-mail and Web services, and database services (database size is approximately 150 GB) to a variety of application programs. Assume further that Microsoft products such as Windows Server and Microsoft SQL Server will be the primary software used on the server.

   Obtain product information from IBM (www.ibm.com), Hewlett-Packard (www.hp.com), and Dell Computer Corporation (www.dell.com). Determine which of their products best meet the server requirements.

3. Table 2-2 probably will be out of date before this text is in print. Access the manufacturer Web sites (www.ibm.com and www.hp.com) or contact a sales representative and update Table 2-2 with current representative models of each computer class. Access the student support site for this textbook (http://www.averia.mgt.unm.edu) and download copies of Table 2-2 from earlier textbook editions. What is the rate of change in CPU speed, memory and disk capacity, and cost?

Chapter 3

Data Representation

► Describe numbering systems and their use in data representation

► Compare and contrast various data representation methods

► Describe how nonnumeric data is represented

► Describe common data structures and their uses

Computers manipulate and store a variety of data such as numbers, text, sound, and pictures. This chapter describes how data is represented and stored within computer hardware. It also describes how simple data types are used as building blocks to create more complex data structures such as arrays and records. Understanding data representation is key to understanding hardware and software technology.

DATA REPRESENTATION AND PROCESSING

People can understand and manipulate data represented in a variety of forms. For example, we can understand numbers represented symbolically as Arabic numerals (for example, 8714), Roman numerals (for example, XVII), and simple lines or tick marks on paper (for example, ||| to represent a value of 3). We can understand words and concepts represented using pictorial characters (电 脑), alphabetic characters ("computer" and компьютер, Cyrillic text of the Russian word for computer), or directly as sound waves (a word spoken in a particular language). People also extract data from visual images (photographs and motion pictures) and from the senses of taste, smell, and touch. Much of the brain's processing power and flexibility are due to the rich variety of data representations it can recognize and understand.

To be manipulated or processed by the brain, external data representations such as printed text must be converted to an appropriate internal format and transported to the brain's "processing circuitry." Our sensory organs convert inputs such as sight, smell, taste, sound, and skin sensation into electrical impulses that are transported through the nervous system to the brain. Processing within the brain occurs as networks of neurons exchange data electrically.

Any data and information processor, whether organic, mechanical, electrical, or optical, must be capable of:

► Recognizing external data and converting it to an appropriate internal format
► Storing and retrieving data internally
► Transporting data among internal storage and processing components
► Manipulating data to produce desired results or decisions

Note that these capabilities correspond roughly to computer system components described in Chapter 2—I/O units, primary and secondary storage, system bus, and the CPU.

AUTOMATED DATA PROCESSING

Current computer systems represent data electrically and process it with electrical switches. Two-state (on-off) electrical switches are well suited for representing data that can be expressed in a binary (1 or 0) format, as you'll see in the next chapter. Electrical switches are combined to form processing circuits that are in turn combined to form processing subsystems and entire CPUs. This can be seen as an equation:

$$A + B = C$$

where data inputs A and B, represented as electrical current, are transported through processing circuits (see Figure 3-1). The electrical current that emerges from the circuit represents a data output, C. Automated data processing thus combines physics (electronics) and mathematics.

Figure 3-1 ▶

Two electrical inputs on the left flow through processing circuitry that generates their sum on the right

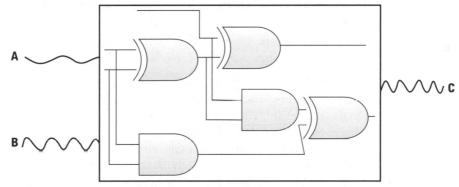

The physical laws of electricity, optics, and quantum mechanics are described by mathematical formulas. If the behavior of a device is based on well-defined, mathematically described laws of physics, then, in theory, that device can implement a processor to perform the equivalent mathematical function. This relationship between mathematics and physics underlies all automated computation devices from mechanical clocks (using the mathematical ratios of various gears) to electronic microprocessors (using the mathematics of electrical voltage and resistance). As you saw in the previous chapter, in the case of quantum mechanics, we understand the mathematical laws but don't yet know how to build reliable and cost-effective computing devices based on those laws.

Basing computer processing on mathematics and physics has limits. Processing operations must be based on mathematical functions such as addition and equality comparison, use numerical data inputs, and generate numerical outputs. From a human perspective, computers are limited because they can represent and process only numeric data. Concepts such as "mother," "friend," "love," and "hate" have no obvious numerical representation, so there is no direct way to process them using computer circuitry.

Although many researchers are studying ways to represent nonnumeric data numerically so it can be processed by computers, as well as ways to implement nonnumeric processing devices, the vast majority of modern computers perform only numerical computing. Even simple nonnumeric data like characters or tiny components of pictures can be processed only by first converting them to numeric form.

BINARY DATA REPRESENTATION

A decimal (base 10) number is a number in which each digit can have one of 10 possible values (0, 1, 2, 3, 4, 5, 6, 7, 8, or 9). A **binary number** is a number in which each digit can have only one of two possible values (0 or 1). Computers represent data using binary numbers for two reasons:

► Binary numbers represented as electrical signals can be reliably transported among computer system components.

► Binary numbers represented as electrical signals can be processed by two-state electrical devices that are relatively easy to design and fabricate.

Reliable data transport is important for computer applications to produce accurate outputs. The relationship between binary numbers, binary electrical signals, and reliable transport is fully described in Chapter 8. Given current technology, binary signals and processing devices represent the most cost-efficient tradeoff among capacity, accuracy, reliability, and cost. Future technological developments could fundamentally alter this tradeoff.

Binary numbers also are well-suited to computer processing because they correspond directly with values in **Boolean logic**. This form of logic is named for nineteenth-century mathematician George Boole, who developed methods of reasoning and logical proof using sequences of statements that could be evaluated only as true or false. Similarly, a computer can perform logical comparisons of two binary data values to determine whether one data value is greater than, equal to, less than, less than or equal to, not equal to, or greater than or equal to another value. As discussed in the previous chapter, this primitive logical capability is the means by which a computer can exhibit intelligent behavior.

Both computers and humans can combine multiple digits to form a single data value to represent and manipulate larger numbers. Decimal and binary notation are alternate forms of a positional numbering system. In a positional numbering system, numeric values are represented as groups, or strings, of digits. The symbol used to represent a digit and the digit position within a string determine its value. The value of the entire string is the sum of the values of all digits within the string.

For example, in the decimal number system, the number 5,689 is interpreted as follows:

$$(5 \times 1000) + (6 \times 100) + (8 \times 10) + 9 =$$
$$5,000 + 600 + 80 + 9 = 5,689$$

The same series of operations can be represented in columnar form, or with positions of the same value aligned in columns:

$$
\begin{array}{r}
5,000 \\
600 \\
80 \\
+\quad 9 \\
\hline
5,689
\end{array}
$$

For whole numbers, values are accumulated from right to left. In the above example, the digit 9 is in the first position, 8 is in the second position, 6 is in the third, and 5 is in the fourth.

The maximum value, or weight, of each position is a multiple of the weight of the position to its right. In the decimal number system, the first (rightmost) position is that of the ones (10^0), and the second position is 10 times that of the first position (10^1). The third position is 10 times the second position (10^2 or 100). The fourth is 10 times the third position (10^3 or 1,000), and so on. In the binary numbering system, each position is two times the previous position. Thus position values for whole numbers are 1, 2, 4, 8, and so forth. The multiplier that describes the difference between one position and the next is the **base** of the numbering system. Another term for base is **radix**. The base, or radix, of the decimal number system is 10. The base, or radix, of the binary numbering system is 2.

The fractional part of a numeric value is separated from the whole part by a period, though in some countries a comma is used instead of a period. In the decimal numbering system, the period or comma is called a **decimal point**. In numbering systems other than decimal, the term **radix point** is used for the period or comma. An example of a decimal value with a radix point is:

$$5,689.368$$

The fractional portion of this real number is .368 and its value is interpreted as follows:

$$(3 \times .1) + (6 \times .01) + (8 \times .001) =$$
$$0.3 \quad + \quad 0.06 \quad + \quad 0.008 \quad = 0.368$$

Proceeding toward the right from the radix point, the weight of each position is a fraction of the position to its left. In the decimal (base 10) numbering system, the first position to the right of the decimal point represents tenths (10^{-1}), the second position represents hundredths (10^{-2}), the third represents thousandths (10^{-3}), and so forth.

In the binary numbering system the first position to the right of the radix point represents halves (2^{-1}), the second position represents quarters (2^{-2}), the third represents eighths (2^{-3}), and so forth. As with whole numbers, each position has a weight that is 10 (or 2) times that of the position to its right. Table 3-1 shows a comparison between decimal and binary representations of the numbers 0 through 10.

Table 3-1 ►

Binary and decimal notations for the values 0 through 10

Place Values	Binary system (Base 2)					Decimal system (Base 10)			
	2^3 8	2^2 4	2^1 2	2^0 1		10^3 1000	10^2 100	10^1 10	10^0 1
	0	0	0	0	=	0	0	0	0
	0	0	0	1	=	0	0	0	1
	0	0	1	0	=	0	0	0	2
	0	0	1	1	=	0	0	0	3
	0	1	0	0	=	0	0	0	4
	0	1	0	1	=	0	0	0	5
	0	1	1	0	=	0	0	0	6
	0	1	1	1	=	0	0	0	7
	1	0	0	0	=	0	0	0	8
	1	0	0	1	=	0	0	0	9
	1	0	1	0	=	0	0	1	0

The number of digits required to represent a particular value depends on the base of the numbering system. The number of digits required to represent a value increases as the base of the numbering system decreases. Therefore, values that can be represented in a relatively compact format in decimal notation may require lengthy sequences of binary digits. For example, the decimal value 99 requires two decimal digits but seven binary digits. Table 3-2 summarizes the number of binary digits required to represent various decimal values.

Table 3-2 ▶

Binary data
representations
for up to
16 positions

Number of bits (n)	Number of values (2^n)	Numeric range (decimal)
1	2	0...1
2	4	0...3
3	8	0...7
4	16	0...15
5	32	0...31
6	64	0...63
7	128	0...127
8	256	0...255
9	512	0...511
10	1024	0...1023
11	2048	0...2047
12	4096	0...4095
13	8192	0...8191
14	16384	0...16383
15	32768	0...32767
16	65536	0...65535

To convert a binary value to its decimal equivalent, use the following procedure:[1]

1. Determine each position weight by raising 10 to the number of positions left (+) or right (-) of the radix point.
2. Multiply each digit by its position weight.
3. Sum all the values calculated in step 2.

Figure 3-2 shows how the binary number 101101.101 is converted to its decimal equivalent, 45.625.

---

[1] The standard Windows calculator can convert among binary, octal, decimal, and hexadecimal. (In Windows XP, access the calculator by clicking Start, All Programs, Accessories, Calculator.) For example, to convert a binary number to decimal, click the View menu and select Scientific, click the Bin (for Binary) option button, enter the binary number, then click the Dec (Decimal) option button to display the number in decimal format.

Figure 3-2 ▶

Computing the
decimal
equivalent of a
binary number

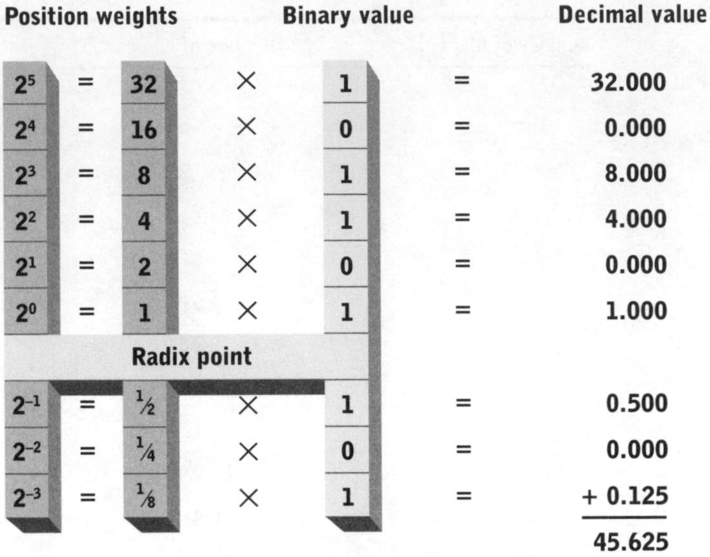

Position weights			Binary value		Decimal value
2^5	=	32	× 1	=	32.000
2^4	=	16	× 0	=	0.000
2^3	=	8	× 1	=	8.000
2^2	=	4	× 1	=	4.000
2^1	=	2	× 0	=	0.000
2^0	=	1	× 1	=	1.000
Radix point					
2^{-1}	=	½	× 1	=	0.500
2^{-2}	=	¼	× 0	=	0.000
2^{-3}	=	⅛	× 1	=	+ 0.125
					45.625

In computer terminology, each digit of a binary number is called a **bit**. A group of bits that describe a single data value is called a **bit string**. The leftmost digit, which has the greatest weight, is called the **most significant digit**, or **high-order bit**. Conversely, the rightmost digit is the **least significant digit**, or **low-order bit**. A string of eight bits is called a **byte**. A byte generally is the smallest unit of data that can be read from or written to a storage device.

The following mathematical rules define addition of positive binary digits:

$$0 + 0 = 0$$
$$1 + 0 = 1$$
$$0 + 1 = 1$$
$$1 + 1 = 10$$

To add two positive binary bit strings, you first must align their radix points as follows:[2]

$$\begin{array}{r} 101101.101 \\ + \ 10100.0010 \\ \hline \end{array}$$

The values in each column are added separately, starting with the least significant, or rightmost, digit. If a column result exceeds one, then the excess value must be carried to the next column and added to the values in that column.

---

[2] The standard Windows calculator can add and subtract binary integers. From the View menu, select Scientific, then click Bin (binary). You can also select Digit grouping from the View menu to place the digits in groups of four for easier readability.

The result of adding the two numbers above is:

```
 111       1
 101101.101
+ 10100.0010

 1000001.1100
```

The result is the same as that obtained by adding the values in base 10 notation:

	Binary		Real fractions		Real decimal
	101101.101	=	45⅝	=	45.625
+	10100.0010	=	20⅛	=	20.125
	1000001.1100	=	65¾	=	65.750

Binary numbers usually contain many digits and are difficult for people to remember and manipulate without error. Compilers and interpreters for higher-level programming languages such as C and Java automatically convert decimal numbers into binary numbers when generating CPU instructions and data values. However, programmers sometimes must deal directly with binary numbers, such as when they program directly in machine language or for some operating system utilities. To minimize errors, and to make dealing with binary numbers easier, numbering systems that are based on even multiples of two are sometimes used. These number systems include:

► Hexadecimal
► Octal

Hexadecimal Notation

Hexadecimal numbering uses 16 as its base or radix (hex = 6, and decimal = 10). There are not enough numeric symbols (Arabic numerals) to represent 16 different values, so English letters represent the larger values (see Table 3-3).

Table 3-3 ►

Base 16 digit	Decimal value		Base 16 digit	Decimal value
0	0		8	8
1	1		9	9
2	2		A	10
3	3		B	11
4	4		C	12
5	5		D	13
6	6		E	14
7	7		F	15

The primary advantage of **hexadecimal notation**, as compared to binary notation, is its compactness. Large numeric values expressed in binary notation require four times as many digits as those expressed in hexadecimal notation. For example, the data content of a byte requires eight binary digits (such as 11110000) but only two hexadecimal digits (such as F0). This compact representation greatly reduces programmer error.

Hexadecimal numbers often designate memory addresses. For example, a 64-kilobyte (KB) memory region contains 65,536 bytes (64 × 1,024 bytes/KB). Each byte is identified by a sequential numeric address. The first byte is always address 0. Therefore, the range of possible memory addresses is 0 to 65,535 in decimal numbers, 0 to 1111111111111111 in binary numbers, and 0 to FFFF in hexadecimal numbers. As you can see from this example, hexadecimal addresses are more compact than decimal or binary addresses due to the higher radix of the number system.

When reading a numeric value in written text, the base of the number may not be obvious. For example, when reading an operating system error message or a hardware installation manual, should the number 1000 be interpreted in base 2, 10, 16, or something else? In mathematical expressions the base of a number usually is specified with a subscript. For example,

$$1001_2$$

indicates to the reader that 1001 should be interpreted as a binary number and

$$6044_{16}$$

indicates that 6044 should be interpreted as a hexadecimal number.

The base of a written number can be made explicit by placing a letter at the end. For example,

$$1001B$$

indicates a binary number and

$$6044H$$

indicates a hexadecimal number. No letter is normally used to indicate a decimal (base 10) number. Some programming languages such as Java and C++ use the prefix 0x to indicate a hexadecimal number (0x1001 is equivalent to 1001_{16}).

Unfortunately, these conventions are not observed consistently. Often it is left to the reader to guess the correct base by the content of the number or the context in which it appears. A value that contains a numeral other than a 0 or 1 cannot be binary. Similarly, the use of letters A through F indicates that the contents are expressed in hexadecimal. Bit strings are usually expressed in binary, and memory addresses are usually expressed in hexadecimal.

Octal Notation

Some operating systems and machine programming languages use **octal notation**. Octal notation uses the base 8 numbering system and has a range of digits from 0 to 7. Large numeric values expressed in octal notation are one-third the length of corresponding binary notation and double the length of corresponding hexadecimal notation.

GOALS OF COMPUTER DATA REPRESENTATION

Although all modern computers internally represent data with binary digits, they don't necessarily represent larger numeric values with positional bit strings. Positional numbering systems are convenient for people to interpret and manipulate, because the sequential processing of digits in a string parallels the way the human brain functions and because people traditionally are taught to perform computation in a linear fashion. For example, positional numbering systems are well-suited to addition and subtraction of numbers in columns using pencil and paper. Computer processors, however, operate differently from a human brain. Data representation tailored to human capabilities and limitations may not be best suited to computers.

Any representation format for numeric data represents a balance among several factors, including:

▶ Compactness
▶ Accuracy
▶ Range
▶ Ease of manipulation
▶ Standardization

As with many computer design decisions, alternatives that perform well in one dimension or category often perform poorly in others. For example, a data format with a high degree of accuracy and a large range of representable values is usually difficult and expensive to manipulate because it is not compact.

Compactness

The term compactness (or size) describes the number of bits used to represent a numeric value. Compact representation formats use relatively few bits to represent a value, but they are limited in the range of values they can represent. For example, the largest binary integer that can be stored in 32 bits is 2^{32}, or $4,294,967,296_{10}$. Halving the number of bits to make the representation more compact decreases the largest possible value to 2^{16}, or $65,535_{10}$.

Computer users and programmers usually prefer a large numeric range. For example, would you be happy if your bank's computer limited your maximum checking account balance to 65,535 pennies? But the extra bit positions required to provide greater numeric range have a cost. Primary and secondary storage devices must be larger and, thus, more expensive. Additional and more expensive capacity is required for data transmission among devices within a computer system or across computer networks. CPU processing circuitry becomes more complex and expensive as additional bit positions are added. In sum, the more compact a data representation format, the less expensive it is to implement in computer hardware.

Accuracy

Although compact data formats minimize the complexity and cost of hardware, they can do so at the expense of accurate data representation. The accuracy, or precision, of representation increases with the number of data bits that are used.

It is possible for routine calculations to generate quantities that are too large or too small to be contained within the finite circuitry of a machine (that is, within a fixed number of bits). For example, the fraction 1/3 cannot be accurately represented within a fixed number of bits because it is a nonterminating fractional quantity (0.333333333 with an infinite number of 3s). In such cases, the quantities must be manipulated and stored as approximations. Each approximation introduces a degree of error. If approximate results are used as inputs for other computations, errors can be compounded. In some cases, the extent of error can be significant. Therefore, it is possible for a computer program to be without apparent logical flaws and yet produce inaccurate results.

If all data types were represented in the most compact form possible, then approximations would introduce unacceptable margins of error. If a large number of bits were allocated to each data value instead, machine efficiency and performance would be sacrificed and hardware cost would be increased substantially. The most favorable tradeoffs can be achieved by using an optimum coding method for each type of data and/or for each type of operation that must be performed. This is the main reason for the variety of data representation formats used in modern CPUs.

Ease of Manipulation

When discussing computer processing, *manipulation* refers to executing processor instructions such as addition, subtraction, and equality comparison. *Ease* refers to machine efficiency. The efficiency of a processor depends on its complexity (that is, the number of its primitive components and the complexity of the wiring that binds them together). Efficient processor circuits perform their function quickly due to a relatively small number of components and the short distance that electricity must travel. More complex devices require more time to perform their functions.

Data representation formats vary in their ability to implement processor circuits efficiently. For example, consider the difficulty most people have performing computations with fractions, compared to working with decimal numbers. The decimal format is more efficiently processed by people than the fractional format.

Unfortunately, there is no one best representation format for all types of computation operations. For example, representing large numeric values as logarithms simplifies multiplication and division for people and computers because $\log A + \log B = \log (A \times B)$ and $\log A - \log B = \log (A \div B)$. Logarithms complicate other operations such as addition and subtraction, and they can substantially increase a number's length (for example, $\log 99 \approx 1.99563519459754991534025577775033$).

Standardization

Data must be communicated among devices in a single computer system and among other computer systems via computer networks. To ensure correct and efficient data transmission, data formats must be suitable for a wide variety of devices and computer systems. This especially is true for alphanumeric data. For this reason, various organizations have created standard data encoding methods. Adhering to such standards provides computer users with the flexibility to combine hardware from different vendors with minimal data communication problems.

CPU DATA TYPES

The CPUs of most modern computers can represent and process five primitive data types:

► Integer
► Real number
► Character
► Boolean
► Memory address

Some CPUs include variants of these five types for specialized application software. The representation format for each data type balances compactness, accuracy, ease of manipulation, and standardization.

Integers

An **integer** is a whole number—a value that does not have a fractional part. For example, the values 2, 3, 9, and 129 are integers while the value 12.34 is not. Integer data formats can be signed or unsigned. Most CPUs provide an **unsigned integer** data type, which stores positive integer values as ordinary binary numbers. The value of an unsigned integer is always assumed to be positive.

A signed integer uses one bit to represent whether the value is positive or negative. The choice of which bit value (zero or one) represents which sign (positive or negative) is arbitrary. The sign bit is normally the high-order bit in a numeric data format. In most data formats, the sign bit is one for a negative number and zero for a nonnegative number (note that zero is a nonnegative number).

The sign bit occupies a bit position that would otherwise be available to store part of the data value. Using a sign bit reduces the largest positive value that can be stored in any fixed number of bit positions. For example, the largest positive value that can be stored in an 8-bit unsigned integer is 255, or 2^8-1. If a bit is used for the sign, then the largest positive value that can be stored is 127 or 2^7-1.

With unsigned integers, the lowest value that can be represented always is zero. With signed integers, the lowest value that can be represented is the negative of the highest value that can be stored (for example, -127 for 8-bit signed binary). The choice between signed and unsigned numeric data formats determines the largest and smallest values that can be represented.

Excess Notation   One format that can be used to represent signed integers is **excess notation**. Excess notation always uses a fixed number of bits with the leftmost bit representing the sign. The value zero is represented by a bit string with a one in the leftmost digit and zeros in all of the other digits. As shown in Table 3-4, all nonnegative values have a one as the high-order bit, and negative values have a zero in this position. In essence, excess notation divides a range of ordinary binary numbers in half and uses the lower half for negative values and the upper half for nonnegative values.

Table 3-4 ▶

Excess notation using four bits

Bit string	Decimal value	
1111	7	
1110	6	
1101	5	
1100	4	
1011	3	Nonnegative values
1010	2	
1001	1	
1000	0	
0111	-1	
0110	-2	
0101	-3	
0100	-4	
0011	-5	Negative values
0010	-6	
0001	7	
0000	-8	

To represent a specific integer value in excess notation you must know how many storage bits are to be used, whether the value fits within the numeric range of excess notation for that number of bits, and whether the value to be stored is positive or negative. For any number of bits, the largest and smallest values in excess notation will be $2^{(n-1)} -1$ and $-2^{(n-1)}$, where n is the number of available storage bits. For example, consider storing a signed integer in 8 bits using excess notation. Because the leftmost bit is a sign bit, the largest positive value that can be stored is 2^7-1, or 127_{10}, and the smallest negative value that can be stored is -2^7, or -128_{10}. The range of positive values appears to be smaller than the range of negative values because zero is considered a positive (nonnegative) number in excess notation. Attempting to represent larger positive or smaller negative values will result in errors because the leftmost (sign) bit may be overwritten with an incorrect value.

Now consider how +9 and -9 are represented in 8-bit excess notation. Both values are well within the numeric range limits for 8-bit excess notation. The ordinary binary representation of $+9_{10}$ in 8 bits is 00001001. Recall that the excess notation representation of zero is always a leading one-bit followed by all zero-bits—10000000 for 8-bit excess notation. Since +9 is nine integer values greater than zero, we can calculate the representation of +9 by adding its ordinary binary representation to the excess notation representation of zero as follows:

$$
\begin{array}{r}
10000000 \\
+ \quad 00001001 \\
\hline
10001001
\end{array}
$$

To represent negative values, we use a similar method based on subtraction. Since -9 is nine integer values less than zero, we can calculate the representation of -9 by subtracting its ordinary binary representation from the excess notation representation of zero as follows:

$$
\begin{array}{r}
10000000 \\
- \quad 00001001 \\
\hline
01110111
\end{array}
$$

Two's Complement Notation   In the binary number system, the complement of 0 is 1 and the complement of 1 is 0. The complement of a bit string is formed by substituting 0 for all values of 1 and 1 for all values of 0 (the complement of 1010 is 0101). This transformation is the basis of **two's complement** notation. In this notation, nonnegative integer values are represented as ordinary binary values. For example, a two's complement representation of 7_{10} using four bits is: 0111.

Bit strings for negative integer values are determined by the following transformation:

Complement of positive value + 1 = negative representation

Parentheses are a common mathematical notation for showing the complement of a value; if A is a numeric value, (A) represents its complement. In 4-bit two's complement representation, -7_{10} is calculated as:

$$
\begin{aligned}
(0111) + 0001 &= \\
1000 + 0001 &= \\
1001 &= -7_{10}
\end{aligned}
$$

As another example, consider the two's complement representation of $+35_{10}$ and -35_{10} in eight bits. The ordinary binary equivalent of $+35_{10}$ is 00100011, which is also the two's complement representation. To determine the two's complement representation of -35_{10}, use the previous formula with eight-bit numbers:

$$
\begin{aligned}
(00100011) + 00000001 &= \\
11011100 + 00000001 &= \\
11011101 &= -35_{10}
\end{aligned}
$$

Two's complement notation is awkward for most people, but it is highly compatible with digital electronic circuitry for the following reasons:

► The leftmost bit represents the sign.

► A fixed number of bit positions are used.

► Only two logic circuits are required to perform addition on single-bit values.

► Subtraction can be performed as addition of a negative value.

The latter two reasons enable CPU manufacturers to build processors with fewer components than are required for other integer data formats, which saves money and increases computational speed. For these reasons, all modern CPUs represent and manipulate signed integers using two's complement format.

Range and Overflow Most modern CPUs use either 32 or 64 bits to represent a two's complement value (we'll assume 32 bits for the remainder of this text unless otherwise specified). A small positive value like one occupies 32 bits even though, in theory, only two bits are required (one for the value and one for the sign). The unneeded bits are padded with leading zeros. Processor and data communication circuitry are more efficiently implemented for fixed-width formats. Although people can easily deal with numeric values of varying lengths, computer circuitry is not nearly as flexible. The additional complexity required to process variable-length data formats results in unacceptably slow performance.

The **numeric range** of a two's complement value is $-(2^{n-1})$ to $(2^{n-1}-1)$, where n is the number of bits used to store the value. The exponent is n-1 because one bit is used for the sign. For 32-bit two's complement format, the numeric range is $-2,147,483,648_{10}$ to $2,147,483,647_{10}$. With any fixed-width data storage format, it is possible that the result of executing a computation will be too large to fit within the format. For example, adding $2^{31}-1$ to any other positive number generates a result that is too big to be stored in 32 bits. This condition is referred to as **overflow** and is treated as an error by the CPU. Executing a subtraction instruction can also result in overflow, for example, $-(2^{31}) - 1$. Overflow occurs when the absolute value of a computational result contains too many bits to fit into a fixed-width data format.

As with most other aspects of CPU design, data format length is one design factor that needs to be balanced with others. Large formats reduce the chance of overflow by increasing the maximum absolute value that can be represented, but many bits are wasted (padded with leading zeros) when smaller values are stored. If bits were free then there would be no tradeoff. But extra bits increase processor complexity and storage requirements, which, in turn, increase computer system cost. A CPU designer chooses a data format width by balancing numeric range; the chance of overflow during program execution; and the complexity, cost, and speed of processing and storage devices.

To avoid overflow and to increase accuracy, some computers and programming languages define additional numeric data types called **double precision** data formats. A double precision data format combines two adjacent fixed-length data items to hold a single value. Double precision integers sometimes are called **long integers**.

Overflow also can be avoided by careful programming. If the programmer anticipates that overflow is possible, the units of measure for program variables can be made larger. For example, calculations on centimeters could be converted to meters or kilometers, as appropriate. Converting to different units of measure may not be practical if data values are restricted to whole numbers. If numbers with fractional parts are required, then a non-integer data format is required.

Real Numbers

A **real number** can contain both whole and fractional components. The fractional portion is represented by digits to the right of the radix point. For example, the following computation uses real number data inputs and generates a real number output:

$$18.0 \div 4.0 = 4.5$$

The equivalent computation in binary notation is:

$$10010 \div 100 = 100.1$$

Representing a real number within computer circuitry requires some way to separate the whole and fractional components of the value (that is, the computer equivalent of a written radix point). A simple way to accomplish this is to define a storage format in which a fixed-length portion of the bit string holds the whole portion and the remainder of the bit string holds the fractional portion. Figure 3-3 illustrates such a format with a sign bit and fixed radix point.

Figure 3-3 ▶

32-bit storage format for real numbers using a fixed radix point

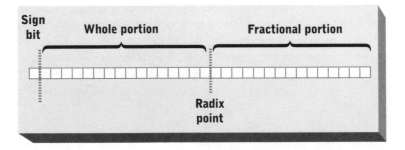

Data Representation

The format illustrated in Figure 3-3 is structurally simple due to the fixed location of the radix point. The advantage of this simplicity is simpler and faster CPU processing circuitry. Unfortunately, that processing efficiency is gained by limiting numeric range. Although the sample format uses 32 bits, its numeric range is substantially less than 32-bit two's complement. Only 16 bits are allocated to the whole portion of the value. Thus the largest possible whole value is $2^{16}-1$, or 65,535. The remaining bits store the fractional portion of the value, which can never be greater than or equal to one.

We could increase the numeric range of the format by allocating a larger number of bits to the whole portion (shifting the radix point in Figure 3-3 to the right). If the total size of the format is fixed at 32 bits, however, then the reallocation would reduce the number of bits used to store the fractional portion of the value. Reallocating bits from the fractional portion to the whole portion reduces the precision of fractional quantities. Reduced precision, in turn, reduces computational accuracy.

Floating Point Notation   One way of dealing with the tradeoff between range and precision is to abandon the concept of a fixed radix point. To represent extremely small (precise) values, move the radix point far to the left. For example, the value:

$$0.0000000013526473$$

has only a single digit to the left of the radix point. Similarly, very large values can be represented by moving the radix point far to the right, as in:

$$1352647300000000.0$$

Note that both examples have the same number of digits. The first example trades range of the whole portion for increased fractional precision, and the second example trades fractional precision for increased whole range by having the radix point float left or right. Values can be either very large or very small (precise) but not both at the same time.

People tend to commit errors when manipulating long strings of digits. To minimize errors, people often write large numbers in a more compact format called scientific notation. In scientific notation the two numbers on the previous page are represented as $13,526,473 \times 10^{-16}$ and $13,526,473 \times 10^8$. Note that the base of the number system (10) is part of the multiplier. The exponent can be interpreted as the number and direction of positional moves of the radix point, as illustrated in Figure 3-4. Negative exponents indicate movement to the left, and positive exponents indicate movement to the right.

Figure 3-4 ▶

Conversion of
scientific
notation to
decimal notation

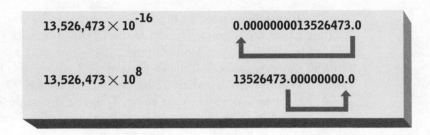

Real numbers are represented in computers using **floating point notation.**
Floating point notation is similar to scientific notation except that 2 (rather than 10)
is the base. A numeric value is derived from a floating point bit string according to
the following formula:

$$value = mantissa \times 2^{exponent}$$

The mantissa holds the bits that are interpreted to derive the digits of the
real number. By convention, the mantissa is assumed to be preceded by a radix
point. The exponent field indicates the position of the radix point.

Many CPU-specific implementations of floating point notation are possible.
Differences among these implementations may include the length and coding formats of the mantissa and exponent and the location of the radix point within the
mantissa. Although two's complement can be used to code either or both fields,
other coding formats may offer better design tradeoffs. Prior to the 1980s there
was little compatibility in floating point format among different CPUs, which
made it difficult or impossible to transport floating point data among different
computers.

The Institute of Electrical and Electronics Engineers (IEEE) addressed this
problem by defining standard representation formats for floating point data. IEEE
Standards 754 and 854 define 32-bit and 64-bit floating point coding formats,
respectively. Both formats have been adopted by computer and microprocessor
manufacturers. For the remainder of this chapter all references to floating point
representation will refer to the IEEE 32-bit format unless otherwise specified.

The IEEE 32-bit floating point format is illustrated in Figure 3-5. The leading sign bit applies to the mantissa, not the exponent, and is one if the mantissa
is negative. The 8-bit exponent is coded in excess notation (this implies that its
first bit is a sign bit). The 23-bit mantissa is coded as an ordinary binary number.
It is assumed to be preceded by a binary 1 and the radix point. This extends the
precision of the mantissa to 24 bits, although only 23 actually are stored.

Figure 3-5 ▶

IEEE 32-bit
floating point
format

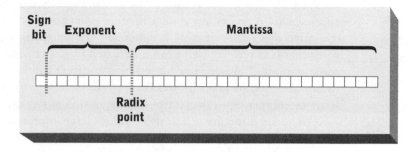

Range, Overflow, and Underflow The number of bits in a floating point string and the formats of the mantissa and exponent fields impose limits on the range of values that can be represented. The number of digits in the mantissa determines the number of significant (that is, nonzero) digits in the largest and smallest values that can be represented. The number of digits in the exponent determines the number of possible bit positions to the right or to the left of the radix point.

Using the number of bits assigned to each field, the largest absolute value of a floating point value appears to be:

$$1.11111111111111111111111 \times 2^{11111111}$$

Exponents containing all zeros and all ones, however, represent special data values in the IEEE standards. Thus, the usable exponent range is reduced and the decimal range for the entire floating point value is approximately 10^{-45} to 10^{38}.

Floating point numbers with large absolute values have large positive exponents. When overflow occurs, it always occurs within the exponent. Floating point representation is also subject to a related error condition called **underflow**. Very small numbers are represented by negative exponents. Underflow occurs when the absolute value of a negative exponent is too large to fit within the bits allocated to store it.

Precision and Truncation Recall that scientific notation, including floating point notation, trades numeric range for accuracy. Accuracy is reduced as the number of digits available to store the mantissa is reduced. The 23-bit mantissa used in floating point format represents approximately seven decimal digits of precision. However, many useful numbers contain more than seven nonzero decimal digits. One example of such a number is the decimal equivalent of the fraction 1/3:

$$1/3 = 0.33333333 \ldots$$

The number of digits to the right of the decimal point is infinite. Only a limited number of mantissa digits are available, however.

Numbers such as 1/3 are stored in floating point format by **truncation**. The numeric value is stored in the mantissa, starting with its most significant bit, until all available bits are consumed. The remaining bits are discarded. An error or approximation occurs any time a floating point value is truncated. However, the truncated digits are relatively insignificant compared to the significant, or large, value that is stored. Problems can arise when truncated values are used as input to computations. The error introduced by truncation can be magnified as truncated values are used to generate inaccurate results. The error resulting from a long series of computations that starts with truncated inputs can be relatively large.

An added difficulty is that more values have nonterminating representations in the binary system than in the decimal system. For example, the fraction ¹⁄₁₀ is nonterminating in binary notation. The representation of this value in floating point notation is a truncated value. Such problems can usually be avoided through careful programming. As a rule of thumb, programmers reserve floating point calculations for quantities that may vary continuously over wide ranges, such as measurements made by scientific instruments.

When possible, programmers use integer data types to avoid truncation. One common example is financial applications. Most monetary systems have at least one fractional monetary unit, such as pennies and quarters—fractions of a U.S. dollar. Novice programmers sometimes assume that monetary amounts should be stored and manipulated as floating point numbers. Truncation errors due to nonterminating representations of tenths and other fractions inevitably will occur. Cumulative errors will mount as truncated numbers, or approximations, are input to subsequent calculations.

Careful programmers use integer arithmetic for accounting and financial applications. To do so, a programmer stores and manipulates monetary amounts in the smallest possible monetary unit, for example, U.S. pennies or Mexican pesos. Small denominations are converted to larger ones only when needed for output, such as printing a dollar amount on a check or account statement.

Processing Complexity   The difficulty that we encounter when learning to use scientific and/or floating point notation is understandable. The formats are far more complex than integer data formats, which affects both people and computers. Although floating point formats are optimized for processing efficiency, they still require relatively complex processing circuitry. The simpler two's complement format used for integers requires much less complex computational circuitry.

The difference in processing circuitry complexity translates directly to a difference in speed when performing calculations. The magnitude of the difference depends on a number of factors, including the computation being performed and the exact details of the processing circuitry. As a general rule, simple computational operations like addition and subtraction take at least twice as long for floating point numbers than for integers. The difference is even greater for operations like division and exponentiation.

For this reason and for reasons of accuracy, careful programmers never use a real number when an integer will suffice. This is particularly important for data items that are frequently updated.

Character Data

English and many other languages use alphabetic letters, numerals, punctuation marks, and a variety of other special-purpose symbols (such as $ and &) in their written form. Each of these symbols is a **character**. A sequence of characters that forms a meaningful word, phrase, or other useful group is a **string**. In most programming languages single characters are surrounded by single quotation marks ('c') and strings are surrounded by double quotation marks ("computer").

Character data cannot be directly represented or processed within a computer because computers are designed to process only digital data (bits). Character data can be represented indirectly by defining a table that assigns numeric values to individual characters. For example, the integer values 0 through 9 could be used to represent the characters (numerals) '0' through '9', the uppercase letters 'A' through 'Z' could be represented as the integer values 10 through 36, and so forth.

A table-based substitution of one set of symbols or values for another is one example of a coding method. All coding methods share several important characteristics, including:

▶ All users must use the same coding and decoding method.

▶ The coded values must be capable of being stored or transmitted.

▶ The specific coding method represents a tradeoff among compactness, ease of manipulation, accuracy, range, and standardization.

The following sections describe some of the more common coding methods for character data.

EBCDIC Extended Binary Coded Decimal Interchange Code (EBCDIC) is a character coding method developed by IBM in the 1960s. All IBM computers used to use EBCDIC but, today, only the S/390 mainframe still uses it. Characters are encoded as strings of eight bits.

ASCII The **American Standard Code for Information Interchange** (ASCII) is a widely used coding method in data communication that has been adopted in the United States. The international equivalent of this character set is **International Alphabet Number 5 (IA5)**, a standard promulgated by the **International Standards Organization** (ISO). Virtually all other computers, including IBM's non-mainframe computers, use ASCII or its newer relative, Unicode.

ASCII is a 7-bit format because most computers and peripheral devices transmit data in bytes and because parity checking is used widely for detecting transmission errors. Parity checking and other error detection and correction methods are discussed in Chapter 8. For now, the important characteristic of parity checking is that it requires one extra bit per character. Thus, one of every eight bits is not part of the data value, leaving only seven bits for data representation.

The standard version of ASCII used for data transfer is sometimes called ASCII-7 to emphasize its 7-bit format. This coding table has 128, or 2^7, defined characters. Computers that use 8-bit bytes are capable of representing 256, or 2^8, different characters. In most computers the ASCII-7 characters are included within an 8-bit character coding table as the first, or lower, 128 table entries. The additional, or upper, 128 entries are defined by the computer manufacturer and typically are used for graphic characters such as line drawing characters or multinational characters such as á, ñ, Ö, and Ω. This encoding method is sometimes called ASCII-8. The term is a misnomer, as it implies that the entire table (all 256 entries) is standardized. In fact, only the first 128 entries are defined by the ASCII standard.

Table 3-5 shows portions of the ASCII and EBCDIC coding tables.

Table 3-5 ▶

Partial listing of ASCII and EBCDIC codes

Symbol	ASCII	EBCDIC
0	0110000	11110000
1	0110001	11110001
2	0110010	11110010
3	0110011	11110011
4	0110100	11110100
5	0110101	11110101
6	0110110	11110110
7	0110111	11110111
8	0111000	11111000
9	0111001	11111001
A	1000001	11000001
B	1000010	11000010
C	1000011	11000011
a	1100001	10000001
b	1100010	10000010
c	1100011	10000011

Device Control   When text is printed or displayed on an output device, it is often formatted in a particular way. For example, a customer record can be displayed on a video display terminal in a manner similar to a printed form. Text output to a printer is normally formatted into lines and paragraphs as it is in this textbook.

Certain text can be highlighted when printed or displayed using methods such as underlining, boldface font, or reversed background and foreground color.

At the time ASCII was defined, data was commonly transmitted one character at a time over a single data communication wire. ASCII defines a number of device control codes (see Table 3-6) that can be used for text formatting by sending them immediately before or after the character(s) they modify. Among the simpler codes are carriage return, which moves the print head or insertion point to the beginning of a text line; line feed, which moves the print head or insertion point down one line; and bell, which rings a bell or generates a short sound such as a beep. In ASCII, each of these functions is assigned a numeric code and a short character name, for example, CR for carriage return, LF for line feed, and BEL for bell.

Some ASCII device control codes are used to control data transfer. For example, ACK is sent to acknowledge correct receipt of data and NAK is sent to indicate that an error has been detected. ASCII defines a total of 33 device control codes that occupy the first 32 table entries (numbered 0 through 31) and the last table entry (number 127).

Software and Hardware Support Because characters are usually represented within the CPU as unsigned integers, there is little or no need for special character processing instructions. Instructions that move and copy unsigned integers behave the same whether the content being manipulated is an actual numeric value or an ASCII encoded character. Similarly, an equality or inequality comparison instruction that works for unsigned integers also works for values that represent characters.

The results of non-equality comparisons are less straightforward. The assignment of numeric codes to characters follows a specific order called a **collating sequence**. A greater-than comparison with two character inputs (for example, 'a' less than 'z') returns a result based on the numeric comparison of the corresponding ASCII codes; that is, whether the numeric code for 'a' is less than the numeric code for 'z'. If the character set has an order and if the coding method follows the order, then less-than and greater-than comparisons usually produce expected results.

However, using numeric values to represent characters can produce some unexpected or unplanned results. For example, the collating sequence of letters and numerals in ASCII follows the standard alphabetic order for letters and numeric order for numerals, but uppercase and lowercase letters are represented by different codes. As a result, an equality comparison between uppercase and lowercase versions of the same letter returns false because the numeric codes aren't identical (for example, 'a' does not equal 'A', as shown in Table 3-5). Punctuation symbols also have a specific order in the collating sequence, though there is no widely accepted ordering for such symbols.

Table 3-6 ▶

ASCII control
codes

Decimal code	Control character	Description
000	NUL	Null
001	SOH	Start of Heading
002	STX	Start of Text
003	ETX	End of Text
004	EOT	End of Transmission
005	ENQ	Enquiry
006	ACK	Acknowledge
007	BEL	Bell
008	BS	Backspace
009	HT	Horizontal Tabulation
010	LF	Line Feed
011	VT	Vertical Tabulation
012	FF	Form Feed
013	CR	Carriage Return
014	SO	Shift Out
015	SI	Shift In
016	DLE	Data Link Escape
017	DC1	Device Control 1
018	DC2	Device Control 2
019	DC3	Device Control 3
020	DC4	Device Control 4
021	NAK	Negative Acknowledge
022	SYN	Synchronous Idle
023	ETB	End of Transmission Block
024	CAN	Cancel
025	EM	End of Medium
026	SUB	Substitute
027	ESC	Escape
028	FS	File Separator
029	GS	Group Separator
030	RS	Record Separator
031	US	Unit Separator
127	DEL	Delete

ASCII Limitations ASCII's designers could not foresee the long lifetime (approaching one-half century) of the code or the revolutions in I/O device technologies that would occur during that lifetime. Modern I/O device characteristics, such as color, bitmapped graphics, and selectable fonts were never envisioned by the ASCII designers. Unfortunately, ASCII does not have sufficient range to define a set of control codes large enough to account for all of the formatting and display capabilities in modern I/O devices.

ASCII also is an English-based coding method. This is not surprising, given that at the time ASCII was defined, the United States accounted for most computer use and virtually all computer production. ASCII has a heavy bias toward Western languages in general and American English in particular, which became a significant limitation as computer use and production proliferated worldwide.

Recall that 7-bit ASCII has only 128 table entries, 33 of which are used for device control. Only 95 printable characters can be represented. This is enough for a usable subset of the characters commonly employed in American English text. But that subset does not include any modified Roman characters, such as ç and á, or those from other alphabets, such as Σ.

The International Standards Organization (ISO) partially addressed the problem by defining many different 256-entry tables based on the ASCII model. One of these, called **Latin-1**, contains the ASCII-7 characters in the lower 128 table entries and most of the characters used by Western European languages in the upper 128 table entries. The upper 128 entries are sometimes called the **multinational characters**. The number of available character codes in a 256-entry table, however, is still much too small to represent the full range of printable characters in world languages.

Further complicating matters is the fact that some printed languages are not based on characters in the Western sense. Chinese, Japanese, and Korean written text consists of ideographs, which are pictorial representations of words or concepts. Ideographs are, in turn, composed of individual graphic elements, sometimes called strokes, that number in the thousands. Other written languages, such as Arabic, present similar though less severe coding problems.

Unicode

The **Unicode** Consortium (see www.unicode.org) was founded in 1991 to develop a multilingual character encoding standard encompassing all of the world's written languages. The original members were Apple Computer Corporation and Xerox Corporation, but many computer companies soon joined. The effort enabled software and data to cross international boundaries. Major Unicode standard releases are coordinated with standard 10646 of the International Standards Organization (ISO). As of this writing, the latest standard is Unicode 4.01, published in March 2004.

Like ASCII, Unicode is a coding table that assigns nonnegative integers to represent individual printable characters. The ISO Latin-1 standard, which includes ASCII-7, is incorporated into Unicode as the first 256 table entries. Thus, ASCII is a subset of Unicode. The primary difference between ASCII and Unicode is the size of the coding table. Unicode uses 16-bit code providing 65,536 table entries numbered 0 through 65,535.

The additional table entries are primarily used for characters, strokes, and ideograms of languages other than English and its Western European siblings. Unicode includes many other alphabets such as Arabic, Cyrillic, and Hebrew, and thousands of Chinese, Japanese, and Korean ideograms and characters. Some extensions to the ASCII device control codes also are provided. As currently defined, Unicode can represent written text from all modern languages. Approximately 50,000 characters are defined in Unicode 4.01. Approximately 6,000 characters are reserved for private use.

Unicode is widely supported in modern software, including most operating systems and word processors. Because most CPUs represent characters as 32-bit unsigned integers, there is no real need to upgrade processor hardware to support Unicode. The primary impact of Unicode on hardware is in storage and I/O devices. Because the size of the numeric code is double that of ASCII, pure text files are twice as large. This appears to be a problem at first glance, but the impact is reduced by the use of custom word-processing file formats and the ever-declining cost of secondary storage. Most documents are not stored as ASCII or Unicode text files. Instead, they are stored in a format that intermixes text and formatting commands. Because formatting commands also occupy file space, the file size increase experienced when switching from ASCII to Unicode is generally far less than 100 percent.

Devices designed for character I/O have traditionally used ASCII by default and have implemented vendor-specific methods or used older ISO standards to process character sets other than Latin-1. The typical method is to maintain an internal set of alternate coding tables where each table contains a different alphabet or character set. Device-specific commands switch from one table to another. Unicode unifies and standardizes the content of these various tables. It offers a standardized method for processing international character sets. That standard has been widely adopted by I/O device manufacturers. Backward compatibility with ASCII is not a problem because Unicode includes ASCII.

Boolean Data

The Boolean data type has only two data values—true and false. Most people do not think of Boolean values as data items, but the primitive nature of CPU processing requires the ability to store and manipulate Boolean values. Recall that processing circuitry acts to physically transform input signals into output signals. If the input signals represent numbers and the processor is performing a computational function, such as addition, then the output signal represents the numerical result.

When the processing function is a comparison operation like greater than or equal to, the output signal represents a Boolean result of true or false. This result is stored in a register (as is any other processing result) and may be used by other instructions as input (for example, a conditional branch or GOTO). The Boolean data type is potentially the most concise coding format because only a single bit is required. For example, binary 1 can represent true and 0 can represent false. Most CPU designers seek to minimize the number of different coding formats in use to simplify processor design and implementation. CPUs generally use an integer coding format for integers to represent Boolean values. When coded in this manner, the integer value zero corresponds to false and all nonzero values correspond to true.

To conserve memory and storage, programmers will sometimes "pack" many Boolean values into a single integer by using each individual bit to represent a separate Boolean value. Although this conserves memory, it generally slows program execution due to the complicated instructions required to extract and interpret individual bits from an integer data item.

Memory Addresses

As described in Chapter 2, primary storage is a series of contiguous bytes of storage. Conceptually, each memory byte has a unique identifying or serial number. The identifying numbers are called addresses, and their values start with zero and continue in sequence (with no gaps) until the last byte of memory is addressed.

In some CPUs this conceptual view is also the physical method of identifying and accessing memory locations. That is, memory bytes are identified by a series of nonnegative integers. This approach to assigning memory addresses is called a flat memory model. In CPUs that use a flat memory model, it is logical and typical to use two's complement or unsigned binary as the coding format for memory addresses. The advantage of this approach is that it minimizes the number of different data types and the complexity of processor circuitry.

Some CPU designers choose not to use simple integers as memory addresses in order to maintain backward compatibility with earlier generations of CPUs that didn't use a flat memory model. For example, the Intel Pentium family of CPUs maintains backward compatibility with the 8088 CPU used in the first IBM personal computer. Earlier generations of processors typically used an alternate approach to memory addressing called a segmented memory model. In a segmented memory model, primary storage is divided into a series of equal-sized (for example, 64 KB) segments called pages. Pages are identified by sequential nonnegative integers. Each byte within a page is also identified by a sequentially assigned nonnegative integer. Each byte of memory has a two-part address: the first part identifies the page, and the second part identifies the byte within the page.

Because a segmented memory address contains two integer values, the data coding method for single signed or unsigned integers cannot be used. Instead, a specific coding format for memory addresses must be defined and implemented. For this reason, CPUs with segmented memory models have an extra data type, or coding format, for storing memory addresses.

Technology
Focus

Intel Memory Address Formats

Intel microprocessors have been used in personal computers since the first IBM personal computer was sold in 1981. The original IBM PC used an Intel 8088 microprocessor. Later generations of Intel processors (the 8086, 80286, 80386, 80486, and Pentium family of processors) have all maintained backward compatibility with the 8088. That is, each subsequent processor can execute CPU instructions designed for the 8088. Backward compatibility has been considered an essential ingredient in the success and proliferation of personal computers based on the Intel 80X86 family of CPUs.

Early Intel microprocessors were designed and implemented with speed of processor operation as a primary goal. Because a general-purpose CPU uses memory addresses in nearly every instruction execution cycle, it is important to make memory access as efficient as possible. Processor complexity rises with the number of bits that must be processed simultaneously. Large coding formats for memory addresses and other data types require complicated and relatively slow processing circuitry. Intel designers wanted to minimize the complexity of processor circuitry associated with memory addresses. They also needed to balance the desire for efficient memory processing with the need to maintain a relatively large range of possible memory addresses.

Intel designers made two significant design decisions for the 8088 to achieve this balance:

► A 20-bit size for memory addresses

► A segmented memory model

The 20-bit address format limited usable (addressable) memory to 1 megabyte (MB) (2^{20} addressable bytes). This did not seem like a significant limitation because few computers (including most mainframes) had that much memory at the time. The 20-bit format is divided into two parts: a 4-bit segment identifier and a 16-bit segment offset. The 16-bit segment offset identifies a specific byte within a 64 (2^{16})-KB memory segment. Because 4 bits are used to represent the segment, there are 16, or 2^4, possible segments.

Intel designers anticipated that most programs would fit within a single 64-KB memory segment. Further, they knew that 16-bit memory addresses could be processed much more efficiently than larger addresses. Therefore, they defined two types of address-processing functions—those that used the 4-bit segment portion of the address and those that ignored it. Memory could be accessed rapidly when the processor ignored the segment identifier. Memory access was much slower when the segment identifier was used. Intel designers assumed that most memory accesses would not require the segment identifier.

Both the 64-KB segment size and the 1-MB total memory limit soon became significant constraints. Although early programs for the IBM personal computer were generally smaller than 64 KB, later versions of those same programs were larger. Because the processor design imposed a performance penalty when the segment identifier was used, later versions ran much more slowly than earlier versions. In addition, both the operating system and the computer hardware were becoming more complex and consuming more memory resources. As a result, computer designers, operating systems designers, and users chafed under the 1 MB limit.

Intel designers first addressed these constraints in the 80286. They increased the segment identifier to 8 bits, increasing total addressable memory to 16 MB. The 80386 went a step further by providing two methods of addressing. The first, called real mode, was a method compatible with that of the 8088. The second, called protected mode, was an entirely new method based on 32-bit memory addresses. Protected mode addressing eliminated the performance penalty for programs larger than 64 KB. The Pentium and Xeon family of microprocessors continue to use both addressing methods.

This discussion, and the earlier discussion of ASCII, illustrates a few of the pitfalls for computer designers in choosing and implementing data representation methods and formats. A fundamental characteristic of any data representation method is that it is a tradeoff. For Intel memory addresses, the tradeoff was made among processor cost, speed of program execution, and the memory requirements of typical PC software. The tradeoff may have been optimal or, at least, reasonable, given the state of all these variables in 1981. But neither time nor technology stands still. The optimality of any CPU design decision, including data representation formats, is quickly compromised as technology changes. Attempts to maintain backward compatibility with previous CPU designs can be difficult, expensive, and may severely limit future design choices.

DATA STRUCTURES

The previous sections outlined the data types that a CPU can manipulate directly: integer, real number, character, Boolean, and memory address. The data types directly supported by a CPU are sometimes called **primitive data types**, or **machine data types**. Computers can also process more complex data such as strings, arrays, text files, databases, and digital representations of audio and image data such as MP3, JPEG, and MPEG files. Audio and image data representation and related hardware devices are discussed in Chapter 7. The remainder of this chapter will concentrate on other complex data formats that are commonly manipulated by system and application software.

It would be difficult to develop programs of any kind if only primitive data types were available. Most application programs need to combine primitive data items to form useful aggregations. A common example is a character or text string. Most application programs define and use character strings, but few CPUs provide instructions that directly manipulate them. Application development is simplified if character strings can be defined and manipulated (that is, read, written, and compared) as a single unit instead of one character at a time.

A **data structure** is a related group of primitive data elements that is organized for some type of common processing. Data structures are defined and manipulated within software. Computer hardware cannot manipulate data structures directly, but must deal with them in terms of their primitive components such as integers, floating point numbers, single characters, and so on. Software must translate operations on data structures into an equivalent set of machine instructions that operate on individual primitive data elements. For example, Figure 3-6 shows a comparison of two strings decomposed into comparison operations on each character.

Figure 3-6 ▶

Software decomposes operations on data structures into operations on primitive data components

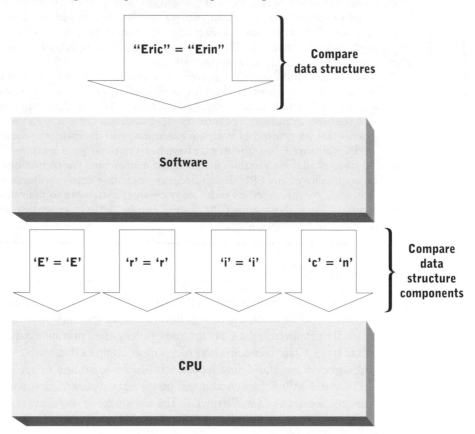

The complexity of data structures is limited only by the imagination and skill of programmers. As a practical matter, certain data structures are useful in a wide variety of situations and are commonly implemented. Examples of such data structures include character strings or arrays, records, and files. System software often provides application services to manipulate these commonly used data structures. For example, an operating system normally provides services for reading and writing to and from files.

Other data structures are less commonly supported by system software. Examples of these include numeric arrays, indexed files, and complex database structures. Indexed files are supported in some, but not all, operating systems. Numeric arrays are supported within most programming languages but not within most operating systems. Database structures normally are supported by a database management system. Most programming languages support direct manipulation of character strings.

Data structures have an important role in system software development. For example, linked lists are commonly used by operating systems to keep track of memory blocks allocated to programs and disk blocks allocated to files and directories. Indexes are widely used within database management systems to speed search and retrieval operations. Programmers who develop system software generally take an entire course in data structures and efficient methods for manipulating them.

Pointers and Addresses

Whether implemented within system or application software, virtually all data structures make extensive use of pointers and addresses. A **pointer** is a data element that contains the address of another data element. An **address** is the location of some data element within a storage device. Addresses vary in content and representation depending on the storage device being addressed. Secondary storage devices are normally organized as a sequence of data blocks. A block is a group of bytes that is read or written as a unit. For example, disk drives on personal computers usually read and write data in 512-byte blocks. For block-oriented storage devices, an address can normally be represented as an integer containing the sequential position of the block. Integers also can be used to represent the address of an individual byte within a block.

As described earlier, memory addresses can be complex if a segmented memory model is used. For the purpose of discussing data structures, we will assume that a flat memory model is used and that memory addresses are represented by nonnegative integers.

Arrays and Lists

Many types of data can be grouped into lists. A list is a set of related data values. In mathematics, a list is considered unordered. That is, no specific list element is designated as the first, second, or last element. When writing software, a programmer usually prefers to impose some ordering upon the list. For example, a list of the days of the week may be ordered sequentially, starting with Monday.

An **array** is an ordered list in which each element can be referenced by an index to its position. An example of an array for the first five letters of the English alphabet appears in Figure 3-7. Note that the index values are numbered starting at zero, a common (although not universal) practice in computer programming. Although the index values appear in Figure 3-7, they are not stored. Instead, they are inferred from the location of the data value within the storage allocated to the array. In a high-level programming language, individual array elements are normally referenced by the name of the array and the index value. For example, the third letter of the alphabet stored in an array might be referenced as:

alphabet[2]

where alphabet is the name of the array and 2 is the index value (numbered from zero).

Figure 3-7 ▶

Array elements
in contiguous
storage locations

Index	Value
0	A
1	B
2	C
3	D
4	E

Figure 3-8 shows a character array, or string, stored sequentially in contiguous memory locations. In this example, each character of the name John Doe is stored within a single byte of memory and the characters are ordered in sequential byte locations starting at byte 1000. An equivalent organization may be used to store the name on a secondary storage device. The address of an individual array element can be calculated with the starting address of the array and the index of the desired element. For example, if we wish to retrieve the third character in the array, we can compute its address as the sum of the address of the first element plus the index value, assuming index values start at zero.

　　　　　　　　　　　　　　　　　　　　　　　　　　　　　Data Representation

Figure 3-8 ▶

Character array
in contiguous
storage locations

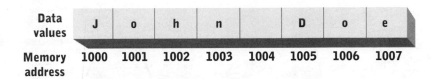

Data values	J	o	h	n		D	o	e
Memory address	1000	1001	1002	1003	1004	1005	1006	1007

Using contiguous storage locations, especially in secondary storage devices, complicates the allocation of storage locations. For example, it may be difficult to add new elements to the end of an array if those storage locations are already allocated to other data items. Because of this, contiguous storage allocation is generally used only for arrays of fixed length.

For a variety of reasons, it may be desirable to store individual elements of the array in widely dispersed storage locations. A **linked list** is a data structure that uses pointers so list elements can be scattered among nonsequential storage locations. Figure 3-9 shows a generic example of a **singly linked list**. Each list element occupies two storage locations. The first location holds the data value of a list element. The second storage location holds the address of the next element value. Figure 3-10 shows a character string stored within a linked list. Note that individual characters are scattered among nonsequential, or noncontiguous, storage locations.

Figure 3-9 ▶

Value and pointer fields of a singly linked list

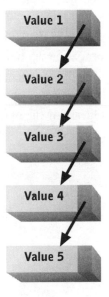

Figure 3-10 ▶

Character array
stored in
noncontiguous
memory
locations

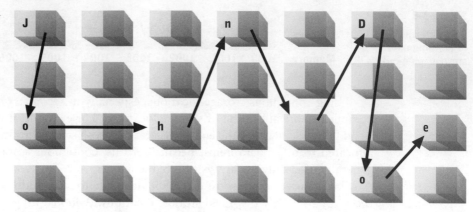

More storage locations are required for a linked list than for an array with equivalent content because both data and pointers must be stored. Using pointers also complicates the task of locating individual array elements. References to specific array elements must be resolved by following the chain of pointers, starting with the first array element. This can be very inefficient if the number of array elements is large. For example, accessing the 1000th element requires reading and following the pointers within the first 999 elements.

Linked lists are easier to expand or shrink than arrays. The procedure to add a new element is as follows (see Figure 3-11):

1. Allocate storage for the new element.

2. Copy the pointer from the element preceding the new element into the pointer field of the new element.

3. Write the address of the new element into the pointer field of the preceding element.

Figure 3-11 ►

Insertion of a
new element into
a singly
linked list

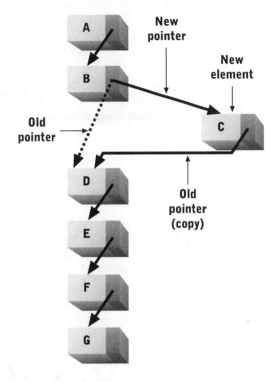

In contrast, insertion of an element into a list stored in contiguous memory can be very time-consuming. The procedure is as follows (see Figure 3-12):

1. Allocate a new storage location to the end of the list.
2. For each element past the insertion point, copy the element value to the next storage location, starting with the last element and working backward to the insertion point.
3. Write the new element value in the storage location at the insertion point.

Inserting an element near the beginning of the array is highly inefficient due to the large number of required copy operations.

Figure 3-12 ►

Insertion of a new element into an array stored in contiguous memory locations

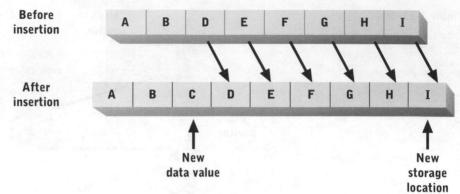

Before insertion

| A | B | D | E | F | G | H | I |

After insertion

| A | B | C | D | E | F | G | H | I |

New data value   New storage location

Figure 3-13 depicts a more complicated linked list called a **doubly linked list**. Each element of a doubly linked list has two pointers. One pointer points to the next element in the list and the other points to the previous element in the list. The primary advantage of doubly linked lists is that they can be traversed in either direction with equal efficiency. The primary disadvantages are that more pointers must be updated each time an element is inserted into or deleted from the list, and more storage locations are required to hold the extra set of pointers.

Figure 3-13 ►

A doubly linked list

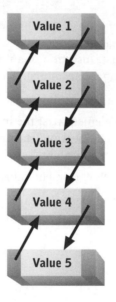

Value 1

Value 2

Value 3

Value 4

Value 5

Records and Files

A **record** is a data structure composed of other data structures or primitive data elements. Records are commonly used as a unit of input and output to files or databases. For example, consider the following data items:

Account-Number   Street-Address
Last-Name   City
First-Name   State
Middle-Initial   Zip-Code

This set of data may be the contents of the data structure for a customer record as shown in Figure 3-14. Each component of the record either is a basic data element (for example, Middle-Initial) or another data structure (for example, a character array for Street-Address). To speed input and output, records normally are stored in contiguous storage locations. This restricts the record's array components to a fixed length.

Figure 3-14 ▶

Record data structure

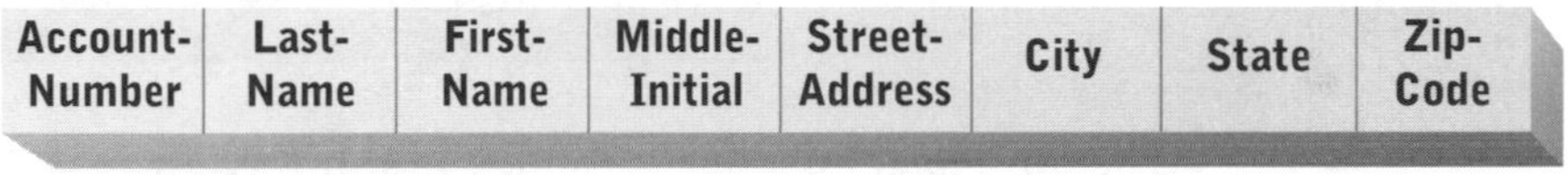

Account-Number	Last-Name	First-Name	Middle-Initial	Street-Address	City	State	Zip-Code

A sequence of records on secondary storage is called a file. A sequence of records stored within main memory is normally called a table, although its structure is essentially the same as a file. Files can be organized in many different ways, the most common being sequential and indexed.

In a sequential file, the records are stored in contiguous storage locations. As with arrays stored in contiguous storage, accessing a specific record is relatively simple. The address of the nth record in a file can be computed as:

$$address\text{-}of\text{-}first\text{-}record + ((n\text{-}1) \times record\text{-}size)$$

If the first byte of the first record is at address 1 and the size of a record is 200 bytes, the address of the fourth record is 601.

Sequential files suffer the same problems as contiguous arrays when records are being inserted and deleted. A copy procedure similar to that shown earlier in Figure 3-11 must be executed to add a record to a file. The procedure is even less efficient for files than for arrays because of the relatively large size of the records that must be copied.

One method of solving this problem is to use linked lists. With files, the data elements of a linked list are entire records instead of basic data elements. The methods for searching, record insertion, and record deletion essentially are the same as previously described.

Another method of organizing files uses an index. An **index** is an array of pointers to records. The pointers can be ordered in any sequence desired by the user. For example, a file of customer records may be ordered by ascending account number as shown in Figure 3-15.

Figure 3-15 ▶

An indexed file

The advantage of using an index lies in the efficiency of record insertion, deletion, and retrieval. When a record is added to a file, storage is allocated for the record and the data is placed in the storage location. The index then is updated by inserting the address of the new record. The index update is accomplished by the same procedure as an array update. Because the array contains only pointers, it is relatively small in size and fast to update.

Classes and Objects

Up to this point, we have discussed data and programs as two fundamentally different things. Programs contain instructions, which transform data inputs into data outputs when executed. Data items are held within storage devices, moved to CPU registers when needed, transformed into data outputs by executing instructions, and sent to output or storage devices once again. Under this view of computer systems and software behavior, data items are relatively static, and programs are the active means for transforming and updating data items.

In the 1980s and 1990s, computer researchers developed an alternative view of computer and software behavior that combined program instructions and data into a single data structure. A **class** is a data structure that contains both traditional, or static, data elements and programs that manipulate that data. The programs in a class are **methods**. A class combines related data items in much the same way as a record, but it extends the record to include methods that manipulate the data items.

Consider the customer record in Figure 3-14 as a starting point for defining a Customer class. The record contains some of the primitive data items and data structures that describe various features of a customer (others, such as account balance, could be added to create a more complete representation of a customer).

To turn the customer record into a Customer class, methods must be added that manipulate or modify data elements in the record. For example, you could add a simple method for updating the content of an individual data element, called Update-First-Name. You could check for legal values, such as that the value supplied for Zip-Code is a valid U.S. Zip Code within the customer's state. Functions such as Print-Mailing-Label, or methods to modify the customer's account balance to reflect transactions, such as Apply-Payment, could be added. One possible set of data elements and methods for the Customer class is shown in Figure 3-16.

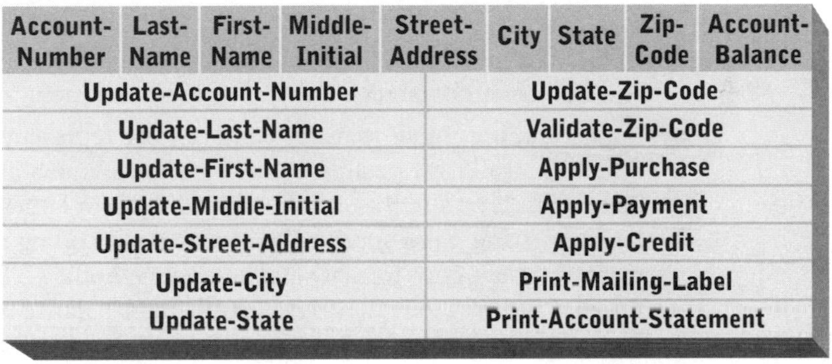

Figure 3-16 ▶

A Customer class containing traditional data elements (darker boxes) and methods (lighter boxes)

Account-Number	Last-Name	First-Name	Middle-Initial	Street-Address	City	State	Zip-Code	Account-Balance
Update-Account-Number					Update-Zip-Code			
Update-Last-Name					Validate-Zip-Code			
Update-First-Name					Apply-Purchase			
Update-Middle-Initial					Apply-Payment			
Update-Street-Address					Apply-Credit			
Update-City					Print-Mailing-Label			
Update-State					Print-Account-Statement			

An **object** is one instance, or variable, of the class. Each person who is a customer would be represented by one variable or object of the Customer class. Each object could be stored within a storage device, and each object's data elements would occupy a separate part of the storage device.

Viewed as just another data structure, an object differs little from a record. Much as the data element Street-Address could be represented with a character array, a method could be represented as an array of CPU instructions. Methods could be represented in other ways, including linked lists of instructions, or pointers to files containing sequential sets of instructions. In essence, the only new primitive data type required to represent a method within an object is an instruction. As we'll see in the next chapter, instructions, like data, have specific representation and storage formats.

SUMMARY

► Data can be represented in many ways. To be processed by any device, data must be converted from its native format into a form suitable for the processing device. Modern computers represent data as electronic signals and implement processing devices using electronic circuitry. Electronic processing devices exploit the physical laws of electricity. Because the laws of electricity can be stated as mathematical equations, electronic devices can perform the computational functions embedded within those equations. Other ways of implementing processors, such as mechanics and optics, exploit similar mathematically stated physical laws.

► All data, including nonnumeric data, are represented within a modern computer system as strings of binary digits, or bits. Bits are used because zero and one can be encoded in clearly recognizable electrical states, such as high and low voltage. Binary electrical states can be processed and stored by electrical devices that are reliable and cheap. A handful of primitive data types are represented and processed by a CPU, including integers, real numbers, characters, Boolean values, and memory addresses.

► Numeric values other than zero and one are represented by combining multiple bits into larger groups, called bit strings, much as multiple decimal digits such as 2, 5, and 9 can be combined to form larger values such as 592. Each bit string has a specific data format and coding method. There are many different data formats and coding methods. CPU designers choose formats and methods that represent the best balance among compactness, ease of manipulation, accuracy, range, and standardization. The optimal balance can vary depending on the type of data and the intended uses for that data.

► Integers have no fractional component and can use relatively simple data formats. Two's complement is the most common integer format, though excess notation sometimes is used. Real numbers have both whole and fractional components, and their complex structure requires a complex data format called floating point notation. Floating point notation represents a numeric value as a mantissa multiplied by a positive or negative power of two. A value can have many digits of precision in floating point notation for very large or very small magnitudes, but not both large and small magnitudes at the same time. Floating point formats used in modern computers typically follow a standard promulgated by the IEEE. Floating point format is less accurate and more difficult to process than two's complement format.

► Character data is not inherently numeric. Characters are converted to numbers by means of a coding table. Each character occupies a sequential position within the table. A character is represented by converting it to its integer table position. Characters are extracted from integers by the reverse process. Many different tables are possible. Most modern computers either use the ASCII or Unicode coding tables. ASCII is an older standard geared toward the English language. Unicode is a more modern table that is large enough to encompass ASCII and all of the world's written languages.

► A Boolean data value must be true or false. Memory addresses can be simple or complex numeric values depending on whether the CPU uses a flat or segmented memory model. Flat memory addresses can be represented as a single integer. Segmented memory addresses require multiple integers. Many CPUs also provide double precision numeric data types, which double the number of bits used to store a value.

► Programs often define and manipulate data in larger and more complex units than primitive CPU data types. A data structure is a related group of primitive data elements that is organized for some type of common processing. Data structures are defined in software. To enable a CPU to manipulate a data structure, software decomposes operations on the data structure into an equivalent set of operations on its primitive components. Commonly used data structures include arrays, linked lists, records, tables, files, indices, and objects. Many data structures use pointers, which are stored memory addresses, to link primitive data components.

In this chapter you've learned the various ways that data is represented within a modern computer system, focusing particularly on the relationship between data representation and the CPU. This chapter and the preceding chapters introduced basic concepts and terminology of systems architecture. The understanding you've gained in the first three chapters lays the foundation for the more detailed discussion of the hardware implementations of data processing, storage, and communication in upcoming chapters.

Key Terms

address
American Standard Code for
   Information Interchange
   (ASCII)
array
base
binary number
bit
bit string
Boolean data type

Boolean logic
byte
character
class
collating sequence
data structure
decimal point
double precision
doubly linked list
excess notation

Extended Binary Coded
   Decimal Interchange Code
   (EBCDIC)
flat memory model
floating point notation
hexadecimal notation
high-order bit
index
integer
International Alphabet
   Number 5 (IA5)

International Standards
  Organization (ISO)
Latin-1
least significant digit
linked list
long integer
low-order bit
machine data type
method
most significant digit

multinational character
numeric range
object
octal notation
overflow
pointer
primitive data type
radix
radix point
real number

record
segmented memory model
singly linked list
string
truncation
two's complement
underflow
Unicode
unsigned integer

—————————————— Vocabulary Exercises ——————————————

1. An element in a(n) _____ contains pointers to both the next and previous list elements.

2. _____ notation encodes a real number as a mantissa multiplied by a power (exponent) of two.

3. A(n) _____ is an integer stored in double the normal number of bit positions.

4. Increasing the size (number of bits) of a numeric representation format increases the _____ of values that can be represented.

5. Assembly (machine) language programs for most computers use _____ notation to represent memory address values.

6. A(n) _____ is a data item composed of multiple primitive data items.

7. In some IBM mainframe computers, characters are encoded according to the _____ coding scheme.

8. A(n) _____ is the address of another data item or structure.

9. In a positional numbering system, the _____ separates digits representing whole number quantities from digits representing fractional quantities.

10. A(n) _____ is an array of characters.

11. Most Intel CPUs implement a(n) _____ where each memory address is represented by two integers.

12. A set of data items that can be accessed in a specified order using a set of pointers is called a(n) _____.

13. A(n) _____ contains eight _____.

14. A(n) _____ list stores one pointer with each list element.

15. The result of adding, subtracting, or multiplying two integers may result in overflow, but never _____ or _____.

16. A(n) _____ is a sequence of primitive data elements stored in sequential storage locations.

17. A(n) _____ is a group of data elements that usually describe a single entity or event.

18. A(n) _____ data item can contain only the values true or false.

19. A(n) _____ is an array of data items, each of which contains a key value and a pointer to another data item.

20. Many computers implement _____ numeric data types to provide greater accuracy and prevent overflow and underflow.

21. Unlike ASCII and EBCDIC, _____ is a 16-bit character coding table.

22. The _____ is the bit of lowest magnitude within a byte or bit string.

23. _____ occurs when the result of an arithmetic operation exceeds the number of bits available to store it.

24. Within a CPU, _____ arithmetic generally is easier to implement than _____ arithmetic due to a simpler data coding scheme and data manipulation circuitry.

25. Under the _____, memory addresses consist of a single integer.

26. The _____ has defined a character coding table called _____, which combines the ASCII-7 coding table with an additional 128 Western European multinational characters.

27. Portability of data represented in _____ is guaranteed if each computer's CPU represents real numbers using an IEEE standard notation.

28. The ordering of characters within a coding table is called a(n) _____.

29. A(n) _____ is a data structure that contains both static data and methods.

30. A(n) _____ is one instance or variable of a class.

1. What is the binary representation of the decimal number 10? What is the octal representation? What is the hexadecimal representation?

2. Why is binary data representation and signaling the preferred method of computer hardware implementation?

3. What is excess notation? What is two's complement notation? Why are they needed (in other words, why can't integer values be represented by ordinary binary numbers)?

4. What is the numeric range of a 16-bit two's complement value? A 16-bit excess notation value? A 16-bit unsigned binary value?

5. What is overflow? What is underflow? How can the probability of their occurrence be minimized?

6. How and why are real numbers more difficult to represent and process than integers?

7. Why may a programmer choose to represent a data item in IEEE 64-bit floating point format instead of IEEE 32-bit floating point format? What additional costs may be incurred at run time (when the application program executes) as a result of using the 64-bit instead of the 32-bit format?

8. Why doesn't a CPU evaluate the expression 'A' = 'a' as true?

9. What are the differences between ASCII and EBCDIC?

10. What primitive data types normally can be represented and processed by a CPU?

11. What is a data structure? List several types of common data structures.

12. What is an address? What is a pointer? For what are they used?

13. How is an array stored in main memory? How is a linked list stored in main memory? What are their comparative advantages and disadvantages? Give an example of data that would be best stored as an array. Give an example of data that would be best stored as a linked list.

14. How does a class differ from other data structures?

Problems and Exercises

1. Develop an algorithm or program to implement the following function:

    insert_in_linked_list (element,after_pointer)

    The parameter "element" is a data item to be added to a linked list. The parameter "after_pointer" is the address of the element after which the new element will be inserted.

2. Develop an algorithm or program to implement the following function:

insert_in_array (element,position)

The parameter "element" is a data item to be added to the array. The parameter "position" is the array index at which the new element will be inserted. Make sure to account for elements that must be moved over to make room for the new element.

3. Consider the following binary value:

1000 0000 0010 0110 0000 0110 1101 1001

What number (base ten) is represented if the value is assumed to represent a number stored in two's complement notation? Excess notation? IEEE floating point (32-bit) notation?

4. How are the values 515_{10} and -515_{10} represented as ordinary binary numbers? How are they represented as octal and hexadecimal numbers? How are they represented in 16-bit excess notation? How are they represented in 16-bit two's complement notation?

5. How is the binary value $101101 \times 2^{-101101}$ represented in 32-bit IEEE floating point notation?

Research Problems

1. Choose a commonly used microprocessor such as the Intel Pentium (www.intel.com) or IBM PowerPC (www.ibm.com). What data types are supported? How many bits are used to store each data type? How is each data type internally represented?

2. Most personal and office productivity applications, such as word processors, are marketed internationally. To minimize development costs, software producers develop a single version of the program, but have separate configurations for items like menus and error messages that vary from language to language. Investigate a commonly used development environment for application tools such as Microsoft Visual Studio (www.microsoft.com). What tools and techniques (for example, Unicode data types and string tables) are supported for developing multilingual programs?

3. Object-oriented programming has been adopted widely due to its inherent ability to reuse code. Most application development software provides class libraries and extensive support for complex data structures, including various types of linked lists. Investigate one of these libraries, such as Microsoft Foundation Classes (www.microsoft.com) or the Java 2 Platform (java.sun.com) application program interface (API). What data structures are supported by the library? What types of data are recommended for use with each data structure object? Which classes contain which data structures, and what methods does the library provide?

Chapter 4

Processor Technology and Architecture

Chapter 2 gave a brief overview of computer processing, including the function of a processor, general-purpose and special-purpose processors, and the components of a central processing unit (CPU). This chapter explores CPU operation, instructions, components, and implementation (see Figure 4-1). It also takes a look at future trends in processor technology and architecture.

Figure 4-1 ▶

Topics covered in this chapter

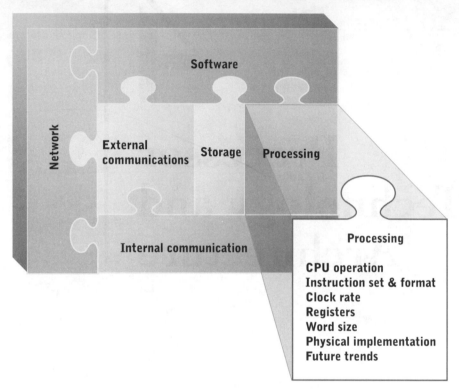

CPU OPERATION

Recall from Chapter 2 that a CPU has three primary components—the control unit, the arithmetic logic unit (ALU), and a set of registers (see Figure 4-2). The control unit moves data and instructions between main memory and registers. The ALU performs all computation and comparison operations. Registers are storage locations that hold inputs and outputs for the ALU.

Figure 4-2   ►

C P U
components

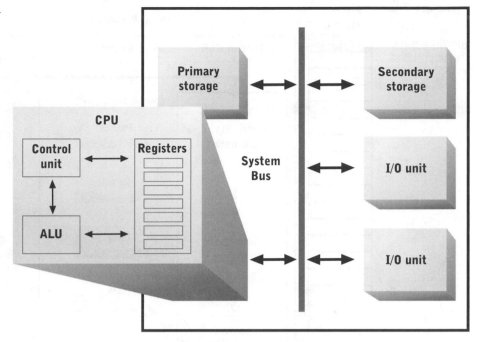

A complex chain of events occurs when the CPU executes a program. To start, the control unit reads the first instruction from primary storage. The control unit then stores the instruction in a register and, if necessary, reads data inputs from primary storage and also stores them in registers. If the instruction is a computation or comparison instruction, the control unit signals the ALU what function to perform, where the input data is located, and where to store the output data. Instructions to move data to memory, I/O devices, or secondary storage are executed by the control unit itself. When the first instruction has been executed, the next instruction is read and executed. The process continues until the final instruction of the program has been executed.

The actions performed by the CPU can be divided into two groups—the **fetch cycle** (or **instruction cycle**) and the **execution cycle**. During the fetch cycle, data inputs are prepared for transformation into data outputs. During the execution cycle, the transformation takes place and data output is stored. The CPU constantly alternates between instruction and execution cycles. Figure 4-3 shows the flow between instruction and execution cycles (denoted by solid arrows) and data and instruction movement (denoted by dashed arrows).

Figure 4-3 ▶

Control and data
flow during the
instruction and
execution cycles

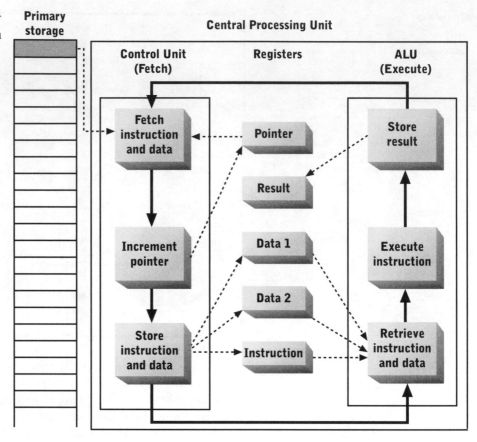

During the fetch cycle, the control unit:

► Fetches an instruction from primary storage

► Increments a pointer to the location of the next instruction

► Separates the instruction into components—the instruction code (or number) and the data inputs to the instruction

► Stores each component in a separate register

During the execution cycle, the ALU:

► Retrieves the instruction code from a register

► Retrieves data inputs from registers

► Passes data inputs through internal circuits to perform the addition, subtraction, or other data transformation

► Stores the result in a register

At the conclusion of the execution cycle a new fetch cycle is started. The control unit keeps track of the next program instruction location by incrementing a pointer after each fetch. The second program instruction is retrieved during the second fetch cycle, the third instruction is retrieved during the third fetch cycle, and so forth.

INSTRUCTIONS AND INSTRUCTION SETS

An **instruction** is a command to the CPU to perform one of its primitive processing functions on specific data inputs. It is the lowest-level command that software can direct a processor to perform. As stored in memory, an instruction is merely a bit string that is logically divided into a number of components. The first group of bits represents the unique binary number of the instruction, commonly called the **op code**. Subsequent groups of bits hold the input values for the instruction, called **operands**. The operand can contain a data item (such as an integer value) or the location of a data item (such as a memory address, a register address, or the address of a secondary storage or I/O device).

Figure 4-4 ▶

An instruction containing one op code and two operands

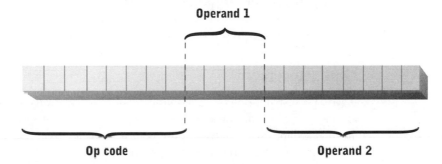

An instruction directs the CPU to route electrical signals representing data input(s) through a predefined set of processing circuits that implement the desired function. Data inputs are accessed from storage or extracted directly from the operands and stored in one or more registers. For computation and logic functions, the ALU accesses the registers and sends the corresponding electrical signals through the appropriate processing circuitry. This circuitry transforms the input signals into output signals representing the processing result. This result is stored in a register in preparation for movement to a storage device or I/O device, or for use as input to another instruction.

The control unit executes some instructions without assistance from the ALU, including instructions for moving or copying data, as well as simple tasks like halting or restarting the CPU.

The collection of instructions that a CPU can process is called the CPU's **instruction set**. Instruction sets vary among CPUs in the following ways:

► Number of instructions
► Size of individual instructions, op codes, and operands
► Supported data types
► Number and complexity of processing operations performed by individual instructions

Instruction set variations reflect differences in design philosophy, processor fabrication technology, class of computer system, and type of application software. CPU cost and speed depend on these characteristics.

The full range of processing operations expected of a modern computer can be implemented with approximately one dozen instructions. Such an instruction set can perform all of the computation, comparison, data movement, and branching functions for integer and Boolean data types. Computation and comparison of real numbers can be accomplished by software with complex sequences of integer instructions operating separately on the whole and fractional parts. Small instruction sets were common in early CPUs and microprocessors, but most modern CPUs have much larger instruction sets. Details of the minimal instruction set are described in the following sections.

Data Movement

A **MOVE** instruction copies data bits to storage locations. MOVE can copy data among any combination of registers and primary storage locations. A **load** operation is a data transfer from main memory into a register. A **store** operation is a data transfer from a register into primary storage.

MOVE tests the bit values in the source location and places copies of those values in the destination location. The former bit values in the destination are overwritten. At the completion of the MOVE, both sending and receiving locations hold identical bit strings. The name "move" is a misnomer because the data content of the source location is unchanged.

MOVE also is used to access storage and I/O devices. An input or storage device writes to a specific memory location, and the CPU retrieves the input by reading that memory location and copying its value into a register. Similarly, data is output or stored by writing to a predefined memory address or range of addresses. The output or storage device continually monitors the content of its assigned memory address(es) and reads newly written data for storage or output.

Data Transformations

The most primitive data transformation instructions are based on Boolean logic:

► NOT
► AND
► OR
► XOR

These four Boolean instructions and the ADD and SHIFT instructions (which are discussed in detail in the following sections) are the basic building blocks of all numeric comparisons and computations. They are summarized in Table 4-1.

Table 4-1 ►

Primitive data transformation instructions

Instruction	Function
NOT	Each result bit is the opposite of the operand bit
AND	Each result bit is true if both operand bits are true
OR	Each result bit is true if either or both operand bits are true
XOR	Each result bit is true if either, but not both, operand bits are true
ADD	Result is the arithmetic sum of operands
SHIFT	Move all bit values left or right as specified by operand

NOT   A NOT instruction transforms the Boolean value true (1) into false (0) and the value false into true. The rules that define the output of NOT on single bit inputs are:

$$NOT\ 0 = 1$$
$$NOT\ 1 = 0$$

With bit strings, NOT treats each bit in the bit string as a separate Boolean value. For example, executing NOT on the input 10001011 produces the "opposite" result—01110100. Note that NOT has only one data input, whereas all other Boolean instructions have two.

AND   An AND instruction generates the result true if both of its data inputs are true. The following rules define the result of AND on single bit data inputs:

$$0\ AND\ 0 = 0$$
$$1\ AND\ 0 = 0$$
$$0\ AND\ 1 = 0$$
$$1\ AND\ 1 = 1$$

The result of AND with two bit string inputs is shown in the following example:

```
         10001011
AND   11101100
         10001000
```

OR There are two types of OR operations in Boolean logic. An **inclusive OR** instruction (the word inclusive usually is omitted) generates the value true if either or both data inputs are true. The rules that define the output of inclusive OR on single bit inputs are:

```
0 OR 0 = 0
1 OR 0 = 1
0 OR 1 = 1
1 OR 1 = 1
```

The result of inclusive OR with two bit string inputs is shown in the following example:

```
        10001011
OR   11101100
        11101111
```

An **exclusive OR,** also called **XOR,** instruction generates the value true if either, but not both, data inputs are true. The rules that define the output of XOR on single bit inputs are:

```
0 XOR 0 = 0
1 XOR 0 = 1
0 XOR 1 = 1
1 XOR 1 = 0
```

Note that if either operand is one, the result is the complement of the other operand. Specific bits within a bit string can be inverted by XORing the bit string with a string containing zeros in all positions except the positions to be negated. For example, the following XOR inverts only the right four bit values and leaves the left four bit values unchanged:

```
          10001011
XOR   00001111
          10000100
```

Every bit in a bit string can be inverted by XORing with a string of ones:

```
          10001011
XOR   11111111
          01110100
```

Note that XORing any input with a string of ones produces the same result as executing NOT. Thus, NOT isn't required in a minimal instruction set.

ADD   An **ADD** instruction accepts two numeric inputs and produces their arithmetic sum. For example, the following ADD shows the binary addition of two bit strings:

$$
\begin{array}{r}
10001011 \\
\underline{\text{ADD} \quad 00001111} \\
10011010
\end{array}
$$

Note that the mechanics of the addition operation are the same regardless of what the bit strings represent. In this example, if the bit strings represent unsigned binary numbers, then the operation is:

$$139_{10} + 15_{10} = 154_{10}$$

If the bit strings represent signed integers in two's complement format, then the operation is:

$$-117_{10} + 15_{10} = -102_{10}$$

Binary addition does not work for complex data types such as floating point and double precision numbers. If the CPU supports complex numeric data types, then a separate ADD instruction must be implemented for each type.

SHIFT   The effect of a **SHIFT** instruction is shown in Figure 4-5. In Figure 4-5(a), the value 01101011 occupies an 8-bit storage location. Bit strings can be shifted to the right or left, and the number of positions shifted may be greater than one. Typically, the second operand holds an integer value that indicates the number of bit positions by which the value will be shifted. Positive or negative operand values indicate shifting to the left or right.

Figure 4-5 ▶

Original data byte (a) SHIFTed two bits to the right (b)

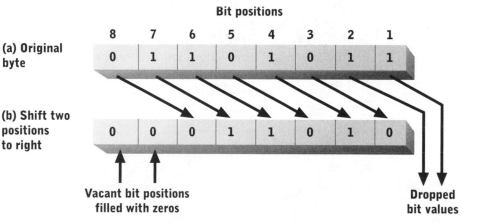

Figure 4-5(b) shows the result of shifting the value two positions to the right. The resulting value is 00011010. In this case, the values in the two least significant positions of the original string have been lost and the vacant bit positions are filled with zeroes.

Figure 4-5 is an example of a **logical SHIFT**. A logical SHIFT can extract a single bit from a bit string. Figure 4-6 shows how shifting an 8-bit value (a) to the left by four bits (b) and then to the right by seven bits creates a result (c) with the third bit of the original string in the rightmost position. Since all other bit positions contain zeros, the entire bit string can be interpreted as true or false. Shifting an 8-bit two's complement value to the right by seven positions is a simple way to extract and test the sign bit.

Figure 4-6 ▶

Extracting a single bit with logical SHIFT instructions

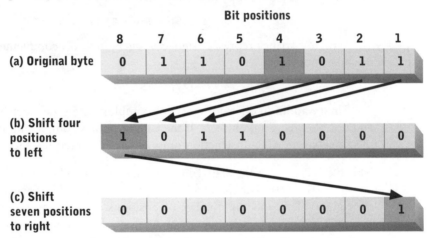

An **arithmetic SHIFT** instruction performs multiplication or division, as illustrated in Figure 4-7. If a bit string contains an unsigned binary number (Figure 4-7(a)), then shifting to the left by one bit (b) multiplies the value by two, and shifting the original bit (a) to the right by two bits (c) divides by four. Arithmetic SHIFT instructions are more complex when applied to two's complement values because bits must be shifted "around" the sign bit. Most CPUs provide a separate arithmetic SHIFT instruction that preserves the sign bit of a two's complement value.

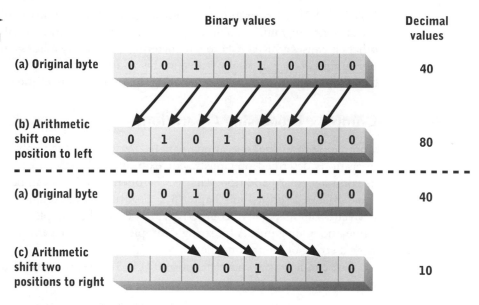

Figure 4-7 ▶

Multiplying and dividing binary values using SHIFT instructions

Binary values

Decimal values

(a) Original byte | 0 | 0 | 1 | 0 | 1 | 0 | 0 | 0 | 40

(b) Arithmetic shift one position to left | 0 | 1 | 0 | 1 | 0 | 0 | 0 | 0 | 80

(a) Original byte | 0 | 0 | 1 | 0 | 1 | 0 | 0 | 0 | 40

(c) Arithmetic shift two positions to right | 0 | 0 | 0 | 0 | 1 | 0 | 1 | 0 | 10

Sequence Control

Sequence control operations alter the flow of instruction execution in a program. These operations include:

▶ Unconditional branch

▶ Conditional branch

▶ Halt

Branch Operations A BRANCH (or JUMP) instruction causes the processor to depart from sequential instruction order. Recall that the control unit fetches the next instruction from memory at the conclusion of each execution cycle. The control unit consults a register to determine where the instruction is located in primary storage. BRANCH has one operand containing the memory address of the next instruction. BRANCH is actually implemented as a MOVE instruction. The BRANCH operand is loaded into the register that the control unit uses to fetch the next instruction.

In an **unconditional BRANCH,** the processor always departs from the normal execution sequence. In a **conditional BRANCH,** the BRANCH occurs only if a specified condition is met, such as the equivalence of two numeric variables. The condition must be evaluated and the Boolean result stored in a register. The conditional BRANCH instruction checks the content of that register and only branches if the value contained there is true.

HALT   A HALT operation suspends the normal flow of instruction execution in the current program. In some CPUs, it causes the CPU to cease all operations. In others it causes a BRANCH to a predetermined memory address. A portion of the operating system is typically stored at this address, effectively transferring control to the operating system and terminating the previously executing program.

Complex Processing Operations

Complex processing operations can be performed by combining the simpler operations described previously. For example, subtraction can be implemented as complementary addition. That is, the operation A - B can be implemented as A + (-B). As described in Chapter 3, a negative two's complement value can be derived from its positive counterpart by taking the complement of the positive value and adding one. A bit string's complement can be generated by XORing it with a string of binary one digits.

For example, the complement of 0011 (3_{10}), represented as a two's complement value, can be derived as:

$$XOR(0011,1111) + 0001 = 1100 + 0001 = 1101 = -3_{10}$$

This result is added to implement a subtraction operation. For example, the result of subtracting 0011 from 0111 can be calculated as:

$$7_{10} - 3_{10} = ADD(ADD(XOR(0011,1111),0001),0111)$$
$$= ADD(ADD(1100,0001),0111)$$
$$= ADD(1101,0111)$$
$$= 10100$$

Because 4-bit values are used, the result of 10100 is truncated from the left, resulting in a value of 0100.

Comparison operations can be implemented in much the same way as subtraction. A comparison operation generates a Boolean output value. Typically, an integer value of zero is interpreted as false and any nonzero value is interpreted as true. The comparison A ≠ B can be implemented by generating the complement of B and adding it to A. If the two numbers are equal, the result of the addition will be a string of zeros (interpreted as false). An equality comparison can be implemented by negating the Boolean result of an inequality comparison.

Greater-than and less-than comparisons also can be performed with subtraction followed by extraction of the sign bit. For the condition A < B, subtracting B from A generates a negative result if the condition is true. In two's complement notation, a negative value always has a one in the leftmost position (that is, the sign bit). SHIFT can be executed to extract the sign bit. For example, the two's complement value 10000111 is a negative number. The sign bit is extracted by shifting the value 7 bits to the right, resulting in the string 00000001. The SHIFT result is interpreted as a Boolean value (one, or true, in this case).

For example, the comparison:

$$0111 < 0011$$

can be evaluated as:

$$SHIFT(ADD(0111,ADD(XOR(0011,1111),0001)),0011)$$
$$SHIFT(ADD(0111,ADD(1100,0001)),0011)$$
$$SHIFT(ADD(0111,1101),0011)$$
$$SHIFT(0100,0011)$$
$$0000$$

The second operand of the SHIFT instruction is a binary number representing the direction and number of bit positions to shift (+3, or right 3, in this example). The result is zero and is interpreted as the Boolean value false.

A Short Programming Example[1]

Consider the following high-level programming language statement:

```
IF (BALANCE < 100) THEN
        BALANCE = BALANCE - 5
ENDIF
```

Such a computation might be part of a program that applies a monthly service charge to checking or savings accounts with a balance below $100. A program that implements this computation using only the previously defined low-level CPU instructions is shown in Figure 4-8. Table 4-2 shows the register contents after each instruction is executed when the account balance is $64. The coding format for all numeric data is 8-bit two's complement.

---

[1] You may wish to review the discussion of two's complement notation in Chapter 3 before studying this example.

Figure 4-8 ▶

A simple
program using
primitive CPU
instructions

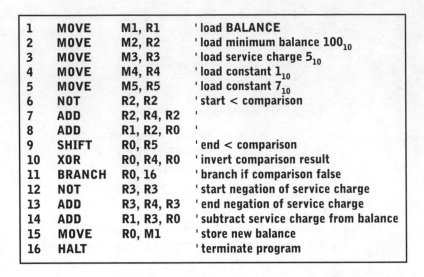

1	MOVE	M1, R1	'load BALANCE
2	MOVE	M2, R2	'load minimum balance 100_{10}
3	MOVE	M3, R3	'load service charge 5_{10}
4	MOVE	M4, R4	'load constant 1_{10}
5	MOVE	M5, R5	'load constant 7_{10}
6	NOT	R2, R2	'start < comparison
7	ADD	R2, R4, R2	'
8	ADD	R1, R2, R0	'
9	SHIFT	R0, R5	'end < comparison
10	XOR	R0, R4, R0	'invert comparison result
11	BRANCH	R0, 16	'branch if comparison false
12	NOT	R3, R3	'start negation of service charge
13	ADD	R3, R4, R3	'end negation of service charge
14	ADD	R1, R3, R0	'subtract service charge from balance
15	MOVE	R0, M1	'store new balance
16	HALT		'terminate program

Table 4-2 ▶

Register con-
tents after
executing each
instruction in
Figure 4-8
when account
balance in M1
is $64

Instruction	R0	R1	R2	R3	R4	R5
1		01000000				
2			01100100			
3				00000101		
4					00000001	
5						00000111
6			10011011			
7			10011100			
8	11011100					
9	00000001					
10	00000000					
11						
12				11111010		
13				11111011		
14	00111011					
15						
16						

Instructions 1 through 5 load the account balance, minimum balance, service charge, and needed binary constants from memory locations M1 through M5. A less-than comparison is performed in instructions 6 through 9. The right side of the comparison is converted to a negative value by executing a NOT instruction (instruction 6) and adding 1 to the result (instruction 7). The result is added to the account balance (instruction 8), and the sum is shifted seven places to the right to extract the sign bit (instruction 9). At this point register R0 holds the Boolean result of the less-than comparison.

For this example, all BRANCH instructions are assumed to be conditional on the content of a register. The BRANCH is taken if the register holds a Boolean true value and otherwise ignored. To jump beyond the code that implements the service charge if the account balance is above the minimum, we must invert the Boolean result of the condition prior to branching. Instruction 10 inverts the sign bit stored in the rightmost bit of R0 by XORing it against 00000001 (stored in R5). Instruction 11 executes a conditional BRANCH. Because the original sign bit was one, the inverted value is zero. Thus the BRANCH is ignored and processing proceeds with instruction 12.

Instructions 12 through 14 subtract the service charge stored in register R3 from the account balance. Instructions 12 and 13 convert the positive value to a negative value, and instruction 14 adds it to the account balance. Instruction 15 saves the new balance in memory. Instruction 16 halts the program.

Instruction Set Extensions

The instructions described to this point are sufficient for a simple general-purpose processor. All of the more complex functions normally associated with computer processing can be implemented by combining those primitive building blocks. Most CPUs provide a much larger set of instructions, including advanced computation operations such as multiplication and division, negation of two's complement values (NOT followed by ADD 1), and testing a sign bit and other single bit manipulation functions. Such instructions sometimes are referred to as **complex instructions** because they represent combinations of primitive processing operations.

Complex instructions represent a tradeoff between processor complexity and programming simplicity. For example, consider the three-step process of NOT - ADD - ADD that performs subtraction in Figure 4-8, instructions 12 through 14. Because subtraction is a commonly performed operation, most CPU instruction sets include one or more subtraction instructions. This complicates the CPU by requiring extra processing circuitry for the additional instruction, but simplifies machine language programs.

Complex instructions also represent a tradeoff between CPU complexity and program execution speed. Multistep instruction sequences execute faster if they are executed within hardware as a single instruction, avoiding the overhead of fetching multiple instructions and accessing intermediate results in registers. Other efficiencies may also be realized by hardwiring the steps together. However, these efficiencies have limits, as described later.

Additional instructions are required when new data types are added. For example, most CPUs provide instructions for adding, subtracting, multiplying, and dividing integers. If double-precision integers and floating point numbers are supported, additional computation instructions must be included for those data types. It is not unusual to see a half dozen different ADD instructions in an instruction set to support integers and real numbers in single and double precision as well as signed and unsigned "short" integers.

Some instruction sets also include instructions that combine data transformations with data movement. The primitive instructions in Figure 4-8 use registers for both data input and data output. In some CPUs, data transformation instructions allow one or more operands to be a memory address. In essence, this combines data movement with data transformation.

INSTRUCTION FORMAT

Recall that an instruction consists of an op code (instruction number) and zero or more operands (representing data values or storage locations). An **instruction format** is a template that specifies the number of operands and the position and length of the op code and operands. Instruction formats vary among CPUs in many ways, including op code size, meaning of specific op code values, data types used as operands, and the length and coding format of each type of operand. The term instruction format erroneously implies that each CPU uses a single format. Because instructions vary in the number and type of operands they require, CPUs must support multiple instruction formats. Different formats represent various combinations of operand types.

Figure 4-9 shows an instruction format consisting of an op code and three operands with a total length of 20 bits. The sample format is typical of instructions that use register inputs and outputs. The op code occupies the first 8 bits. Most CPUs represent the op code as an unsigned binary number, which provides 256 possible instructions numbered 0 through 255. Each operand is a 4-bit unsigned binary number identifying one of 16 possible registers.

Figure 4-9 ►

An instruction format with three register operands

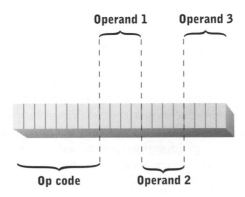

Operand 1   Operand 3

Op code   Operand 2

Figure 4-10 shows another instruction format for load and store instructions. The 32-bit format contains three operands. The first two operands store register numbers. The first operand contains the number of a register that holds data to be transferred to or from memory. The second register contains a partial memory address.

Figure 4-10 ►

An instruction format that includes a segmented memory address

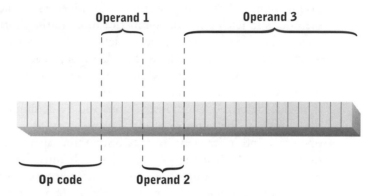

Recall from the previous chapter that some CPUs use a segmented memory addressing scheme in which a memory address consists of two parts—a memory segment identifier and a segment offset. For segmented memory schemes, the second operand identifies a register that holds the segment identifier, and the third operand is the segment offset. The control unit uses the contents of the register identified in the second operand and the content of the third operand to construct a complete memory address.

Instruction Length

Instruction formats within a single CPU can be fixed or variable in length. **Fixed length instructions** simplify the instruction-fetching process within the control unit. If the instruction format is fixed length, then the amount by which the instruction pointer must be incremented after each fetch is a constant. This increment is the length of an instruction.

Instruction format lengths can be equalized by "padding" shorter instruction formats with trailing zero bits. For example, the format in Figure 4-9 could be padded with 12 zero bits to increase its length to match the format in Figure 4-10. The CPU ignores the extra bits when processing instructions that use the padded format.

With a **variable length instruction**, the amount by which the instruction pointer is incremented after a fetch is the length of the most recently fetched instruction. The control unit must check the op code of each fetched instruction to determine the proper increment value. Conceptually, you might think of a table with op code values on the left and format lengths on the right. The control unit would look up the op code in the table and extract the appropriate value to add to the instruction pointer.

Variable length instructions also complicate instruction fetching because the number of bytes to be fetched is not known in advance. One method of addressing this uncertainty is to always fetch the number of bytes in the longest instruction format, but this can result in many unnecessary memory accesses. Another method is to fetch only the number of bytes in the shortest format. Then the op code is examined to determine the length of the instruction and the number of additional bytes to fetch, if any. In either case, extra computer resources are used in the fetch operation that would not have been used if the instruction length were fixed and known before the fetch.

Although fixed length instructions and fields simplify control unit functions, they do so at the expense of efficient memory use. Some instructions have no operands and others have one, two, or three operands. If fixed length instructions are used, then the instruction length must be the length of the longest instruction, for example, an instruction with two or three large operands. Smaller instructions stored in memory must be padded with empty bit positions to extend their length to the maximum. The memory used to pad short instructions is wasted, and programs require more memory during execution and more time to load from secondary storage.

Reduced Instruction Set Computing

Reduced Instruction Set Computing (RISC) is a philosophy of processor and computer system design that first was employed in the late 1980s. The primary architectural feature of a RISC processor is the absence of some, but not all, complex instructions from the instruction set. In particular, RISC processors avoid instructions that combine data transformation and data movement operations. For example, a non-RISC processor might provide a data transformation instruction of the form:

Transform(Address1,Address2,Address3)

Address1 and Address2 are the memory addresses of data inputs, and Address3 is the memory address to which the result is stored. In a RISC processor, transformation operations always use register inputs and outputs. The single complex instruction shown above requires four separate RISC instructions:

MOVE(Address1,R1)
MOVE(Address2,R2)
Transform(R1,R2,R3)
MOVE(R3, Address3)

Each instruction is loaded and executed independently, consuming at least four execution cycles.

Although the lack of many complex instructions is the primary distinguishing feature of a RISC processor, other differences typically include fixed length instructions, short instruction length, and a large number of general-purpose registers. To describe these differences fully, we must first contrast RISC with its opposite design philosophy—**Complex Instruction Set Computing** (CISC).

Complex instruction sets were developed because early computers had limited memory and processing power. Memory was very expensive and many computers barely had enough to hold an entire application program. To compensate for limited memory, many CPU designers provided complex instructions that would do more work per instruction. As a result, programs required less memory, and complex operations executed more quickly.

For example, assume that a floating point addition operation, implemented as a single complex instruction, can be executed in one processor cycle. Assume further that an equivalent operation can be performed with five different integer math instructions, each of which requires one processor cycle. In this example, a direct implementation of the complex floating point instruction saves four processor cycles each time it is executed.

If complex instructions are so beneficial, then why would anyone want to eliminate them? The most important reason is that the benefits of complex instructions are subject to the law of diminishing returns. Each complex instruction provides benefits if measured in isolation, but as more and more complex instructions are added, the instruction set becomes large and complex, creating inefficiencies that affect the entire CPU.

A large instruction set creates two problems. First, it complicates the control unit because there are more instructions to interpret and usually more instruction formats and data types to deal with. Also, a large set of complex instructions often goes hand in hand with variable length instruction formats. This adds even more complexity to the job of fetching and decoding instructions, which in turn increases fetching and decoding time. Because every simple or complex instruction must be fetched before it is executed, a performance penalty is applied to every instruction, even the simple ones.

The second problem arising from a large instruction set is microprocessor size. As discussed later in this chapter, speed improvements in microprocessors are primarily achieved by miniaturization. The simpler the processor, the easier the task of shrinking it and the more reliably smaller versions can be fabricated. Because CISC processors are much more complex than RISC processors, they are more difficult to fabricate reliably in smaller sizes. RISC processor design reduces the size and complexity of the instruction set to increase raw speed in fetching and executing instructions. RISC design follows a "less is more" strategy and extends the instruction set only when the benefits are very high.

The primary disadvantage of RISC CPUs is the extra memory required for program storage and execution. Because many complex instructions are not present, programs must use multiple simple instructions in their place. The equivalent sequences of simple instructions occupy more storage than their complex equivalents and increase program memory requirements. Although this is a disadvantage for RISC, it is not very significant at present because memory cost has fallen more rapidly than CPU cost.

RISC processors also are relatively inefficient at executing programs that do many of the functions for which complex instructions are designed, for example, transforming data items stored in memory and immediately storing the results back to memory. Although this is a potential disadvantage, detailed studies of typical program behavior have shown that many complex instructions are used infrequently. Typical programs spend most of their time executing relatively primitive instructions such as load, store, add, and compare. In many cases, the speed advantage of complex instructions is not realized frequently enough to make up for the performance penalty applied to every instruction in a more complex instruction set.

CPU performance depends heavily on the construction of the programs being executed, regardless of whether the CPU is RISC, CISC, or a hybrid. Software should be optimized to take advantage of the presence or absence of specific CPU capabilities. In modern software development, most of the optimization effort takes place when a program written in a higher-level language such as C or Java is compiled to produce CPU instructions. Compilation is discussed in Chapter 10.

Despite the inherent advantages of RISC design, CISC CPUs from Intel including the Pentium, Xeon, and Itanium dominate desktop computing and have made significant inroads into workstation and server computing. But how can Intel CISC CPUs compete against newer and supposedly faster RISC CPUs? And why does Intel stick with "old" CISC technology?

One part of the answer is backward compatibility. Early Intel desktop CPUs were developed in an era when CISC was clearly dominant. Software developed for those CPUs was typically sold to customers in the form of CPU instructions (for example, MS-DOS or Windows .EXE program files). Since customers didn't want to abandon their old software when upgrading to new CPUs, newer Intel CPUs needed to execute the same complex instructions supported by older CPUs. Although there are multiple ways to accomplish this backward compatibility, one of the easiest is to continue and expand on older CISC designs in newer chips.

Another part of the answer is Intel's dominance in CPU fabrication technology. In essence, Intel can economically build more complex CPUs because it has more advanced fabrication technology and a large market share over which to spread the costs of that technology. By using the latest fabrication technology, Intel can produce CISC CPUs with clock speeds that rival their RISC competitors despite their more complex circuitry.

Yet another part of the answer is a set of advanced RISC techniques used by Intel and most other microprocessor manufacturers. These techniques will be covered later in this chapter. For now, it's sufficient to say that just because a CPU supports complex instructions doesn't mean that it can't employ RISC design principles "under the hood." In a similar vein, few processors that claim to be RISC completely avoid all CISC characteristics. The terms RISC and CISC are overused in marketing literature to paint stark contrasts among CPU designs. But the realities are seldom as far apart as a strict interpretation of those adjectives might imply.

CISC and RISC are different ends of a design continuum representing tradeoffs among several design factors. The economics and technology of program design and behavior, processor fabrication, and memory cost made CISC an optimal solution in the early days of computing. In the 1990s, as memory became cheaper and CPU fabrication technology improved, the optimal tradeoff shifted toward RISC. At the moment there is no clear "best" approach to processor design, partly because processors of each type borrow heavily from the best features of the other type. Future changes in technology and in how computers are used may favor RISC, CISC, or a completely different approach to CPU design.

CLOCK RATE

The system clock is a digital circuit that generates timing pulses, or signals, and transmits the pulses to other devices within the computer (note: this is not the clock used to keep track of the current date and time). The clock is generally an entirely separate device with a dedicated communication line that is monitored by all devices within the computer system. All actions, especially the instruction and execution cycles of the CPU, are timed according to this clock. Storage and I/O devices are timed by the clock signal, and all devices within a computer system coordinate their activities with the system clock.

The frequency at which the system clock generates timing pulses is the **clock rate** of the system. Each "tick" of the clock begins a new **clock cycle**. CPU and computer system clock rates are expressed in **hertz (Hz)**. One hertz corresponds to one clock cycle per second. Modern CPUs and computer systems have clocks that generate millions or billions of timing pulses per second. The frequency of these clocks is measured in **megahertz (MHz)**, meaning millions of cycles per second, or **gigahertz (GHz)**, billions of cycles per second.

The inverse of the clock rate is called the CPU **cycle time**. In most CPUs, the cycle time is the time required to fetch and execute the simplest instruction in the instruction set. For example, assume that the CPU clock rate is 5 GHz and that NOT is the simplest instruction. The time required to fetch and execute a NOT instruction can be computed as the inverse of the clock rate:

$$\text{cycle time} = \frac{1}{\text{clock rate}} = \frac{1}{5,000,000,000} = 0.0000000002 \text{ second} = 0.2 \text{ nanosecond}$$

Clock rate and cycle time are important CPU performance measures. However, they tell only part of the performance story for a CPU or computer system. CPU clock rate is frequently misinterpreted by equating it with:

► Instruction execution rate
► Overall computer system performance

From the perspective of program execution speed, the most important CPU performance consideration is the rate at which instructions are executed. That rate is generally stated in units called **millions of instructions per second** (**MIPS**). MIPS are assumed to measure CPU performance when manipulating single precision integers. CPU performance when manipulating single precision floating point numbers is measured in **millions of floating point operations per second** (**MFLOPS**).

If all instructions, regardless of function or data type, were executed within a single clock cycle, then clock rate, MIPS, and MFLOPS would be equivalent. However, execution time varies with the complexity of the processing function. Instructions for simple processing functions such as Boolean functions, equality comparison, and integer addition execute quickly. Instructions for functions like multiplication and division execute more slowly.

Most complex instructions require more than one clock cycle to execute. Execution of a complex instruction continues through one or more successive fetch and execute cycles. A new instruction fetch does not begin until the beginning of the next clock cycle after the previous instruction completes execution. Complex integer computations typically require two to four clock cycles, and floating point computations typically require three to ten clock cycles.

The number of instructions executed in a given time interval depends on the mix of simple and complex instructions in a program. Assume, for example, that a program executes 100 million integer instructions. Also assume that 50 percent are simple instructions requiring a single clock cycle and 50 percent are complex instructions requiring an average of three clock cycles. The average program instruction requires two clock cycles, in this case, $(1 \times 0.50) + (3 \times 0.50)$, and the CPU MIPS rate is 50 percent of the clock rate when executing this program. Different mixes of simple and complex instructions and different mixes of integer and floating point data will result in different ratios of clock rate to MIPS and MFLOPS. For all but the most trivial programs, MIPS and MFLOPS will be much smaller than the CPU clock rate.

The previous MIPS calculation assumes that nothing hinders the CPU in fetching and executing instructions. But the CPU relies on slower devices to keep it supplied with instructions and data. For example, main memory is typically two to ten times slower than the processor. That is, the time required to complete a main memory read or write operation is typically two to ten CPU clock cycles. Accessing secondary storage is thousands or millions of times slower than the CPU. The CPU may be idle while waiting for access to storage and I/O devices. Each clock cycle that the CPU spends waiting for a slower device is called a **wait state**.

Unfortunately, a CPU can spend much of its time in wait states, during which no instructions are being executed. The effective MIPS rate of a computer system can be much lower than the MIPS rate of the CPU measured in isolation because of the delays imposed by waiting for storage and I/O devices. Various methods of minimizing wait states are discussed in Chapter 6.

MIPS and MFLOPS are poor measurements for comparing computer system performance because processor instruction sets vary so widely and because CPU performance depends so heavily on memory access speed and other aspects of computer design. A better measure of comparative computer system performance is how quickly a specific program or set of programs is executed.

A **benchmark** is a measure of CPU or computer system performance when executing one or more specific tasks. A widely used benchmark for comparing CPU performance is the Standard Performance Evaluation Corporation (SPEC) CPU2004 program suite (www.specbench.org). CPU2004 has two component benchmarks: CINT2004, which measures integer computation performance, and CFP2004, which measure floating point computation performance. Both component benchmarks are suites of approximately one dozen programs that are as similar as possible to typical computationally intensive programs.

The numeric scores computed as benchmark results are not direct measures of instructions executed. Rather, they are numbers that are only meaningful in comparison. For example, it is fair to say that a CPU with a CINT2004 score of 2500 is twice as fast performing typical integer computations as another CPU with a CINT2004 score of 1250.

CPU REGISTERS

Registers play two primary roles in CPU operation. First, they provide a "scratch-pad" for the currently executing program, holding data that is needed quickly or frequently by the program. Second, registers store information about the currently executing program and about the status of the CPU, for example, the address of the next program instruction, error messages, and signals from external devices.

General-Purpose Registers

General-purpose registers are used only by the currently executing program. General-purpose registers are typically used to hold intermediate results or to hold data values that will be used frequently, such as loop counters or array indices. Register accesses are fast because registers are implemented within the CPU. In contrast, storing and retrieving from primary storage is much slower. Using registers to store data needed immediately or frequently increases program execution speed.

Adding general-purpose registers increases execution speed, but only up to a point. There are a limited number of intermediate results or frequently used data items in any given process or program, so CPU designers try to find the optimal tradeoff among the number of general-purpose registers, the extent to which those registers will be used by a typical process, and the cost of implementing those registers. As the cost of producing individual registers has decreased, their number has increased. Current CPUs typically provide several dozen general-purpose registers.

Special-Purpose Registers

Every processor has registers that are used by the CPU for specific purposes. Some of the more important special-purpose registers are:

► Instruction register
► Instruction pointer
► Program status word

When the control unit fetches an instruction from memory, it stores it in the **instruction register**. The control unit then extracts the op code and operands from the instruction and performs any additional data movement operations that are needed to prepare for execution. The process of extracting the op code and operands, loading data inputs, and signaling the ALU is called instruction **decoding**.

The **instruction pointer** can also be called the **program counter**. Recall that the CPU alternates between the instruction (fetch and decode) and execution (data movement or transformation) cycles. At the end of each execution cycle, the control unit starts the next fetch cycle by retrieving the next instruction from memory. The address of the instruction retrieved is stored in the instruction pointer. The instruction pointer is incremented by the control unit either during or immediately after each fetch cycle.

The CPU deviates from sequential execution only if a BRANCH instruction is executed. A BRANCH is implemented by overwriting the value of the instruction pointer with the address of the instruction to which the branch is directed. An unconditional BRANCH instruction is actually a MOVE from the branch operand, which contains the branch address, to the instruction pointer.

The **program status word** (**PSW**) contains data that describes the status of the CPU and the currently executing program. Each bit within the PSW is a separate Boolean variable, sometimes called a **flag**, that represents one data item. The content and meaning of the flags vary widely from one CPU to another. In general, PSW flags have three primary uses:

► Store the result of a comparison operation
► Control conditional branch execution
► Indicate actual or potential error conditions

The sample program shown earlier in Figure 4-8 performs comparison using the XOR, ADD, and SHIFT instructions. The result was stored in a general-purpose register and was interpreted as a Boolean value. This method was used because the instruction set was limited.

Most CPUs provide one or more COMPARE instructions. COMPARE takes two operands and determines whether the first is less than, equal to, or greater than the second. Because there are three possible conditions, the result cannot be stored in a single Boolean variable. Most CPUs use two PSW flags to store the result of a COMPARE. One flag is set to true if the operands are equal and the other flag indicates whether the first operand is greater than or less than the second. If the first flag is true, then the second flag is ignored.

To implement program branches based on a COMPARE result, two additional conditional BRANCH instructions are provided. One conditional BRANCH is based on the equality flag and the other is based on the less-than or greater-than flag. Using COMPARE with related conditional BRANCH instructions simplifies machine language programs, speeding up their execution.

Other PSW flags represent status conditions resulting from instruction execution by the ALU. Conditions such as overflow, underflow, or an attempt to perform an undefined operation such as dividing by zero are each represented by a PSW flag. After each execution cycle, the control unit tests PSW flags to determine whether an error has occurred. They can also be tested by an operating system or an application program to determine appropriate error messages and corrective actions.

WORD SIZE

A **word** is a unit of data that contains a fixed number of bytes or bits. A word can be loosely defined as the amount of data that a CPU processes at one time. Depending on the CPU, processing may include arithmetic, logic, fetch, store, and copy operations. Word size normally matches the size of general-purpose registers. Word size is a fundamental CPU design decision with implications for most other computer system components.

In general, a CPU with a large word size can perform a given amount of work faster than a CPU with a small word size. For example, a processor with a 64-bit word size can add or compare 64-bit integers by executing a single instruction because the registers that hold the operands and the ALU circuitry are 64 bits wide.

Now consider manipulating 64-bit data in a CPU with a 32-bit word size. Because the operands are larger than the word size, they must be partitioned and the operation(s) carried out on the pieces. For example, in a comparison operation, the CPU compares the first 32 bits of the operands and then, in a second execution cycle, compares the second 32 bits. The process is inefficient because multiple operands are loaded from or stored to memory, and multiple instructions are executed to accomplish what is logically a single operation.

Because of these inefficiencies, a 64-bit CPU usually is more than twice as fast as a 32-bit processor when processing 64-bit data values. Inefficiencies are compounded as the complexity of the operation increases. For example, division and exponentiation may be four or five times slower on a 32-bit processor than on a 64-bit processor.

CPU word size also has implications for system bus design. Maximum CPU performance is achieved when the bus width is at least as large as the CPU word size. If the bus width is smaller, every load and store operation requires multiple transfers to or from primary storage. For example, moving 8 bytes of character data to contiguous memory locations requires two separate data movement operations on a 32-bit bus, even if the word size of the processor is 64 bits. Similarly, fetching a 64-bit instruction requires two separate transfers across the bus.

The physical implementation of memory is likewise affected by word size. Although the storage capacity of memory is always measured in bytes, data movement between memory and the CPU is generally performed in multiples of the word size. For a 64-bit CPU, memory should be organized to read or write at least 64 contiguous bits in a single access operation. Lesser capabilities will leave the CPU idle pending completion of memory accesses.

As with many other CPU design parameters, increases in word size are subject to the law of diminishing returns. Two primary issues are the usefulness of the extra bits and their cost. The extra bits are useful if the larger word is less than or equal to the size of data items that application programs normally manipulate.

The added benefit of increased word size drops off sharply past 64 bits. Recall from the previous chapter that integers are typically stored in 32-bit two's complement format and that real numbers are typically stored in either 32-bit or 64-bit IEEE floating point format. Of what use are the extra 32 bits of a 64-bit CPU when adding together two 32-bit data items? They are of no use at all. The extra 32 bits simply store zeros that are carried through the computational process and provide benefit only when manipulating 64-bit data items. But many current word processors, financial applications, and other programs never manipulate 64-bit data items. Only a few types of application programs can use the additional bits. Examples include numerical-processing applications that need very high precision (such as navigation), numerical simulations of complex phenomena, some database- and text-processing applications, and programs that manipulate continuous streams of audio or video data. For other programs, the increased computational power of the larger word size is wasted.

The waste of extra word size wouldn't be a major concern if cost were not an issue. However, doubling word size generally increases the number of CPU components by 2.5 to 3 times, increasing CPU cost and complicating microprocessor fabrication. Costs tend to rise at nonlinear rates when approaching the limits of current fabrication technology. Until recently, fabrication technology limits made 64-bit CPUs an expensive luxury—generally used only in larger and more expensive computer systems. Now technology has reached the point where placing hundreds of millions of transistors on a single chip is cost effective. The Intel Itanium® and Xeon and IBM POWER CPUs feature a 64-bit word size, now needed by many Internet and multimedia applications.

ENHANCING PROCESSOR PERFORMANCE

Modern CPUs use a number of advanced techniques to improve performance, including:

► Memory caching
► Pipelining
► Branch prediction and speculative execution
► Multiprocessing

Memory caching will be discussed in Chapter 5. The remaining topics are all forms of parallel processing and are discussed in the following sections.

Pipelining

Refer back to Figure 4-3 to review the various steps in fetching and executing an instruction:

1. *Fetch* from memory
2. *Increment* and store instruction pointer (IP)
3. *Decode* instruction and store operands and instruction pointer
4. *Access* ALU inputs
5. *Execute* instruction within the ALU
6. *Store* ALU output

Note that each function in the above list is performed by a separate portion, or stage, of the CPU circuitry. All computation and comparison instructions must pass through each stage of this circuitry in sequence.

Pipelining is a method of organizing CPU circuitry to enable multiple instructions to be in different stages of execution at the same time. A pipelining processor operates similarly to an automobile assembly line. Automobiles are assembled by moving them sequentially through dozens or hundreds of production stages. Each stage does a small part of the assembly process (for example, install a door, engine, or dashboard) and then passes the automobile onto the next stage. As each automobile passes to the next stage, another arrives from the previous stage.

Assume for the moment that a typical computation or comparison instruction requires 6 CPU cycles to pass through all six stages—one cycle for each stage. In a traditional processor, each instruction must pass through all six stages before the next instruction can enter the first stage. Thus, a sequence of ten instructions requires $10 \times 6 = 60$ CPU cycles to complete execution. But what if it were possible to overlap the processing stages in a manner similar to an automobile assembly line (see Figure 4-11)? For example, the first instruction could be in the store stage while the second instruction was in the execute stage, the third instruction was in the access stage, and so forth. In theory, allowing instructions to overlap in different processor stages would reduce execution time for all ten instructions to 15 CPU cycles—a 400% improvement!

The theoretical improvement shown in Figure 4-11 is unlikely to be realized for several reasons. One reason is that it would be nearly impossible to build a processor in which all six stages take exactly one clock cycle. Operations such as fetching the next instruction from main memory and performing a computation within the ALU usually take longer than simpler operations such as incrementing the instruction pointer. The efficiency of pipelining is reduced if some stages take longer than others.

Figure 4-11 ►

Overlapped instruction execution via pipelining

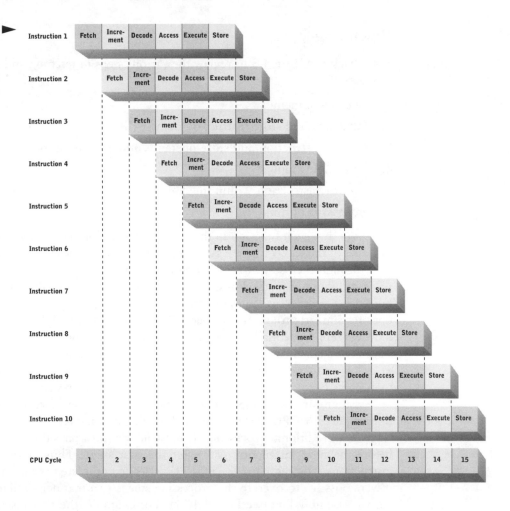

Processor Technology and Architecture

Another reason that a 400% performance improvement is unlikely is that instructions are not always executed sequentially. When the instruction sequence is altered by a conditional branch instruction, the CPU empties the content of the pipeline and starts over with the first instruction of a new sequence. This problem and various methods of addressing it are described next.

Branch Prediction and Speculative Execution

Recall from Chapter 2 that one feature that enables a general-purpose processor to emulate intelligent behavior is its ability to execute algorithms that alter their behavior based on different data inputs. The example used in that chapter was a progressive tax computation where the tax rate increases as income increases. The processor applies the progressive tax algorithm by executing a series of comparison operations followed by conditional branches corresponding to a program such as that shown in Figure 4-12.

Figure 4-12 ▶

The income tax program from Chapter 2

```
10  INPUT INCOME
20  IF INCOME > 7150 THEN GOTO 50
30  TAX = INCOME * 0.10
40  GOTO 180
50  IF INCOME > 29050 THEN GOTO 80
60  TAX = 715.00 + (INCOME - 7150) * 0.15)
70  GOTO 180
80  IF INCOME > 70350 THEN GOTO 110
90  TAX = 4000.00 + (INCOME - 29050) * 0.25)
100 GOTO 180
110 IF INCOME > 146750 THEN GOTO 140
120 TAX = 14325.00 + (INCOME - 70350) * 0.28)
130 GOTO 180
140 IF INCOME > 319100 THEN GOTO 170
150 TAX = 35717.00 + (INCOME - 146750) * 0.33)
160 GOTO 180
170 TAX = 92592.50 + (INCOME - 319100) * 0.35)
180 OUTPUT TAX
190 END
```

Consider the six-stage pipeline shown in Figure 4-11. If instruction 4 is a conditional branch and the condition that controls a conditional branch is true, the branch will overwrite the instruction pointer in CPU cycle 9 with a new value. At that point in time, instructions 5-9 have all been fetched based on incrementing an IP value that is now invalid. Because the branch condition is true, all work on those instructions must be abandoned and the processor must start anew to fill the pipeline and produce useful processing results. The opportunity for parallelism in completing instructions after instruction 4 has been lost and it will be six more CPU cycles before another processing result is stored.

This situation is similar to deciding immediately to switch models being built on an auto assembly line because customers no longer want the current model. The assembly line must be stopped and all automobiles currently in production must be removed. In addition, no new models will be available until one automobile proceeds through every stage of the assembly process from start to finish.

There are various approaches to dealing with the problem of conditional branches. One approach is to prefetch several instructions ahead of the current instruction and examine them to see if any are conditional branch instructions. By analyzing the order of pending instructions, the CPU may determine that it can evaluate the branch condition early. This approach is complicated and not all branches can be evaluated out of sequence. But if it succeeds often enough, it can improve processor speed by redirecting the CPU's efforts to the "right" set of instructions.

Under another approach, called **branch prediction**, the CPU guesses whether a branch condition will be true or false based on past experience. To provide a performance improvement, the guess must be made during or soon after the conditional branch instruction is loaded, before it is completely executed. Branch prediction is best applied to conditional branch instructions embedded within a program loop. For example, branch prediction could be used to compute withholding tax for all employees of a single filing status, that is, using a loop to compute the tax for all employees whose tax is computed with the same algorithm.

Using the payroll program in Figure 4-12, and assuming that most single employees have incomes greater than $26,250 but less than or equal to $63,550, the CPU will guess that the branch condition in line 20 will be true and that the branch condition in line 50 will be false. After the CPU loads and examines the conditional branch in line 20, it will immediately load the conditional branch instruction from line 50, bypassing machine instructions for lines 30 and 40. After the CPU loads and examines the conditional branch instruction in line 50, it will predict false for the branch condition and begin work on the instructions that implement line 60 in the next CPU cycle.

As the CPU executes conditional branch instructions within the loop it "keeps score," tracking how often the condition for each branch instruction has been true or false. Based on the scores, the CPU determines whether to load the next sequential instruction past the conditional branch or the instruction at the address of the branch operand. The program path chosen by the CPU is speculative—the CPU won't know whether its guess is correct until the conditional branch instruction proceeds through all CPU stages. Thus, the term **speculative execution** describes instructions executed after the guess but before the final result is known with certainty.

The processing needed to keep track of conditional branch instructions and their scores can add considerable overhead to early CPU stages, which may slow the entire processor. If the CPU's guess is later found to be incorrect, instructions in intermediate stages of processing must be flushed from the pipeline.

Another approach that avoids the complexities associated with branch prediction and speculative execution is simultaneous execution (multiprocessing) of both program paths after a conditional branch. Rather than guessing which path will be taken, the CPU assumes that either path is possible and executes instructions from both paths until the conditional branch is finally evaluated. At that point, further work on the incorrect path is abandoned and all effort is directed to the correct path.

Simultaneous execution of both paths requires redundant CPU stages and registers. The necessary processing redundancy can be achieved with parallel stages within a single CPU or with multiple CPUs sharing memory and other resources.

Multiprocessing

The term **multiprocessing** describes any CPU architecture in which duplicate CPUs or processor stages can execute in parallel. There is a range of possible approaches to implementing multiprocessing, including:

▶ Duplicate circuitry for some or all processing stages within a single CPU
▶ Duplicate CPUs implemented as separate microprocessors sharing main memory and a single system bus
▶ Duplicate CPUs on a single microprocessor that also contains main memory caches and a special bus to interconnect the CPUs

The approaches are listed from less difficult to more difficult implementation. The first approach was made possible by microprocessor fabrication technology improvements in the late 1980s and early 1990s. The second became cost-effective with fabrication technology advances in the late 1990s and early 2000s. The latter approach has recently emerged as a feasible and cost-effective approach.

Duplicate processing stages of a single CPU enable some, but not all, instructions to execute in parallel. For example, consider the formulas:

1. $$((a+b) \times c) - d) \div e$$

2. $$((a+b) \times (c+d)) - ((e+f) \div (g+h))$$

Formula 1 provides no opportunities for parallel execution because the computation operations must be implemented in left to right order to produce the correct result. But in formula 2, all four addition operations can be performed in parallel before the multiplication and division operations, which can also be performed in parallel before performing the subtraction operation. Assuming that each operation requires a single ALU cycle, a processor with a single ALU stage consumes seven ALU cycles to calculate the result of formula 2. If two ALUs are available, formula 2 can be calculated in four ALU cycles by performing two additions in parallel, followed by two more additions in parallel, followed by the multiplication and division in parallel, followed by the subtraction.

The performance improvement assumes a sufficient number of general-purpose registers to hold all of the intermediate processing results. Earlier processing stages must also be duplicated to keep the multiple ALUs supplied with instructions and data. As with pipelined processing, conditional branch instructions limit the ability of the CPU to execute nearby instructions in parallel. But the performance gains are substantial for complex algorithms such as computing bank account interest and image processing.

Embedding multiple CPUs in a single computer system and sharing resources such as main memory among them presents a greater range of possibilities for executing parallel instructions. Since a typical computer system executes multiple programs at the same time, multiple CPUs enable instructions from two separate programs to execute simultaneously. However, the CPUs must cooperate to ensure that they don't interfere with one another. For example, the CPUs shouldn't try to write to the same memory location, send a message over the system bus, or access the same I/O device at the same time. To keep that from happening, the processors must continually "watch" one another to avoid interference. That "watching" adds complexity to each processor.

The latest microprocessor fabrication technology enables multiple CPUs to be placed on the same chip. The coordination complexities are similar to those of CPUs on separate chips, but implementing the CPUs on the same chip enables them to communicate at much higher speeds. It also enables shared memory and other common resources to be placed on the same chip. This topic is further discussed in Chapter 6.

Technology
Focus

Intel Pentium Processor Family

The Pentium processor is a direct descendent of the Intel 8086 and 8088 processors used in the original IBM personal computer. It maintains backward compatibility with those processors in all areas, including instruction set, data types, and memory addressing. The Pentium processor was introduced in 1993. The Pentium has been updated several times through generations named the Pentium Pro, Pentium MMX (MMX stands for MultiMedia eXtensions), Pentium II, Pentium III, Pentium 4, and Xeon. This discussion will concentrate on features common to the entire Pentium family and the MMX streaming multimedia extensions.

Processor Technology and Architecture

Data Types

The Pentium supports a rich variety of data types (see Table 4-3). For purposes of describing data types, a word is considered to be 16 bits in length (as it was in the 8086/8088). Data types are of 5 sizes, including 8 (byte), 16 (word), 32 (doubleword), 64 (quadword), and 128 bits. Two variable length data types also are defined.

Table 4-3 ▶

Pentium data type summary

Data Type	Length (Bits)	Coding Format
Unsigned integer	8, 16, or 32	Ordinary binary
Signed integer	8, 16, or 32	Two's complement
Real	32, 64, or 80	IEEE floating point standards
Bit field	32	32 binary digits
Bit string	Variable up to 2^{32} bits	Ordinary binary
Byte string	Variable up to 2^{32} bytes	Ordinary binary
MMX (integer)	8×8, 4×16, 2×32, 1×64	Packed two's complement
MMX (real)	4×32	Packed IEEE 32-bit floating point

No special data types are provided for Boolean or character data, which are assumed to use appropriate integer data types. A bit field data type consists of up to 32 individually accessible bits. Two variable length data types also are defined. A bit string is a sequence of 1 to 2^{32} bits. This data type is commonly used for large binary objects such as compressed audio or video files. A byte string consists of 1 to 2^{32} bytes. Elements of a byte string can be bytes such as ASCII characters, words such as Unicode characters, or doublewords (32 bits).

Four 64-bit MMX data types are provided. Three of these are packed data types, in which multiple integers are "packed" within a single 64 bit unit. The packed integers may be 8, 16, or 32 bits in length. The fourth data type is a quadword (64-bit) integer. Packed data types allow MMX instructions to operate on each component of a packed data type in parallel. An MMX instruction can execute the same instruction on eight data inputs at once. Streaming multimedia instructions can use all of the MMX data types and an additional 128-bit data type containing four 32-bit floating point numbers.

Four different memory address formats are provided. Two of these are segmented memory addresses that are backward-compatible with earlier 80x86 processors. The third is a 32-bit flat memory model address compatible with the 80386 and later processors. The fourth is a 36-bit flat address model that was introduced with the Pentium Pro. Segmented memory addresses are offset with respect to predefined segment registers. Six different segment registers are available and are assumed to point to specific portions of program address space. Flat 32-bit and 36-bit addresses are the norm for most software developed since the early 1990s.

Instruction Set

The Pentium instruction set contains 28 system instructions, 92 floating point instructions, 52 MMX instructions, 61 streaming multimedia instructions, and 164 integer, logical, and other instructions. This is one of the largest instruction sets among modern microprocessors. As expected, such a large instruction set has many instruction formats of varying length. Instructions can have from zero to three operands. Legal operands include words, registers, and 32-bit memory addresses.

A detailed discussion of the Pentium instruction set fills an entire volume of several hundred pages of Intel documentation. The large number of instructions represents both the rich variety of supported data types and the desire to implement as many processing functions as possible in hardware. This clearly places the Pentium in the CISC camp of CPU design.

Word Size

There is no clear definition of word size for the Pentium. For backward compatibility with the 8088 and 8086 microprocessors, a word is defined as 16 bits. All registers in the integer ALUs are 32 bits in length, the upper and lower 16 bits can be accessed separately for backward compatibility. The two internal processing paths of the integer ALUs also are 32 bits wide. The external bus interface is 64 bits wide and the on-chip memory cache interface is 128 bits wide.

The Pentium has a separate floating point ALU with its own registers and internal processing paths. Registers within this ALU are 80 bits wide as are the internal processing paths, which enables single instruction operations to be performed on the largest floating point data type. Although this is a powerful floating point processing capability, most application programs seldom use it.

Clock Rate

The original Pentium debuted at a clock rate of 60 MHz and eventually reached 200 MHz. The fastest Pentium 4 processor at this writing has a 3.6 GHz clock rate with an 800 MHz interface to the system bus.

64-Bit Processing

In 2004, Intel released a Xeon processor with a 64-bit word size, which closely mirrored the capabilities of a competing processor released by AMD the year before. Intel called the processor extensions Extended Memory 64 Technology® (EM64T). EM64T increases the size of general-purpose registers and ALU data paths to 64 bits, which enables full 64-bit computations on integers and memory addresses.

The primary benefit of EM64T is the ability to access terabytes of memory, far more than the 4 gigabyte ceiling imposed by 32-bit architecture. Many current server applications and a few workstation applications can benefit from that ability. Typical desktop computers are rapidly approaching 4 gigabyte memory capacity. Intel plans to release Pentium processors with EM64T in 2005 or 2006.

A modern CPU is a complex system of interconnected electrical switches. Early CPUs typically contained several hundred to a few thousand switches. Modern CPUs contain millions of switches. In this section, we will look first at how these switches perform basic processing functions and then at how these switches and circuits have been implemented physically.

Switches and Gates

The basic building blocks of computer processing circuits are electronic switches and gates. Electronic switches control electrical current flow in a circuit, and are implemented as transistors (described in more detail in a few pages). Switches can be interconnected in various ways to build gates. A **gate** is a circuit that can perform a processing function on an individual binary electrical signal, or bit. One and two bit processing functions performed by gates include the logical functions AND, OR, XOR, and NOT. Figure 4-13 shows the electrical component symbols for NOT, AND, OR, XOR, and NAND (NOT AND). A NOT gate is also called a signal inverter, because it transforms a value of zero or one into its inverse or opposite.

Processing circuits for more complex operations are constructed by combining logic gates. One example is the XOR gate. Although it is represented by a single symbol as shown in Figure 4-13(d), it is actually constructed by combining NOT, AND, and OR gates.

Figure 4-13 ►

Electrical component symbols for signal inverter or NOT gate (a); AND gate (b); OR gate (c); XOR gate (d); and NAND gate (e)

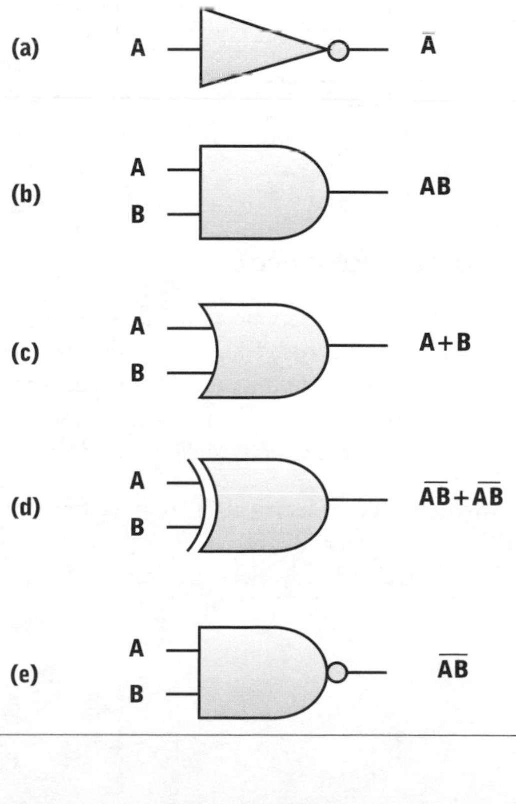

Figure 4-14 shows the components of half adder and full adder circuits for single bit inputs. Addition circuits for multiple bit inputs are constructed by combining a half adder for the least significant bit position with full adder circuits for the remaining bit positions. More complex processing functions require more complicated interconnections of gates. Modern CPUs have millions of gates to perform a wide variety of processing functions on a large number of bits.

Figure 4-14 ▶

Circuit diagrams for half adder (a) and full adder (b)

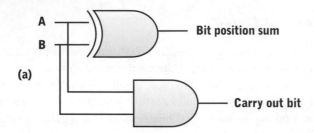

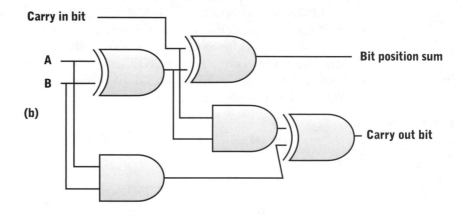

Electrical Properties

Gates and their interconnections carry and transform electrical signals representing binary ones and zeros. The construction of these switches and their connections is important in determining the CPU's speed and reliability. The speed and reliability of a CPU are affected not only by the materials used in its fabrication, but also by the properties of electricity such as conductivity, resistance, and heat.

Conductivity Electrical current is the flow of electrons from one place or device to another. An electron requires a sufficient energy input to excite it to move. Once excited, electrons move from place to place using molecules as stepping stones. Conductive molecules are typically arranged in straight lines generically called **wires** or **traces**. Because each switch is usually interconnected with many other switches, a CPU contains many more traces than switches. The traces are as important as the switches in determining CPU speed and reliability.

The ability of an element or substance to enable electron flow is called **conductivity**. Substances that electrons can flow through are called **conductors**. Electrons travel through a perfect conductor with no loss of energy. With less than perfect conduction, energy is lost as electrons pass through. If enough energy is lost, the electrons cease to move and the flow of electrical current is halted. Commonly used conductors include aluminum, copper, and gold.

Resistance   The loss of electrical power that occurs within a conductor is called **resistance**. A perfect conductor would have no resistance. Unfortunately, all substances have some degree of resistance. Conductors with relatively low resistance include some well-known and valuable metals such as silver, gold, and platinum. Fortunately, some cheaper materials such as copper serve nearly as well.

The laws of physics tell us that energy is never really lost but merely converted from one form to another. Electrical energy isn't really lost due to resistance. Instead, it is converted to heat, light, or both, depending on the specific conductive material. The amount of generated heat depends on the amount of electrical power being transmitted and the resistance of the conductor. Higher power and/or higher resistance increase the amount of generated heat.

Heat   Heat has two negative effects on electrical conductivity. The first is physical damage to the conductor. To reduce heat and avoid physical destruction of a switch or trace, we must use a very low-resistance material, reduce the power input, or increase the size of the switch or trace.

The second negative effect of heat is that it changes the inherent resistance of the conductor. The resistance of most materials increases as their temperature increases. To keep operating temperature and resistance low, some method must be provided to remove or dissipate heat. This can be accomplished in many ways. The simplest way is to provide a cushion of moving air around the device. Heat migrates from the surface of the device or its coating to the air and then is transported away by fans or through ventilation openings.

A **heat sink** (see Figure 4-15) is an object specifically designed to absorb heat and rapidly dissipate it via air or water movement. A heat sink is placed in direct physical contact with an electrical device to dissipate heat from the device, exposing a large surface area to the moving air to allow more rapid dissipation.

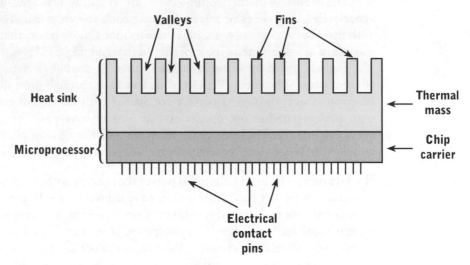

Figure 4-15 ▶

A heat sink
attached to a
surface-mounted
microprocessor

Valleys   **Fins**

Heat sink

**Thermal
mass**

Microprocessor

**Chip
carrier**

**Electrical
contact
pins**

Speed and Circuit Length To perform processing, electrons must move through the traces and switches of a processing circuit. In a vacuum, electrical energy travels at the (constant) speed of light, which is approximately 180,000 miles per second. Electricity travels through a trace at approximately 70% of the speed of light. There is a fundamental relationship between circuit length and processing speed. The time required to perform a processing operation is a function of the length of the circuit and the speed of light.

Because circuit length is the only variable, the path to faster processing is clear: reduce circuit length. Shorter circuits require smaller switches and shorter and narrower traces. Miniaturization has been the primary basis for CPU speed and clock rate improvement since the first electrical computer. The first IBM personal computer used an Intel 8088 microprocessor with a clock rate of 4.77 MHz. As of this writing, Intel microprocessors have clock rates as high as 3.6 GHz.

Processor Fabrication

Reliable, efficient, and cost-effective electrical circuits must balance power requirements, resistance, heat, size, and cost. The earliest computers were constructed with ordinary copper wire and vacuum tube switches. They were unreliable due to the heat generated by the vacuum tubes. They also were quite large, typically filling an entire room with processing circuitry less powerful than that found today in a cheap calculator. Improvements in materials and fabrication techniques have vastly increased the performance and reliability of processors.

Transistors and Integrated Circuits In 1947, researchers at Bell Laboratories discovered a class of materials called **semiconductors**. The conductivity of these materials varies in response to the electrical inputs applied. **Transistors** are made of semiconductor material that has been treated, or doped, with chemical impurities to enhance the semiconducting effects. Silicon and germanium are basic elements with resistance characteristics that can be controlled or enhanced through the use of chemicals called dopants.

In the early 1960s, photolithography techniques were developed to fabricate miniature electronic circuits from multiple layers of metals, oxides, and semiconductor materials. This new technology made it possible to fabricate several transistors and their interconnections on a single chip to form an **integrated circuit** (IC). Integrated circuits reduced manufacturing cost per circuit because many chips could be manufactured in a single sheet, or wafer. Combining multiple gates on a single chip also reduced the manufacturing cost per gate and created a compact, modular, and reliable package. As fabrication techniques improved, it became possible to put hundreds, then thousands, and today, hundreds of millions of electrical devices on a single chip. The term **microchip** was coined to refer to this new class of electronic devices.

Microchips and Microprocessors A **microprocessor** is a microchip that contains all of the circuits and connections that implement a CPU. The first microprocessor (see Figure 4-16) was designed by Ted Hoff of Intel and introduced in 1971. Microprocessors ushered in a new era in computer design and manufacture. The most important part of a computer system could now be produced and purchased as a single package. Computer system design was simplified because computer designers didn't have to construct processors from smaller components. Microprocessors also opened an era of standardization as a small number of microprocessors became widely used. The personal computer revolution would not have been possible without standardized microprocessors.

Figure 4-16 ▶

The Intel 4004
microprocessor
containing 2,300
transistors

Courtesy of Intel Corporation

Processor Technology and Architecture

Current Technology Capabilities and Limitations

Gordon Moore, an Intel founder, made an observation during a speech in 1965 that has come to be known as **Moore's Law**. Moore observed that the rate of increase in transistor density on microchips had increased steadily, roughly doubling every 18 to 24 months. He further observed that each doubling was achieved with no increase in unit cost. Moore's Law has proven surprisingly durable. Figure 4-17 shows that increases in transistor density of Intel microprocessors have followed Moore's Law. Software developers and users have come to expect CPU performance to double every couple of years with no price increase.

Figure 4-17 ▶

Increases in transistor density for Intel microprocessors

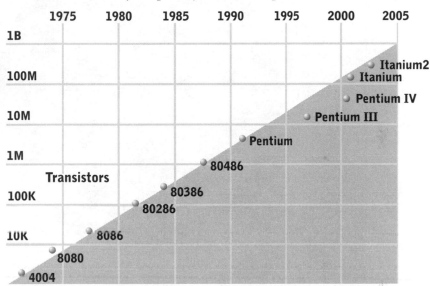

Arthur Rock, a venture capitalist, made a short addendum to Moore's Law. **Rock's Law** states that the cost of fabrication facilities for the latest chip generation doubles every four years. This law also has proven quite durable. A fabrication facility using the latest production processes currently costs at least 10 billion dollars. Improvements in chip fabrication have been achieved by ever more expensive and exotic techniques. The added expense hasn't led to price increases, because newer fabrication methods produce larger numbers of chips and the demand for chips exceeds the production capacity of any single factory.

Current fabrication technology is capable of squeezing over 300 million transistors onto a wafer of silicon approximately one square centimeter in size (see Figure 4-18). The process starts with a wafer, which is a flat disk of pure silicon. A thin layer of conductive or semiconductive material is spread over the surface of the wafer. The material then is exposed to ultraviolet light focused through a patterned map similar to an overhead transparency. An etching chemical is then applied which removes the portion of the layer that was exposed. The portion that remains has a very specific pattern, or map, corresponding to one layer of the microprocessor circuitry. Additional layers are added, then etched, gradually building up complete transistors and wires.

Figure 4-18 ▶

A memory chip containing 330 million transistors

Courtesy of Intel Corporation

The wafer contains hundreds or thousands of microprocessors, which are cut apart, encased in individual carriers, and tested. It is not unusual for 70 to 80 percent of the microprocessors to fail initial testing. Testing failures can be the result of impurities in materials, contamination of layers as they are laid down on the wafer, errors in etching, and variations in any number of production process parameters. Those that survive initial tests are then subjected to further tests at successively higher clock rates. A microprocessor that fails to perform reliably at a higher clock rate still may be usable and rated for sale at a lower clock rate.

Increases in the number of circuits and packing density on a chip have been achieved by shrinking the size of transistors and the traces connecting them, and by increasing the number of layers added to the wafer. Shrinking the size of transistors and traces is accomplished by an etching process that leaves thinner and thinner lines of conductive and semiconductive material on the wafer, while increasing the number of layers puts additional components one on top of another on the chip.

Increased packing density causes some problems, however. As the size of the devices and wiring decreases, the ability to transmit electrical power is reduced. This requires reduced operating voltage (currently as low as 1.3 volts, which is down from 5 volts in the 1980s). Low voltage circuits are more susceptible to damage from voltage surges and from static electricity. Also, low voltage signals are more susceptible to contamination and noise from nearby circuits and external interference.

Electrical resistance is also a problem because circuits must have uniform resistance properties throughout the chip. A slight increase in resistance in one part of the chip may prevent sufficient electrical power from reaching another part of the chip. Uniform resistance requires very high and uniform purity of the conductive and semiconductive material in the chip. Resistance can also cause problems due to heat. Because a chip is a sealed package, there is no way to circulate air or any other substance among its circuits. Heat must migrate through the chip carrier to be dissipated by a fan or heat sink. Heat generation per circuit is minimized by their small size and low-resistance materials. But each device and trace does generate heat. Higher clock rates generate more heat as more electrons are forced through the circuits. Inadequate or barely adequate cooling can substantially shorten a microprocessor's operating life.

Business
Focus

Server CPUs

Beginning with the Pentium line of CPUs, Intel has developed products along two parallel development paths, one targeted toward "ordinary" desktop computers and the other targeted toward workstations and small servers. The history of both development paths is summarized in Table 4-4. In essence, each release of a Pentium processor has been shortly followed by a similar, but more powerful CPU targeted to a different market.

Table 4-4 ▶

Intel desktop and server CPUs based on Pentium technology

Desktop CPU	Server CPU
Pentium (1993)	Pentium Pro (1995)
Pentium II (1997)	Pentium II Xeon (1998)
Pentium III (1999)	Pentium III Xeon (1999)
Pentium 4 (2000)	Xeon (2001)
	Xeon MP (2002)
Pentium 4 EM64T (2005)	Xeon EM64T (2004)

Intel Xeon CPUs share much technology with their Pentium counterparts including their instruction sets, integer and floating point execution units, similar approaches to managing memory caches, and similar manufacturing processes. However, Xeon CPUs provide extra features not found in the Pentium, including:

▶ Larger memory caches

▶ The ability to access more main memory

▶ The ability to share access to main memory, the system bus, and other resources among four CPUs

Although the Xeon fills a significant niche in the small server market, it isn't sufficiently powerful to perform processing duties in many large-scale servers. Large-scale servers, especially those that host database management systems, generally use CPUs with higher computational and memory performance, on-chip error checking, and support for more than four CPUs per server. IBM, Sun Microsystems, Hewlett-Packard, and Digital Equipment Corporation (DEC) dominated this market in the 1990s. DEC was purchased by Compaq, which merged with Hewlett-Packard in 2002.

In the mid-1990s, Intel teamed with Hewlett-Packard to develop a new generation of CPUs. The new CPU architecture they developed is now called Explicitly Parallel Instruction Computing (EPIC). Characteristics of the architecture include 64-bit word size, pipelined execution, prefetching of instructions and operands, and speculative execution of multiple processing paths (for example, beginning to execute both paths of an If-Then-Else while the If condition is being evaluated).

Intel's first CPU based on the EPIC architecture, called the Itanium, was released in 2001. The Itanium2 was released in 2002. Over the last few years, the Itanium and IBM POWER processors have leapfrogged each other for the claim to world's fastest mass-produced microprocessor, as measured by the SPEC CFP2000/2004 benchmarks. In addition to raw speed, the Itanium also provides on-chip support for the Pentium instruction set, which allows it to run many legacy software applications.

Although the Itanium is well suited to large-scale servers, computer manufacturers have been slow to build systems based on it. IBM POWER CPUs have been more successful, largely due to IBM's entrenched position in the large-scale server market and its long history of developing and supporting server operating systems.

To match IBM's success, Intel will have to convince computer manufacturers to develop new mainframe class servers with the Itanium and convince large-scale server customers to buy them and adopt compatible operating systems. Itanium market penetration is expected to be slow given the time and cost of mainframe development and the significant costs and risks inherent in purchasing and operating large-scale servers.

Questions

1. Assume that you're purchasing new servers for a medium-sized network where the primary server functions are file and printer sharing. Which CPU—Pentium, Xeon, Itanium, or POWER—is the best choice?

2. Assume that you're purchasing new servers for a server farm. The servers host e-commerce Web sites and supporting software such as an Oracle DBMS. Which CPU is the best choice?

3. The Itanium2 and POWER5 CPUs currently operate at about one half the clock rate of the fastest 32-bit CPUs. Why is there such a significant clock rate difference? How can the Itanium2 and POWER5 deliver improved performance compared to a 32-bit CPU with double the clock rate?

FUTURE TRENDS

Will Moore's Law continue to hold for the foreseeable future? Will transistor size and cost continue to shrink at the present rate? Many industry experts don't think so. They predict that we will hit a technological "wall" between 2010 and 2020 when the limits of current microprocessor design and fabrication will be reached.

Beyond a certain point, further miniaturization will be more difficult to achieve because of the nature of the etching process and the limits of semiconducting materials. The size of traces and devices depends on the etching beam wavelength, but there are limits to our ability to generate and focus short wavelength lasers and X-rays. The width of chip traces also is rapidly approaching the molecular width of the materials from which the traces are constructed; a trace can't be less than one molecule wide.

Subsequent improvements will require fundamentally different approaches to microprocessor design and fabrication. These approaches may be based on a number of technologies including optics and quantum physics. None of these technologies has yet been proven on a commercial scale.

Optical Processing

In an electrical computer, computation is performed by forcing electricity to flow through millions of electrical switches and wires. Optical computing could eliminate interconnection and simplify fabrication problems because photon pathways can cross without interfering with one another. Although fiber-optic wires are usually used to connect optical devices, it is possible to connect them without wires. If the sending and receiving devices are aligned precisely, then light signals sent between them can travel through free space. Eliminating wires would vastly improve fabrication cost and reliability.

Commercially available and affordable optical computers have yet to become a reality. As long as semiconductors continue to get faster and cheaper, there isn't sufficient economic incentive to develop optical technologies. There is economic incentive to pursue specialized processing applications in the areas of telecommunications and networking. Practical optical processors will probably appear first as dedicated communication controllers and later evolve into full-fledged computer processors.

Electro-Optical Processing

Electro-optical transistors and other devices hold promise as the interface between semiconductor processors and purely optical memory and storage devices. To date, the most commonly used material for such devices is **gallium arsenide**. Like silicon and other materials, it can be used to fabricate conventional semiconductors. Unlike silicon, it has both optical and electrical properties. Gallium arsenide can be used to create transistors with optical controls; a gallium arsenide transistor can enable electron flow when an optical input is applied.

Gallium arsenide transistors already are used in telecommunications switching equipment. They also are used in some supercomputers to allow electrical CPUs to interface directly with an optical bus. They provide sufficient data transfer capacity to supply dozens or hundreds of processors with instructions and data. Technologies such as holographic storage have the potential to deliver gigabytes of data in a matter of seconds. Optical communication channels will be required to provide processor access to all of that data.

Intel has recently demonstrated another approach to interfacing electrical and optical components. A silicon-based semiconductor device is used to encode data in externally generated laser light. The device accepts electrical pulses representing bits and modifies the phase characteristics of a split laser to encode the bit values at a rate of one billion bits per second. Other advances, such as silicon-based optical receivers need to be developed to fully realize the promise of silicon-based optics. If Intel is successful, they will be able to combine high-speed optical and electrical components on a single chip at relatively low cost compared to more exotic materials such as gallium arsenide.

Quantum Processing

Quantum computing was introduced as a Technology Focus section in Chapter 2. Quantum computing uses quantum states to simultaneously encode two values per bit, called a qubit, and uses quantum processing devices to perform computations.

In theory, quantum computing is well-suited to solving problems that require massive amounts of computation, such as factoring large numbers and calculating private encryption keys.

Current computers solve many different types of problems, only a few of which require massive computation. As currently envisioned, quantum computing provides no benefit for the other problem classes as compared to conventional electrical processors. Quantum processors may never replace conventional processors in the vast majority of computer systems, or perhaps they will develop in ways we can't imagine. Much as Bell Labs researchers could not have imagined 4 GHz microprocessors when they developed the first transistor, current researchers probably can't foresee the quantum devices that will be built later in the twenty-first century nor how they'll be applied to present and future computing problems.

SUMMARY

▶ The CPU continuously alternates between the instruction, or fetch, cycle and the execution cycle. During the fetch cycle, the control unit fetches an instruction from memory, separates the op code and operands, stores the operands in registers, and increments a pointer to the next instruction. During the execution cycle, either the control unit or the ALU executes the instruction. The ALU executes the instruction if it is an arithmetic or logical operation. The control unit executes all other instruction types.

▶ Primitive CPU instructions can be classified into three types: data movement, data transformation, and sequence control. Data movement instructions copy data among registers, primary storage, secondary storage, and I/O devices. Data transformation instructions implement simple Boolean operations (NOT, AND, OR, and XOR), addition (ADD), and bit manipulation (SHIFT). Sequence control instructions control the next instruction to be fetched or executed. More complex processing operations are implemented by appropriate sequences of primitive instructions.

▶ An instruction format is a template describing the op code position and length and the position, type, and length of each operand. Most CPUs support multiple instruction formats. CPUs can support fixed length or variable length instruction formats. Fixed length formats are simpler for the control unit to fetch and decode. Variable length instructions use primary and secondary storage more efficiently. RISC CPUs use fixed length instructions and generally avoid complex instructions, particularly those that combine data movement and data transformation. RISC CPUs are simpler than CISC CPUs, but they are less efficient than CISC CPUs when performing some complex operations.

- The CPU clock rate is the number of instruction and execution cycles potentially available in a fixed time interval. Typical CPUs have clock rates measured in thousands of MHz (1000 MHz = 1 GHz). A CPU generally executes fewer instructions than are implied by its clock rate. The rate of actual or average instruction execution is measured in MIPS or MFLOPS. Many instructions, especially floating point operations, require multiple execution cycles to complete. The CPU also may incur wait states pending storage or I/O device accesses.

- CPU registers are of two types—general and special purpose. Programs use general-purpose registers to hold intermediate results and frequently needed data items. General-purpose registers are implemented within the CPU so their contents can be read or written quickly. Within limits, increasing the number of general-purpose registers decreases program execution time. The CPU uses special-purpose registers to track processor and program status. Important special-purpose registers include the instruction pointer and program status word.

- Word size is the number of bits that a CPU can process simultaneously. Within limits, CPU efficiency increases with word size because inefficient piece-by-piece operations on large data items are avoided. For optimal performance, other computer system components such as the system bus should match or exceed CPU word size.

- Techniques to enhance processor peformance include pipelining, branch prediction, speculative execution, and multiprocessing. Pipelining improves performance by allowing multiple instructions to execute simultaneously in different stages. Branch prediction and speculative execution ensure that the pipeline is kept full while executing conditional branch instructions. Multiprocessing provides multiple CPUs to enable simultaneous execution of different processes or programs.

- CPUs are electrical devices implemented as silicon-based microprocessors. Like all electrical devices, they are subject to basic laws and limitations of electricity including resistance, heat generation, and speed as a function of circuit length. Microprocessors use small circuit size, low-resistance materials, and heat dissipation to ensure fast and reliable operation. Microprocessors are fabricated using expensive processes based on ultraviolet or laser etching and chemical deposition. Those processes, and semiconductors themselves, are approaching fundamental physical size limits that will stop further improvements. Candidate technologies that may move performance beyond semiconductor limitations include optical processing, hybrid optical-electrical processing, and quantum processing.

Now that we've described the inner workings of computer processors, we need to move on to the other computer system components. In the next chapter you will learn the characteristics and inner function of primary and secondary storage devices, including magnetic and optical storage technology—the bases of all modern storage devices.

Key Terms

ADD
AND
arithmetic SHIFT
benchmark
BRANCH
branch prediction
clock cycle
clock rate
complex instruction
Complex Instruction Set
    Computing (CISC)
conditional BRANCH
conductivity
conductor
cycle time
decoding
exclusive OR (XOR)
execution cycle
fetch cycle
fixed length instruction
flag
gallium arsenide
gate
gigahertz (GHz)

HALT
heat sink
hertz (Hz)
inclusive OR
instruction
instruction cycle
instruction format
instruction pointer
instruction register
instruction set
integrated circuit (IC)
JUMP
load
logical SHIFT
megahertz (MHz)
microchip
microprocessor
millions of floating point
    operations per second
    (MFLOPS)
millions of instructions per
    second (MIPS)
Moore's Law

MOVE
multiprocessing
NOT
op code
operand
pipelining
program counter
program status word (PSW)
Reduced Instruction Set
    Computing (RISC)
resistance
Rock's Law
semiconductor
SHIFT
speculative execution
store
trace
transistor
unconditional BRANCH
variable length instruction
wait state
wire
word

Vocabulary Exercises

1.   The _____ time of a processor is one divided by the clock rate (in Hz).

2.   A CPU typically implements multiple _____ to account for differences among instructions in the number and type of operands.

3.   _____ generates heat within electrical devices.

4.   _____ is a semiconducting material with optical properties.

5.   A(n) _____ is an electrical switch built of semiconducting materials.

6. A(n) _____ improves heat dissipation by providing a thermal mass and a large thermal transfer surface.

7. One _____ is one cycle per second.

8. Applying a(n) _____ OR transformation to input bit values one and one generates true. Applying a(n) _____ OR transformation to those same inputs generates false.

9. When first fetched from memory, an instruction is placed in the _____ and then _____ to extract its components.

10. Using _____ instructions simplifies the process of instruction fetching and decoding.

11. A(n) _____ is an electrical circuit that implements a Boolean or other primitive processing function on single bit inputs.

12. A microchip containing all of the components of a CPU is called a(n) _____.

13. A(n) _____ instruction transforms the bit pairs 1/1, 1/0, and 0/1 into 1.

14. The address of the next instruction to be fetched by the CPU is held in the _____.

15. The contents of a memory location are copied to a register while performing a(n) _____ operation.

16. A(n) _____ or _____ contains multiple transistors or gates in a single sealed package.

17. A(n) _____ instruction always alters the instruction execution sequence. A(n) _____ instruction alters the instruction execution sequence only if a specified condition is true.

18. A(n) _____ processor does not directly implement complex instructions.

19. A(n) _____ instruction copies data from one memory location to another.

20. The CPU incurs one or more _____ when it is idle, pending the completion of an operation by another device within the computer system.

21. A(n) _____ is the number of bits processed by the CPU in a single instruction. It also describes the size of a single register.

22. In many CPUs, a register called the _____ stores condition codes, including those representing processing errors and the results of comparison operations.

23. The components of an instruction are its _____, and one or more _____.

24. Two one-bit values generate a one result value when a(n) _____ instruction is executed. All other input pairs generate a zero result value.

25. A(n) _____ CPU typically uses variable length instructions and has a large instruction set.

26. A(n) _____ operation transforms a zero-bit value to one and a one-bit value to zero.

27. _____ predicts that transistor density and processor power will double every two years or less.

28. A(n) _____ is a measure of CPU or computer system performance when executing one or more specific tasks.

29. _____ is a CPU design technique in which instruction execution is broken into multiple stages and different instructions can execute in different stages simultaneously.

Review Questions

1. Describe the operation of a MOVE instruction. Why is the name "move" a misnomer?

2. Why does program execution speed generally increase as the number of general-purpose registers increases?

3. What are special-purpose registers? Give three examples of special-purpose registers and explain how each is used.

4. What are the advantages and disadvantages of fixed length instructions as compared to variable length instructions? Which type generally are used in a RISC processor? Which type are generally used in a CISC processor?

5. Define word size. What are the advantages and disadvantages of increasing word size?

6. What characteristics of the CPU and of primary storage should be balanced to obtain maximum system performance?

7. How does a RISC processor differ from a CISC processor? Is one processor type better than the other? Why or why not?

8. What factor(s) account for the dramatic improvements in microprocessor clock rates over the last three decades?

9. What potential advantages are provided by optical processors as compared to electrical processors?

10. Which is the better measure of relative computer system performance — a benchmark such as CPU2004 or a processor speed measure such as GHz, MIPS, or MFLOPS? Why?

11. How does pipelining improve the efficiency of a CPU? What is the potential effect of executing a conditional branch instruction on the efficiency gained by pipelining? What technique(s) can be employed to make pipelining more efficient when executing conditional branch instructions?

12. How does multiprocessing improve the efficiency of a computer system?

Problems and Exercises

1. Develop a program consisting of primitive CPU instructions to implement the following procedure:

```
integer i,a;

i=0;
while (i < 10) do
  a=i*2;
  i=i+1;
endwhile
```

2. Assume that a microprocessor has a cycle time of 0.5 nanoseconds. What is the processor clock rate? Also, assume that the fetch cycle is 40 percent of the processor cycle time. What memory access speed is required to implement load operations with zero wait states? What memory access speed is required to implement load operations with two wait states?

3. Processor R is a 64-bit RISC processor with a 2 GHz clock rate. The average instruction requires one cycle to complete, assuming zero wait state memory accesses. Processor C is a CISC processor with a 1.8 GHz clock rate. The average simple instruction requires one cycle to complete, assuming zero wait state memory accesses. The average complex instruction requires 2 cycles to complete, assuming zero wait state memory accesses. Processor R can't directly implement the complex processing instructions of Processor C. Executing an equivalent set of simple instructions requires an average of 3 cycles to complete, assuming zero wait state memory accesses.

   Program S contains nothing but simple instructions. Program C executes 70 percent simple instructions and 30 percent complex instructions. Which processor will execute program S more quickly? Which processor will execute program C more quickly? At what percentage of complex instructions will the performance of the two processors be equal?

4. Assume the existence of a CPU with a 4.8 GHz clock rate. The instruction and execution cycles are each 50 percent of the clock cycle. The average instruction requires 0.5 nanoseconds to complete execution. Main memory access speed for a single instruction is 2 nanoseconds on average. What is the expected average MIPS rate for this CPU? What modern microprocessor architectural features might be added to the CPU to improve its MIPS rate?

Research Problems

1. Investigate the instruction set and architectural features of a modern RISC processor such as the Sun UltraSparc (www.sun.com), Apple/Motorola G5 (www.apple.com/g5processor/), or IBM POWER5 (www.chips.ibm.com). In what ways does it differ from the architecture of the Intel Pentium and Itanium processor families? Which processor(s) employ advanced techniques such as pipelining, speculative execution, and multiprocessing?

2. AMD (www.amd.com) produces microprocessors that execute CPU instructions designed for the Intel Pentium processor family. Investigate their current product offerings. Do they offer true Pentium compatibility? Is their performance comparable to that of Intel processors?

3. Until recently, microprocessor transistor interconnections were implemented with aluminum. Some manufacturers are now implementing transistor interconnections with copper using a process developed by IBM (www.ibm.com). Investigate the advantages of chips fabricated with copper. Will copper extend the lifetime of semiconductor microprocessor technology? Why or why not?

Chapter 5

Data Storage Technology

Chapter Goals

- ▶ Describe the distinguishing characteristics of primary and secondary storage

- ▶ Describe the devices used to implement primary storage

- ▶ Describe memory allocation schemes

- ▶ Compare and contrast secondary storage technology alternatives

- ▶ Describe factors that determine storage device performance

- ▶ Choose appropriate secondary storage technologies and devices

In Chapter 2, you were briefly introduced to the topic of storage, including the role of storage in a computer system and the differences between primary and secondary storage. In this chapter, you will explore storage devices and their underlying technologies in depth (see Figure 5-1). The chapter outlines characteristics common to all storage devices, and then explains the technology, strengths, and weaknesses of primary and secondary storage.

Figure 5-1 ▶

Topics covered in this chapter

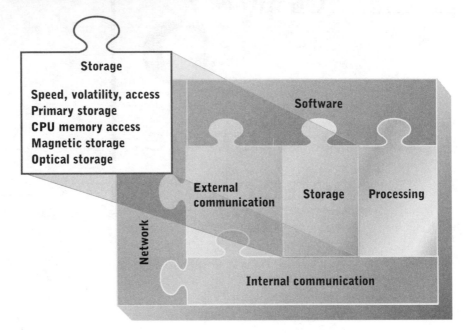

STORAGE DEVICE CHARACTERISTICS

A storage device consists of a read/write mechanism and a storage medium. The **storage medium** is the device or substance that actually holds data. The **read/write mechanism** is the device used to read or write data to and from the storage medium. A **device controller** provides the interface between the storage device and system bus. Device controllers are discussed in detail in Chapter 6.

In some storage devices, the read/write mechanism and storage medium are a single unit using the same technology. For example, most types of primary storage use electrical circuits implemented with semiconductors for both the read/write mechanism and storage medium. In other storage devices, the read/write mechanism and the storage medium use fundamentally different technologies. For example, tape drives use an electro-mechanical device for the read/write mechanism and a magnetic storage medium composed of polymers and metal oxides.

A typical computer system has many storage devices (see Figure 5-2) because no one device or technology can cost-effectively meet all storage needs for a single computer system or user. Different devices are used for different storage purposes because there is no single storage device that is best in all the primary characteristics that distinguish storage devices, which include:

► Speed
► Volatility
► Access method
► Portability
► Cost and capacity

Each characteristic is discussed in detail in the sections that follow.

Figure 5-2 ►

Primary and secondary storage and their component devices

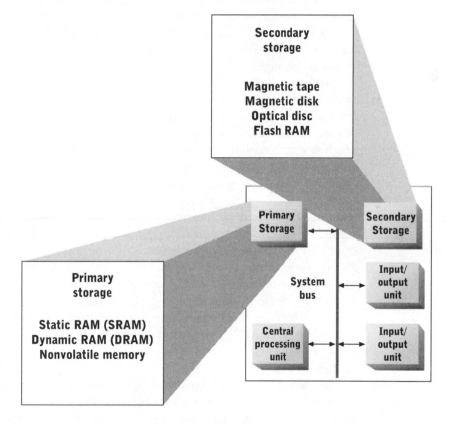

Speed

Speed is the most important characteristic that differentiates primary and secondary storage. Primary storage extends the limited capacity of CPU registers. The CPU continually moves data and instructions between registers and primary storage. CPU instruction and execution cycles are extremely short, a few nanoseconds at most.

If primary storage devices cannot be read or written within a single CPU cycle, then the CPU must wait for the read/write operation to complete. A **wait state** is a CPU cycle spent waiting for a storage access or I/O operation to finish. Wait states reduce CPU and computer system performance. To minimize wait states, all or part of primary storage is implemented with the fastest available storage devices. With current technology, primary storage speed is typically faster than secondary storage speed by a factor of 10^5 or more.

Speed is also an important issue for secondary storage. Many information system applications require access to large databases to support ongoing processing. Program response time in such systems depends on secondary storage access speed, which also affects overall computer performance in other ways. Before a program can be executed, its executable code is copied from secondary to primary storage. The delay between a user request for program execution and the first prompt for user input depends on the speed of both primary and secondary storage.

Storage device speed is called access time. **Access time** is the time required to execute one complete read or write operation. Access time is assumed to be the same for both reading and writing unless otherwise stated. Access time for some storage devices such as random access memory (RAM) is the same regardless of which storage location is accessed. For other storage devices such as disks, access time varies with storage location. For such devices, access time typically is expressed as an average of access times for all storage locations called **average access time** (access times are described in more detail later in this chapter). Access times of primary storage devices are generally expressed in nanoseconds (billionths of a second). Access times for secondary storage devices are typically expressed in milliseconds (thousandths of a second).

By itself, access time is an incomplete measure of storage device speed. A complete measure consists of access time and the unit of data transfer to or from the storage device. Data transfer unit size varies from one storage device to another. The data transfer unit for primary storage devices is usually a word. Depending on the CPU, a word may represent 2, 4, 8, or more bytes, 4 and 8 bytes being the most common word sizes. Data transfer unit size sometimes is used as an adjective, as in "64-bit memory."

Secondary storage devices read or write data in units larger than words. **Block** is a generic term that describes secondary storage data transfer units. Block size is normally stated in bytes. It also varies widely among storage devices and can even vary within a single storage device.

The term **sector** describes the data transfer unit for magnetic and optical disk drives. As with blocks, sector size is generally stated in bytes and can vary from one device to another. A 512-byte block is the most common data transfer unit for magnetic disks.

A storage device's **data transfer rate** is computed by dividing 1 by the access time (expressed in seconds) and multiplying the result by the unit of data transfer (expressed in bytes). For example, the data transfer rate for a primary storage device with 15-nanosecond access time and a 32-bit word data transfer unit can be computed as:

$$\frac{1 \text{ second}}{15 \text{ nanoseconds}} \times 64 \text{ bits} = \frac{1}{0.000000015} \times 8 \text{ bytes} \approx 533,333,333 \text{ bytes/second}$$

Using average access time, the same formula can be used to compute an average data transfer rate for devices with variable access times.

Volatility

A storage device or medium is **nonvolatile** if it holds data without loss over long periods of time. A storage device or medium is **volatile** if it cannot reliably hold data for long periods. Primary storage devices are generally volatile. Secondary storage devices are generally nonvolatile.

Volatility is actually a matter of degree and conditions. For example, RAM is nonvolatile as long as external power is continually supplied. But it is generally considered a volatile storage device because continuous power cannot be guaranteed under normal operating conditions. Magnetic tape and disk are considered nonvolatile storage media. But data held on magnetic media is often lost after a few years due to natural magnetic decay and other factors. Data stored on nonvolatile media may be lost due to obsolescence of compatible read/write devices, as happened with 12-inch laser discs and many older tape formats.

Access Method

The physical structure of a storage device's read/write mechanism and storage medium determines the way(s) in which data can be accessed. Three broad classes of access are recognized:

► Serial access
► Random access
► Parallel access

A single device can use multiple access methods.

Serial Access   A **serial access** storage device stores and retrieves data items in a linear, or sequential, order. Magnetic tape is the only widely used form of serial access storage. Data is written to a linear storage medium in a specific order and can be read back only in that same order. For example, to play the last song on a cassette or digital audio tape requires playing or fast-forwarding past all of the songs that precede it. Similarly, if there are *n* digital data items stored on a magnetic tape, then the nth data item can be accessed only after the preceding $n-1$ data items have been accessed.

Serial access time depends on the current position of the read/write mechanism and on the position of the desired data item within the storage medium. If both positions are known, then access time can be computed as the difference between the current and target positions multiplied by the amount of time required to move from one position to the next.

Serial access devices are not used for frequently accessed data due to their inefficient access method. Even when data items are accessed in the same order as written, serial access devices are much slower than other forms of storage. Serial access storage is primarily used to hold backup copies of data stored on other storage devices, for example, in a weekly tape backup of user disk files.

Random Access A random **access device**, also called a **direct access device**, is not restricted to any specific order when accessing data. Rather, it can directly access any storage location. All primary storage devices and disk storage devices are random access devices.

Access time for a random access storage device may or may not be a constant. It is a constant for most primary storage devices. It is not a constant for disk storage because of the physical, or spatial, relationships of the read/write mechanism, storage media, and storage locations on the media. This issue is discussed fully in a later section.

Parallel Access A **parallel access** device can simultaneously access multiple storage locations. Although RAM is considered to be a random access storage device, it is also a parallel access device. The confusion arises from differences in the definition of the unit of data access. If one considers the unit of data access to be a bit, then access is parallel. That is, random access memory circuitry can access all of the bits within a byte or word at the same time.

Parallel access can also be achieved by subdividing data items and storing the component pieces on multiple storage devices. For example, some operating systems can store file content on several disk drives. Different parts of the same file can be read at the same time by issuing parallel commands to each disk drive.

Portability

Data can be made portable by storing it on a removable storage medium or device. For example, compact disc and tape storage are portable because the storage media are removable and their storage formats are standardized. In devices such as external disk drives and USB flash memory, the storage media and read/write mechanism are integrated into single portable device that can be quickly connected to or disconnected from a computer system.

Portable secondary storage devices and devices with removable storage media typically have slower access speed than permanently installed devices and those with non-removable media. High-speed access requires tight control of environmental factors. For example, high-speed access for magnetic disks is achieved, in part, by sealed enclosures for the storage media, thus minimizing or eliminating dust and air density variations. Removable storage media come in contact with a larger variety of damaging environmental influences than do isolated media.

Cost and Capacity

Each of the storage device attributes described so far is directly related to device cost (see Table 5-1). Improving speed, volatility, or portability increases cost if all other factors are held constant. Cost also increases as an access method moves from serial to random to parallel. Primary storage is generally expensive compared to secondary storage due to its high speed and its combination of parallel and random access methods.

Secondary storage devices are usually thought of as having much greater capacity than primary storage devices. In fact, the capacity difference between primary and secondary storage in most computer systems results from a compromise between cost and other device characteristics. If cost were not a factor, most users would opt for nonvolatile semiconductor-based memory devices instead of disk drives to implement secondary storage. However, most users need gigabytes of secondary storage, and that much nonvolatile semiconductor-based memory costs more than most users can afford. So users sacrifice speed and parallel access to gain sufficient capacity at an acceptable cost.

Table 5-1 ►

Storage device
characteristics
and their rela-
tionship to cost

Characteristic	Description	Cost
Speed	Time required to read or write a bit, byte, or larger unit of data	Cost increases as speed increases
Volatility	Ability to hold data indefinitely, particularly in the absence of external power	For devices of similar type, cost decreases as volatility increases
Access method	Can be serial, random, or parallel; parallel devices are also serial or random access	Serial is the least expensive; random is more expensive than serial; parallel access is more expensive than non-parallel access
Portability	Ability to easily remove and reinstall the storage media from the device or the device from the computer	For devices of similar type, portability increases cost; if all other characteristics are held constant
Capacity	Maximum data quantity held by the device or storage medium	Cost usually increases in direct proportion to capacity

Memory-Storage Hierarchy

A typical computer system has a variety of primary and secondary storage devices. The CPU and a small amount of high-speed RAM usually occupy the same chip. Slower RAM on separate chips comprises the bulk of primary storage. One or more magnetic disk drives are usually complemented by an optical disk drive and at least one form of removable magnetic storage.

The range of storage devices within a single computer system forms a memory-storage hierarchy, as illustrated in Figure 5-3. Cost and access speed generally decrease as one moves down the hierarchy. Due to lower cost, capacity tends to increase as one moves down the hierarchy. A computer system designer or purchaser attempts to find an optimal mix of cost and performance for a particular purpose.

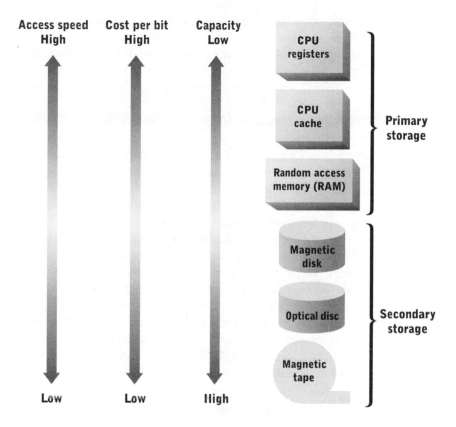

Figure 5-3 ▶

Comparison of storage devices in terms of cost and access speed

Access speed	Cost per bit	Capacity
High	High	Low
Low	Low	High

CPU registers

CPU cache

Random access memory (RAM)

Primary storage

Magnetic disk

Optical disc

Magnetic tape

Secondary storage

PRIMARY STORAGE DEVICES

The critical performance characteristics of primary storage devices are access speed and data transfer unit size. Primary storage devices must closely match CPU speed and word size to avoid wait states. The evolution of memory devices has closely mirrored the evolution of processing devices. Technologies applied to the construction of processors have, in general, been applied simultaneously to the construction of primary storage devices.

Storing Electrical Signals

Data are represented within the CPU as digital electrical signals. Digital electrical signals are also the basis of data transmission among all devices attached to the system bus. Any storage device or controller must accept electrical signals as input and generate electrical signals as output.

Electrical power can be stored directly by various devices including batteries and capacitors. Unfortunately, there is a tradeoff between access speed and volatility. Batteries are quite slow to accept and/or regenerate electrical current. They also lose their ability to accept charge with repeated use. Capacitors can charge or discharge with much greater speed than batteries. However, small capacitors lose their charge at a fairly rapid rate and require a fresh injection of electrical current at regular intervals (hundreds or thousands of times per second).

An electrical signal can be stored indirectly by using its energy to alter the state of a device such as a mechanical switch, or a substance such as a metal. An inverse process regenerates an equivalent electrical signal. For example, electrical current can generate a magnetic field. The strength of the magnetic field can induce a permanent magnetic charge in a nearby metallic compound. The stored magnetic charge can generate an electrical signal equivalent to the one used to create the original magnetic charge. Magnetic polarity, which is either positive or negative, can be used to represent the values zero and one.

Early computers implemented primary storage as rings of ferrous material (iron and iron compounds), a technology called **core memory**. The rings, or cores, are embedded in a two-dimensional wire mesh. Electricity sent through two wires induces a magnetic charge in one metallic ring. The polarity of the charge depends on the direction of electrical flow through the wires.

Modern computers use memory implemented with semiconductors. Basic types of semiconductor memory include random access memory (RAM) and read-only memory (ROM). There are many variations of each memory type, described in the following sections.

Random Access Memory

Random access memory (RAM) is a generic term describing primary storage devices with the following characteristics:

► Microchip implementation using semiconductors

► Ability to read and write with equal speed

► Random access to stored bytes, words, or larger data units

RAM is fabricated in the same manner as microprocessors. One might assume that microprocessor clock rates are well matched to RAM access speeds. Unfortunately, this is not the case for many reasons, including:

► Reading and writing many bits in parallel requires additional circuitry.

► When RAM and microprocessors are on separate chips, delays are encountered when moving data from one chip to another.

There are two basic RAM types and several variations on each. **Static RAM (SRAM)** is implemented entirely with transistors. The basic storage unit is a flip-flop circuit (see Figure 5-4). Each flip-flop circuit uses two transistors to store one bit. Additional transistors (typically two or four) perform read and write operations. A flip-flop circuit is an electrical switch that remembers its last position. One position represents zero and the other represents one. Flip-flop circuits require a continuous supply of electrical power to maintain their positions. SRAM is volatile unless a continuous supply of power can be guaranteed.

Figure 5-4 ▶

A flip-flop circuit composed of two NAND gates: the basic component of SRAM and CPU registers

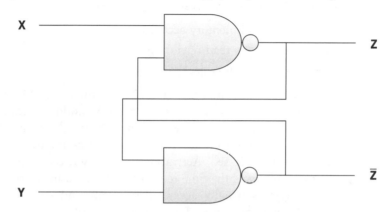

Dynamic RAM (DRAM) uses transistors and capacitors. The capacitors are the *dynamic* element. DRAM circuitry is less complex than SRAM circuitry. However, because the capacitors quickly lose their charge, they require a fresh infusion of power thousands of times per second. DRAM chips include circuitry that automatically performs refresh operations. Each refresh operation is called a **refresh cycle**. Unfortunately, a DRAM chip cannot perform a refresh operation at the same time it performs a read or write operation.

SRAM is approximately 10 times more expensive to fabricate than DRAM because of its greater complexity. Because DRAM circuitry is simpler, more memory cells can be packed into each chip (in other words, DRAM has higher density). Despite its simpler circuitry, DRAM is slower than SRAM due to its required refresh cycles and less efficient circuitry for accessing individual bits. With current fabrication technology, typical access times are 50 nanoseconds for DRAM and 5 nanoseconds for SRAM (a 10:1 difference). Improvements in fabrication technology can decrease both access times but cannot change the relative performance difference.

Note that neither RAM type can match current microprocessor clock rates, which range from 3 to 4 GHz at this writing. For zero wait state memory accesses, those clock rates require the following memory access speeds:

$$\frac{1}{3 \text{ GHz}} = \frac{1}{3,000,000,000} \approx 0.00000000033 \approx 0.33 \text{ nanoseconds}$$

$$\frac{1}{4 \text{ GHz}} = \frac{1}{4,000,000,000} = 0.00000000025 = 0.25 \text{ nanoseconds}$$

The fastest DRAM is at least 50 times slower than modern microprocessors. SRAM is 5 to 10 times slower.

A number of technologies are used to bridge the performance gap between memory and microprocessors:

► Read-ahead memory access
► Synchronous read operations
► On-chip memory caches

Computer memory contains more than just SRAM or DRAM circuits. Memory circuits are grouped in memory modules containing tens or hundreds of megabytes. Each module contains additional circuitry to implement read and write operations to random locations—circuitry that must be activated before data can be accessed. Activating the read/write circuitry consumes time, which in turn lengthens the time required to access a memory location.

Programs usually access instructions and data items sequentially. Read-ahead memory access takes advantage of this tendency by activating the read/write circuitry for location *n+1* during or immediately after an access to memory location *n*. If the CPU subsequently issues an access command for location *n+1*, then the memory module already has performed part of the work needed to complete the access.

Synchronous DRAM (SDRAM) is a read-ahead RAM that uses the same clock pulse as the system bus. Read and write operations are broken into a series of simple steps, each of which can be completed in one bus clock cycle (this is similar to pipelined processing as described in Chapter 4). Several clock cycles are required to complete a single random access. But if memory is being written or read sequentially, then accesses occur once per system bus clock tick after the first access has completed.

Current SDRAM can match bus clock rates over 500 MHz. The latest generations of SDRAM technology, called double data rate (DDR) and DDR2, perform two read or write stages per bus clock cycle.

Cached DRAM, also called **enhanced DRAM (EDRAM)**, consists of a small amount of SRAM placed in each DRAM device. When a read request for a specific word is received, that word and several words that follow are retrieved. The extra words are stored in the SRAM cache in anticipation of future sequential read requests. If those requests are made, the data can be more quickly accessed from the SRAM cache. Other uses of caching are discussed in Chapter 6.

Nonvolatile Memory

Memory manufacturers have worked for decades to develop semiconductor and other forms of random access memory with long-term or permanent data retention. The generic term for such a memory device is **nonvolatile memory (NVM)**. Of course, manufacturers and consumers would like NVM to cost the same or less than conventional SRAM and DRAM and have similar or faster read/write access times. Thus far, those goals have proven elusive, but some current and emerging memory technologies show considerable promise.

Even though it currently lacks the access speed of RAM, NVM has many applications in computing and other industries including storage of programs and data in portable devices such as handheld computers and cell phones, portable secondary storage in devices such as digital cameras, and permanent program storage in computer motherboards and peripheral devices. One of the oldest uses in computers is to store programs such as computer system boot subroutines like the system BIOS. These instructions can be loaded at high speed into main memory from NVM. Software that resides in NVM is called **firmware**.

Early NVM technology has evolved through several generations of devices. **Read-only memory (ROM)** is the earliest type of NVM with data content written permanently during manufacture. **Erasable programmable ROM (EPROM)** is manufactured blank, is written (programmed) with a special EPROM writer, and erased by exposure to ultraviolet light. The latest (and only currently used) form of ROM is **electronically erasable programmable ROM (EEPROM)**. An EEPROM can be programmed, erased, and reprogrammed by signals sent from a CPU. The primary drawbacks of EEPROM technology are low density, relatively high cost, and write speed that is much too slow to enable EEPROMs to serve as primary storage devices.

The most common NVM in use today is called **flash RAM**. It is competitive with DRAM in storage density (capacity) and read performance. Unfortunately, write performance is much slower than DRAM. Also, each write operation is mildly destructive, resulting in cell destruction after 100,000 or more write operations. Because of its relatively slow write speed and limited number of write cycles, flash memory is primarily used for secondary storage and for firmware that isn't frequently updated.

Other NVM technologies under development could overcome some of the shortcomings of flash RAM. Two promising candidates are ferroelectric RAM and polymer memory. **Ferroelectric RAM** embeds a crystal of a metallic compound within the bit storage/access circuitry. The center atom can be moved from one side of the crystal to the other to represent bit values. Read and write operations are similar to conventional DRAM or SRAM at comparable speeds. Ferroelectric RAM access speeds are much faster than flash RAM but repeated read operations can destroy cell contents. Research is continuing to develop nondestructive read capability and low-cost manufacturing methods.

Polymer memory uses a special plastic with electrical resistance that can be increased or decreased by an electrical field. The advantages of polymer memory include low-cost materials, nondestructive read access, and individual memory cells with no transistors. Current polymer memory research is focused on efficient methods for accessing individual memory cells and low-cost fabrication methods.

Memory Packaging

Memory packaging is similar to microprocessor packaging. Memory circuits are embedded within microchips, and groups of chips are packed on a small circuit board that can be installed or removed easily from a computer system. Early RAM and ROM circuits were packaged in **dual in-line packages** (DIPs), as shown in Figure 5-5. Installing a DIP on a printed circuit board is a tedious and precise operation. Also, single DIPs mounted on the board surface occupy a relatively large portion of the total surface area.

Figure 5-5 ►

Dual in-line package (DIP) chip and single in-line memory module (SIMM)

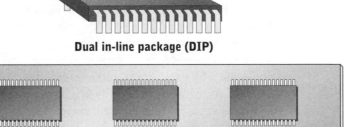

Dual in-line package (DIP)

Single in-line memory module (SIMM)

In the late 1980s, memory manufacturers adopted the **single in-line memory module (SIMM)** as a standard RAM package (see Figure 5-5). Each SIMM incorporates multiple DIPs on a tiny printed circuit board. The edge of the circuit board has a row of electrical contacts and the entire package is designed to lock into a SIMM slot on a computer motherboard. The **double in-line memory module (DIMM)** is a newer packaging standard. It is essentially a SIMM with independent electrical contacts on both sides of the module.

Current microprocessors include a small amount of on-chip memory; this technology is fully described in Chapter 6. As fabrication techniques improve, the amount of memory that can be packaged with the CPU on a single chip will grow. The logical extension of this trend is a CPU and all of its primary storage implemented on a single chip. This would virtually eliminate the current gap between microprocessor clock rates and memory access speeds.

CPU MEMORY ACCESS

Although main memory is not currently implemented as a part of the CPU, the CPU's need to load instructions and data from memory and to store processing results requires close coordination between both devices. Specifically, the physical organization of memory, the organization of programs and data within memory, and the method(s) of referencing specific memory locations are critical design issues for both primary storage devices and processors.

Physical Memory Organization

The main memory of any computer can be regarded as a sequence of contiguous, or adjacent, memory cells, as shown in Figure 5-6. Addresses of these memory locations are assigned sequentially so that available addresses proceed from zero (low memory) to the maximum available address (high memory).

Figure 5-6 ▶

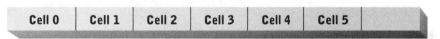

The sequential physical organization of memory cells

Data values and instructions generally occupy multiple bytes of storage. For example, an integer typically occupies 4 consecutive bytes, or 32 bits, of storage. When written in a program, the bits and bytes of a numeric value are typically ordered from highest position weight on the left to lowest position weight on the right. If a 32-bit integer is written as:

$$00110010\ 11010011\ 01001100\ 01101001$$

it is assumed that the left position weight is the largest (2^{31}) and the right position is the smallest (2^0). When considered as a byte sequence, the leftmost byte is called the **most significant byte** and the rightmost byte is called the **least significant byte**.

In many CPU and memory architectures, the least significant byte of a data item is placed at the lower memory address. For example, the 32-bit word $02468ACE_{16}$ would be stored in memory with CE at the lowest memory address, 8A at the next address, 46 at the address after that, and 02 at the highest address. Other CPU and memory architectures store bytes in the reverse order. **Big endian** describes architectures that store the most significant byte at the lowest memory address. **Little endian** describes architectures that store the least significant byte at the lowest memory address.

The **addressable memory** of a CPU or computer system is the highest numbered storage byte that can be represented. Addressable memory is usually determined by the number of bits used to represent an address. For example, if 32 bits are used to represent a memory address, then addressable memory is 2^{32} = 4,294,967,296 bytes, or 4 GB. A computer system's **physical memory** is the actual number of memory bytes that physically are installed in the machine. Physical memory is usually smaller than addressable memory.

Memory Allocation and Addressing

Memory allocation describes the assignment of specific memory addresses to system software, application programs, and data. The simple allocation scheme shown in Figure 5-7 supports an operating system and a single program stored in contiguous memory blocks. The operating system occupies the lowest memory block and the program occupies the next block of memory. Memory allocation diagrams such as Figure 5-7 show physical memory as an array of addresses with the lowest memory address at the bottom and the highest memory address at the top.

Figure 5-7 ►

A simple memory allocation scheme

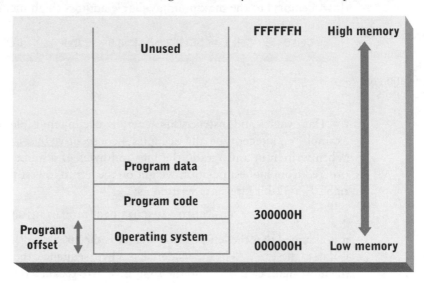

In Figure 5-7, the operating system occupies the lowest range of memory addresses, and the first instruction of the program is one address higher than the last byte of the operating system. The difference between the first address in physical memory and the address of the first program instruction is called the program's offset. In Figure 5-7, the offset is 300000_{16} - 0 = 300000_{16}.

A programmer could explicitly include the program offset in any memory address operands when writing or compiling a program. For example, assume that the program offset is 100,000 bytes. If the tenth program instruction is a BRANCH to the first instruction, then the tenth instruction's operand must be 100,000. Note that the programmer must know the program offset to correctly specify any memory address operands in BRANCH, MOVE, and other instructions.

Data Storage Technology

The term **absolute addressing** describes memory address operands that refer to actual physical memory locations.

Generally, a programmer doesn't know exactly where a program will be loaded in memory when it is executed, or the exact value of the offset. Even if the programmer did know the offset when writing the program, the offset value might later change for many reasons including:

▶ Upgrading the operating system, thus changing its size and the program offset

▶ Reconfiguring the operating system, such as allocating more I/O buffers or adding new device drivers

▶ Loading and executing multiple programs

If a program uses absolute addressing, then it must be rewritten every time the program offset is changed.

Because a program's offset generally can't be known in advance, programs are written as though their first instruction will always be loaded in memory address 0. The CPU automatically converts memory address operands into physical memory addresses as the program executes. To do so, the operating system calculates and stores the program offset in a register when the program is first loaded into memory. During program execution, the CPU automatically adds the program offset to all memory address operands before accessing memory. This method of automatically computing physical memory addresses is called **indirect addressing** or **relative addressing**. The register that holds the offset value is called an **offset register**.

Memory addressing and allocation schemes become much more complex when multiple programs are loaded into memory. Figure 5-8 shows an operating system and two programs stored in memory simultaneously. Note that the starting address and offset value of each program is different (300000_{16} for Program 1, 900000_{16} for Program 2). The operating system stores the offset value for each program currently loaded into memory. Before CPU control is passed from one program to another, the operating system loads the appropriate offset value(s) into the offset register(s).

Figure 5-8 ▶

Memory
allocation for
multiple
programs

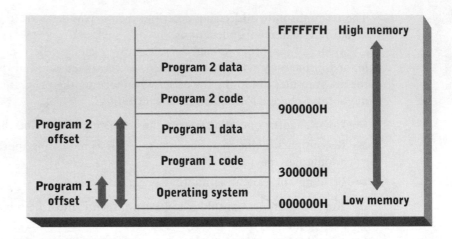

MAGNETIC STORAGE

Magnetic storage devices exploit the duality of magnetism and electricity. That is, electrical current can generate a magnetic field and a magnetic field can generate electricity. A magnetic storage device converts electrical signals into magnetic charges, captures the magnetic charge on a storage medium, and later regenerates an electrical current from the stored magnetic charge. The polarity of the magnetic charge represents the bit values zero and one.

Figure 5-9 illustrates a simple magnetic storage device. A wire is coiled around a metallic **read/write head**. To perform a write operation, an electrical current is passed through the wire, which generates a magnetic field across the gap in the read/write head. The direction of current flow through the wire determines the polarity of the field, that is, the position of the positive and negative poles of the magnetic field. Reversing the direction of the current reverses the polarity.

A magnetic storage medium is placed next to the gap. The portion of the medium surface closest to the magnetic field is permanently charged. The stored charge polarity is identical to the polarity of the generated magnetic field. The strength of the stored charge is directly proportional to the strength of the magnetic field. The strength of the magnetic field is determined by several factors including the number of coils in the wire, the strength of the electrical current, and the mass of the metallic read/write head. The storage medium must be constructed of or coated with a substance capable of accepting and storing a magnetic charge, such as a metallic compound.

Figure 5-9 ►

Principles of
magnetic data
storage

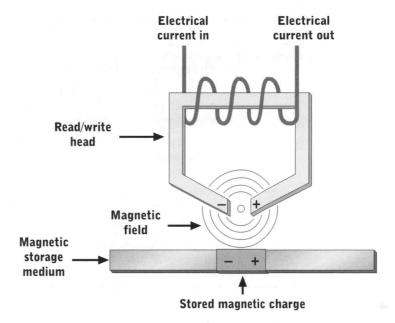

A read operation is the inverse of a write operation. The portion of the storage medium with the desired data is placed near the gap of the read/write head. The stored charge radiates a small magnetic field. When the magnetic field comes in contact with the read/write head, a small electrical current is induced in the read/write head and the wire wrapped around it. The current direction depends on the polarity of the stored charge. Electrical switches at either end of the coiled wire detect the direction of current flow and sense either a zero or one bit value.

Magnetic storage devices must control or compensate for some undesirable characteristics of magnetism and magnetic storage media including:

► Magnetic decay

► Magnetic leakage

► Minimum threshold current for read operations

► Storage medium coercivity

► Long-term storage medium integrity

A magnetic storage device must carefully balance all of these factors to achieve cost-effective and reliable storage. Different storage device requirements, such as speed vs. cost, dictate different trade-offs. Table 5-2 summarizes the factors and measures that protect data against loss.

Table 5-2 ▶

Factors that lead to data loss in magnetic storage devices

Factor	Description
Magnetic decay	Natural charge decay over time—data must be written at higher power than the read threshold to avoid data loss
Magnetic leakage	Cancellation of adjacent charges of opposite polarity and migration of charge to nearby areas—data must be written at higher power than the read threshold to avoid data loss
Coercivity	Ability of storage medium to hold charge—medium must be of sufficient mass and coercivity to hold charges strong enough to counteract decay and leakage
Areal density	Coercible material per bit decreases as areal density increases—higher areal density makes stored data more susceptible to loss due to decay and leakage
Media integrity	Stability of coercible material and its attachment to substrate—physical stress and temperature/humidity extremes must be avoided to prevent loss of coercible material

Magnetic Decay and Leakage

The tendency of magnetically charged particles to lose their charge over time is called **magnetic decay**. Magnetic decay is constant over time and proportional to the power of the charge. The electrical switches used in a read operation require a minimum, or threshold, current level. Because induced current power is proportional to magnetic charge strength, a successful read operation requires magnetic charge above a certain threshold, called the read threshold.

Over a sufficient period of time, magnetic decay will cause the stored charge power to fall below the read threshold. At that point, the data content of the storage medium is effectively lost. This phenomenon is the primary reason that data stored on disks and tapes is usually unreadable after a few years. To minimize the potential for such data loss, data bits must be written with sufficient charge to compensate for decay. Magnetic storage devices write data at a substantially higher charge than the read threshold, thus ensuring long-term readability. But there are limits to how much charge a read/write head can generate and how much the storage medium can hold.

The strength of individual bit charges can also decrease due to **magnetic leakage** from adjacent bits. This phenomenon can be heard in audio recordings made on analog magnetic tape (refer to Problem 3 at the end of this chapter). Any charged area continuously generates a magnetic field that may affect nearby bit areas. If the polarity of adjacent bits is opposite, then their magnetic fields will tend to cancel out the charge of both areas. The strength of both charges will fall below the read threshold.

Data Storage Technology

Areal Density

Coercivity is the ability of a substance or magnetic storage medium to accept and hold magnetic charge. Coercivity varies widely among elements and compounds. In general, metals and metallic compounds offer the highest coercivity at reasonable cost. For any specific material, coercivity is directly proportional to mass. In magnetic storage media, mass is a function of the surface area in which a bit is stored, the thickness of the medium or its coercible coating, and the density of the chargeable material within the medium or coating. Larger, thicker, and more dense areas can hold more charge due to their higher coercible mass.

Most users want to store as much data as possible on a storage medium. The simplest way to do this is to reduce the surface area used to store a single bit value, thus increasing the total number of bits that can be stored on the medium. For example, the areal density of a two-dimensional medium can be quadrupled by halving the length and width of each bit area (see Figure 5-10). The surface area allocated to a bit is called the **areal density, recording density,** or **bit density**. Areal density is typically expressed in bits, bytes, or tracks per inch, abbreviated as bpi, Bpi, and tpi, respectively.

Figure 5-10 ▶

Areal density is a function of the length and width of an individual bit area (a). Density may be quadrupled by halving the length and width of bit areas (b)

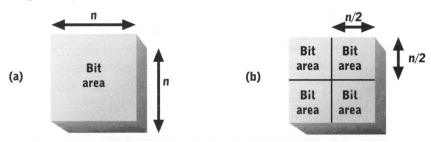

Some magnetic storage devices such as hard disks have fixed areal density, but others such as tape have variable areal density. Because reductions in area per bit reduce coercible mass per bit, the problems of magnetic decay and leakage increase as areal density increases. Designers and purchasers of magnetic media and devices must find a suitable balance between high areal density and storage longevity.

Media Integrity

The integrity of a magnetic storage medium depends on the nature of its construction and the environmental factors to which it is subjected. Some magnetic storage media, such as tape and floppy disk, use thin coatings of coercible material layered over a plastic or other substrate. Age and environmental stress can loosen the bond between the coating and substrate causing the coating to wear away. Physical stress on the medium from fast-forwarding and rewinding tape can accelerate the process. Temperature and humidity extremes can also accelerate the process.

Loss of coercible coating represents a loss of strength in stored magnetic charges. As with magnetic leakage and decay, data becomes unreadable when the remaining charge falls below the threshold of readability. To extend the life of magnetic media they must be protected from physical abuse and temperature and humidity extremes. Removable magnetic media are sometimes stored in climate-controlled vaults to ensure long-term media integrity.

Magnetic Tape

A **magnetic tape** is a ribbon of plastic with a coercible (usually metallic oxide) surface coating. Tapes are mounted in a tape drive for reading and writing. The **tape drive** contains motors that wind and unwind the tape. It also contains one or more read/write heads and related circuitry. Tape drives are relatively slow serial access devices. Tapes are primarily used to make backup copies of data stored on faster secondary storage devices and to physically transport large data sets.

Tapes and tape drives come in a variety of shapes and sizes. Older tape drives used wide tapes of up to 1-inch width wound on large open reels up to 10 inches in diameter. Modern tape drives use much narrower tapes permanently mounted in plastic cassettes or cartridges (see Figure 5-11). When a cassette is inserted into a tape drive, an access door automatically opens and the read/write mechanism is positioned around or next to the exposed tape. Tape widths have shrunk over time because technology improvements have increased areal density.

Figure 5-11 ►

Components of a typical cassette or cartridge tape

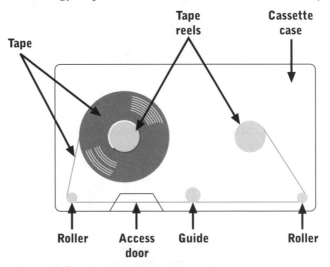

Tapes compound the magnetic leakage because the tape is wound upon itself. Leakage can occur from adjacent bit positions on the same area of the tape as well as from the layer of tape wound above or below on the reel. Tapes also are susceptible to problems arising from stretching, friction, and temperature variations. As a tape is wound and unwound it tends to stretch. If enough stretching occurs, the distance between bit positions can be altered, making individual bits difficult or impossible to read.

There are two basic approaches to recording data onto a tape surface. **Linear recording** places bits along parallel tracks that run along the entire length of the tape, as shown in Figure 5-12(a). Areal density depends on bit spacing within each track and on the number of tracks. **Helical scanning**, as shown in Figure 5-12(b), reads and writes data to or from a tape by rotating the read/write head at an angle to the tape and moving from tape edge to tape edge. This method requires much more complex read/write mechanisms.

Figure 5-12 ▶

Data recorded in linear parallel tracks (a) and with helical scanning (b)

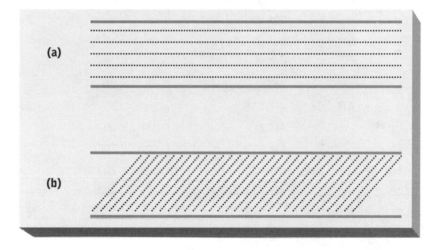

Helical scanning was originally developed for video recording, such as VHS tapes. For equal cost drives, helical scanning can pack more data into the same tape area than linear recording. The primary disadvantages of helical scanning are relatively slow tape motor speeds and direct contact of the read/write heads with the tape. Searching for specific data on a tape is slower with helical scanning drives due to slower tape motor speeds. Helically scanned tapes are subject to wear due to friction as the tape passes over the read/write head. Linear recording drives usually maintain a small gap between the read/write head and the tape to eliminate friction.

Recording data to tape at high speed and density requires precise alignment of the tape with internal drive components. Reliable tape storage requires careful control of environmental factors such as heat, dust, and dirt. Environmental control and precise alignment can be implemented within the cartridge, drive, or both. High-precision cartridges are relatively expensive, but cheaper drive components can be used. If high precision and environmental control are designed into the tape drive, then its price increases but cheaper cartridges can be used. If an application calls for high capacity, speed, and reliability, then both the cartridge and the drive must employ relatively expensive components.

There are at least five competing magnetic tape technology/format combinations and several variations on each (see Table 5-3). The tape storage industry has had relatively little cooperation on standards since the early 1990s. Modern standards are proprietary and subject to licensing fees. For the consumer, the result has been a maze of competing technologies with lots of hype, lots of misinformation, and little compatibility.

Table 5-3 ▶

Comparison of modern tape formats

Format Family	Standard	Uncompressed Capacity (GB)	Manufacturers
DDS (DAT)	DDS-1	2	Hewlett-Packard
	DDS-2	4	Sony
	DDS-3	12	
	DDS-4	20	
	DAT72	36	
AIT	AIT-1	35	Sony
	AIT-2	50	
	AIT-3	100	
	AIT-4	200	
	AIT-5 (proposed)	400	
Mammoth	Mammoth 1	20	Exabyte
	Mammoth 2	60	
	VXA-1	33	
	VXA-2	80	
DLT	DLT4000	20	Quantum
	DLT7000	35	
	DLT8000	40	
	Super DLTtape 1	160	
	Super DLTtape 2	300	
	Super DLTtape 3 (proposed)	600	
LTO	Ultrium-1	100	Hewlett-Packard
	Ultrium-2	200	IBM
	Ultrium-3 (proposed)	400	Seagate
	Ultrium-4 (proposed)	800	

Data Storage Technology

Technology
Focus

Magnetic Tape Formats and Standards

Prior to the mid-1990s, there were several widely used tape storage standards. Mainframe tape drives typically used 0.5- or 1-inch open reel tape. IBM set standards for mainframe tapes because it made the majority of mainframe computers. Tape drives for smaller computers either followed IBM's standards or the open standards of the **Quarter Inch Committee (QIC)**, as shown in Table 5-4. In those days, IBM's dominance helped to create an "us vs. IBM" mentality that perhaps explains the cooperative spirit that led to the QIC.

Table 5-4 ►

Sample QIC tape format specifications

Format	Year	Cartridge Size (Inches)	Capacity (GB)	Tracks	Areal Density (bpi)
QIC-80	1988	4 x 6	.08	28	14,700
QIC-120	1991	4 x 6	0.125	15	10,000
QIC-525	1992	4 x 6	0.525	26	20,000
QIC-2100	1993	4 x 6	2.1	30	50,800
QIC-3095	1995	3.25 x 2.5	4	72	67,733
QIC 3220	1997	3.25 x 2.5	10	108	106,400

The explosion in the market for PCs, small servers, and minicomputers in the 1980s and 1990s led to a corresponding surge in the tape market. Many companies entered the business and many new technologies were developed. To gain a competitive edge, many vendors closely guarded their technology. Cooperation became the exception, and when companies did cooperate, they created proprietary standards and charged substantial licensing fees for other companies to produce compatible media and drives. Technology quickly outpaced the QIC standards and no vendors were willing to contribute proprietary technology to update them.

The **Digital Data Storage (DDS)** standards were developed by Hewlett-Packard and Sony from an earlier technology called **Digital Audio Tape (DAT)**. All DDS tapes are 4 mm wide and all DDS drives use helical scanning. DDS tapes are relatively cheap, but they aren't designed for precise alignment. DDS drives compensate for cheap tape cassettes with elaborate technology for tape and environmental control, making the drives relatively expensive. DDS storage is used widely in small servers. The DAT72 format will probably will probably be the last DDS standard since both Sony and Hewlett-Packard are vigorously pursuing other tape standards and technologies.

Digital Linear Tape (DLT) is a standard developed by Quantum. DLT tape is 0.5 inches wide and the cartridge contains only one reel. The other reel is located in the drive, which uses an elaborate mechanism to grab a plastic tape leader from the cartridge and wind it onto the takeup reel. Longer tape lengths and higher capacity per tape can be achieved with the single reel cartridge. DLT records in parallel linear tracks, one at a time, from one end of the tape to another. At the end of the tape, the drive motors reverse direction and the next track is recorded. The end-to-end track format has advantages when searching for specific data, but it requires many passes to fill the tape completely.

Sony and Exabyte breathed new life into DAT with Sony's **Advanced Intelligent Tape (AIT)** and Exabyte's **Mammoth** standards. Both standards use helical scanning on 8-mm tapes, and employ a more expensive and precisely manufactured tape cartridge. Both standards also improve the tape drive technology to pack more data onto a single tape. AIT includes a small RAM cache within each cartridge, which stores directory information in order to speed searching and data access.

Linear Tape Open (LTO) is a standard developed by Hewlett-Packard, IBM, and Seagate. Despite its name and claims, LTO is a proprietary standard. Actually, LTO is two standards—Accelis and Ultrium. LTO uses linear recording and employs a number of technology improvements in tape cartridges, coercible materials, read/write heads, and tape control. All of this technology comes at a substantial price; LTO is primarily designed for enterprise-wide servers supporting mission-critical applications. The Accelis standard maximizes retrieval speed at the expense of per-tape capacity. The Ultrium standard maximizes capacity at the expense of access speed. No Accelis-based products have been released and it's possible that none ever will because the declining cost and increasing capacity of magnetic disks have reduced the market niche for Accelis.

Magnetic Disk

Magnetic disk media are flat, circular **platters** with metallic coatings that are rotated beneath read/write heads (see Figure 5-13). Multiple platters can be mounted and rotated on a common spindle. Data are normally recorded on both sides of a platter. A **track** is one concentric circle of a platter, or the surface area that passes under a read/write head when its position is fixed. A **cylinder** is the set of all tracks at an equivalent distance from the edge or spindle on all platter surfaces.

A sector is a fractional portion of a track (see Figure 5-14). The number of sectors per track ranges from eight for low-density floppy disks to several dozen for high-capacity hard disks. A single sector, or block, usually holds 512 bytes, which is also the normal data transfer unit to or from the drive.

Figure 5-13 ►

Primary
components of a
typical disk drive

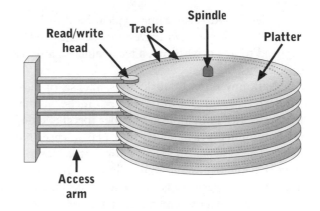

Figure 5-14 ►

Organization of
tracks and
sectors on one
surface of a disk
platter

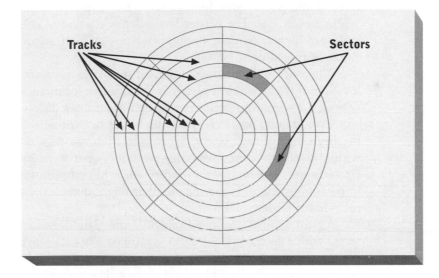

There is a read/write head for each side of each platter. Read/write heads are mounted on the end of an **access arm**. Access arms of equal length are attached to a positioning servo. The servo moves the access arms so that the read/write heads can be positioned anywhere between the outermost track, which is closest to the outer edge of the platter, and the innermost track, which is closest to the center of the disk. Read/write heads normally do not make contact with the recording surface of a platter. Rather, the head floats above the platter on a thin cushion of air as the disk spins.

A **hard disk** is a magnetic disk medium with a rigid metal base, or substrate. Typical platter diameter for hard disks is between three and five inches. Multiple platters are spun at relatively high speeds of up to 15,000 revolutions per minute. Drive capacity depends on the number of platters, platter size, and areal density. At this writing, typical hard disk drive capacities range from 40 to 400 GB. Multiple hard drives can be enclosed in a single storage cabinet and referred to as a **drive array**, a term which is also used to describe arrays of optical disk drives.

A **floppy disk**, or **disk**, uses a base of flexible or rigid plastic material. Modern disks range from 2.5 to 3.5 inches in diameter. A disk contains a single platter coated with an iron or other metallic compound on both sides. The platter is mounted permanently in a plastic case that can be removed from a disk drive. The case has an access door that is closed when the disk is removed from the drive and opens automatically when the disk is inserted into the drive. Disks are spun at lower speeds than hard disks. Disk capacities range up to 5 MB.

Larger-capacity removable magnetic disks are externally similar to floppy disks. Examples include Iomega REV and Zip disks. Platter sizes are usually 3.5 inches or less. The platter has a dense coating of coercible material. Spin rates are higher than for floppies but not as high as for hard disks. Storage capacity ranges up to 35 GB.

When originally manufactured, all magnetic disks are blank. Although the location of tracks within a platter is fixed, sector locations within a track are not. One of the operations performed when formatting a disk is to fix the location of sectors within tracks. This is accomplished by writing synchronization data at the beginning of each sector. Synchronization information can also be written within a sector during formatting or subsequent write operations. Read/write circuitry uses synchronization information to compensate for minor variations in rotation speed and other factors that might disturb the precise timing needed for reliable reading and writing.

A read or write command identifies the platter, track within the platter, and sector within the track. Controller circuitry switches activate the read/write head for the platter. The positioning servo moves the access arms over the track. The controller circuitry then waits for the appropriate sector to rotate underneath the read/write head. As the sector starts to rotate beneath the read/write head, the read or write operation is started. The operation is completed just as the last portion of the sector passes the read/write head. The access arms are typically left in their last position until the next read or write command is received.

Disk access time depends on several factors including time required to:

► Switch among read/write heads

► Position the read/write heads over a track

► Wait for the desired sector to rotate beneath the read/write heads

Disk drives share one set of read/write circuits among all read/write heads. The read/write circuitry must be electronically attached, or switched, to the appropriate read/write head before a sector can be accessed. The time required to perform this switching is no more than a few nanoseconds and is referred to as **head-to-head switching time**. Switching occurs in sequence among the read/write heads, and multiple switching operations may be required to activate a desired head. For example, if there are 10 read/write heads, switching from the first head to the last requires 9 separate switching operations (first to second, second to third, and so forth). Head-to-head switching time is the least significant component of access time.

The positioning servo that moves the access arms and read/write heads is a bidirectional motor that can be started and stopped very quickly to position the read/write heads precisely over a desired track. The time required to move from one track to another is called **track-to-track seek time**, typically measured in milliseconds. It is an average number due to variations in the time required to start, operate, and stop the positioning servo.

Once the read/write heads are positioned over the desired track, the controller waits for the desired sector to rotate beneath the heads. The controller keeps track of the current sector position by continually scanning the sector synchronization information written on the disk during formatting. When the appropriate sector is sensed, the read or write operation is started. The time that the disk controller must wait for the proper sector to rotate beneath the heads is called **rotational delay**. Rotational delay depends on platter rotation speed. Increasing spin rate reduces rotational delay.

Both track-to-track seek time and rotational delay vary from one read or write operation to another. For example, if two consecutive read requests access adjacent sectors on the same track and platter, then rotational delay and track-to-track seek time for the second read operation will be zero. If the second read access is to the sector preceding the first, then rotational delay will be high; almost a full platter rotation. If one read request references a sector on the outermost track and the other request references the innermost track, then track-to-track seek time will be high.

The amount of track-to-track seek time and rotational delay cannot be known in advance because the location of the next sector to be accessed can't be known in advance. Access time for a disk drive is not a constant. Instead, it is usually stated using a variety of performance numbers including raw times for head-to-head switching and rotation speed. The most important performance numbers are average access time, **sequential access time**, and **sustained data transfer rate**.

Average access time is computed by assuming that two consecutive accesses are sent to random locations. Given a large number of random accesses, the expected value of head-to-head switching time will correspond to switching over half the number of recording surfaces. The expected value of track-to-track seek time will correspond to movement over half the tracks. The expected value of rotational delay will correspond to one-half of a platter rotation. If head-to-head (HTH) switching time is 5 nanoseconds, there are 10 read/write heads, and track-to-track (TTT) seek time is 5 microseconds, 1,000 tracks are on each recording surface, and the platters are rotated at 7,500 RPM, the average access delay is computed as:

average access delay =   HTH switch   +   TTT seek   + rotational delay

$$= \left(5 \text{ nanoseconds} \times \frac{10}{2} \right) + \left(\frac{1000}{2} \times 5 \text{ microseconds} \right) + \left(\frac{60 \text{ seconds}}{7500 \text{ RPM}} \div 2 \right)$$

=   .000000025   + $(500 \times .000005)$ + $(.008 \div 2)$ seconds

=   .000000025   +   .0025   +   .004 seconds

≈   6.5 milliseconds

Average access time is the sum of average access delay and the time required to read a single sector. The time required to read a single sector depends entirely on the rotational speed of the disk and the number of sectors per track. Using the previous figures, and assuming there are 24 sectors per track, the average access time is computed as:

$$\text{average access time} = \text{average access delay} + \left(\frac{60 \text{ seconds}}{7500 \text{ RPM}} \div 24 \right)$$

=   .0065   + $(.008 \div 24)$

≈   6.83 milliseconds

Sequential access time is the time required to read the second of two adjacent sectors on the same track and platter. Head-to-head switching time, track-to-track seek time, and rotational delay are all zero for such an access. The only component of access time is the time required to read a single sector, that is, the second half of the above equation, or 0.33 milliseconds in that example.

Because sequential access time is so much faster than average access time, disk performance is improved dramatically if related data are stored in sequential sectors. For example, loading a program from disk to memory is relatively fast if the entire program is stored in sequential locations on the same track. If the program won't fit in a single track, then performance is maximized by storing the program in equivalently positioned tracks on multiple platters.

If portions of the program are scattered around the disk in random sectors, then performance is reduced due to switching read/write heads, positioning the access arm, and waiting for the desired sectors to rotate beneath the heads. File storage in sequential sectors improves performance only if file contents are accessed sequentially. If file contents are accessed in random order, then sequential storage will not provide significant performance improvements.

File contents tend to become scattered, or fragmented, in many nonsequential sectors over time. Fragmentation is a normal by-product of adding, deleting, modifying, and moving files and directories. Most operating systems provide a utility program that performs an operation called **disk defragmentation**. A disk defragmentation utility reorganizes disk content so that all contents of an individual file are stored in sequential sectors, tracks, and platters.

Effective disk drive performance also depends on the speed at which data can be moved through the disk controller circuitry to and from memory or the CPU. Communication channel capacity is rarely a restriction on the data transfer rate of a single disk drive but it can limit the effective throughput of a drive array.

The disk drive data transfer rate is a summary performance number that combines the physical aspects of data access with the electronic aspects of data transfer to the disk controller or system. As with access times, there are generally two different numbers stated. Maximum data transfer rate is the fastest rate that the drive can support and assumes sequential access to sectors. Because mechanical delays are minimal with sequential accesses, this number is quite high, typically several megabytes per second. The maximum data transfer rate for the previously described performance figures, assuming no controller delays, is:

$$\frac{1}{0.00033 \text{ seconds}} \times 512 \text{ bytes per sector} \approx 1.5 \text{ megabytes per second}$$

Sustained data transfer rate is computed with much less optimistic assumptions about data location. The most straightforward method is to use average access time. For the previous example, sustained data transfer rate is:

$$\frac{1}{0.00683 \text{ seconds}} \times 512 \text{ bytes per sector} \approx 75 \text{ kilobytes per second}$$

Because operating systems typically allocate disk space to files sequentially, the sustained data transfer rate is generally much higher than the above computation would imply. Table 5-5 provides performance statistics for various hard disk drive models from Seagate Technology, Inc. and Maxtor Corporation.

Table 5-5 ►

Hard disk drive performance statistics

Manufacturer	Model	Platters	Capacity (GB)	Rotation Speed (RPM)	Average Access Time (Milliseconds)
Seagate	ST336752	8	36.7	15,000	3.6
	ST373405	8	73.4	10,000	5.1
	ST1181677	24	181.6	7,200	7.4
Maxtor	8C073L0	8	73.4	15,000	3.2
	KU018L2	2	18.4	10,000	4.5
	6L060L3	2	60	7,200	8.5

Another factor that complicates performance calculations is that most manufacturers pack more sectors in the outer tracks of a platter. Look closely at Figure 5-14 and note that the platter surface area of sectors increases as you move toward the outer edge. Coercible material per sector is greater at the platter edge than in the center. To increase capacity per platter, disk manufacturers divide the tracks into two or more zones and vary the sectors per track in each zone (see Figure 5-15).

Figure 5-15 ▶

A platter divided into two zones with more sectors per track in the outer zone

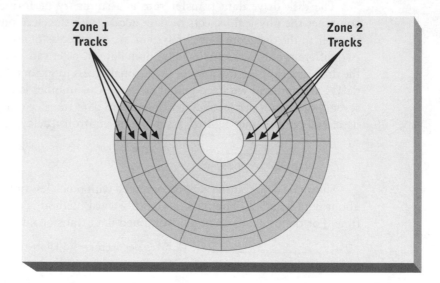

Computing average access time is more complex when sectors are more densely packed on the outer portions of the platter because the assumption that an average access requires moving the read/write head over half the tracks is no longer valid. For that reason, most average access time computations assume movement over a quarter to a third of the tracks.

Technology
Focus

Pocket-Size Data Storage

Floppy disk drives were a standard component of nearly every personal computer until the early 2000s. Their attractiveness, particularly in 3.5-inch hard case format, was obvious—cheap portable storage that can fit in a shirt pocket. Data transport via floppy disk was sometimes called SneakerNet, because carrying floppies between computers was a common way to share data before networks were widely implemented.

The last widely adopted improvement to the personal computer floppy disk standard came in late 1989—the 1.4 MB 3.5-inch standard floppy still used today. After that time, Intel and Microsoft were largely silent on the issue of improved floppy disk drives. The standard stagnated, in a manner similar to that of QIC tapes. Portable storage technology did continue to evolve, but in a myriad of proprietary ways.

Iomega Corporation developed several successful removable magnetic media including Zip, Jaz, and REV disks. Other companies developed various forms of removable optical discs. Because of the limitations of the various formats and the alphabet soup of competing standards, no magnetic or rewritable optical media or device has filled the gap left by the decline of the floppy disk.

The situation took a significant turn for the better in the early 2000s when two technologies were combined to fill the gap. The first technology was the Universal Serial Bus (USB), a standard that defined a simple interface for connecting a variety of external devices to a computer including keyboards, printers, network interface units, audio/video interfaces, and secondary storage devices. Operating system vendors developed standard device drivers to interface with external disk drives via a USB port. Many manufacturers of external and removable magnetic and optical drives developed USB products.

The second technology was flash memory, which matured rapidly in the late 1990s and early 2000s. By 2003, 64MB flash memory chips were being mass produced for less than $10 each. Several memory and secondary storage vendors saw the advantages of combining flash memory with the USB interface. The result was a portable device with flash memory storage, a USB connector, and a logic chip that makes the flash memory appear to the operating system and hardware as an ordinary disk drive. Though the device is standardized, it lacks a standard name. Commonly used names include memory key, USB drive, flash drive, and jump drive. Figure 5-16 shows a USB drive compared to a floppy disk.

Figure 5-16 ▶

A 1.4 MB floppy disk and a 512 MB flash RAM portable storage device

OPTICAL MASS STORAGE DEVICES

Optical storage came of age in the 1990s. Its primary advantages, as compared to magnetic storage, are higher areal density and longer data life. Typical optical recording densities are at least 10 times greater than for magnetic storage devices. Higher density is achieved with tightly focused lasers that can access a very small storage medium area. Magnetic fields cannot be focused on such a small area without overwriting surrounding bit positions.

Most optical storage media can retain data for decades. Optical media are not subject to magnetic decay and leakage. Optical storage is popular due to standardized and relatively inexpensive storage media. The compact disc (CD) format used for musical recordings is also supported by most optical storage devices. The cost to manufacture CDs is considerably less per bit than for removable magnetic storage technologies.

Optical storage devices store bit values as variations in light reflection (see Figure 5-17). The storage medium is a surface of highly reflective material. The read mechanism consists of a low-power laser and a photoelectric cell. The laser is focused onto one bit location of the storage medium at a specific angle. The photoelectric cell is positioned at a complementary angle to intercept reflected laser light. A highly reflective surface spot will cause the photoelectric cell to generate a detectable electrical current. A poorly reflective surface spot will not reflect enough light to cause the photoelectric cell to fire. The current, or lack of current, produced by the photoelectric cell is interpreted as a binary zero or one.

Figure 5-17 ▶

Optical disc read operations for a one bit (a) and a zero bit (b)

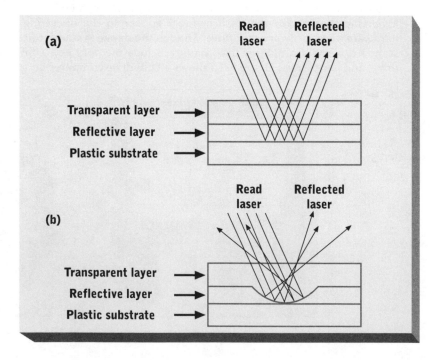

There are multiple ways to implement bit areas with low reflectivity. In Figure 5-17, low reflection is a result of a concave "dent" or pit in the reflective layer. Other ways of achieving low reflectivity bit areas include burned areas, dyes, and substances that change from crystalline to amorphous states.

An optimal surface material could be rapidly and easily changed from highly to poorly reflective and then back again an unlimited number of times. The search for such a material and a reliable and efficient method of altering its reflectivity has proven problematic. Some materials change quickly in one direction but not another, like photosensitive sunglass materials. The ability of most materials to change reflectivity degrades over time and with repeated use. There are several different optical write technologies currently in use or under development, but there is currently no clearly dominant technology.

Current optical storage devices use a disk storage medium. (Note that in current usage, optical storage media are often called *discs* while magnetic storage media are called *disks*.) Because the recording surface is a disk, there are many similarities to magnetic disk storage including the use of a read/write mechanism mounted on an access arm, a spinning disk, and performance limitations due to rotational delay and movement of the access arm.

Although magnetic and optical devices both use disk technology, their performance is quite different. Optical disc storage is slower for a number of reasons, including the use of removable media, the use of a single spiral track, and the inherent complexity of an optical write operation. The performance difference has narrowed considerably in recent years and is approximately 10 to 1 at present. Magnetic hard disks typically have access times under 5 milliseconds. Read access time for optical drives is generally greater than 50 milliseconds. Write access time is typically two or three times longer than read access time. The gap is expected to shrink as optical technology matures and as magnetic technology reaches its physical limits.

Magnetic and optical storage are not direct competitors at present because they have very different cost/performance tradeoffs. Magnetic hard disks are still the cost performance leader for general-purpose on-line secondary storage. Optical storage is favored for read-only storage with low performance requirements, for applications with extremely high capacity requirements, and where portability in a standardized format is needed. The two technologies will compete more directly as optical access times improve and as optical write technology develops and matures. Optical storage devices are available in a wide variety of storage formats and write technologies, summarized in Table 5-6. These technologies and media formats are discussed in detail following the table.

Table 5-6

Technologies
and storage for-
mats for optical
and magneto-
optical storage

Technology/Format	Writable?	Description
CD-ROM	No	Adaptation of musical compact disc technology, 650 MB capacity
CD-R	One time only	CD-ROM format with a dye reflective layer that can be written with a low-power laser
CD-RW	Yes	CD-ROM format with a phase change reflective layer, can be written up to 1000 times
Magneto-optical	Yes	Combination of optical and magnetic technology, expensive and outdated, rapidly giving way to CD-RW and writeable DVD
DVD-ROM	No	Adaptation of DVD video technology, similar to CD-ROM but more advanced, 4.7 GB capacity
DVD-R	One time only	DVD-ROM format, same basic technology as CD-R with performance and capacity improvements
DVD-RAM	Yes	Format similar but not completely compatible with DVD-ROM, updated CD-RW phase change technology with performance and capacity improvements
DVD-RW	Yes	A competitor to DVD-RAM
DVD+RW	Yes	A competitor to DVD-RAM

CD-ROM

Compact disc technology was originally developed by Sony and Philips for storing and distributing music in a format known as **compact disc digital audio (CD-DA)**. **Compact disc read-only memory (CD-ROM)** is compatible with CD-DA but includes additional formatting information to store directory and file information. Both CD-DA and CD-ROM are read-only storage technologies and formats because data are permanently embedded in the disc during manufacture.

A compact disc is 120 millimeters (approximately 4.75 inches) in diameter, is made largely of durable polycarbonate, and holds approximately 650–700 MB, depending on the format. Bit values are represented as flat areas called *lands* and concave dents called *pits* in the reflective layer. Data are recorded in a single continuous track that spirals outward from the center of the disc. Error correction data written along the track at regular intervals helps the CD drive read mechanism to recover from errors introduced by dust, dirt, and scratches.

CD-ROM is a popular medium for distributing software and large data sets such as maps, census data, telephone directories, and other large databases. Its standardized format, high density, and cheap manufacturing cost make it well suited for these purposes. Its primary drawbacks are that discs cannot be rewritten, and their capacity is limited to 700 MB.

CD-R

Compact disc-recordable (CD-R) drives use a laser that can be switched between high and low power and a laser-sensitive dye embedded in the CD-R disc. The dye is stable when scanned at low power during a read operation, but changes its reflective properties when scanned at higher power during a write operation. The dyes are sensitive to heat and light, so CD-R discs should be stored in a dark location at room temperature. Also, because the dyes degrade over time, data may be lost after a few decades. Recordable DVDs use similar dyes and have similar data retention problems.

CD-R discs and drives are relatively cheap. CD-R drives can read standard CDs and most computer CD readers can read CD-R discs. CD-R drives and discs are commonly used in several areas including:

▶ Creating music CDs on home computers

▶ Backing up data from other storage devices

▶ Creating archives of large data sets, such as photo albums

▶ Manufacturing small quantities of identical CDs for distribution, such as parts catalogs and conference proceedings

Magneto-Optical

A **magneto-optical** drive uses a laser and reflected light to sense bit values. The variations in reflectivity of the disk surface are achieved by applying a magnetic charge to the bit area. Reading is based on the polarity of the reflected laser light, not its strength. The magnetic charge shifts the polarity of the reflected laser light. The direction of the shift is determined by the polarity of the magnetic charge. This polarity shift is known as the Kerr effect.

The disk surface is a material that resists magnetic charge at normal temperatures. At higher temperatures, approximately 150 degrees Celsius, the material can be charged by a magnetic field in much the same way as purely magnetic media. To bring the material to a chargeable temperature, a high-power laser is focused on a bit position. A magnetic write head positioned below the media surface then generates the appropriate charge. Because the laser beam focuses on a single bit position, only that bit position is raised to a sufficient temperature. Surrounding bits are unaffected by the magnetic field from the write head because they are cooler.

Magneto-optical technology peaked in the mid-1990s. Few new magneto-optical drives have been sold since then due to the introduction of cheaper

CD-RW drives. Magneto-optical technology still has advantages over CD-RW in access speed and capacity, but those advantages are rapidly fading as phase-change technology matures and as newer DVD phase-change formats are developed and deployed. There is still a market for magneto-optical drives because many organizations created data archives on magneto-optical disks and still need access to those archives. Thus, the technology will probably not completely fade away for several more years.

Phase-Change Optical Discs

Phase-change optical technology enables nondestructive writing to optical storage media. The technology is based on materials that can change state easily from noncrystalline, or amorphous, to crystalline, and then back again. The reflective characteristics of the materials are quite different in their amorphous and crystalline states. The difference is less than with traditional CDs, but sufficient to be detected by newer optical scanning technologies.

The reflective layer of a phase-change disc is a compound of tellurium, selenium, and tin. It changes from an amorphous state to a crystalline state when heated to a precise temperature. Heating the material to its melting point changes it back to an amorphous state. The melting point is relatively low so high-power lasers aren't required. However, multiple passes are usually required to generate sufficient heat, so write times are substantially longer than read times. The reflective layer loses its ability to change state with repeated heating and cooling. Current CD-RW media wear out after approximately 1000 write operations.

Compact disc-Rewritable (**CD-RW**) is a standard for phase-change optical discs that are the same size as standard CDs. Most CD-RW drives can read and write CD-R and CD-ROM discs. Older CD-R and CD-ROM drives cannot read CD-RW discs because they are substantially less reflective than CD-R and CD-ROM discs. Access time is comparable to CD-ROM and CD-R drives, typically between 50 and 100 milliseconds.

DVD

DVD is an acronym for both **digital video disc** and **digital versatile disc** (the two terms are equivalent). A consortium of companies developed DVD as a standard format for distributing movies and other audiovisual content. Like CD-ROM, **DVD-ROM** is an adaptation of DVD in which DVD disks can store computer data. Most DVD-ROM drives can also read CD, CD-R, and CD-RW discs.

DVD improves on CD and CD-RW technology in several ways:

► Increased track and bit density due to smaller wavelength lasers and more precise mechanical control
► Improved error correction
► Multiple recording sides and layers

DVD technology now can store between 4 and 12 times as much data as a CD on the same size disc, depending on the specific format. DVD-ROM is the current standard format for read-only computer data. A single-sided, single-layer DVD-ROM disc holds 4.7 GB. The DVD-ROM standard also defines discs with capacities of 8.5 GB (double-sided) and 17 GB (double-sided, two recording layers per side). Virtually all DVD, DVD-R, DVD-RW, and DVD-RAM drives can read DVD-ROM discs. Also, most DVD drives can read all CD formats.

DVD discs have more densely packed data and are spun at higher speed than CDs. Sustained data transfer rates are several times those of CD formats. However, because they use similar disc access technology, DVD access times approximate those of CD formats (50 to 100 milliseconds). Access times should improve as the technology matures.

DVD-R drives and media are similar in most ways to CD-R drives and media. First-generation DVD-R discs had a 3.9 MB capacity, which was later expanded to 4.7 MB to ensure compatibility with DVD-ROM. As of this writing (late 2004), dual-layer recordable DVD drives are just coming to market. A standard called DVD+R DL has already been ratified and a competing DVD-R DL standard is nearly completed. Both disc types can store 8.5 GB on a single disk in two separate recording layers.

There are three competing standards for rewritable DVDs: DVD-RAM, DVD-RW, and DVD+RW. **DVD-RAM** devices were released over a year before DVD-RW and DVD+RW devices. Lack of compatibility with standard home DVD players and DVD-ROM readers hampered adoption. The DVD-RW and DVD+RW standards arrived later than DVD-RAM, but offered superior compatibility and capacity.

The **DVD-RW** standard is supported by the DVD Forum (an industry trade association) and Panasonic, Toshiba, Apple Computer, Hitachi, NEC, Pioneer, Samsung, and Sharp. The competing **DVD+RW** standard has the backing of several large companies including Hewlett-Packard, Philips, and Sony. Both standards rely on similar technology and provide similar performance and capacity. Because of their similarities, and the lack of a clear market leader, most drive manufacturers have developed products that support both standards.

Business
Focus

Which Storage Technology?

The Ackerman, Holt, Sanchez, and Trujillo (AHST) law firm has thousands of clients, dozens of partners and associates, and a costly document storage and retention problem. Prior to 1990, AHST archived legal documents and billing records on paper. As in-office file cabinets filled, the paper was boxed, labeled, and transported to a warehouse for storage. AHST retains all records for a period of at least 20 years and longer in some cases.

In the late 1980s, AHST adopted automated procedures for archiving documents. Most documents were stored as computer files in their original format (WordPerfect, at that time). Documents with signatures were scanned and stored as JPEG files. Several possible storage media were considered including various forms of magnetic tape and removable magneto-optical (MO) disks. The firm chose MO drives and disks manufactured by Sony based on the following selection criteria, listed roughly in order of importance:

▶ Permanence (25 year minimum document recoverability)

▶ Cost

▶ High storage capacity

▶ Ease of use

▶ Stability and market presence of the technology and selected vendor(s)

Over the years AHST has used several different MO media including 600 MB, 1.3 GB, and 2.3 GB 5.25-inch disks and several different drives. Document formats have also changed over the years to include Microsoft Word and Adobe Acrobat files. Starting last year, AHST began recording depositions on a digital video camera and archiving them to MPEG files stored on 2.3 GB MO disks. AHST estimates its current annual storage requirements for document and video deposition archives at 500-750 GB, approximately 90% of which is video.

Electronic document and secondary storage technology have evolved considerably since AHST chose magneto-optical storage. Many new tape formats have been developed and optical formats such as CD and DVD have become commonplace. At the same time, magneto-optical technology has evolved even though its popularity has waned. Current Sony MO drives and media can store up to 9.1 GB per disk. However, drives and media are becoming harder to find and costs are high relative to more modern alternatives. For those reasons, the firm is revisiting the choice of technology for permanent document storage.

Questions

1. Describe the pros and cons of the removable secondary storage technologies described in this chapter with respect to the firm's selection criteria.

2. What technology(ies) should AHST adopt for future document and video storage? Why?
3. Assume that AHST chooses the technology(ies) that you recommend. How and when might they become obsolete?
4. Assume that AHST chooses the product(s) that you recommend. How and when might they become obsolete?

SUMMARY

► A typical computer system has multiple storage devices filling a variety of needs. Primary storage devices support the immediate execution of programs, and secondary storage devices provide long-term storage of programs and data. The characteristics of speed, volatility, access method, portability, cost, and capacity vary among storage devices, forming a memory-storage hierarchy. Primary storage tends to be at the top of the hierarchy with faster access speeds and higher costs per bit of storage, while secondary storage tends to be in the lower portion of the hierarchy, with slower access speeds and lower costs.

► The critical performance characteristics of primary storage devices are their access speed and the number of bits that can be accessed in a single read or write operation. Modern computers use memory implemented with semiconductors. Basic types of memory that are built from semiconductor microchips include random access memory (RAM) and nonvolatile memory (NVM). The primary RAM types are static and dynamic. Static RAM (SRAM) is implemented entirely with transistors. Dynamic RAM (DRAM) is implemented using transistors and capacitors. The more complex circuitry of SRAM is more expensive to implement but provides faster access times. Neither type of RAM can match current microprocessor clock rates. NVM is usually relegated to specialized roles and secondary storage due to its slower write speeds and limited number of rewrites.

► Programs are created as though they occupied contiguous primary storage locations starting at the first location. Programs are not usually placed in low memory because system software or other programs reside there. Indirect addressing reconciles the difference between where a program "thinks" it is located within memory and where it actually is located. The actual address of a program's first instruction is stored in an offset register and this address is added to each address reference made by the program. Indirect addressing enables multiple programs to reside in memory and allows programs to be moved during execution.

► Magnetic secondary storage devices store data bits as magnetic charges. Magnetic tapes are ribbons of plastic coated with a coercible coating. Data is written to (or read from) tape by passing it over the read/write head of a tape drive. Magnetic disks are platters coated with coercible material. Platters are rotated within a disk drive and read/write heads access data at various locations on the platter(s). Magnetic disk drives are random access devices because the read/write head can be freely moved to any location on a disk platter.

► Optical discs store data bits as variations in light reflection. An optical disc drive reads data bits by shining a laser beam onto a small disc location. High and low reflections of the laser are interpreted as ones and zeros. The cost per bit of optical storage is less than magnetic storage, at the expense of slower access speed. Types of optical discs (and drives) include various forms of CDs and DVDs. CD-ROM and DVD-ROM are written during manufacture. CD-R and DVD-R are manufactured blank and can be written to once. CD-RW and various types of rewritable DVDs use phase-change technology to write and rewrite the media.

Now that CPU, primary storage, and secondary storage have been covered in some detail, it's time to turn our attention to how those devices are integrated in a computer system. Chapter 6 describes how processing, I/O, and storage devices interact, and discusses methods of improving overall computer system performance. Chapter 7 describes various types of input/output devices.

Key Terms

absolute addressing
access arm
access time
addressable memory
Advanced Intelligent Tape (AIT)
areal density
average access time
big endian
block
CD-DA (compact disc digital audio)
CD-R
CD-ROM (compact disc read-only memory)
CD-RW
coercivity

core memory
cylinder
data transfer rate
device controller
Digital Audio Tape (DAT)
Digital Data Storage (DDS)
Digital Linear Tape (DLT)
direct access
disk defragmentation
disk
double in-line memory module (DIMM)
drive array
dual in-line package (DIP)
DVD
DVD-RAM

DVD-ROM
DVD-RW
DVD+RW
dynamic RAM (DRAM)
electronically erasable programmable read-only memory (EEPROM)
enhanced DRAM (EDRAM)
erasable programmable read-only memory (EPROM)
ferroelectric RAM
firmware
flash memory
flash RAM
floppy disk
hard disk

Data Storage Technology

head-to-head switching time
helical scanning
indirect addressing
least significant byte
linear recording
Linear Tape Open (LTO)
little endian
magnetic decay
magnetic leakage
magnetic tape
magneto-optical
Mammoth
memory allocation
most significant byte
nonvolatile
nonvolatile memory (NVM)

offset register
parallel access
physical memory
platter
polymer memory
Quarter Inch Committee (QIC)
random access
random access memory (RAM)
read-only memory (ROM)
read/write head
read/write mechanism
refresh cycle
relative addressing
rotational delay
sector
segment register

sequential access time
serial access
single in-line memory module
  (SIMM)
static RAM (SRAM)
storage medium
sustained data transfer rate
synchronous DRAM
  (SDRAM)
tape drive
track
track-to-track seek time
volatile
wait state

Vocabulary Exercises

1. Dynamic RAM requires frequent _____ to maintain its data content.

2. _____ rate is the speed at which data can be moved to/from a storage device over a communication channel.

3. Two standard optical storage media that are written only during manufacture are called _____ and _____.

4. _____, _____, and _____ are competing standards for rewritable DVD discs.

5. The _____ of a hard disk drive generates or responds to a magnetic field.

6. Data stored on magnetic media for long periods of time may be lost due to _____ and _____.

7. A(n) _____ stores data in magnetically charged areas on a rigid platter.

8. The _____ endian storage format places the _____ byte of a word in the lowest memory address. The _____ endian storage format places the _____ byte of a word in the highest memory address.

9. The contents of most forms of RAM are _____, making them unsuitable for long-term data storage.

10. _____ and _____ are outdated technologies for nonvolatile primary storage. _____ and _____ are promising new technologies for implementing NVM.

11. The CPU can incur one or more _____ when it accesses slower storage devices.

12. Under relative addressing, the content of a(n) _____ is added to calculate the corresponding physical memory address.

13. The number of bits used to represent a memory address determines the amount of a CPU's _____.

14. _____ is typically stated in milliseconds for secondary storage devices and nanoseconds for primary storage devices.

15. The three components of average access time for a disk drive are _____, _____, and _____.

16. _____ storage is slower, cheaper, and less volatile than _____ storage.

17. In a magnetic disk drive, a read/write head is mounted on the end of a(n) _____.

18. The access method for RAM is _____ or _____ if words are considered the unit of data access. The access method is _____ if bits are considered the unit of data access.

19. DDS and AIT both use _____ to record bits and tracks on a magnetic tape.

20. _____ and _____ are two modern standards/formats for linear recording on magnetic tapes.

21. Under _____, program memory references correspond to physical memory locations. Under _____, the CPU must calculate the physical memory location that corresponds to a program memory reference.

22. A(n) _____ uses a combination of optical and magnetic storage technologies.

23. A(n) _____ is a series of sectors stored along one concentric circle on a platter.

24. The data transfer rate of a secondary storage device can be calculated by dividing one by its access time and multiplying the result by _____ size.

25. _____, _____, and _____ are storage formats, originally designed for music or video recording, that have been applied to computer data storage.

26. Tape drives are _____ devices. _____ are random or direct access devices.

27. Average access time can usually be improved by _____ files stored on a disk.

28. Modern personal computers generally use memory packaged on small standardized circuit boards called _____.

29. The _____ of a magnetic or optical storage medium is the ratio of bits stored to a unit of medium surface area.

30. For most disk drives the unit of data access and transfer is a(n) _____ or _____ .

31. Software programs permanently stored in ROM are called _____ .

32. Many open standards for cartridge tape storage have been defined by the _____ .

33. Competing standards for rewritable DVDs include _____ , _____ , and _____ .

Review Questions

1. What factor(s) limit the speed of an electrically based processing device?

2. What is/are the difference(s) between static and dynamic RAM?

3. What improvements are offered by synchronous DRAM as compared to ordinary DRAM?

4. Why isn't flash RAM commonly used to implement primary storage?

5. Describe current and emerging nonvolatile RAM technologies. What advantages are potentially offered by the emerging technologies as compared to current flash RAM technology?

6. Describe serial, random, and parallel access. What types of storage devices use each method?

7. What is direct addressing? What is relative addressing?

8. What are the costs and benefits of indirect addressing?

9. How is data stored and retrieved on a magnetic mass storage device?

10. Describe the factors that contribute to a disk drive's average access time. Which of these factors are improved if spin rate is increased? Which are improved if areal density is increased?

11. What problems contribute to read/write errors on magnetic tapes? Are these problems also present with other magnetic storage media/devices?

12. What is/are the advantage(s) and disadvantage(s) of helical scanning as compared to linear recording?

13. Why is the areal density of optical discs higher than the areal density of magnetic disks? What factor(s) limit this areal density?

14. Describe the processes of reading from and writing to a phase change optical disc. How does the performance and areal density of these discs compare to magnetic disks?

15. List and briefly describe the various standards for recordable and rewritable CDs and DVDs. Are any of the standards clearly superior to their competitors?

Problems and Exercises

1. Assume that a magnetic disk drive has the following characteristics:

    ▶ 10,000 RPM spin rate

    ▶ 2 nanosecond head-to-head switching time

    ▶ 3 microsecond average track-to-track seek time

    ▶ 5 platters, 1,024 tracks/platter side recorded on both sides, 50 sectors per track on all tracks

    Questions:

    A. What is the storage capacity of the drive?

    B. How long will it take to read the entire contents of the drive sequentially?

    C. What is the serial access time for the drive?

    D. What is the average (random) access time for the drive (assume movement over half the tracks and half the sectors within a track)?

2. Assume that a CPU has a clock rating of 2.4 GHz. Assume further that half of each clock cycle is used for fetching and the other half for execution.

    Questions:

    A. What access time is required of primary storage for the CPU to execute with zero wait states?

    B. How many wait states per fetch will the CPU incur if all of primary storage is implemented with 5 nanosecond static RAM?

    C. How many wait states per fetch will the CPU incur if all of primary storage is implemented with 10 nanosecond SDRAM?

3. Try the following experiment. Find an analog cassette tape that was recorded at least a year ago. Start the playback in the blank section preceding a musical selection with the volume turned up fairly high. You will hear the song start at a very low volume just before it actually starts at normal volume. What causes this phenomenon? What are its implications for digital data stored on magnetic tapes?

Research Problems

1. Select a handful of computers at work or school covering a range of ages and types (e.g., ordinary PCs, workstations, and servers). Visit a Web site such as www.kingston.com or www.memory4less.com and locate a 1 GB memory expansion for each computer. Note the prices and memory type for each upgrade. What factors account for the variation in memory price and type for each computer?

2. IBM has invested heavily in research and development for holographic storage systems. Investigate this research to determine the nature of holographic storage, its applicability to primary and secondary storage, and the likelihood of seeing products based on holographic technology in the near future.

3. Assume that you are the chief information officer of a rapidly growing company. You manage several dozen small- and medium-size servers and are about to acquire your first mainframe to implement a data warehouse. You have a variety of DDS and AIT tape drives. You are concerned about tape drive capacity, reliability, and cost, and are considering switching to another tape format/technology. Which of the modern tape formats provides the greatest capacity at the lowest cost? Which is most reliable? Which is least likely to become obsolete five years from now? Which should you buy?

4. Assume that you are a home computer user and amateur photographer with almost 20 GB of digital photographs stored on your hard disk. You are running short of disk space and you want to archive current and future photographs onto removable optical discs. Investigate the capacity and cost per various recordable and rewritable CD and DVD media. Which media type offers the lowest cost? Which offers the best performance? Which would you choose and why?

Chapter 6

System Integration and Performance

Chapter Goals

- ► Describe the system bus and bus protocol

- ► Describe how the CPU and bus interact with peripheral devices

- ► Describe the purpose and function of device controllers

- ► Describe how interrupt processing coordinates the CPU with secondary storage and I/O devices

- ► Describe how buffers, caches, and data compression improve computer system performance

In earlier chapters we discussed the processing and storage components of a computer system. Now it's time to describe how those components communicate with one another. This chapter explores the system bus and device controllers and describes how the CPU uses them to communicate with secondary storage and input/output devices. The chapter also describes bus protocols, interrupt processing, buffering, caching, and compression, and how these elements impact system performance (see Figure 6-1).

Figure 6-1 ▶

Topics covered in this chapter

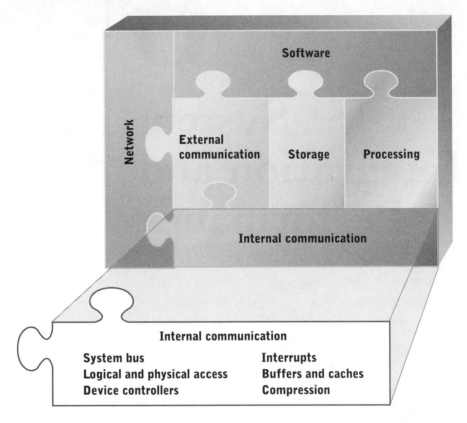

SYSTEM BUS

A **bus** is a set of parallel communication lines that connect two or more devices. The **system bus** connects the CPU with main memory and other system components, as shown in Figure 6-2. Devices other than the CPU and primary storage are usually called **peripheral devices**. Each bus line can carry a single bit value during any bus transfer operation. Computer systems have subsets of bus lines, including the data bus, the address bus, and the control bus, which are dedicated to carrying specific types of information.

System Integration and Performance

Figure 6-2 ▶

Components of
the system bus
and attached
devices

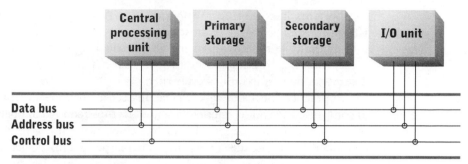

System bus

As its name implies, the **data bus** moves data among computer system components. The number of lines is typically the same as, or a multiple of, the CPU word size. For example, a computer system with a 64-bit CPU would have 64 or 128 bus data lines.

The **address bus** carries the bits of a memory address. Modern address buses have at least 32 lines. When a peripheral device transfers data to main memory, it transmits the data item via the data bus while simultaneously transmitting the address of that data item via the address bus. The address bus is used only for bus transfers in which main memory is the target, or destination, device.

The **control bus** carries commands, command responses, status codes, and similar messages. Computer system components coordinate their activities by sending appropriate signals over the control bus.

Bus Clock and Data Transfer Rate

One or more control bus lines carry the **bus clock** pulse, a common timing reference for all attached devices. The frequency of bus clock pulses is measured in megahertz (MHz). It may be equivalent to the CPU clock rate or a fraction of the clock rate. Each clock pulse marks the start of a new opportunity to transmit data or a control message.

The time interval from one clock pulse to the next is called a **bus cycle**. Bus cycle time is the inverse of the bus clock rate. For example, if the bus clock rate is 400 MHz then the duration of each bus cycle is:

$$\text{bus cycle time} = \frac{1}{\text{bus clock rate}}$$

$$= \frac{1}{400,000,000 \text{ Hz}}$$

$$= 2.5 \text{ nanoseconds}$$

By its very nature, the bus must be relatively long because it connects many different computer system components. Miniaturization of all computer system

---

components has reduced the length of a typical system bus, but it is still 10 centimeters or longer in most computer systems. A bus cycle cannot be any shorter than the time required for an electrical signal to traverse the bus from end to end. Bus length imposes a theoretical maximum on bus clock rate and a theoretical minimum on bus cycle time.

In practice, bus clock rate is generally set far below its theoretical maximum to compensate for noise, interference, and the time required to operate interface circuitry in peripheral devices. Slower bus clock rates also give computer component manufacturers the ability to increase peripheral device reliability while holding cost to acceptable levels. As an analogy, consider the speed limit on a road. Even though the road itself might be able to support a maximum speed of 150 miles per hour, the speed limit is not set that high for several reasons. Vehicles that could travel that fast would be expensive, and reliable and safe transport would require high standards of road and vehicle maintenance, skilled drivers, and perfect weather and visibility.

The maximum capacity of a bus or any other communication channel is the product of clock rate and data transfer unit. For example, the theoretical capacity of a bus with 64 data lines and a 400 MHz clock rate is:

$$\text{bus capacity} = \text{data transfer unit} \times \text{clock rate}$$

$$= 64 \text{ bits} \times 400 \text{ MHz}$$

$$= 8 \text{ bytes} \times 400{,}000{,}000 \text{ Hz}$$

$$= 3{,}200{,}000{,}000 \text{ bytes per second}$$

Such a measure of communication capacity is called a **data transfer rate**. Data transfer rates are expressed in bits or bytes per second, for example, 100 MB per second. See the Appendix for further discussion of data transfer rate units of measure.

There are only two ways to increase the maximum bus data transfer rate: increase the clock rate or increase the data transfer unit size (the number of data lines). As discussed, bus length, conservative engineering, and peripheral device costs limit maximum clock rates. Cost and engineering considerations also limit the number of data bus lines. But the number of data bus lines should always be at least equal to CPU word size. This helps to avoid CPU wait states for bus data transfers. Data bus width larger than CPU word size can provide performance improvements if the CPU can effectively use multiple-word inputs.

Bus Protocol

The **bus protocol** governs the format, content, and timing of data, memory addresses, and control messages sent across the bus. Every peripheral device must follow the protocol rules. In the simplest sense, a bus is just a set of communication lines. In a larger sense, a bus is a combination of communication lines, a bus protocol, and devices that implement the bus protocol.

The bus protocol has two important effects on maximum data transfer rate. First, control signals sent across the bus consume bus cycles, reducing the cycles available to transmit data. For example, a disk drive transfers data to main memory as a result of an explicit CPU command. Sending the command requires a bus cycle. In some bus protocols the command is followed by an acknowledgment from the disk drive and later by a confirmation that the command was carried out. Each signal (command, acknowledgment, and confirmation) consumes a separate bus cycle.

An efficient bus protocol consumes a minimal number of bus cycles for commands, maximizing bus availability for data transfers. Unfortunately, efficient bus protocols tend to be complex, increasing the complexity and cost of the bus and all peripheral devices. The SCSI bus protocol, which is covered in a Technology Focus section later in this chapter, is an example of an efficient but complex bus protocol.

The bus protocol regulates bus access to prevent devices from interfering with one another. If two peripheral devices attempt to send a message at the same time, the messages collide and produce electrical noise. A bus protocol avoids collisions by one of two access control approaches: a master-slave approach or a peer-to-peer approach.

In traditional computer architecture, the CPU is the focus of all computer activity. As part of this role, the CPU is also the **bus master** and all other devices are **bus slaves**. No device other than the CPU can access the bus except in response to an explicit instruction from the CPU. There are no collisions as long as the CPU waits for a response from one device before issuing a command to another device. In essence, the CPU plays the role of bus traffic cop.

Because there is only one bus master, the bus protocol is simple and efficient. However, overall system performance is significantly reduced because transfers among devices, such as from disk to memory, must pass through the CPU. All transfers consume at least two bus cycles; one to transfer data from the source device to the CPU and another to transfer data from the CPU to the destination device. The CPU cannot execute computation and logic instructions while transferring data among other devices.

Performance is improved if storage and I/O devices can transmit data among themselves without explicit CPU involvement. There are two commonly used approaches to implementing such transfers: direct memory access and peer-to-peer buses. Under **direct memory access (DMA)**, a device called a **DMA controller** is attached to the bus and to main memory. The DMA controller assumes the role of bus master for all transfers between memory and other storage or I/O devices. While the DMA controller manages bus traffic, the CPU is free to execute computation and data movement instructions.

In a **peer-to-peer bus,** any device can assume control of the bus, or act as a bus master for transfers to any other device. A single master must be chosen when multiple devices want to become a bus master at the same time. A **bus arbitration unit** is a simple processor attached to a peer-to-peer bus that decides which device(s) must wait when multiple devices want to become a bus master. Peer-to-peer bus protocols are substantially more complex and expensive than master-slave bus protocols, but their complexity is offset by more efficient use of the CPU and the bus.

LOGICAL AND PHYSICAL ACCESS

In most computer systems, the system bus is physically implemented on a large printed circuit board with attachment points for various devices (see Figure 6-3). Some devices, such as bus and memory controller circuitry, are permanently mounted to the board. Others, such as the CPU, memory modules, and some peripheral device controllers, are physically inserted into bus ports or dedicated sockets.

Figure 6-3 ►

A typical personal computer motherboard

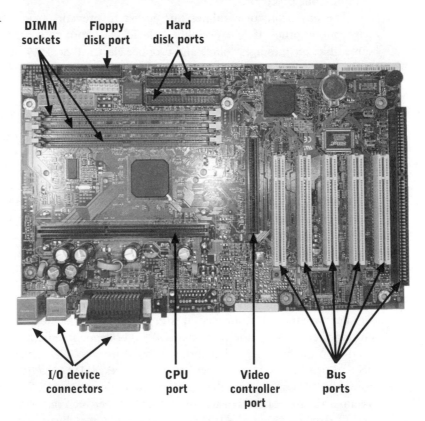

DIMM sockets   Floppy disk port   Hard disk ports

I/O device connectors   CPU port   Video controller port   Bus ports

An **I/O port** is a communication pathway from the CPU to a peripheral device. In most computers, an I/O port is a memory address, or a set of contiguous memory addresses, that can be read or written by the CPU and a single peripheral device. Each peripheral device can have multiple I/O ports and use them for different purposes, for example, transmitting data, commands, and error codes. The CPU communicates with a peripheral device by moving data to or from an I/O port's memory address(es). Dedicated bus interface circuitry monitors changes to I/O port memory content and automatically copies updated contents to the proper bus port.

An I/O port is more than a memory address or data conduit; it is also a logical abstraction used by the CPU and bus to interact with each peripheral device in a similar way. I/O ports enable the CPU and bus to interact with a keyboard in exactly the same way they interact with a disk drive or video display. I/O ports simplify the CPU instruction set and the bus protocol because special instructions and bus circuitry aren't required for each different peripheral device. The computer system is also more flexible, because new types of peripheral devices can be incorporated into an older computer system simply by allocating new I/O ports.

Despite similar interfaces to the CPU and bus, peripheral devices differ in significant ways, including storage capacity (if any), data transfer rate, internal data coding methods, and whether the device is a storage or I/O device. Physical device details, such as how a disk read/write head is positioned, how a certain color is displayed on a video display, or the font and position of a printed character, are not directly dealt with by the simple interface between peripheral device and CPU described thus far.

In essence, the CPU and bus interact with each peripheral device as if it were a storage device containing one or more bytes stored in sequentially numbered addresses. A read or write operation from this hypothetical storage device is called a **logical access**. The set of sequentially numbered storage locations is called a **linear address space**. Logical accesses to a peripheral device are similar to access to memory. One or more bytes are "read" or "written" as each I/O instruction is executed. The bus address lines carry the position within the linear address space being read or written, and the bus data lines carry data. Complex commands and status signals also can be encoded and sent via the data lines.

As an example of a logical access, consider a disk drive physically organized into sectors, tracks, and platters. When the CPU executes an instruction to read a specific sector, it does not transmit the platter, track, and sector number as parameters of the read command. Instead, it "thinks" of the disk as a linear sequence of storage locations, each holding one sector of data, and sends a single number to identify the location it wants to read.

To physically access the appropriate sector, the location within the assumed linear address space must be converted into the corresponding platter, sector, and track, and the disk hardware must be instructed to access that specific location. Linear addresses can be assigned to physical sectors in any number of ways, one of which is shown in Figure 6-4. The disk drive itself or, as described later, its device controller, translates the linear sector address into the corresponding physical sector location on a specific track and platter.

Figure 6-4 ▶

An example of
assigning logical
sector numbers
to physical
sectors on disk
platters

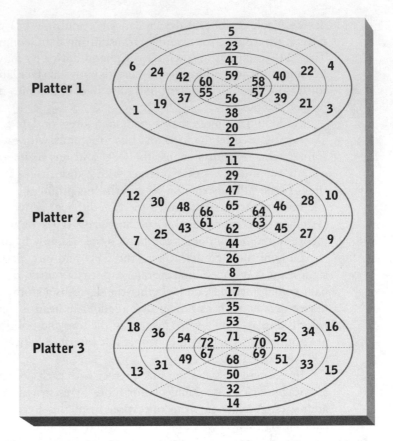

With tape drives, translating logical addresses to physical addresses is much more straightforward because blocks on a tape are physically organized in a linear sequence. A logical access to a storage (block) location is translated into commands to position the appropriate part of the tape physically under the read/write head.

Some I/O devices such as keyboards and sound cards communicate only in terms of sequential streams of bytes that fill the same memory location. With these devices, the concept of an address or location is irrelevant. From the CPU's point of view, the I/O device is a storage device with a single location that is read or written repeatedly.

Other I/O devices do have storage locations in the traditional sense. For example, the individual character or pixel positions of a video display or printed page are logical storage locations. Each position is assigned an address within a linear address space, and those addresses are used to manipulate individual pixels or character positions. The device or device controller translates logical write operations into the physical actions necessary to illuminate the corresponding video display pixel or place ink at the corresponding position on the page.

DEVICE CONTROLLERS

Storage and I/O devices are normally connected to the system bus through a **device controller** as shown in Figure 6-5. Device controllers perform the following functions:

► Implement the bus interface and access protocols

► Translate logical accesses into physical accesses

► Enable several devices to share access to a bus connection

Figure 6-5 ►

Secondary storage and I/O device connections using device controllers

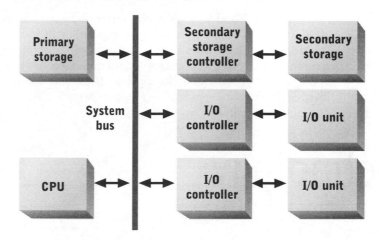

A device controller performs all bus interface functions for its attached peripheral device(s). Device controllers monitor the bus control lines for signals to peripheral devices and translate those signals into appropriate commands to the storage or I/O device. Similarly, device controllers translate data and status signals from the device into appropriate bus control and data signals.

Device controllers perform some or all of the conversion between logical and physical access commands. Device controllers "know" the physical details of the attached devices and issue specific instructions to the device based on that knowledge. For example, a disk controller converts a logical access to a specific disk sector within a linear address space into a command to read from a specific head, track, and sector.

Device controllers can connect multiple peripheral devices to share a single bus port. This function is important because the number of bus ports on the system bus usually is limited to 16 or fewer. Computer systems, especially midrange computers and mainframes, may have several dozen storage and I/O devices. Most storage and I/O devices can't sustain the system bus data transfer rate for extended periods. Sharing a single bus connection among multiple slower devices efficiently allocates the relatively large communication capacity of a bus port to many lower-capacity devices.

Mainframe Channels

In many mainframe computer systems, a device controller can be a dedicated special-purpose computer called an **I/O channel**,[1] or simply a **channel**. The distinction between an I/O channel and a device controller is not clear cut. It is a function of power and capability in several key areas, including:

▶ Number of devices that can be controlled
▶ Variability in type and capability of attached devices
▶ Maximum communication capacity

Typical I/O and storage device controllers can control only a few devices of similar type. For example, the disk controller in most desktop computers can control one or two disk drives. In contrast, a secondary storage channel in a mainframe computer can control several dozen secondary storage devices of various types, such as magnetic disks, optical discs, and magnetic tape drives. Another channel in the same mainframe might control up to 100 video display terminals or point-of-sale I/O devices.

Technology
Focus

SCSI

The acronym **SCSI** (pronounced "skuzzy") stands for **Small Computer System Interface**. It is a family of standard buses designed primarily for secondary storage devices. There are nine different SCSI standards representing nine generations of technology: SCSI-1, SCSI-2, SCSI-3, Ultra SCSI, Ultra2 SCSI, Ultra3 SCSI, Ultra320 SCSI, Ultra640 SCSI, and Serial Attached SCSI (SAS). There also are multiple variations of several SCSI standards resulting in a blizzard of confusing terminology, such as fast, wide, differential, high voltage differential, and low voltage differential. To keep things simple, this section will concentrate primarily on SCSI-1 and SCSI-2. More details on SAS are presented in Chapter 8.

A SCSI bus implements both a low-level physical I/O protocol and a high-level logical device control protocol. The bus can connect up to 16 devices, each of which is assigned a unique device identification number of 0 through 15. The bus can be as long as 25 meters. Logically, the bus can be divided into two subsets—the control and data buses. The control bus transmits control and status signals. The data bus transmits data and device identification numbers. In SCSI-1, the data bus is 8 bits wide. Later SCSI standards use a 16-bit data bus.

---

[1] The term "channel" originally was coined by IBM to describe a specific component of its 7000 series mainframe computers and it has since gained the generic meaning described in this section. Vendors of other mainframe computer systems often use different terms, for example, **peripheral processing unit** and **front end processor,** to describe functionally similar components.

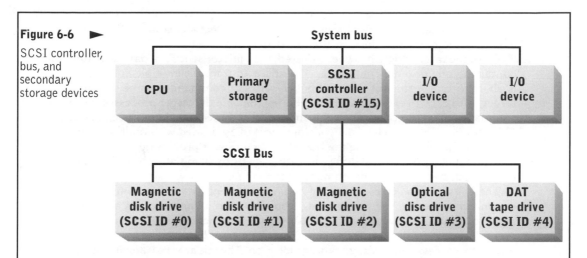

Figure 6-6 ▶

SCSI controller, bus, and secondary storage devices

A SCSI controller translates signals between the system bus and the SCSI bus. This architecture is shown in Figure 6-6. A partial listing of the SCSI-1 control signals is shown in Table 6-1.

Table 6-1 ▶

A subset of the SCSI-1 control signals

Name	Description
Acknowledge	Controls asynchronous data transfer between devices in combination with the Request signal
Attention	Instructs a target to place a message to the initiator on the bus
Busy	Sent continuously by the initiator and/or target while a bus transaction is in progress
Control/Data	Sent by the target to indicate whether the data bus contains data or a control signal
Input/Output	Indicates the direction of data transfer on the bus (target to initiator or initiator to target)
Message	Sent by the target to inform the initiator that a message is being sent on the data bus
Request	Controls asynchronous data transfer between devices in combination with the Acknowledge signal
Select	Selects a particular device as the target; the identification number of the target is simultaneously placed on the data bus

When a device needs to initiate a transaction, it asks for control of the bus by sending a busy signal and placing its identification number on the data bus. If multiple devices send the busy signal simultaneously, the device with the highest identification number becomes the bus master and is designated as the initiator. The initiator then selects a target device by sending a select signal and placing the identification number of the target on the data bus.

Once the target device has been selected, the initiator sends it a data structure containing one or more commands. The ability to transmit multiple commands is an important feature of the SCSI bus. A series of commands can be sent to a target, with control of the bus released during the time the target processes those commands. The target queues the commands and executes them in sequence. This process avoids much of the control overhead associated with multiple commands. Other devices can use the bus while the target is processing commands.

A target can send several different status signals to an initiator, including OK, busy, check condition, condition met, intermediate.

The OK signal indicates successful receipt and intent to satisfy a command or command sequence. The target sends a busy signal if it can't process a request at the current time, for example, because its queue of pending commands is full. A check condition signal tells the initiator that an error has occurred, such as an attempt to access a nonexistent storage location. The initiator must explicitly ask the target to send a message indicating the nature of the error. The condition met status signal is sent when a search command is executed successfully. SCSI defines several different search commands including searches based on data content. An intermediate signal is sent to indicate the successful completion of one command in a command sequence.

Data transfer between target and initiator can be implemented either synchronously or asynchronously. In a synchronous transfer, the target sends a Request signal simultaneously with the data. The initiator simultaneously sends an Acknowledge signal, but the target does not wait for that signal to begin sending the next data item. Synchronous data transfer uses the full capacity of the bus, which can be up to 20 MB per second with SCSI-2, for data transfer. In asynchronous data transfer, the target sends a Request signal and data simultaneously. The target doesn't send the next data item until the initiator explicitly acknowledges the receipt of the previous data item. Asynchronous data transfer can proceed no faster than half the data transfer rate of the bus because every other cycle is used by the target to send an acknowledge signal.

A SCSI bus has several desirable characteristics including:

▶ Non-proprietary standard

▶ High data transfer rate

▶ Peer-to-peer capability

▶ High-level (logical) data access commands

▶ Multiple command execution

▶ Interleaved command execution

Any company can freely manufacture and distribute SCSI-compatible devices. Data transfer rates can be as high as 3 GB per second with SAS, which is comparable to a mainframe channel. Peer-to-peer capability attains efficient bus control with minimal control overhead. Target devices can become initiators (bus masters) in order to transfer data to a requestor. The ability to combine multiple commands and to interleave command sequences increases the likelihood that most of the bus capacity will be used to transfer data rather than transfer control and status signals.

The SCSI command set is rich and high-level. A SCSI device hides its physical organization from other SCSI devices and other computer system components. Disk devices appear as a linear address space to the CPU and/or operating system. Defective storage blocks are automatically detected by SCSI disk drives and eliminated from the linear address space. A wide variety of devices can be connected to a SCSI bus, including magnetic and optical disk drives, tape drives, and data communication devices.

INTERRUPT PROCESSING

Secondary storage and I/O devices are much slower than the CPU in terms of data transfer rates. Table 6-2 lists access times and data transfer rates for various types of secondary storage devices. Table 6-3 shows data transfer rates of various I/O devices. Slow access times and data transfer rates are primarily due to mechanical limitations, for example, disk rotation speed. The time interval between a request by the CPU for input and the moment that input is received can span thousands, millions, or billions of CPU cycles.

Table 6-2 ▶

Performance characteristics of typical memory and storage devices

Storage Device	Maximum Data Transfer Rate	Access Time
SDRAM	2 GB/second	0.5 nanoseconds
Hard disk	10 MB/second	4 milliseconds
Floppy disk	500 KB/second	200 milliseconds
CD-ROM	250 KB/second	100 milliseconds
DAT tape	1 MB/second	-

Table 6-3 ▶

Typical speeds for various I/O devices

Device	Speed
Video display	25 to 200 MB/second
Ink-jet printer	3 to 25 pages per minute
Laser printer	4 to 120 pages per minute
Network interface	100 to 1000 Mbits/second

If the CPU waits for a device to complete an access request, the CPU cycles that could have been (but weren't) devoted to instruction execution are called **I/O wait states**. To prevent such inefficient use of the CPU, peripheral devices communicate with the CPU using interrupt signals. In a logical sense, an **interrupt** is a signal to the CPU that some event has occurred that requires the CPU to execute a specific program or process. In a physical sense, an interrupt is an electrical signal sent by a peripheral device over the control bus.

A portion of the CPU, separate from the components that fetch and execute instructions, continuously monitors the bus for interrupt signals and copies them to an **interrupt register**. The interrupt signal is a numeric value called an **interrupt code**, usually equivalent to the bus port number of the peripheral device sending the interrupt. At the conclusion of each execution cycle, the control unit checks the interrupt register for a nonzero value. If one is present, the CPU suspends execution of the current process, resets the interrupt register to zero, and proceeds to process the interrupt. When the interrupt has been processed, the CPU resumes executing the suspended process.

Coordinating peripheral device communication with interrupts allows the CPU to do something useful while it is waiting for an interrupt. If the CPU is executing only a single process or program, then there is no performance gain. If the CPU is sharing its processing cycles among many processes, the performance improvement is substantial. When one process requests data from a peripheral device, the CPU suspends it and starts executing another process's instructions. When an interrupt is received indicating that the access is complete, the CPU retrieves the data, suspends the process it is currently executing, and returns to executing the process that requested the data.

Interrupt Handlers

Interrupt handling is more than just a hardware feature; it is also a method of calling system software programs and processes. The operating system provides a set of processing routines, or service calls, to perform low-level processing operations such as reading data from a keyboard or writing data to a file stored on disk. An interrupt is a request by a peripheral device for operating system assistance in transferring data to or from a program. For example, an interrupt signal indicating that requested input is ready is actually a request to the operating system to retrieve the data and place it where the program that requested it can access it, such as in a register or in memory.

There is one operating system service routine, called an **interrupt handler**, to process each possible interrupt. Each interrupt handler is a separate program stored in a separate part of primary storage. In order to process an interrupt, the CPU must load and execute the first instruction of the correct interrupt handler.

An interrupt table stores the memory address of the first instruction of each interrupt handler. When the CPU detects an interrupt, it executes a master interrupt handler program called the supervisor. The **supervisor** examines the interrupt code stored in the interrupt register and uses it as an index to the interrupt table. The supervisor extracts the corresponding memory address and transfers control to the interrupt handler at that address.

Multiple Interrupts

The interrupt handling mechanism just described seems adequate until one considers the possibility of an interrupt arriving while the CPU is busy processing a previous interrupt. Which interrupt has priority? What is done with the interrupt that doesn't have priority?

Interrupts can be roughly classified into the following categories:

► I/O event
► Error condition
► Service request

The interrupt examples discussed thus far have been I/O events used to notify the operating system that an access request has been processed and that data is ready for transfer. Error condition interrupts are used to indicate errors that occur during normal processing. Error interrupts can be generated by software (for example, when attempting to open a non-existent file) or by hardware (for example, when attempting to divide by zero or when battery power in a portable computer is nearly exhausted).

Application programs generate interrupts to request operating system services. An interrupt code is assigned to each service program, and an application program requests a service by placing the corresponding interrupt number in the interrupt register. The interrupt code is detected at the conclusion of the execution cycle, the requesting process is suspended, and the service program is executed.

An operating system groups interrupts by their importance or priority. For example, error conditions normally are given higher priority than other interrupts. Critical hardware errors such as a power failure are given the highest priority. Interrupt priorities determine whether an interrupt that arrives while another interrupt is being processed is handled immediately or is delayed until current processing is finished. For example, if a hardware error interrupt code is detected while an I/O interrupt is being processed, the I/O processing is suspended and the hardware error is processed immediately.

Stack Processing

Consider the following situation. While executing instructions in an application program, an interrupt is received from a pending I/O request. The interrupt is detected and the appropriate interrupt handler is called. As it executes, the interrupt handler overwrites several general-purpose registers. When the interrupt handler terminates, processing of the application program resumes, but an error occurs because the interrupt handler altered a value in a general-purpose register that was needed by the application program.

How could this error have been prevented? How did the CPU know which instruction from the application program to load after the interrupt handler terminated? The operating system needs to be able to restart a suspended program at exactly the position it was interrupted. When the program resumes execution, it needs all register values to be in the same state as when it was interrupted. The mechanism that allows a program to resume execution in exactly the same state as before interruption is called a stack.

A **stack** is an area of storage that is accessed in a last-in, first-out (LIFO) basis. Access is similar to a stack of plates or trays in a cafeteria. Items can be added to or removed only from the top of the stack. Accessing the item at the bottom of the stack requires removing all items above it.

In a computer system, the stack is a primary storage area that holds register values of interrupted processes or programs. When a process is interrupted, values in CPU registers are added to the stack in an operation called a **push**. The saved register values are sometimes called the **machine state**. When an interrupt handler finishes executing, the CPU removes values on the top of the stack and loads them back into the appropriate registers. This operation is called a **pop**.

It is not always necessary to push the values of all registers onto the stack. At minimum, the current value of the instruction pointer must be pushed because the instruction at that address will be needed to restart the interrupted process where it left off. Think of the instruction pointer as a bookmark in an executing program. If indirect addressing is in use, then the offset register also must be pushed. The general-purpose registers are pushed because they may contain intermediate results that are needed for further processing. They also might contain values awaiting output to a storage or I/O device.

Multiple machine states can be pushed onto the stack as interrupts of high precedence occur while processing interrupts of lower precedence. It is possible for the stack to fill to capacity, in which case further attempts to push values onto the stack result in a **stack overflow** error. The size of the stack limits the number of processes that can be interrupted or suspended.

The stack is a reserved area of main memory. A special-purpose register called the **stack pointer** always points to the next empty address in the stack. The stack pointer is automatically incremented or decremented each time the stack is pushed or popped. Most CPUs provide one or more instructions to push and pop the stack.

Performance Effects

The sequence of events that occurs when processing an interrupt is summarized in Figure 6-7. The sequence is complex and consumes many CPU cycles. The most complex parts are the push and pop operations, which copy many values between registers and main memory. Wait states may occur if the stack is not implemented within a cache. (Caching is described in the next section.)

Figure 6-7 ►

Interrupt
processing

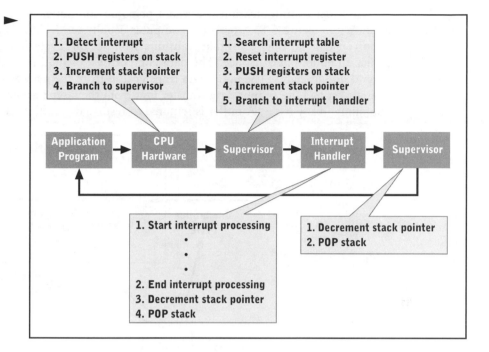

Supervisor execution also consumes CPU cycles. The table lookup procedure is relatively fast but still consumes at least 10 CPU cycles. There are other steps in the process that consume additional CPU cycles; these will be discussed in a later chapter. In total, processing an interrupt typically consumes at least 100 CPU cycles in addition to the cycles consumed by the interrupt handler itself.

BUFFERS AND CACHES

Access to I/O and storage devices is inherently slow. Mismatches in data transfer rate and data transfer unit size are addressed in part by interrupt processing, which consumes substantial CPU resources. RAM can be employed to overcome the mismatches in two distinct ways: buffering and caching. The primary goal of both buffering and caching is to improve overall computer system performance.

Buffers

A **buffer** is a small storage area (usually DRAM or SRAM) that holds data in transit from one device to another. Buffers resolve differences in data transfer rate or data transfer unit size. A buffer is required when there is a difference in data transfer unit size. A buffer isn't required when data transfer rates differ, but using one generally improves performance.

Consider communication between a personal computer and a laser printer as shown in Figure 6-8. A personal computer usually sends data to a printer via a Universal Serial Bus (USB) that transmits 1 bit per bus cycle. The laser printer prints an entire page at once. The data content of a printed page can be up to several million bytes. The input data transfer unit from the personal computer is a single bit and the output data transfer unit from the laser printer is a full page containing many bytes. A buffer is required to resolve this difference.

Figure 6-8 ▶

A buffer resolves differences in data transfer unit size between a personal computer and a laser printer

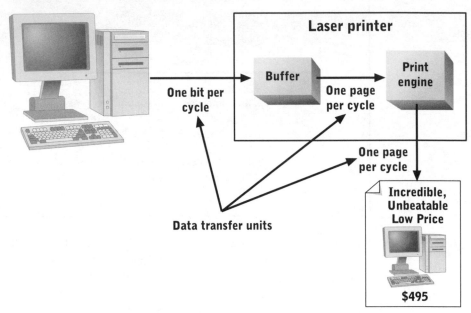

As single bits are transferred over the USB, they are added to the buffer in sequential order. When all bits of a page have been received, they are removed from the buffer and transferred to the print engine as a single unit. Buffer size must be at least as large as the data output unit. If the buffer is not large enough to hold a full page, an error called a **buffer overflow** occurs. A buffer for an I/O device is typically implemented within the device, as it is in a laser printer.

A buffer can also improve system performance when two devices have different data transfer rates. Consider, for example, communication between a personal computer and an Internet service provider (ISP) via an analog modem. The modem transmits data to and from the ISP at a relatively slow rate, such as 56 Kbps. Data is sent to the modem from primary storage or the CPU at the data transfer rate of the system bus, such as 400 Mbps. There is a data transfer rate difference of approximately 10^4.

Assume for the moment that the modem has a four-character buffer and that data is being transmitted from the computer to the ISP. The four-character buffer limits data transfer to the modem to four characters at a time. Once the modem places four characters in the buffer, it sends an interrupt to indicate that no further data should be sent. This message prevents more characters from arriving and causing buffer overflow before the modem can transmit the current buffer content to the ISP. Once the buffer content has been sent to the ISP, the modem transmits another interrupt to indicate that it is again ready to receive input.

How much system bus data transfer capacity is used to transmit 64 KB (65,536 bytes) via the modem? The answer is 49,152 bus cycles. Each four bytes transmitted result in two interrupts being sent across the bus; one to stop further transmission and another to restart it. The two interrupts are not data, but they do consume bus cycles. Further, the CPU must process those interrupts. If we assume there are 100 CPU cycles per interrupt, then 3,276,800 CPU cycles are consumed to process interrupts while transferring 64 KB to the modem.

Now consider the same scenario with a 1,024-byte buffer in the modem. A single transfer from main memory to the modem would send as many bytes per bus cycle as could be carried on the data bus, for example, 32 bits or 4 bytes. After 256 bus transfers (1,024 bytes at 4 bytes per bus cycle) the modem would send an interrupt to stop transmission. After the buffer content was transmitted to the ISP, the modem would transmit another interrupt to restart data transfer from memory.

Computer system performance improves dramatically with the larger buffer (see Table 6-4). Two interrupts are generated for each block of 1,024 bytes transferred. The total number of bus cycles consumed is:

$$16,384 \text{ data transfers} + \left(\frac{65,536 \text{ bytes}}{1024 \text{ bytes per transfer}} \times 2 \text{ interrupts} \right) = 16,512 \text{ bus cycles}$$

The number of CPU cycles falls to 12,800 ($65,536 \div 1,024 \times 2 \times 100$). Bus and CPU cycles are thus freed for other purposes such as executing programs.

Table 6-4

Bus and CPU cycles consumed for a 64-KB data transfer with various buffer sizes

Buffer Size	Bus Data Transfers	Interrupts	Total Bus Transfers	Improvement	CPU Interrupt Processing Cycles	Improvement
4	16,384	32,768	49,152	n/a	3,276,800	n/a
8	16,384	16,384	32,768	33.33%	1,638,400	50%
16	16,384	8,192	24,576	25%	819,200	50%
32	16,384	4,096	20,480	16.67%	409,600	50%
64	16,384	2,048	18,432	10%	204,800	50%
128	16,384	1,024	17,408	5.56%	102,400	50%
256	16,384	512	16,896	2.94%	51,200	50%
512	16,384	256	16,640	1.52%	25,600	50%
1,024	16,384	128	16,512	0.77%	12,800	50%

Diminishing Returns

Note the rates of change in performance improvement for total bus transfers and CPU cycles shown in Table 6-4. As buffer size increases, CPU cycle consumption decreases at a nearly linear rate, but total bus cycles decrease at a diminishing rate. If we assume that there are no excess CPU cycles, then all buffer size increases shown in the table provide significant benefit. In other words, CPU cycles not used for I/O interrupts are always applied to other useful tasks.

But if we assume that extra CPU cycles are available, in other words, that the CPU is not being used at full capacity, then the only significant benefit of larger buffer size is reduced bus cycles. Because bus cycle improvement drops off rapidly, there is a point at which further buffer size increases provide no significant benefit. If we also assume that buffer cost increases with buffer size (a reasonable assumption), then there is also a point at which the cost of additional RAM is greater than the monetary value of more efficient bus utilization.

If the discussion in the previous two paragraphs sounds vaguely familiar to you, then you've probably studied economics. The underlying economic principle is called the law of diminishing returns. Simply stated, the law says that when multiple resources are required to produce something useful, adding more and more of a single resource produces fewer and fewer benefits. The law is just as applicable to buffer size in a computer system as it is to labor or raw material inputs in a factory. The law has many other applications within computer systems, some of which are described elsewhere in this book.

Table 6-5 shows bus and CPU performance improvements for a 64-byte transfer to our hypothetical modem. Note that the law of diminishing returns affects both bus and CPU performance. Both types of performance improve at the same rate as in Table 6-4 until buffer size exceeds the amount of data being transferred, after which there is no further improvement. Why do increases in buffer size beyond 64 bytes provide no improvement? Because in this example there's nothing to put into the extra buffer space.

Table 6-5 ▶

Bus and CPU cycles consumed for a 64-byte data transfer with various buffer sizes

Buffer Size	Bus Data Transfers	Interrupts	Total Bus Transfers	Improvement	CPU Interrupt Processing Cycles	Improvement
4	16	32	48	n/a	3200	n/a
8	16	16	32	33.33%	1600	50%
16	16	8	24	25%	800	50%
32	16	4	20	16.67%	400	50%
64	16	2	18	10%	200	50%

Buffer Size	Bus Data Transfers	Interrupts	Total Bus Transfers	Improvement	CPU Interrupt Processing Cycles	Improvement
128	16	2	18	0%	200	0%
256	16	2	18	0%	200	0%
512	16	2	18	0%	200	0%
1024	16	2	18	0%	200	0%

The results in Tables 6-4 and 6-5 do more than show the law of diminishing returns in action. They also show the importance of clearly defined assumptions and a clear understanding of the work that a computer system will be asked to do. Accurate computation of improvements requires a full understanding of all affected components, in this case, the CPU, bus, modem, and buffer; clearly stated operational assumptions, for example, whether the CPU is being used at full capacity; and well-understood workload characteristics, in this case, size and frequency of data transfers to/from the modem.

Cache

Like a buffer, a **cache** is a storage area (usually RAM) that improves system performance. However, a cache differs from a buffer in several important ways including:

▶ Data content is not automatically removed as it is used
▶ Cache is used for bidirectional data transfer
▶ Cache is used only for storage device accesses
▶ Caches are usually much larger than buffers
▶ Cache content must be managed intelligently

The basic idea behind caching is simple. Access to data contained within a high-speed cache can occur much more quickly than access to slower storage devices such as magnetic disk. The speed difference is due entirely to the faster access speed of the RAM used to implement the cache. Performance improvements require a sufficiently large cache and the "intelligence" to use it effectively.

Performance improvements are achieved differently for read and write accesses. During a write operation, a cache acts as a buffer. Data first is stored in the cache and then transferred to the storage device. Performance improvements are the same as those of a buffer: reduced bus and CPU overhead due to larger continuous data transfers. Because caches are usually much larger than buffers, the performance improvement is increased. But the law of diminishing returns usually results in only slightly better performance than would be achieved with a typical buffer.

Write caching can result in more significant performance improvement when one write access must be confirmed before another can begin. Some programs, such as transaction updates for banks and retailers, require confirmed writes. In Figure 6-9, when data is written to a cache (1), the confirmation signal is sent immediately (2), before the data is written to the storage device (3).

Figure 6-9 ▶

A storage write operation using a cache

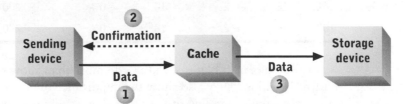

Sending a confirmation before data is written to the secondary storage device can improve program performance because the program can immediately proceed with other processing tasks. If the program is performing a series of write operations, performance improvement ceases as soon as the cache is full. As with a buffer, interrupts are used to coordinate the data transfer activity.

Immediate write confirmations are risky because an error might occur while copying data from the cache to the storage device. The most significant danger is total cache failure due to a power failure or other hardware error. In that case all data, possibly representing many cached write operations, is lost permanently, and there is no way to inform the program(s) that wrote the data of the loss.

Data written to a cache during a write operation is not automatically removed from the cache as it is written to the underlying storage device. This can provide a performance advantage in situations where data is reread shortly after it is written. A subsequent read of the same data item will be much faster unless the data item has been purged from the cache for some other reason, for example, to make room for other data items.

Most of the performance benefits of a cache occur during read operations. In Figure 6-10, read accesses are first routed to the cache (1). If the data is already in the cache then it is accessed from there (2). Performance is improved because access to the cache is much faster than access to the storage device. If the requested data is not in the cache, then it must be read from the storage device (3). A performance improvement is realized only if requested data already is waiting in the cache.

Figure 6-10 ▶

A storage read operation using a cache

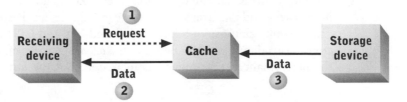

System Integration and Performance

A **cache controller** is a processor that manages cache content. A cache controller can be implemented in:

► A storage device storage controller or communication channel, as a special-purpose processor controlling RAM installed within the controller or channel

► The operating system, as a program that uses part of primary storage to implement the cache

The former approach is more common with primary storage caches, and the latter with secondary storage caches.

A cache controller guesses what data will be requested in the near future and loads that data from the storage device into the cache before it is actually requested. Guessing methods can be simple or complex. One simple method is to assume linear access to storage locations. For example, while servicing a read request for location *n*, the cache controller pre-fetches location *n+1* and possibly several subsequent storage locations. More complex guessing methods require more complex processing circuitry or software. The extra complexity and expense is justified only if the guessing accuracy improves and if those improvements lead to significant overall performance improvement.

When a read operation accesses data already contained within the cache, the access is called a **cache hit**. When the data needed is not in the cache, the access is called a **cache miss**. The ratio of cache hits to read accesses is called the cache's **hit ratio**. A cache miss requires that a **cache swap** to or from the storage device be performed. The cache controller guesses which data items are least likely to be needed in the near future, writes them to the storage device, and purges them from the cache. The requested data is then read from the storage device and placed in the cache.

Caches can be implemented within main memory if an operating system program serves as the cache controller. However, this approach can reduce overall system performance by reducing memory and CPU cycles available to application programs. Modern device controllers can implement a cache with RAM and the cache controller installed on the device controller. In that case, the CPU and system software are unaware that a cache is in use.

A surprisingly small cache can provide significant performance gains. Typical ratios of cache size to storage device capacity range from 10,000:1 to 1,000,000:1. Primary storage caches with a 10,000:1 size ratio typically achieve cache hits more than 90 percent of the time. Actual performance gain depends on cache size and the nature of storage device accesses. Frequent sequential read accesses tend to increase performance. Frequent random or scattered read accesses tend to reduce the hit ratio.

Primary Storage Cache Current processor speeds exceed the ability of dynamic RAM (DRAM) to keep them supplied with data and instructions. Although static RAM (SRAM) more closely matches the processor speed, it is generally too expensive to use for all of the primary storage. When the CPU accesses DRAM, it incurs wait states. For example, a 4-GHz processor incurs 10 wait states each time it reads sequentially from 400 MHz SDRAM.

One way to limit wait states is to use an SRAM cache between the CPU and SDRAM primary storage. The SRAM cache can be integrated into the CPU, located on the same chip as the CPU, or located between the CPU and SDRAM. Modern computers use a combination of approaches; for example, a relatively small (64 KB) cache within the CPU, a larger (512 KB) cache on the CPU chip, and an even larger (2 MB) cache between the microprocessor and SDRAM. When all three cache types are used, the within-CPU cache is called a **level one (L1) cache**, the on-chip cache is called a **level two (L2) cache**, and the off-chip cache is called a **level three (L3) cache**.

Primary storage cache control can be quite sophisticated. The default sequential flow of instructions allows simple guesses based on linear storage access to work some of the time. But accesses to operands and conditional branch instructions are not accounted for by this simple strategy. Modern CPUs use sophisticated methods to improve the hit ratio. For example, instead of guessing whether a conditional branch will be taken, the cache controller may prefetch both subsequent instructions. This consumes more of the cache but guarantees that the next instruction will be in the cache. Another strategy is to examine operands as instructions are loaded into the cache. If an operand is a memory address, then the cache controller prefetches the data at that location into the cache.

Technology
Focus

Itanium® 2 Memory Cache

The Intel Itanium® 2 microprocessor uses three levels of primary storage caching. The CPU, L1 cache, and L2 cache are implemented on a single chip. The L3 cache is implemented on a set of separate chips and can be 3, 4, or 6 MB. All of the chips are combined into a single integrated package that plugs into a computer system motherboard. The integrated package reduces distance between components, so the L3 cache can communicate with the CPU at the CPU's clock rate. Figure 6-11 shows the relationship among the caches, other CPU components, and the system bus.

The L1 cache is divided into two 16-KB segments; one for data and the other for instructions. The L2 cache is 256 KB and holds both instructions and data. L1 data cache contents are swapped automatically with the L2 cache as needed. The instruction fetch and branch prediction unit examines incoming instructions for conditional branches and attempts to prefetch instructions from both branch paths into the L1 instruction cache. The branch prediction unit also interacts with the CPU execution units to determine the likelihood of branch conditions being true or false.

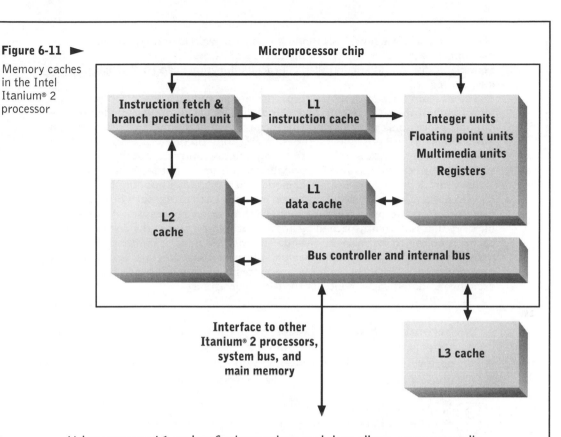

Figure 6-11 ▶

Memory caches in the Intel Itanium® 2 processor

Microprocessor chip

Instruction fetch & branch prediction unit

L1 instruction cache

Integer units
Floating point units
Multimedia units
Registers

L2 cache

L1 data cache

Bus controller and internal bus

Interface to other Itanium® 2 processors, system bus, and main memory

L3 cache

Using separate L1 caches for instructions and data allows some streamlining of processor functions. For the instruction execution units, the L1 instruction cache is a read-only cache. The L1 data cache can be read or written as operands are moved to and from registers. Cache controller functions are optimized for each type of access. However, it is possible for a program to modify its own instructions. In this case, the code section being modified is stored in both L1 caches at the same time. The processor automatically detects this condition and maintains consistency between the two caches.

The bus controller acts as a cache controller, bus controller, and as a bus connecting the CPU chip, including its embedded L1 and L2 caches, the L3 cache, the system bus, and other Itanium® 2 processors. The bus controller coordinates data movement between the L2 and L3 caches and between the L3 cache and main memory. It also coordinates data movement between the caches, main memory, and peripheral devices via the system bus.

The Itanium® 2 can assign various caching modes to 2-KB, 4-KB, or 4-MB regions of main memory. Write-back caching enables full caching of both read and write operations. Main memory is updated only when necessary, such as when a modified cache line must be removed to make room for new content. Write-through caching causes all write operations to update the caches and main memory at the same time. This mode is useful for regions of memory that can be read directly without knowledge of the CPU by other devices on the system bus, for example, a disk or video controller. All caching functions can be disabled for any memory region.

The Itanium® 2 was designed to participate in multiprocessor computer systems. Multiprocessor systems add complexity to cache management because multiple processors can cache the same memory region(s) simultaneously. Program execution errors are possible if a program executing on one processor updates a cached memory location and another program executing on another processor subsequently reads that same memory location from an outdated cache copy.

Itanium® 2 processors use a technique called memory snooping to maintain consistency among their caches and main memory. Each processor's bus controller monitors access to main memory by up to three other processors. An access to memory that is currently cached by another processor causes the corresponding cache contents to be marked invalid. The modified cache contents are automatically reloaded from main memory the next time they are accessed. If a memory region cached by multiple processors is written, all processors directly exchange updated cache contents to ensure consistency.

Secondary Storage Caches Disk caching is common in modern computer systems, particularly in file and database servers. In a file server, disk caching performance can be improved if information about file access is tracked and used to guide the cache controller. Specific strategies include:

► Give frequently accessed files higher priority for cache retention

► Use read-ahead caching for files that are read sequentially

► Give files opened for random access lower priority for cache retention

The operating system is the best source of file access information because it updates the information dynamically as it services file access requests. Because the operating system executes on the CPU, it is difficult to implement access-based cache control if the cache controller is a special-purpose processor within a disk controller.

Many computer system designers now rely on the operating system to implement disk caching instead of using specialized disk controller hardware. They feel that the extra cost of hardware-based disk caching solutions is better spent on more primary storage and faster or additional CPUs. The operating system uses the extra memory and CPU cycles to implement larger caches and more intelligent access-based cache control. This approach makes sense only if the computer system has sufficient CPU capacity to devote to cache management.

PROCESSING PARALLELISM

Many applications are simply too big to be executed by a single CPU or computer system. Examples include:

► Large-scale transaction processing applications—for example, computing monthly social security checks and electronic fund transfers

► Data mining—for example, examining a year of sales transactions at a large grocery store chain to identify purchasing trends

► Scientific applications—for example, producing hourly weather forecasts for a 24-hour period covering all of North America.

Solving problems of this magnitude would require days, months, or years for even the fastest single CPU or computer system. The only way to solve them in a timely fashion is to break the problems into pieces and solve each of the pieces in parallel with separate CPUs. Even for applications that could be performed by a single CPU, parallel processing can improve performance. This section examines various parallel processing architectures and their performance implications.

Multicore Processors

As described in Chapter 5, current semiconductor fabrication techniques are capable of placing hundreds of millions of transistors and their interconnections within a single microchip. A full-featured 64-bit CPU, even one with multiple ALUs and pipelined processing, typically requires less than 100 million transistors. That raises an obvious question—what else can be placed within a single microchip to improve processor and computer system performance?

Until recently, the only answer was cache memory. For example, the Itanium® 2 and most other high-performance CPUs devote more than half the microchip transistor count to cache memory. This architecture yields significant performance enhancements by helping to overcome the speed differences between CPU circuitry and off-chip memory. Current fabrication technology enables multimegabyte caches on the same chip as the CPU. Continuing improvements in fabrication technology could enable even larger caches. However, the performance benefit of those large caches is subject to rapidly diminishing returns.

The latest trend in high-performance CPU design embeds multiple CPUs and cache memory on a single chip—an approach called a **multicore architecture**, where the term *core* describes the logic, computation, and control circuitry of a single CPU. As of this writing, IBM POWER5 CPUs are available in dual core architecture (see Figure 6-12) and Intel plans to release a dual core version of the Itanium in 2005. Future advances in semiconductor fabrication will enable a larger number of cores in a single microprocessor.

Figure 6-12 ▶

The IBM
POWER5
dual core
microprocessor

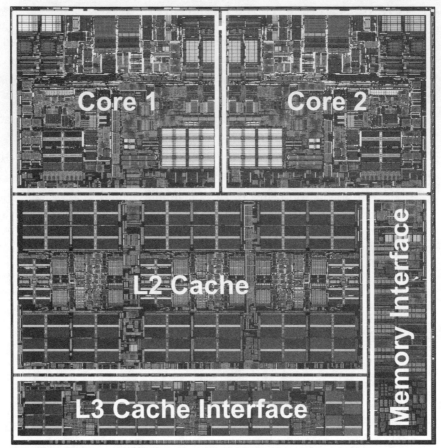

Core 1 Core 2

L2 Cache

Memory Interface

L3 Cache Interface

Courtesy of International Business Machines Corporation.
Unauthorized use not permitted.

Multicore architectures typically share memory cache, memory interface, and off-chip I/O circuitry among the cores. This reduces the total transistor count and cost and provides some synergistic benefits compared to separate CPUs on a single motherboard. In particular, a common data set can be stored in cache memory and each core can work on a different part. This yields substantial performance improvements in many numerical modeling and image processing applications because communication and data exchange among CPUs doesn't have to traverse the relatively slow system bus.

Multi-CPU Architecture

Multi-CPU architecture is a more traditional approach to multiprocessing which employs two or more single core CPUs on a single motherboard or set of interconnected motherboards. When multiple CPUs occupy a single motherboard, they share primary storage and a single system bus. When necessary, they can exchange messages over the system bus and transfer data by reading and writing primary storage. However, the speed at which those exchanges occur is much slower than with a multicore architecture.

Multi-CPU architecture is common in midrange computers, mainframe computers, and supercomputers; the number of CPUs tends to increase with computer class. Multiple programs can execute in parallel and the operating system can move them among CPUs to meet changing needs. Multi-CPU architecture is a cost-effective approach to computer system design when a single computer system executes many different application programs or services. In that scenario, there is little need for CPUs to share data or exchange messages. Thus, the performance penalty associated with those exchanges is seldom incurred.

Multi-CPU architecture is also common in workstations, though it is often less optimal in that environment because workstations frequently execute large computationally-intensive application programs that can benefit from the efficiencies of multicore architecture. However, it is a cost-effective approach to workstation design since a dual core microprocessor is usually more expensive than two single core microprocessors. Also, the similarity in the CPU and I/O power of midrange and workstation computers enables manufacturers to build both types from a common pool of subcomponents such as motherboards and storage subsystems. Multicore CPUs may find wider application in future workstations as manufacturing quantities increase and per unit costs drop.

Scaling Up and Scaling Out

The phrase **scaling up** describes approaches to increasing processing and other computer system power by using ever larger and more powerful computers. Multicore and multi-CPU architecture are both examples of scaling up because they increase the power of a single computer system. The alternative approach is **scaling out**—partitioning processing and other tasks among multiple computer systems. Two of the approaches to multicomputer architecture described in Chapter 2, clusters and grids, are examples of scaling out. Blade servers are a combination of scaling up and scaling out, though some people consider them only as an example of scaling up since all blades are normally housed within a single computer system or rack.

Until the last decade or so, scaling up was almost always a more cost-effective strategy to increase available computer system power because communication between separate computer systems was extremely slow compared to communication among the components of a single computer system. But beginning in the early 1990s, the speed of communication networks began a rapid increase. Gigabit per second speed across networks is common today and much higher speeds are close at hand. As a result, the relative performance penalty of communication between computer systems has diminished. Also, because networking technology is widely deployed, economies of scale have driven high-speed network costs ever lower.

Other changes that have increased the desirability of scaling out include:

► Distributed organizational structures that emphasize flexibility

► Improved software for managing multicomputer configurations

Modern organizations change much more rapidly than their predecessors of a few decades ago. They need the flexibility to deploy and redeploy all types of resources, including computing resources. In general, scaling out enables organizations to distribute computing resources across many locations and to combine disparate resources to solve large problems.

Until recently, the complexity of multicomputer configurations made it very difficult and expensive to administer widely distributed resources. But improvements in management software have made it feasible to manage large and diverse collections of computing resources with minimal labor input. Nonetheless, it is still less expensive to administer a few large computer systems than many smaller ones. So scaling up is still a cost-effective solution when maximal computer power is required and flexibility is not as important. Examples of environments where scaling up is cost-effective include large data processing centers that service the transaction processing needs of multiple organizations such as banks and insurance companies.

HIGH-PERFORMANCE CLUSTERING

The largest computational problems, such as those encountered in modeling three-dimensional physical phenomena, cannot be solved by any single computer system. Such problems are typically addressed by groups of powerful computers organized into a cluster. Each individual computer system, or node, in the cluster usually contains multiple CPUs, and each CPU may be dual-core.

Figure 6-13 shows a dual cluster architecture employed by the European Centre for Medium-Range Weather Forecasts. Each cluster contains a few dozen IBM pSeries 690 32-CPU computers. The computers are organized in groups of four connected by high-speed local connections. Each four-way group is connected to the other groups by two high-speed communication switches. Secondary storage, labeled GPFS (general-purpose file system) in the figure, is controlled by two four-way groups while the other four-way groups are dedicated to computation.

Figure 6-13 ▶

Organization
of two
supercomputing
clusters

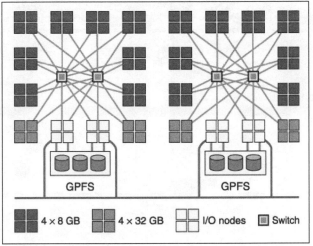

GPFS   GPFS

■■ 4 × 8 GB ■■ 4 × 32 GB ▫▫ I/O nodes ■ Switch

Courtesy European Centre for Medium-Range Weather Forecasts

The cluster organization in Figure 6-13 addresses a common problem in supercomputing—data movement among processing nodes. Consider, for example, a simplified approach to global weather forecasting that breaks up the globe into ten non-overlapping regions and further divides each region into four zones (Figure 6-14).

Each region can be assigned to a four-way computer group in the cluster and each zone can be assigned to a single computer system. The interdependence of forecast calculations within a zone will necessitate substantial data transfer among CPUs in a single computer system. But that data traverses relatively high-speed bus and memory connections within a single motherboard or set of interconnected motherboards. Of equal importance, that data will not traverse connections between computers in a four-way group nor the switches that connect the groups.

Now consider what happens as weather features in the forecast move from zone to zone. For example consider a region consisting of most of North America, Central America, and the western edge of the North Atlantic, and division of that region into four zones, as shown on the left side of Figure 6-14. A low-pressure center or cold front moving from the western to eastern United States moves from one zone to another. That movement creates a shared data dependency between the zone calculations in two computers of the four-way group assigned to the region. Those computers must exchange data to enable forecast calculations to proceed. A similar data exchange among computers assigned to different zones occurs when modeling a hurricane as it moves westward from the Atlantic Ocean into the Caribbean Sea.

Figure 6-14 ►

Sample weather forecast regions (separated by solid lines) and division of the left region into zones (separated by dashed lines)

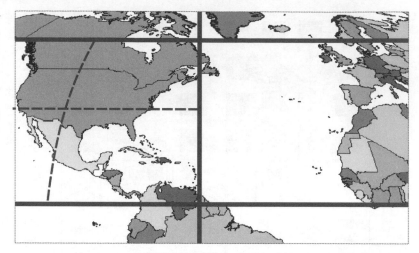

By dividing the forecasting problem into pieces that mirror the cluster organization in Figure 6-13, the problem is hierarchically decomposed into pieces with increasing needs for data exchange. The greatest amount of data is exchanged among CPUs in a single computer system dedicated to a single zone. The next greatest amount of data is exchanged among computer systems in a four-way group, which traverses data connections specific to that group. The least amount of data is exchanged among four-way groups assigned to a region, and that data traverses the switches that connect all four-way groups. Partitioning the problem to match the cluster architecture ensures that most data exchange traverses high-speed paths. Thus, all CPUs are kept as busy as possible by avoiding long waits for data sent from "distant" processors.

COMPRESSION

Compression is a technique that reduces the number of bits used to encode a set of related data items, for example, a file or a stream of motion video images. A **compression algorithm** is a specific mathematical compression technique implemented as a program. Most compression algorithms have a corresponding **decompression algorithm** that restores compressed data to its original or nearly original state. In common use, the term compression algorithm refers simultaneously to both the compression and decompression algorithm.

There are many compression algorithms, which vary in:

► Type(s) of data for which they are best suited

► Whether information is lost during compression

► Amount by which data is compressed

► Computational complexity (CPU cycles required to execute the compression program)

A compression algorithm can be lossless or lossy. With **lossless compression,** the result of compressing and then decompressing any data input is exactly the same as the original input. Lossless compression is required in many applications, for example, accounting records, executable programs, and most stored documents.

With **lossy compression,** the result of compressing and then decompressing a data input is different than, but still similar to, the original input. Lossy compression is usually applied only to audio and video data. The human brain tolerates missing audio and video data and usually can "fill in the blanks." Lossy compression is commonly used to send audio or video streams via low-capacity transmission lines or networks, for example, video conferencing.

The term **compression ratio** describes the ratio of data size in bits or bytes before and after compression. For example, if a 100-MB file is compressed to 25 MB, then the compression ratio is 4:1. Some types of data are easier to compress than others. For example, lossless compression ratios up to 10:1 are easily achieved with word processing documents and ASCII or Unicode text files. Lossless compression ratios of greater than 3:1 are difficult or impossible to achieve with audio and video data. Lossy compression of audio and video can achieve compression ratios of up to 50:1, though at that ratio the lost information is easily discerned by listeners and viewers.

Compression is commonly used to increase the amount of data stored on backup tapes and is sometimes used to reduce disk storage requirements, as shown in Figure 6-15(a). Data sent to the storage device is compressed before it is written. Data read from storage is decompressed before it is sent to the requester. Compression can also increase communication channel capacity, as shown in Figure 6-15(b). Data is compressed as it enters the channel and then decompressed as it leaves the channel. Hardware-based compression is included in all modern standards for analog modems and is often employed for long-distance telephone transmission.

Using data compression alters the balance of processor resources and communication or storage resources in a computer system. Implementing a compression algorithm consumes processor cycles. Algorithms with high compression ratios frequently consume more processing resources than those with low compression ratios. The proper tradeoff between these resources depends on their relative cost, availability, and degree of use. Implementing data compression with CPU processing resources may not be cost effective. Implementing data compression using special-purpose processors is often preferable. Such processors are now commonly available and are frequently embedded in tape backup devices, video display controllers, and high-speed modems.

Figure 6-15 ▶

Data
compression
with a secondary
storage
device (a) and a
communication
channel (b)

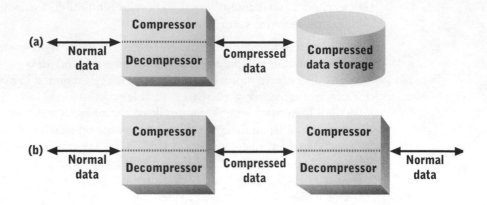

Technology
Focus

MPEG and MP3

The **Motion Picture Experts Group (MPEG)** is an organization that creates and evaluates standards for motion picture recording and encoding technology. The group has created three different standards: MPEG-1, MPEG-2, and MPEG-4. Motion pictures usually contain images and sound, so MPEG standards address recording and encoding formats for both data types. Each standard is divided into layers numbered 1 (systems), 2 (video), and 3 (audio). The audio encoding standard commonly known as **MP3** is actually layer 3 of the MPEG-1 standard. It is a useful encoding method for many types of audio, not just audio tracks within motion pictures.

Analog audio data is converted to digital form by sampling the audio waveform thousands of times per second and then constructing a numerical picture of each sample. On an audio CD, there are 44,100 samples per second per stereo channel. Each sample is represented by a single digital number. Sample quality is improved (due to greater precision and accuracy) if many bits are used to encode each sample. Each sample on an audio CD is encoded in 16 bits. Each second of audio data on a stereo CD requires $44,100 \times 16 \times 2 = 1,411,200$ bits.

A typical motion picture is two hours long and includes far more video than audio data. To encode both data sets within a single storage medium, for example, on a DVD-ROM, all data must be compressed. Normal CD-quality music can be compressed by a ratio of 6:1 and most listeners cannot distinguish the compressed audio data from the original. MP3 is a lossy compression method. It discards significant portions of the original audio data and does not attempt to reconstruct that data for playback. Thus, MP3 has a relatively simple decompression algorithm.

MP3 takes advantage of a few well-known characteristics of human audio perception including:

▶ Sensitivity that varies with audio frequency (pitch)

▶ Inability to recognize faint tones of one frequency simultaneously with much louder tones in nearby frequencies

▶ Inability to recognize soft sounds that occur shortly after louder sounds

These three characteristics interact in complex ways. For example, loud sounds at the extremes of the human hearing range, such as 20 Hz or 20 KHz, more effectively mask nearby soft frequencies than do loud sounds in the middle of the human hearing range. MP3 analyzes digital audio data to determine which sound components are masked by others. For example, a bass drum may be masked for a moment by a bass guitar note. It then compresses the audio data stream by either discarding information about masked sounds or representing them with a reduced number of bits.

Figure 6-16 shows the conceptual components of an MP3 encoder. A frequency separator divides raw audio data into 32 narrow-frequency audio streams. In parallel, a perceptual modeler examines the entire audio signal to identify loudness peaks. The modeler then determines by how much and for how long those loudness peaks can mask nearby audio data. The masking information and the 32 data streams then are sent to an encoder that represents data within each frequency. The encoder uses the masking information from the perceptual modeler to determine what data to discard or to represent with less than 16-bit precision.

Figure 6-16 ▶

MP3 encoding components

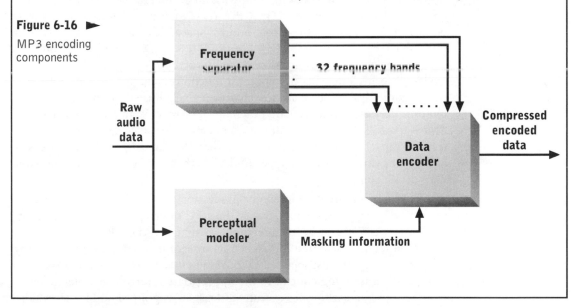

The MP3 standard is widely implemented in MP3 player software, Internet broadcasts, MP3 hardware, and DVD playback devices. There has been much controversy over its widespread use and its effect(s) on music distribution and piracy. Attempts may be made to modify the standard, or legally mandate a variant, that makes copyright infringement more difficult, though there are doubts about whether that is even possible. The standard will definitely need enhancement to address more advanced forms of audio data such as theater-quality surround sound, but the coding method and data format have been proven effective by researchers and the marketplace. That success means MP3 will be with us in some form for many years to come.

SUMMARY

▶ The system bus is the communication pathway that connects the CPU with memory and other devices. The bus is a set of data lines, control lines, and status lines. The number and use of these lines, as well as the procedures for controlling access to the bus, are called the bus protocol. A peer-to-peer bus or direct memory access gives devices other than the CPU temporary control of the bus, resulting in improved computer system performance.

▶ The CPU communicates with peripheral devices through I/O ports. For simplicity, the CPU and bus interact with all peripheral devices using simple data movement instructions. The CPU treats each peripheral device as if it were a storage device with a linear address space. The device, or its controller, translates access commands to the linear address space into whatever physical actions are actually necessary to perform the access. A peripheral device controller performs a number of important functions including logical-to-physical translation, physical bus interface, sharing a single bus connection among multiple devices, and (optionally) caching. A channel is an advanced type of device controller used in mainframe computers. Channels have greater data transfer capacity, a larger maximum number of attached peripheral devices, and greater variability in the types of devices that can be controlled compared with device controllers.

▶ Application programs use interrupt processing to coordinate data transfers to or from peripheral devices, notify the CPU of errors, and call operating system service programs. An interrupt is a signal to the CPU that some condition requires its attention. When an interrupt is detected, the currently executing program is suspended by pushing current register values onto the stack and transferring control to the appropriate interrupt handler. When the interrupt handler finishes executing, the stack is popped and the suspended process resumes execution from the point of interruption.

▶ A buffer is a region of memory that holds a single unit of data for transfer to or from a device. Buffers enable devices with different data transfer rates and unit sizes to efficiently coordinate data transfer. A cache is a large buffer implemented within a device controller or primary storage. When used for input, a cache enables more rapid access if the data being requested already is in the cache. The cache controller guesses what data the CPU will request next and loads that data into the cache before it actually is requested.

▶ Computer system computational capacity can be increased with parallel processing techniques including multicore processors, multi-CPU architecture, and clustering. A multicore processor includes multiple CPUs and shared memory cache in a single microchip. Multi-CPU architecture employs multiple single or multicore processors sharing main memory and the system bus within a single motherboard or computer system. Clusters connect separate computer systems with high-speed interconnections. The computer systems each work on part of a large problem and exchange data as needed.

▶ Compression reduces the number of bits required to encode a data set or stream, effectively increasing the capacity of a communication channel or storage device. A compression algorithm may be lossless or lossy. With lossless compression, data is exactly the same before and after compression and decompression. Lossy compression loses some information during compression and decompression. Data compression requires increased processing resources to implement the compression and decompression algorithms while reducing resources needed for data storage and/or communication.

In Chapter 3 we described various ways in which data are represented. In Chapters 4 and 5 we described CPU, primary storage, and secondary storage hardware. In this chapter we've described how data are transmitted among all computer system hardware devices. Various hardware and software techniques for improving data transfer efficiency were also described and, thus, overall computer system performance. In the next chapter, we'll look at input/output device technology.

Key Terms

address bus	bus protocol	compression
buffer	bus slave	compression algorithm
buffer overflow	cache	compression ratio
bus	cache controller	control bus
bus arbitration unit	cache hit	data bus
bus clock	cache miss	data transfer rate
bus cycle	cache swap	decompression algorithm
bus master	channel	device controller

direct memory access (DMA)
DMA controller
front end processor
hit ratio
I/O channel
I/O port
I/O wait state
interrupt
interrupt code
interrupt handler
interrupt register
level one (L1) cache
level two (L2) cache

level three (L3) cache
linear address space
logical access
lossless compression
lossy compression
machine state
Motion Picture Experts Group
    (MPEG)
MP3
multicore architecture
multi-CPU architecture
peer-to-peer bus
peripheral devices

peripheral processing unit
pop
push
scaling out
scaling up
Small Computer System
    Interface (SCSI)
stack
stack overflow
stack pointer
supervisor
system bus

─────────────── **Vocabulary Exercises** ───────────────

1.   The _____ is the communication channel that connects all computer system components.

2.   A(n) _____ cache is generally implemented on the same chip as the CPU.

3.   The CPU is always capable of being a(n) _____, thus controlling access to the bus by all other devices in the computer system.

4.   A(n) _____ is a small area of memory used to resolve differences in data transfer rate or data transfer unit size.

5.   A(n) _____ is an area of fast memory where data held in a storage device is prefetched in anticipation of future requests for that data.

6.   A cache controller is a hardware device that initiates a(n) _____ when it detects a cache miss.

7.   The _____ transmits command, timing, and status signals between devices in a computer system.

8.   If possible, the system bus _____ rate should equal the speed of the CPU.

9.   The _____ is a register that always contains a pointer to the top of the _____.

10.   The _____ calls an interrupt handler after it looks up its memory address in the _____.

11.   The set of register values placed on the stack while processing an interrupt is also called the _____.

12.   A(n) _____ is a software program that is executed in response to a specific interrupt.

13. During interrupt processing, register values of a suspended process are held on the _____.

14. A(n) _____ is a signal to the CPU or operating system that some device or program requires processing services.

15. A(n) _____ is a hardware device that intervenes when two potential bus masters want control of the bus at the same time.

16. The initiator of a bus transfer assumes the role of a(n) _____. The recipient assumes the role of a(n) _____.

17. The CPU incurs one or more _____ if it is idle pending the completion of an I/O operation.

18. The system bus can be decomposed logically into three sets of transmission lines: the _____ bus, the _____ bus, and the _____ bus.

19. During a(n) _____, one or more register values are copied to the top of the stack. During a(n) _____, one or more values are copied from the top of the stack to registers.

20. The relative size of a data set before and after data compression is described by the compression _____.

21. If the result of compressing and decompressing a file is not exactly the same as the original file, then the compression algorithm is said to be _____. If the result is the same as the original, the compression algorithm is said to be _____.

22. A(n) _____ is a special-purpose processor dedicated to managing the contents of a cache.

23. A(n) _____ is a memory address to which the CPU copies data when writing to an I/O or storage device.

24. The _____ carries interrupts, read commands, and other control signals.

25. Primary storage receives a read command on the _____, the address to be accessed on the _____, and returns the accessed data on the _____.

26. The operating system normally views any storage device as a(n) _____, ignoring the device's physical storage organization.

27. Part of the function of a storage secondary device controller is to translate _____ into physical accesses.

28. _____ is used to transfer data directly between memory and secondary storage devices, without the assistance of the CPU.

29. A(n) _____ is a high capacity device controller used in mainframe computers.

30. An access to primary storage that is found within a cache is called a(n) _____.

31. The _____ defines the control, signals, and other communication parameters of a bus.

32. A(n) _____ processor is a microchip containing two or more CPUs with shared memory cache.

33. The term _____ describes methods of increasing available computer system power that employ ever larger and more powerful individual computer systems.

34. _____ architecture is an efficient approach to increase computer system power when the system will execute many different applications or services at once.

Review Questions

1. What is the system bus? What are its primary components?

2. What is a bus master? What is the advantage of having devices other than the CPU be a bus master?

3. What characteristics of the CPU and of the system bus should be balanced to obtain maximum system performance?

4. What is an interrupt? How is an interrupt generated? How is it processed?

5. What is a stack? Why is it needed?

6. Describe the execution of the push and pop operations.

7. What is the difference between a physical access and a logical access?

8. What functions does a device controller perform?

9. What is a buffer? Why might one be used?

10. How can a cache be used to improve performance when reading data from a storage device? How can a cache be used to improve performance when writing data to a storage device?

11. What is the difference between lossy and lossless compression? For what types of data is lossy compression normally used?

12. Describe how scaling up differs from scaling out. Given the speed difference between a typical system bus and a typical high speed network, is it reasonable to assume that both approaches can yield similar increases in total computational power?

13. What is a multicore processor? What are its advantages compared to multi-CPU architecture? Why have multicore processors only recently become available?

Problems and Exercises

1. Assume that a PC uses a 2-GHz processor, a system bus clocked at 400 MHz, and a 3 megabit per second (Mbps) internal cable modem attached to the system bus. No parity or other error-checking mechanisms are used.

    The modem has a 64-byte buffer. After 64 bytes are received by the modem, it stops accepting data from the network and sends a "data ready" interrupt to the CPU. When the "data ready" interrupt is received, the CPU and operating system perform the following actions:

    a. The supervisor is called.

    b. The supervisor calls the modem's "data ready" interrupt handler.

    c. The interrupt handler sends a command to the modem instructing it to copy its buffer content to main memory.

    d. The modem interrupt handler immediately returns control to the supervisor, without waiting for the copy operation to complete.

    e. The supervisor returns control to the process that was originally interrupted.

    When the modem completes the data transfer, it sends a "transfer completed" interrupt to the CPU and resumes accepting data from the network. In response to the interrupt, the CPU and operating system perform the following actions:

    a. The supervisor is called.

    b. The supervisor calls the "transfer completed" interrupt handler.

    c. The interrupt handler determines whether a complete packet is present in memory. If so, it copies the packet to a memory region of the appropriate application program.

    d. The modem interrupt handler returns control to the supervisor.

    e. The supervisor returns control to the process that was originally interrupted.

    Sending an interrupt requires one bus cycle. A push or pop operation consumes 30 CPU cycles. Incrementing the stack pointer and executing an unconditional branch instruction each require 1 CPU cycle. The supervisor consumes 8 CPU cycles searching the interrupt table before calling an interrupt handler. The "data ready" interrupt handler consumes 50 CPU cycles before returning to the supervisor.

    Incoming packets range in size from 64 bytes to 4096 bytes. The "transfer complete" interrupt handler consumes 30 CPU cycles before returning to the supervisor if it does not detect a complete packet in memory. If it does detect a complete packet in memory, it consumes 30 CPU cycles plus 1 cycle for each 8 bytes of the packet).

Question 1: How long does it take to move a 64-byte packet from it arrival at the cable modem until its receipt in the memory area of the target application program or service? State your answer in elapsed time (seconds or a fraction thereof).

Question 2: Assume that the computer is executing a program that is downloading a large file using the modem. Assume further that all packets are 1024 bytes. What percentage of the computer system CPU capacity is used to manage data transfer from the modem to the application program? What percentage of available bus capacity is used to move incoming data from the modem to the application program? Assume that the bus uses a simple request/response protocol without command acknowledgment.

Question 3: Recalculate your answers to questions 1 and 2 assuming a modem buffer size of 1024 bytes. Assume that all incoming packets are 1024 bytes.

2. A video frame displayed on a computer screen consists of many pixels. Each pixel, or cell, represents one unit of video output. The resolution of a video display is typically specified in horizontal and vertical pixels (such as 800 × 600) and the number of pixels on the screen is simply the product of these numbers (800 × 600 = 480,000 pixels). The data content of a pixel is one or more unsigned integers. For a black and white display, each pixel is a single number typically between 0 and 255 that represents the intensity of the color white. Color pixel data is typically represented as either one or three unsigned integers. When three numbers are used, the numbers are usually between 0 and 255, and each number represents the intensity of a primary color (red, green, or blue). When a single number is used, it represents a predefined color selected from a table or palette of colors.

Motion video is displayed on a computer screen by rapidly copying frames to the video display controller. The video controller continuously converts frame data to an analog RGB signal that is output to the display device. Because video images or frames require many bytes of storage, they are usually copied to the display controller directly from secondary storage. Each video frame is an entire picture, and its data content, as measured in bytes, depends on the resolution at which the image is displayed and the maximum number of simultaneous colors that can be contained within the sequence of frames. For example, a single frame at 800 × 600 resolution with 256 simultaneous colors contains 800 × 600 × 1 = 480,000 bytes of data. Realistic motion video requires a minimum of 20 frames per second to be copied and displayed; 24 or 30 frames per second are common professional standards. Using fewer frames per second results in perceived "jerky" motion because the frames are not being displayed quickly enough to fool the eye and the brain into thinking that they are one continuously changing image.

Assume that the computer system being studied contains a bus mastering disk controller and a video controller that copies data to the video display at least as fast as it can be delivered over the bus. Assume that the system bus can transfer data at a sustained rate of 100 Mbps, as can both of the controllers' bus interfaces. This system will be used to display motion video on a monitor capable of resolutions as low as 640×480 and as high as 1024×768.

Assume that a single disk drive is attached to the disk controller and that it has a sustained data transfer rate of 20 MB per second when reading sequentially stored data. Next, assume that the channel connecting the disk drive to the disk controller has a data transfer rate of 80 Mbps. Finally, assume that the files containing the video frames are stored sequentially on the disk and that copying the content of these files from disk to the display controller is the only activity that the system will perform (no external interrupts, no multitasking, etc.).

Assume that the video display controller contains 2 MB of 50-nanosecond, 8-bit buffer RAM, and that the video image arriving from the bus can be written to the buffer at a rate of 8 bits per 50 nanoseconds. The RAM buffer of the video display can be written from the bus while it is simultaneously being read by the display device, sometimes called dual porting. Finally, data can be received and displayed by the display device as fast as it can be read from the RAM buffer by the video controller.

Question 1: What is the maximum number of frames per second (round down to a whole number) that can be displayed by this system in 256 simultaneous colors at a resolution of 640×480?

Question 2: What is the maximum number of frames per second (round down to a whole number) that can be displayed by this system in 65,536 simultaneous colors at a resolution of 800×600?

Question 3: What is the maximum number of frames per second (round down to a whole number) that can be displayed by this system in 16,777,216 simultaneous colors at a resolution of 1024×768?

3.   An employee listens to an Internet radio station while using his or her office computer. The Internet radio station sends MP3 encoded stereo music through the Internet and company network to the employee's computer system. The music source is compact disc and the music is compressed at an average ratio of 8:1.

The employee's computer has a 100-Mbps Ethernet interface and a 400-MHz 32-bit PCI bus. The employee uses Windows Media Player to listen to the music.

Question 1: How much company network capacity is consumed by the employee while listening to music?

Question 2: What percentage of available bus cycles on the employee's computer are consumed when listening to the Internet broadcast?

Question 3: Does your answer to either Question 1 or Question 2 support forbidding this employee to listen to Internet broadcast music? What if there are 500 employees on the company network?

Research Problems

1. Investigate the implementation of modern PC buses such as the Peripheral Component Interconnect (PCI) bus. How wide are the data and address buses? What is the bus clock rate? What is the maximum data transfer rate? What is the set of commands and command responses? How is peer-to-peer capability implemented? What is the bus arbitration strategy?

2. Fibre Channel (http://www.fibrechannel.org) is a relatively recent bus standard for high-speed connection to storage devices. It is currently used to implement high-speed storage arrays where storage devices are separated by large distances. It can also enable multiple computer systems to share access to a single storage device or array. Investigate the implementation details of the Fibre Channel standard. What are its strengths and weaknesses as compared to Ultra640 and Serial-Attached SCSI? Which would be used in a typical LAN environment? Which would be used in a distributed transaction-processing environment?

3. Most IBM-compatible PCs use secondary storage devices based on one of the IDE/ATA standards. Investigate those standards and compare them to lower-cost SCSI standards such as SCSI-3. Which standard is the best choice for an average home or office workstation? Which standard is the best choice for high-end workstations such as those used by engineers and digital music producers? Which standard is the best choice for a PC-based server used to share files and databases on an office network?

Chapter **7**

Input/Output Technology

Chapter Goals

► Describe common concepts of text and image representation and display including digital representation of grayscale and color, bitmaps, and image compression techniques

► Describe the characteristics and implementation technology of video display devices

► List and describe the three predominant manual input technologies

► Understand printer characteristics and technology

► Describe various types of optical input devices including mark sensors, bar code readers, scanners, and digital cameras

► Identify the characteristics of audio I/O devices, and explain how they operate

People communicate in many different ways, and those differences are reflected in the variety of methods for interacting with computer systems. This chapter describes the concepts, technology, and hardware used in communication between people and computers (see Figure 7-1). Understanding I/O technology is important because that technology expands or limits the ability of computers to assist people in problem solving.

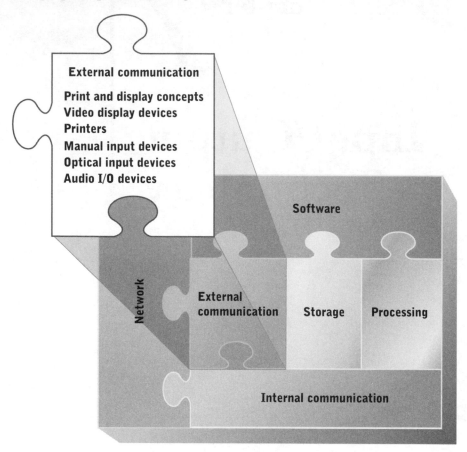

External communication

Print and display concepts
Video display devices
Printers
Manual input devices
Optical input devices
Audio I/O devices

Software

Network

External communication

Storage

Processing

Internal communication

BASIC PRINT AND DISPLAY CONCEPTS

Although printing technology dates from the 15th century, communication via video display devices is less than a century old. However, both share many features, including character representation methods, measurement systems, and methods of generating color. These topics are covered first because they are common to many I/O technologies and devices.

Matrix-Oriented Image Composition

Display surfaces vary widely in size and composition. The most commonly used display surfaces are paper, cathode ray tubes, and flat panel displays. Display surfaces have an inherent background color, usually white for paper and black for video display devices. Display surfaces are divided into rows and columns similar to a large table or matrix. Each cell in the matrix represents one part of an image called a **pixel** (a shortened form of "picture element"). For printed output, a pixel is either empty or contains one or more inks or dyes. For video display a pixel displays either no light or light of a specific color and intensity.

The number of pixels in a display surface depends on:

► Display surface size (height and width)

► Pixel size

For example, a typical 19-inch flat panel display is 28 centimeters (11 inches) high and 37 centimeters (14.5 inches) wide. If display pixels are .25 millimeters square, then the total number of pixels in the display is:

$$280 \text{ mm} \times 370 \text{ mm} \div 0.25 \text{ mm} = 414{,}400$$

The **resolution** of a display is the number of pixels displayed per linear measurement unit. For example, the resolution of the typical 19-inch flat panel display described above is 40 pixels per centimeter or approximately 100 pixels per inch. In the United States, resolution is generally stated in **dots per inch (dpi)**, where a dot is equivalent to a pixel. Higher resolutions correspond to smaller pixel sizes. To an observer, the quality of a printed or displayed image is directly related to the pixel size (see Figure 7-2). Smaller pixel size yields higher print quality because fine details such as smooth curves can be incorporated into the image.

Figure 7-2 ►

Text displayed in two resolutions— 50 dpi (top) and 200 dpi (lower)

Systems Architecture

Systems Architecture

On paper, pixel size corresponds to the smallest drop of ink that can be accurately placed on the page. Decades ago, printers adopted one seventy-second of an inch as a standard pixel size, called a **point**. The term and its use as a printer's measuring system continues, even though modern printing techniques are capable of much higher resolution.

Fonts Written Western languages are based on systems of symbols called characters. Individual characters can be represented as a matrix of pixels, as shown in Figure 7-3. A specific printed character need not exactly match a specific pixel map to be recognizable. For example, the symbols E, E, **E**, ℰ, and *E* are all easily interpreted by people as the letter E, even though their pixel composition varies.

Figure 7-3 ▶

The letters p and A represented within an 8 × 16 pixel matrix

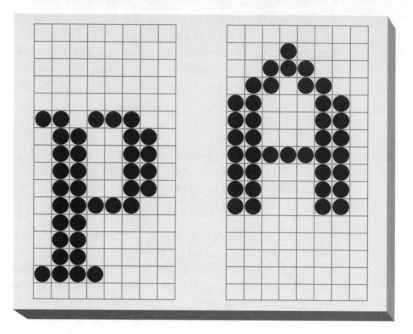

A collection of characters of similar style and appearance is called a **font**. Figure 7-4 shows sample characters from several different fonts. Points are the most common unit for measuring font size. Figure 7-5 shows a single font printed in several point sizes. The point size measurement refers to the height of the characters but not their width, though width is normally scaled to match height. Individual characters vary in height and placement. For example T is taller than a, and the letters p and q extend lower within a text line than the letters a and b. Font point size is the distance between the top of the highest and bottom of the lowest character in the font.

Figure 7-4 ▶

Sample characters printed in various fonts

AaBbCc XxYyZz 123 !@#	News Gothic
AaBbCc XxYyZz 123 !@#	Braggadocio
AaBbCc XxYyZz 123 !@#	Century Schoolbook
AaBbCc XxYyZz 123 !@#	Garamond Condensed
AaBbCc XxYyZz 123 !@#	Tahoma

Input/Output Technology

Figure 7-5 ▶

A sample font
(Times Roman)
printed in various
point sizes

8 point Times Roman
10 point Times Roman
12 point Times Roman
16 point Times Roman
24 point Times Roman

Color   The human eye and brain interpret different light frequencies as different colors. Video display devices generate light of specific color in each pixel. On the printed page, color is the light frequency reflected from the page. Any color in the visible spectrum can be represented as a mixture of varying intensities of primary colors such as red, green, and blue. For example, mixing equal intensities of red and blue produces the color purple. The absence of all three colors is black and full intensity of all three is white.

Different sets of primary colors are used for different purposes. In paint, red, blue, and yellow are used as primary pigments. Video display devices have used red, green, and blue as primary colors since the earliest color televisions. The primary colors for video display are sometimes referred to by their first letters, **RGB**. A video display that generates color using mixtures of red, green, and blue is sometimes called an RGB display.

The printing industry generates color using the inverse of the primary video display colors. In this context, video display colors are called **additive colors** and their inverse colors are called **subtractive colors**. The subtractive colors are cyan (absence of red), magenta (absence of green), and yellow (absence of blue), and they are often referred to by the acronym **CMY**. Black can be generated by combining all three, but a separate black dye, identified by the letter K, is generally used. The four-dye scheme is called **CMYK** color.

Numeric Pixel Content   Because computers are digital devices, pixel content must be described numerically. A stored set of numbers that describes the content of all pixels in an image is called a **bitmap**. The number of bits required to represent a pixel's content depends on the number of different colors the pixel can display.

A **monochrome** display can display one of two colors and thus requires only one bit per pixel. A **grayscale** display can display black, white, and many shades of gray in between. The number of gray shades that can be displayed increases with the number of bits used to represent a pixel. If eight bits per pixel are used, then 254 shades of gray are available in addition to pure black (0) and pure white (255). The number of distinct colors or gray shades that can be displayed is sometimes called the **chromatic depth** or **chromatic resolution**.

Most color display schemes represent each pixel's color with three different numbers. Each number represents the intensity of an additive or subtractive color. Chromatic depth depends on the number of bits used to represent each color's intensity. For example, if each color is represented by an 8-bit number, the chromatic depth is $2^{(3\times8)}$ or approximately 16 million; this is sometimes called **24-bit color**. Each group of 8 bits represents the intensity (0 = none, 255 = maximum) of one color. For example, with RGB color, the numbers 255:0:0 represent bright red, and the numbers 255:255:0 represent bright magenta.

Another approach to representing pixel color using numbers is to define a color palette. A **palette** is simply a table of colors. The number of bits used to represent each pixel determines the table size. If eight bits are used, then the table has 2^8 or 256 entries. Each table entry can contain any RGB color value such as 200:176:40, but the number of different colors that can be represented is limited by the size of the table. Table 7-1 shows the 4-bit color palette used for the Computer Graphics Adapter (CGA) video driver of the original IBM PC/XT.

Dithering is a process that generates color approximations by placing small dots of different colors in an interlocking pattern. If the dots are small enough, the human eye interprets them as being a uniform color representing a mixture of the dot colors. For example, a small pattern of alternating red and blue dots appears to be magenta. A lower percentage of red and higher percentage of blue is interpreted as violet. An alternating pattern of black and white dots looks gray, which can easily be confirmed by examining a gray area of one of the figures in this book with a magnifying glass. Grayscale dithering is usually called **half-toning**.

Dithering is commonly used when a printing or display device has a limited number of display colors or dyes that can't be intermixed. For example, if a printer can apply only four different amounts of a single dye in one pixel area, then only 64 (4^3) different colors can be generated. With dithering, the same printer can simulate several shades between each of those 64 colors at a reduced display resolution.

Table 7-1 ▶

Table Entry Number	RGB Value	Color Name
0	0:0:0	black
1	0:0:191	blue
2	0:191:0	green
3	0:191:191	cyan
4	191:0:0	red
5	191:0:191	magenta
6	127:63:0	brown
7	223:223:223	white
8	191:191:191	gray
9	127:127:255	light blue
10	127:255:127	light green
11	127:255:255	light cyan
12	255:127:127	light red
13	255:127:255	light magenta
14	255:255:127	yellow
15	255:255:255	bright white

Image Storage Requirements

The amount of storage required for an image depends on the number of bits that represent each pixel and on the image height and width in pixels. For example, an 800×600 image has 480,000 pixels. For a monochrome image, 1 bit represents each pixel and the image requires 60,000 bytes of storage (480,000 pixels × 1 bit/pixel ÷ 8 bits/byte). Storing an 800×600 24-bit color image requires 1.44 MB of storage (480,000 pixels × 3 bytes/pixel).

Image storage requirements are the same regardless of whether the image is stored in primary storage, secondary storage, a communication channel, or an I/O device buffer. Image storage requirements can be reduced with bitmap compression techniques. Common bitmap compression formats include **Graphics Interchange Format (GIF)** and **Joint Photographic Experts Group (JPEG)** for single (still) images and **Moving Picture Experts Group (MPEG)** for moving images. However, all of these compression methods are lossy, resulting in some loss of image quality.

Image Description Languages

There are two significant drawbacks to using bitmaps to represent high-quality images. The first is their size. An 8½ × 11-inch, 24-bit color image with 300 dpi resolution requires approximately 25 MB of uncompressed storage. A good compression algorithm might achieve compression ratios of 10:1 or 20:1, reducing the storage requirements to a few megabytes. Files of this size take a great deal of time to transmit over slower communication links such as the parallel printer port of a personal computer or an analog modem.

The second drawback is that there is no standard storage format for raw bitmaps. Bitmap formats vary among software packages and output device types, manufacturers, and models. Image files stored by software must be converted into a different format before they are printed or displayed. The conversion process consumes substantial processing and memory resources, and multiple conversion utilities or device drivers may be required.

An **image description language** (IDL) addresses both drawbacks by storing images compactly. Many, but not all, IDLs are device independent. IDLs use compact bit strings or ordinary ASCII or Unicode text to describe primitive image components. IDLs reduce storage space requirements because a description of a simple image component such as a line or shape is usually much smaller than a bitmap of that same component. Examples of modern IDLs include Hewlett-Packard's Printer Control Language and Adobe's PostScript and Portable Document Format. The internal file formats of many drawing and presentation graphics programs are based on proprietary IDLs (e.g., CorelDRAW and Microsoft PowerPoint).

An IDL can represent image components in several ways:

► Embedded fonts

► Vectors, curves, and shapes

► Embedded bitmaps

Modern I/O devices contain embedded font tables that store symbol bitmaps in ROM or flash RAM. Modern operating systems store font tables in files. An IDL can represent text in a specific font by storing the text in ASCII or Unicode along with a reference to the appropriate font style and size. The CPU, or an embedded processor in an I/O device, can shrink or enlarge a character bitmap to any size according to relatively simple image processing algorithms. Although fonts can consume considerable storage, each symbol is stored only once and reused every time that symbol is displayed or printed.

In graphics, a **vector** is a line segment that has a specific angle and length with respect to a point of origin, as shown in the upper half of Figure 7-6. Vectors can also be described in terms of their weight (line width), color, and fill pattern. Line drawings can be described as a **vector list**, a series of concatenated or linked vectors. Such images resemble "connect the dots" drawings, as shown in the lower half of Figure 7-6. Vector lists can represent many complex shapes such as boxes, triangles, and the inner and outer lines of a table or spreadsheet.

Figure 7-6 ►

Elements of a vector (upper) and two complex shapes built from vectors (below)

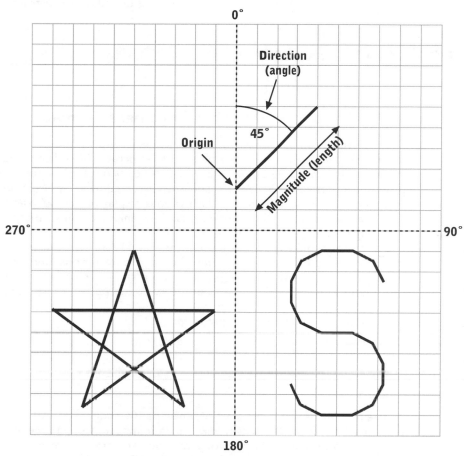

More complex lines and shapes can be constructed with curves. Like a vector, a curve has a point of origin, length, weight, and color. Unlike a vector, a curve has an angle. Curves can be closed and filled to represent objects such as circles and ovals.

Some image components are simply too complex to represent using vectors or curves. An IDL represents complex image components as embedded bitmaps, which may be compressed with GIF, JPEG, or another compression algorithm.

IDLs are a complex form of compression. As with any compression algorithm, storing images with IDLs substitutes processing resources for storage and communication resources. To display or print the desired image, an appropriate program or device driver must process the IDL image description. For video display, the processing may occur in the CPU, in a special-purpose processor embedded in the video controller, or a combination of both. For printed output, processing usually takes place in an embedded processor, though part of the work may be borne by the CPU's execution of device driver instructions.

Technology
Focus

Adobe PostScript and Portable Document Format

PostScript is an image description language designed primarily for printed documents, although it can also be used to generate video display outputs. PostScript is also a programming language. A PostScript file is a program that produces a printed or displayed image when executed by an appropriate interpreter or compiler. Images may include various display objects such as characters, lines, and shapes. Display objects can be manipulated in a variety of ways including rotation, skewing, filling, and coloring.

PostScript commands are composed of normal ASCII characters and may include numeric or textual data, primitive graphic operations, and procedure definitions. Data constants are pushed onto a stack. Graphic operators and procedures pop their input data from the stack and push results, if any, back onto the stack.

Consider the following short program:

```
newpath
400 400 36 0 360
arc
stroke

showpage
```

The first line declares the start of a new path—straight line, curve, or complex line. The second line lists numeric data items (control parameters) to be pushed onto the stack. The control parameters are removed from the stack by the predefined procedure named arc (the third line) as they are used. The left two parameters specify the origin (center) of the arc as row and column coordinates. The center parameter specifies the radius of the arc in points. The remaining parameters specify the starting and ending points of the arc in degrees. In this case, the starting and ending points represent a circle.

Primitive graphic operations are specified in an imaginary drawing space containing a grid (rows and columns) of points. The first three commands specify a tracing within this space. The fourth command states that this tracing should be stroked (drawn). Because no width is specified, the circle is drawn using the current default line width. The final command instructs the output processor to display the contents of the page as defined so far. A PostScript printer or viewing program would execute the program above to generate a single page containing one circle.

Figure 7-7 shows a PostScript program that combines text and graphics. The first two sections define the menu box outline and title separator line with primitive vector and shape drawing commands. The third section defines the text content of the menu and the display font and font size. The last section builds a complex object (the arrow pointer) by defining its outlines, closing the shape, and filling it.

Figure 7-7 ▶

PostScript program that generates the pop-up menu shown on the right

```
% outline box
newpath
245 365 moveto
340 365 lineto
340 265 lineto
245 265 lineto
closepath
stroke

% horizontal line
newpath
245 345 moveto
340 345 lineto
stroke

% text items
/Times Roman findfont 12 scalefont setfont
250 350 moveto (Applications) show
250 330 moveto (Word Processing) show
250 315 moveto (Spreadsheet) show
250 300 moveto (Database) show
250 285 moveto (Graphics) show
250 270 moveto (Games) show

% arrow
newpath
240 305 moveto
-6 6 rlineto
0 -3 rlineto
-20 0 rlineto
0 -6 rlineto
20 0 rlineto
0 -3 rlineto
closepath
fill

showpage
```

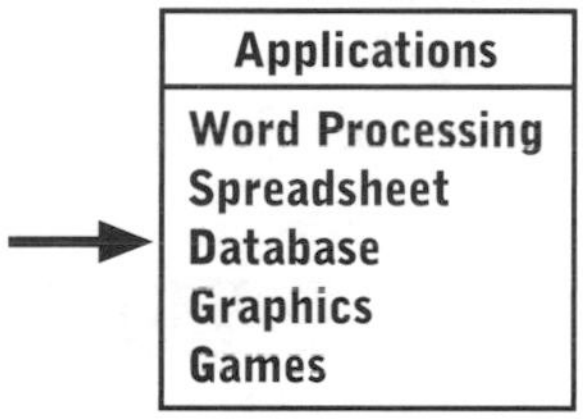

By 1990, PostScript was widely used in the printing and publishing industries. PostScript was also commonly employed as a graphic file interchange format and an embedded printer technology. PostScript printers have special purpose computers to execute PostScript programs and generate printed pages. Some operating systems used PostScript to support graphical user interfaces, though no current operating systems do so.

Although PostScript was successful, it lacked many features that were needed to generate and manage documents as an integrated whole, rather than as a collection of independent images and pages. Missing features included:

▶ Table of contents with page links

▶ Hyperlinks to other documents, Web pages, or active objects such as scripts and executable programs

- ► Annotations and bookmarks
- ► Document properties such as authorship, copyright, and summary
- ► Digital signatures

In the early 1990s, Adobe Systems Incorporated developed the **Portable Document Format (PDF)**, a superset of PostScript. To ensure that PDF would be widely adopted, Adobe developed the free Adobe Reader program that can be installed with a variety of operating systems and Web browsers. Adobe also developed many tools to create PDF documents. Adobe Acrobat Capture enables organizations to convert paper documents to PDF format and store them in searchable archives. Adobe PDF Writer enables users to create PDF documents from desktop word processing and graphics programs such as Microsoft Office and Corel Draw. PDF Writer is installed as a printer driver, and application programs create PDF document files by printing to this driver.

PDF has been widely adopted, especially for documents distributed via the Web. PDF provides a capability that was previously unavailable using standard Web protocols—the ability to distribute compressed documents with complete authorial control over the exact format of the printed and displayed document, regardless of the end user's specific computer, operating system, or printer.

VIDEO DISPLAY

Video display devices have changed greatly since their first widespread use in the middle to late 1960s. Early devices were little more than modified television sets that could display monochrome characters. Modern video displays are much more sophisticated devices. They are larger, have higher resolution, and can display color and graphic images as well as text.

Character-Oriented Video Display Terminals

The first computer video display devices consisted of an integrated keyboard and television screen called a **video display terminal (VDT)** or simply a **terminal**. VDTs were the most common form of video display throughout the 1970s and much of the 1980s until personal computers came into common use. Today, they are used primarily in systems such as retail checkout counters and factory floor environments.

A video display terminal consists of five functional components, as shown in Figure 7-8. The physical keyboard is connected to a keyboard driver. The display generator produces the image actually viewed by a user. Display generators are available in a variety of types and capabilities, which will be described later in this chapter.

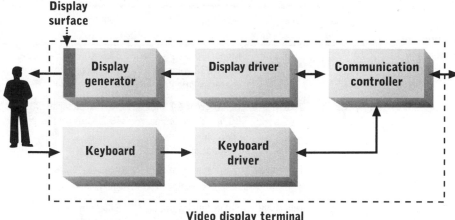

Figure 7-8 ►

Functional components of a video display terminal

Display surface

| Display generator | Display driver | Communication controller |

| Keyboard | Keyboard driver |

Video display terminal

In early VDTs, the display driver contained relatively simple electronic circuitry that translated ASCII or EBCDIC characters into a corresponding pixel matrix, similar to the examples in Figure 7-3, which was then passed to the display generator. In modern VDTs, the display driver is an embedded computer with its own RAM, ROM, and microprocessor. The more sophisticated display driver provides capabilities such as multinational character display, font selection, downloadable fonts, and line drawing.

The communication controller is also a much more sophisticated device. Early VDTs were limited to relatively slow serial transmission standards. Modern VDTs often communicate with computer systems using more advanced standards such as Universal Serial Bus, IEEE Firewire, and Ethernet. Although VDT I/O doesn't generally require the higher I/O throughput of modern communication and network protocols, higher throughput can enable VDTs to communicate with distant computer systems over existing communication infrastructure.

In the 1990s, a new class of VDT appeared, commonly called a **network computer** or **thin client**. A thin client is a hybrid device with a mix of VDT and microcomputer characteristics. Thin clients execute applications within an operating environment such as Java, a Web browser, or Windows Terminal Services. They need sufficient memory and processing power to execute the client end of one or all of those environments, but that power is considerably less than needed by a typical microcomputer.

Operating software is stored in flash RAM or downloaded from a server when the thin client is powered up. All application software and data are stored on a server and accessed via a high-speed communication link or a local area network. Table 7-2 summarizes characteristics of representative VDTs and thin clients manufactured by Wyse Technology.

Table 7-2

Wyse
Technology
VDTs and thin
clients

Model	Type	Display	Communications	Operating Environment
WY-60	VDT	Monochrome, 26 lines, 80 or 132 columns, character-oriented	38.4 Kbps serial with 19.2 Kbps printer port	N/A
WY-325	VDT	Color, 26 or 44 lines, 80 or 132 columns, character-oriented, multinational	38.4 Kbps serial with 19.2 Kbps printer port	N/A
3125SE	Thin client	Color, full-featured graphics display	Ethernet, serial, USB, and printer ports	Windows, Web browser, Java
5455XL	Thin client	Color, full-featured graphics display	Ethernet, serial, USB, PCMCIA, and printer ports	Linux, Web browser, Java, X-terminal

Video Controllers

A modern video display device is called a video monitor or simply a **monitor**. A monitor is connected to a **video controller** that is, in turn, connected to the system bus (see Figure 7-9). Some video controllers are embedded within the monitor, though most are directly attached to a bus port. The video controller accepts commands and data transmitted via a bus from the CPU and generates a TV-style analog video signal, which is transmitted continuously to the monitor.

Figure 7-9

Video controller with monitor and bus connections

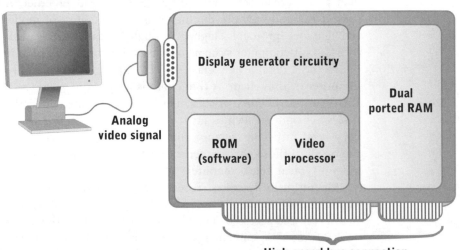

Analog
video signal

Display generator circuitry

Dual ported RAM

ROM (software)

Video processor

High-speed bus connection

Modern video controllers are highly sophisticated devices containing RAM, a microprocessor, and embedded software. In the simplest form of video display, the video controller RAM is a cache for video images in transit from primary storage to the monitor. Software modifies the display by updating the corresponding portions of primary storage and initiating a bus transfer to the video controller. The bus interface circuitry of the controller manages the flow of data from primary storage into the cache.

The controller's display generator circuitry continuously reads the cache contents and generates appropriate analog video signals, which are continuously transmitted to the monitor. Each transfer of a full screen of data from the display generator to the monitor is called a **refresh cycle**. Monitors that require analog video signals need tens of thousands of refresh cycles per second to generate high-quality video images. The number of refresh cycles per second is normally stated in Hertz and called the **refresh rate** (for example, a 75 KHz refresh rate).

Some monitor display technologies can accept digital video signals, which simplifies the controller's display generator circuitry. Rather than being sent continuously, signals are only sent when cache contents change. Monitor and video controller manufacturers are still developing standards for digital video signals.

The rate at which displayed images can be updated is determined by the:

► Data transfer rate of the bus

► Size and performance characteristics of the controller's RAM

► Data transfer rate of the analog or digital video signal

► Number of pixels and chromatic resolution of the monitor

To increase the video data transfer rate, many computer systems have a separate high-speed bus for data transfers from the CPU or primary storage to the video controller.

Video RAM, sometimes called **VRAM**, is different than ordinary RAM because it can be written by the bus interface circuitry or video processor while it is being simultaneously read by display generator circuitry. Simultaneous read/write capability is sometimes called **dual porting**. Modern video controllers contain at least 16 megabytes of dual-ported RAM, which is sufficient to hold four 1280×1024 24-bit color bitmaps.

The microprocessor and embedded software can serve multiple functions. At minimum, they serve as a cache controller, deciding when to save and when to overwrite portions of the cache. High-performance video controllers use more powerful processors, sometimes called **graphics accelerators**, and software to convert incoming IDL commands into appropriate cache contents. As with other IDLs, video controller IDLs can respond to commands to draw simple and complex shapes, fill them with various patterns or colors, and move them from one part of the display to another. Instead of a continuous sequence of large bit maps, controllers transmit a much smaller set of IDL commands, which allows far more video information to be transmitted across the limited capacity of the system bus.

Video Monitors

Video monitors employ a wide variety of technologies to generate displays, with complex tradeoffs among cost and performance characteristics. The following sections describe most common types of video monitors used for computer display.

CRTs   A **cathode ray tube (CRT)** is an enclosed vacuum tube. An electron gun in the rear of the tube generates a stream of electrons that are focused in a narrow beam toward the front surface of the tube. The interior of the display surface is coated with a phosphor. When struck by a sufficient number of electrons, the phosphor emits light. Pulsing the electron beam (rapidly turning it on and off) as it travels across the inside of the screen controls individual pixel brightness. The phosphor's chemical composition determines the display color.

The electron beam is focused and aimed at specific pixels by large electromagnets, which direct the beam to move continually among pixel locations in a left-to-right, top-to-bottom motion. Color images are generated with a grid of three phosphor coatings (red, green, and blue). Different levels of intensity for each color combine to produce a continuous range of color.

The phosphor emits light for only a fraction of a second after the electron beam strikes it. To display light continuously, a pixel area must be refreshed tens of thousands of times per second. Increasing the refresh rate maximizes brightness and perceived image quality while minimizing eyestrain.

Though CRT technology is employed in many computer video monitors sold today, the technology is relatively old and has some significant disadvantages. Physical size and weight increase as the size of the display area increases. The vacuum tube is made of heavy and breakable glass. Larger screen sizes require larger and deeper vacuum tubes. A typical CRT monitor occupies 0.5 cubic meters or more, a substantial portion of the available space on an average desk.

Another drawback of CRT monitors is their power consumption. CRTs consume several hundred watts of power and dissipate much of that as heat. Power requirements and heat generation increase with screen size and refresh rate. The power consumption of a typical workstation CRT monitor is at least half that of the entire computer system.

Many non-CRT video display technologies are under development or in current production. As a group, they are called **flat panel display** technologies due to their characteristic shape. Extensive research and development activities promise that flat panel displays will surpass the cost/performance ratio of CRTs within the next few years.

LCDs A **liquid crystal display (LCD)** contains a matrix of liquid crystals sandwiched between two polarizing filter panels. A liquid crystal twists or untwists when an electrical charge is applied. Light passing through a twisted crystal is rotated 90 degrees. A polarizing filter blocks all light except that which approaches from a specific angle. The front polarizing panel is rotated 90 degrees from the back panel. Light entering through the rear panel cannot pass through the front panel if power is applied to liquid crystal sandwiched between the layers (see Figure 7-10). Color display is achieved with a white light source and a matrix of red, green, and blue color filters layered over or under the front panel.

Figure 7-10 ►

Light entering through one polarized filter cannot pass through the next rotated filter if liquid crystal in the gap is electrically charged (above). Removing the charge returns the liquid crystal to its twisted state, allowing light to pass (below)

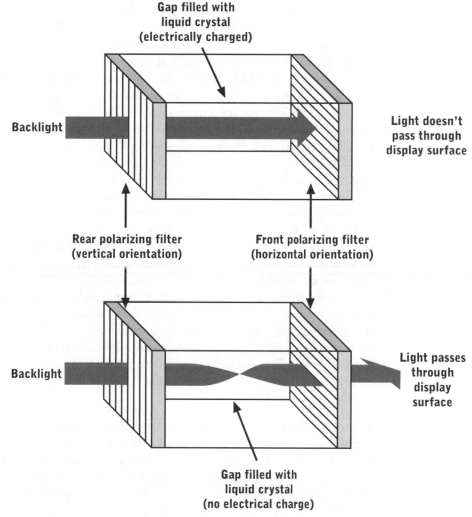

Gap filled with liquid crystal (electrically charged)

Backlight

Light doesn't pass through display surface

Rear polarizing filter (vertical orientation)

Front polarizing filter (horizontal orientation)

Backlight

Light passes through display surface

Gap filled with liquid crystal (no electrical charge)

Applying electrical charge to each display matrix cell requires transistors and wiring. An **active matrix display** uses one or more transistors for every pixel. A **passive matrix display** shares transistors among rows and columns of pixels. Additional transistors enable pixels to be switched on and off more quickly. Also, with transistors dedicated to each pixel, a continuous charge can be generated resulting in a brighter display. However, additional transistors increase the display panel's complexity and cost.

Since the early 1990s, active matrix displays have been manufactured with **thin film transistor** (**TFT**) technology. The wiring and transistors of a TFT display are added in thin layers to a glass substrate. The manufacturing process is similar to that used for semiconductors, with successive layering of wiring and electrical devices added through photolithography. TFT technology enables the reliable manufacture of relatively large displays.

LCD displays of all types have less contrast than CRT displays, because color filters reduce the total amount of light that passes through the front the panel. LCD displays also have a much more limited viewing angle than CRTs. Acceptable contrast and brightness are achieved within a 30-to-60 degree angle. Outside that angle, contrast drops off sharply and colors shift. In contrast, CRT viewing angles usually approach 180 degrees with little loss in contrast or brightness.

The advantages of an LCD display are substantially reduced size, weight, and power consumption as compared with a CRT. Power consumption is generally less than 100 watts for an active matrix display, and much less for a passive matrix display. Weight is relatively low due to small size and the use of only two flat glass or plastic panels. The cost of an LCD display is considerably higher than a CRT of similar size and capability. Because the manufacturing costs of LCD displays have been dropping more quickly than those of CRTs, LCD displays are expected to supplant CRTs within the next few years.

Most current LCDs accept the same analog video signals required by CRTs. However, since individual LCD pixels don't require continuous refresh cycles, the analog signals are converted to digital signals. The conversion process adds unnecessary complexity that can result in slower display updating and reduced image quality. The video display industry has developed standards for digital interfaces between video controllers and LCD monitors that are now appearing in new video controllers.

Plasma Displays A **plasma display** combines elements of CRT and LCD technology. Like LCDs, they're flat panel active matrix devices. Unlike LCDs, they have no backlight and no color filters. Instead, each pixel contains a gas that emits ultraviolet light when electricity is applied (see Figure 7-11). The inner display surface of each pixel is coated with a color phosphor that emits visible light when struck by ultraviolet light.

Input/Output Technology

Figure 7-11 ▶

A plasma
display pixel

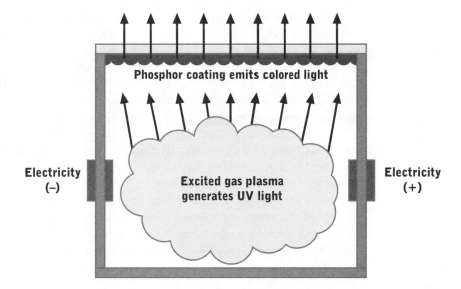

Because they actively generate colored light near the surface of the display, plasma displays have the brightness and viewing angle advantages of a CRT. Plasma displays require more power than LCDs but considerably less than CRTs.

Plasma displays have some shortcomings that limit their use as general-purpose video monitors. The gases and phosphors that generate light have a limited operational lifetime—10,000 to 20,000 hours at present (a year has 8,760 hours). Limited lifetime is a problem with many desktop computer systems and other displays that are used continuously. Also, pixel size is currently larger in plasma displays than in LCDs and CRTs, which reduces comparative image quality when viewed from short distances.

Many companies are investing considerable sums of money in plasma display R&D, driven at present by the market for large-format flat panel televisions. It's likely that some of those R&D results will apply to computer displays, making plasma displays more competitive in the computer monitor marketplace.

PRINTERS

Printer technology can be classified into three common types:

▶ Impact
▶ Laser
▶ Inkjet

Impact technology began with medieval printing presses: a raised image of a printed character or symbol is coated with ink and pressed against paper. The process was updated in the 1880s with the invention of the typewriter, which enabled precise control over paper position and inserted an ink-soaked cloth ribbon between the raised character image and paper. Early printers were little more than electric typewriters that could accept input from a keyboard or a digital communication port.

Later variations on impact printer technology include the line printer and the dot matrix printer. Line printers had multiple disks or rails containing an entire set of raised character images. Each disk or rail occupied one column of a page, enabling an entire row or line to be printed at once. A **dot matrix printer** moves a print head containing a matrix of pins over the paper. A pattern of pins matching the desired character or symbol is forced out of the print head, generating printed images and characters similar to those depicted earlier in Figure 7-3. Some dot matrix printers are still produced, customized for high-speed printing of preprinted multicopy forms.

Inkjet Printers

An **inkjet printer** prints with liquid ink placed directly onto paper. The printer has one or more disposable inkjet cartridges containing one or more large ink reservoirs, a matrix of ink nozzles, and electrical wiring and contact points to enable the printer to control the flow of ink through each nozzle. Each nozzle has a small ink chamber behind it that draws fresh ink from one of the ink reservoirs.

A small drop of ink is forced out of the nozzle in one of two ways:

► Mechanical movement—the back of the chamber is a piezoelectric membrane. When electrical current is applied to the membrane, it bends inward, forcing a drop of ink out of the nozzle. When the current is removed, the membrane returns to its original shape, creating a vacuum in the chamber that draws in fresh ink.

► Heat—the back of the chamber is a resistor that heats rapidly when electrical current is applied (Figure 7-12a). The ink forms a vapor bubble (Figure 7-12b), forcing a drop of ink out of the nozzle. When the current is removed, the vapor bubble collapses (Figure 7-12c), creating a vacuum in the chamber that draws in fresh ink.

Paper is fed from an input hopper and positioned so that the print nozzles are along its top edge. As the print head is drawn across the paper width by a pulley, the print generator rapidly modulates the flow of current to the ink chamber membranes, generating a pattern of pixels across the page. The paper is then advanced and the process is repeated until the entire sheet has been drawn past the moving print head.

Figure 7-12 ▶

Ink drop
formation and
ejection in a
thermal inkjet
chamber

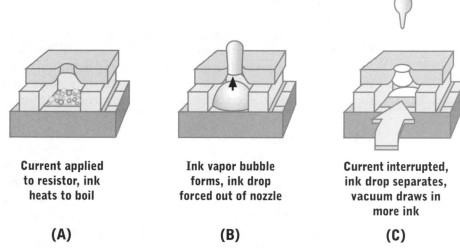

Current applied to resistor, ink heats to boil	Ink vapor bubble forms, ink drop forced out of nozzle	Current interrupted, ink drop separates, vacuum draws in more ink
(A)	**(B)**	**(C)**

Courtesy Hewlett-Packard Company. Reproduced with permission.

Inkjet printer nozzles can produce ink dots as small as one six-hundredth of an inch in diameter (resolution is up to 600 dpi). Color output is generated with CMY inks and alternating rows of nozzles for each ink color. As each row passes a horizontal point on the page, a different color of ink is applied. The inks mix to form printed pixels of various colors. Most color inkjet printers use a separate ink cartridge for black.

Inkjet printer technology was developed in the 1980s and has advanced rapidly to become the most common printing technology. Typical output speeds for inexpensive models are 4-6 pages per minute for black and 2-4 pages per minute for full color. Black and white print quality is good to excellent, and color print quality is often better than that of color laser printers. Inkjet printers are the only type of printer capable of producing high-quality color output within the budget of a typical home computer or small office user.

Printer Communication

Communication between a computer system and an impact printer usually consists of ASCII or Unicode characters, because characters and symbols are the fundamental output unit. An impact printer can print each character as it is received, though small buffers are commonly employed to improve performance.

Inkjet and laser printers use pixels as the fundamental output unit. Thus, more complex communication methods are required. Inkjet and laser printers have relatively large buffers to hold a line, multiple lines, or an entire page of printed output (see Chapter 6, Figure 6-8). An embedded processor and software decide how to store incoming data in the buffer. In the case of bitmaps, the data is simply stored in the appropriate buffer location.

Image description languages are commonly employed to improve printer performance. The processor interprets the IDL shape and line drawing commands, generates the corresponding bitmaps, and stores them in the appropriate part of the buffer. In the case of character-oriented data, the processor generates a bitmap for each printed character according to current settings for font style and size.

Laser Printers

A **laser printer** operates with an electrical charge and the attraction of ink to that electrical charge, as illustrated in Figure 7-13. First, a rotating drum is lightly charged over the width of its surface. The print driver then reads rows of pixel values from the buffer and modulates a tightly focused laser over the width of a metal drum. The drum is then advanced and the process repeated with the next line of pixels. The laser removes the charge wherever it shines on the drum. The drum contains an image of the page with charged areas representing black pixels and uncharged areas representing white pixels. After charging, the drum then passes a station where fine particles of toner, a dry powder ink, are attracted to the charged areas.

Figure 7-13 ►

Components of a laser print engine

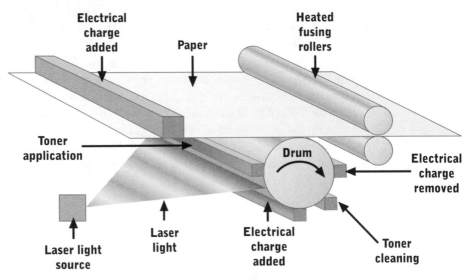

In synchronized motion with the drum, paper is fed through a series of rollers and given a high electric charge. As the paper passes over the drum surface, toner on the drum is attracted to the paper due to its higher charge. The paper and attached toner then are fed through heated rollers that fuse the toner to the paper surface. A light source then removes all charges from the drum and excess toner is removed from the drum by a fine blade and/or vacuum.

Input/Output Technology

Color laser output uses three separate print generators (laser, drum, and laser modulators), one for each color. Paper must be aligned precisely as it passes over the three drums. Color laser printers are much more complex and expensive than their monochrome counterparts. Color can also be generated with three passes over the same print generator with different color toner applied during each pass. This method reduces the printer's complexity, cost, and (unfortunately) maximum speed.

Monochrome laser printers are found in most modern offices. Price reductions have made them affordable for most home and office users. Print speeds range from a few pages per minute to several dozen pages per minute.

Plotters

A **plotter** is a printer that generates line drawings on wide sheets or rolls of paper. Plotters can handle paper widths up to 60 inches, which makes them ideal for producing blueprints and other engineering drawings.

Plotters vary in drawing technology, printing speed, and paper width. Pen-based printing technology was the norm until the mid-1990s. Pen-based technology draws images using one or more moving pens that are raised and lowered onto paper.

Modern plotters use inkjet technology and are little more than large-format inkjet printers that can produce any large output including posters and banners. Comparably sized laser print engines are too expensive to compete with inkjet plotters. The term plotter is rapidly fading in favor of the more descriptive term **large format printer**.

MANUAL INPUT DEVICES

Manual input devices include keyboards, mice, pressure-sensitive pads, and many less familiar but related devices. Until the 1980s, keyboards were the predominant form of manual input. Pointing devices such as mice were introduced in the 1980s, and many newer technologies have arisen in the last decade. This section will concentrate on the most common manual input devices.

Keyboards

Early computer systems accepted input via punched cards or tape. Keypunch machines converted manual keystrokes into punched holes in a cardboard card or in paper tape. Another device called a card reader converted punched cards into electrical inputs that could be recognized by the CPU. Punched cards were passed over a light source within the card reader. Light shining, or not shining, through the punched holes was detected and interpreted as input characters.

Modern keyboard devices translate keystrokes directly into electrical signals, eliminating the need for intermediate storage media such as punched cards or paper tape. Modern keyboards use an integrated microprocessor, also called a **keyboard controller,** to generate bit stream outputs. Pressing a key sends a coded signal to the controller, and the controller generates a bit stream output according to an internal program or lookup table.

A modern keyboard has many special-purpose keys including function keys such as F1, Print Screen, and Escape; display control keys such as ↑, Page Down, and Scroll Lock; and modifier keys such as Shift, Caps Lock, Control, and Alt. There are many valid combinations of two or more keys, such as Control+Alt+Delete. The large number of keys and key combinations makes coding keyboard outputs via ASCII or Unicode impractical. There simply are not enough unused codes in those coding tables to accommodate all keys and key combinations.

A keyboard controller generates output called a **scan code** when one or more keys are pressed. A scan code is a one- or two-byte data element that represents a specific keyboard event. For most keys, the event is simply a key press. For some keys, pressing and releasing the key may represent two different events, generating two different scan codes. The widespread use of IBM-compatible personal computers has created a few de facto scan code standards including those based on keyboard controllers used in IBM personal computers. Most modern keyboards follow the IBM PS/2 standard.

The keyboard can be connected to the computer system in various ways. Most microcomputers connect to keyboards via a wire and connector based on the IBM PS/2 standard. Some computer systems vendors use proprietary interfaces, though most use PS/2 keyboards because they are inexpensive and widely available. Some newer keyboards use Universal Serial Bus (USB) connections, and others use wireless connection standards such as Bluetooth.

Many people have predicted the demise of keyboards over the last few decades under the assumption that a combination of speech recognition, optical scanning, and character recognition technology would soon fill the keyboard's role (those technologies are discussed later in this chapter). Though all of those technologies have advanced considerably, they aren't sufficiently developed to match the accuracy and speed of a well-trained typist. Pending substantial improvements in competing technologies, keyboards will be with us for many years to come.

Pointing Devices

Pointing devices include the mouse, trackball, joystick, and digitizer tablet. The devices all perform a similar function: translating the spatial position of a pointer, stylus, or other selection device into numeric values within a system of two-dimensional coordinates. Pointing devices can be used to enter drawings into the system or to control the position of a **cursor,** or pointer, on a display device.

A **mouse** is a pointing device that is moved on a flat surface such as a table, desk, or rubber pad. The position of the mouse on the surface corresponds to the position of a cursor or pointer on a video display. As the mouse is moved left or right, the pointer on the screen is moved left or right. As the mouse is moved toward or away from the user, the pointer is moved toward the bottom or top of the display.

Mouse position is translated into electrical signals by mechanical or optical components. Most mechanical mice have a roller ball in contact with two wheels mounted on pins. As the mouse moves, the roller ball moves one or both wheels. Wheel movement generates electrical signals that are sent to a device controller and then converted to numeric data describing the direction and distance the mouse has been moved.

An optical mouse uses an optical scanner that constantly scans the surface directly under the mouse at a high resolution. A microprocessor within the mouse compares many scans per second to determine the direction and speed of movement. This approach works on most surfaces including those of apparently uniform color and texture. When scanned at very high resolutions (for example, 600 dpi), most surfaces have discernable texture and color or intensity variation. Optical mice have no moving or exposed parts that can become contaminated by dust or dirt. Some optical mice have two scanners for more accurate tracking.

A mouse typically has one to three buttons mounted on the top. The buttons are clicked to set the position of the cursor in the display, make selections of text or menu items, open programs, or display pop-up or context menus. Some have a scroll wheel on or near the buttons that the user can use to scroll a menu or window. Mouse buttons are simple electromechanical switches. Users select text, menu items, or commands by clicking or double-clicking a single button, or by pressing two buttons at once.

The latest generation of mice and related pointing devices uses embedded gyroscopes and communicates without wires. A gyroscope can detect motion within 3-dimensional space, which frees the pointing device from the confines of a two-dimensional surface such as a desktop. Gyroscopic pointing devices are ideal for lectures and presentations because they provide the user with freedom of movement.

A trackball is essentially an upside-down mechanical mouse. The roller ball is typically much larger than in a mechanical mouse and the selection buttons are mounted nearby. The roller is moved by the fingertips, thumb, or palm of the hand. Trackballs require less desktop space than a mouse. Trackballs are also easy to incorporate into other devices such as keyboards and video games. Because they require minimal finger, hand, and arm dexterity, trackballs are ideal for young children and persons with certain physical disabilities, though ease of use generally requires a relatively large ball.

A **digitizer** consists of a digitizing tablet and a pen, stylus, or both (see Figure 7-14). The tablet is sensitive to the placement of the stylus or pen at any point on its surface. Drafters or artists trace blueprints or drawings on the digitizer tablet or use it for freehand sketching. The stylus enables drawing, tracing, and command functions to be combined in a single device; note the mouse-like buttons on the stylus in Figure 7-14. Tablet PCs use a similar input and display surface but omit the stylus and add handwriting recognition software for text entry.

Figure 7-14 ▶

A digitizing tablet, stylus, and pen

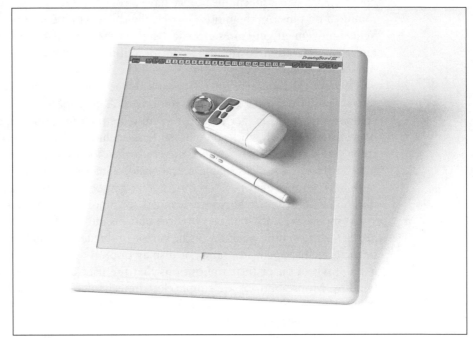

Courtesy GTCO CalComp Peripherals

Digitizing tablets and tablet PCs are examples of a general class of input devices known as **input pads**. Such pads are the basic technology behind other devices such as signature pads, mouse pads, and touch screens. There are three common input pad technologies:

▶ Infrared detector

▶ Photosensor

▶ Pressure-sensitive pad

Infrared sensing is the oldest of these technologies and was commonly used in devices such as touch screens, where the pressure sensing mechanisms must be invisible. An array of infrared beam generators is placed along two adjacent edges of the screen or pad. Arrays of infrared sensors are placed along the other two edges, precisely aligned with the infrared beam generators. When a finger or object touches the pad it interrupts one or more of the beams. Each sensor

Input/Output Technology

generates a binary input to a microprocessor that compares inputs from all sensors to determine where the touch occurs. Infrared technology is fading in favor of alternative TFT-based technologies.

An input pad can also be composed of a two-dimensional array of photosensors. A light pen that generates laser light causes one or more photosensors in the pad to generate a signal when the pen is placed against a point on the pad. Again, a microprocessor compares inputs from all the photosensors to determine the pen position. Photosensing provides relatively high precision and resolution, but is subject to interference from dirt and other contaminants.

Many large input pads contain four separate pressure-sensitive ribbons at each edge of the pad. Each ribbon transmits a separate signal indicating the relative amount of pressure at one edge of the pad. A microprocessor compares the pressure levels for all four ribbons to accurately determine where on the pad pressure from a finger, stylus, or pen is being applied. For example, applying pressure to the center of a square pad will result in equal pressure on all four ribbons. Applying pressure at one corner of the pad will result in equally high pressure on the two ribbons at that corner and equally low pressure on the two ribbons at the opposite corner.

Smaller pads, such as those used for laptop mouse pads or signature pads are manufactured with TFTs. One approach uses an array of sensors that detect interruptions in a weak magnetic field. A finger or a metal stylus interrupts the magnetic field at a specific point, generating a pattern of electrical output in the sensors. Another approach employs an array of pressure-sensitive resistors that vary current flow according to the physical pressure applied by any object. With either approach, a microprocessor constantly compares input from the entire array of sensors or resistors to determine the precise location of the finger or stylus.

IBM defined the most common connection standard for mice and other pointing devices for its PS/2 series of personal computers in the late 1980s. Although the PS/2 computer line is long gone, the standard connector and signaling scheme endures. Within the last few years, other connection standards have emerged, including USB and wireless.

OPTICAL INPUT DEVICES

Optical input devices have come of age in the last decade with the introduction or refinement of devices such as advanced bar-code readers, low-cost optical scanners, and digital still and motion cameras. Optical input devices can be classified into two broad categories based on their application and underlying technology:

▶ Mark and pattern sensors
▶ Image capture devices

Mark and pattern sensors capture input from special-purpose symbols placed on paper or on the flat surfaces of 3-dimensional objects (for example, the shipping label on a box). Specific devices include magnetic mark sensors and bar code readers. Image capture devices, which are relatively new, include digital still and motion video cameras. These devices are based on more complex technology than mark sensors and bar-code readers.

Photosensors are common to both device types. A **photosensor** converts incoming light energy into outgoing electrical energy. Light reflected from a mark, symbol, or object is reflected into a photosensor. High light intensities induce relatively large current outflows from the photosensor. Low intensities produce little or no current. The current outflow of a photosensor is an analog signal representing light intensity. The analog signal is converted to one of a range of digital numbers. Simple devices such as bar-code readers generate a small range of numbers such as 0-7. More complex devices such as digital cameras classify the analog input within a larger range such 0-255, providing a more accurate digital representation of the light input.

Mark Sensors and Bar-Code Scanners

A **mark sensor** scans for light or dark marks at specific locations on a page. Mark sensor input pages are familiar to students who take standardized multiple-choice tests. Input locations are drawn on a page with circles or boxes and are selected by filling them in with a dark pencil. The mark sensor uses preprinted bars on the edge of the page to establish reference points, for example, the row of boxes corresponding to the possible answers to Question #5. The mark sensor then searches for dark marks at specific distances from those reference points. Marks must be of a sufficient size and intensity to be correctly recognized. Older mark sensor devices use magnetic sensing, thus requiring magnetic ink or pencil marks for accurate recognition. Modern mark sensors use fixed arrays of photosensors, which don't require magnetic ink or pencil marks.

A **bar-code scanner** detects specific patterns of bars or boxes. The most common type of **bar code** contains a series of vertical bars of equal length but varying thickness and spacing. The order and thickness of bars is a standardized representation of numeric data. Bar-code readers use one or more scanning lasers to detect the vertical bars. A **scanning laser** sweeps a narrow laser beam back and forth across the bar code. Bars must have precise width and spacing as well as high contrast for accurate decoding.

When a single scanning laser and photosensor are employed, a bar code must be placed at a specific angle in relation to the detector for accurate recognition. Most bar-code scanners employ multiple scanning lasers and photosensors mounted at oblique angles, which enables accurate readings at a variety of physical orientations.

Bar-code readers are typically used to track large numbers of inventory items. Common uses include grocery store inventory and checkout, tracking of packages during shipment, warehouse inventory control, and ZIP code routing of postal mail. The U.S. Postal Service uses a modified form of bar coding with evenly spaced bars of equal thickness but varying height. These appear along the lower edge of an envelope and encode five- and nine-digit ZIP codes (see Figure 7-15).

Figure 7-15 ▶

A sample address with a standard U.S. Postal Service bar code

> The White House
> 1600 Pennsylvania Avenue NW
> Washington, DC 20500
> ‖₁₁‖‖‖₁₁₁‖₁‖₁‖₁‖₁₁‖₁₁₁‖₁‖

Advanced scanning technology can now read two-dimensional bar codes. Figure 7-16 shows a sample 2-D bar code in the PDF417 format. Data is encoded in patterns of small black and white blocks surrounded by vertical bars marking the left and right boundaries of the bar code. PDF417 bar codes can hold approximately 1 KB of data, so descriptive or shipping information can be stored directly on the shipping container or inventory item.

Figure 7-16 ▶

A PDF417 bar code

Optical Scanners

An **optical scanner** generates bitmap representations of printed images. A bright white light shines on the page and reflected light is detected by an array of photosensors. High resolution requires small, tightly packed photosensors. Chromatic resolution is determined by the sensitivity of the photosensors to different light frequencies. Typical desktop scanners have spatial resolution up to 600 dpi and chromatic resolution up to 24 bits. Manual scanners, sometimes called hand scanners, require a user to move the scanning device over a printed page at a steady rate. Automatic scanners use motorized rollers to move paper past a scanning surface or to move the light source and photosensor array under a fixed glass surface.

Optical scanners typically communicate with a computer system using a general-purpose I/O interface such as USB, SCSI, or IEEE Firewire. Those interfaces provide more than adequate data transfer rates for most scanners.

Optical character recognition (OCR) devices combine optical scanning technology with a special-purpose processor or software to interpret bitmap content. Once an image has been scanned, the bitmap is searched for patterns corresponding to printed characters. In some devices, input is restricted to characters in pre-positioned blocks such as the "Amount Enclosed" blocks in Figure 7-17. As with mark sensors, preprinted pre-positioned input blocks provide a reference point for locating individual characters, which simplifies the character recognition process.

Figure 7-17 ▶

Sample form with OCR input fields

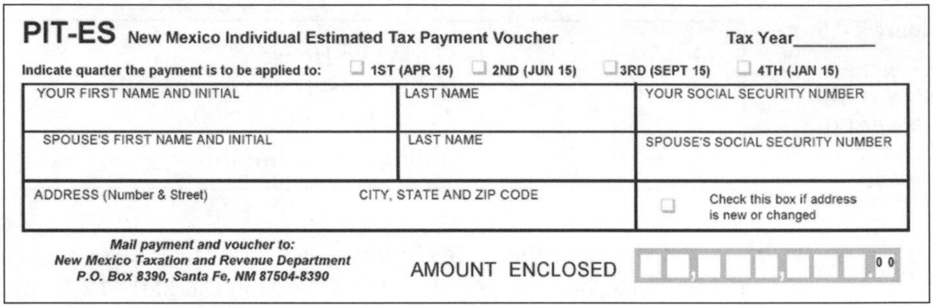

More sophisticated OCR software and hardware place no restriction on the position and orientation of symbols, creating dual problems of locating and recognizing printed symbols. Recognition is most accurate when text is printed in a single font and style with all text oriented in the same direction on the page. If the text is handwritten, contains mixed fonts or styles, is combined with images, or has varying orientations on the page, accurate recognition becomes more difficult. Many inputs are beyond the current capabilities of OCR devices.

Error rates of 10 percent or higher still are common with mixed-font text and even higher with handwritten text. The field has experienced rapid improvement in accuracy and flexibility of input, but much more progress must be made before error rates are acceptable for many types of applications.

Digital Cameras

Digital still and video cameras both employ optical scanning technology to capture images. A two-dimensional photosensor array is placed behind lenses to capture reflected and focused ambient light. The sensors employed in cameras must be more sensitive than those used in optical scanners, since ambient light is usually much dimmer than the bright light employed in scanners. More sensors are required since they must capture an entire image in an instant.

The primary difference between digital still and motion video cameras manifests itself after an image is captured. A digital camera captures one image at a time. When the shutter button is pressed, the camera briefly samples the output of the photosensor array and generates a bitmap. The bitmap is then stored in an internal or removable memory card. To conserve storage space, the user can configure the camera to store images in a compressed format such as JPEG. A digital camera communicates with a computer system over a general-purpose I/O link such as USB or IEEE Firewire.

Both film and digital video cameras capture moving images by rapidly capturing a series of still images called frames. Moving image quality improves as the number of frames per second increases. Digital cameras typically capture 24–30 frames per second. Each frame is stored in a bitmap buffer as it is captured. A separate processor continuously reads buffer content and applies a compression algorithm such as MPEG to the data. The processor can generate better quality compressed images if the buffer is large enough to hold several frames. The compression algorithm can also be configured for various trade-offs between compressed image quality and storage requirements.

Digital video cameras usually have two I/O channels, one for digital output and another for analog TV signals. The digital output channel is usually a standard I/O channel such as USB or IEEE Firewire, though such channels are relatively slow for video data. If the camera has an analog TV output channel, then it also has display generator circuitry to convert the stored image into appropriate television signals. Analog television output quality is generally inferior to digital output since the analog TV standard is nearly half a century old. As with LCD displays, purely digital video connections are under development and will probably supplant analog TV channels within the next decade.

Portable Data Capture Devices

The last decade has seen an explosion of devices that provide rapid data capture for tasks such as warehouse inventory control and package routing and tracking. Most of these devices combine a keyboard, mark or bar-code scanner, and wireless connections to a wired base station or computer system.

Portable data capture devices differ in the degree of computer power embedded within the device itself. Simpler devices function only as an I/O device, sending captured data through a wired or wireless communication channel to a computer system. More complex devices embed computer components such as a CPU and RAM within the device itself. Executable programs are generally developed on a full-featured computer system and then downloaded into the device's flash memory.

AUDIO I/O DEVICES

Sound is an analog signal that must be converted to digital form for computer processing or storage. The process of converting analog sound waves to digital representation is called **sampling**. The content of the audio sound spectrum is analyzed many times per second and converted to a numeric representation. For sound reproduction that sounds natural to people, frequencies between 20 Hz and 20 KHz must be sampled at least 40,000 times per second. The standard for compact disc recordings is 44,100 samples per second.

Sound varies by frequency (pitch) and intensity (loudness). By various mathematical transformations, a complex sound such as a human voice, consisting of many pitches and intensities, can be converted to a single numeric representation. For reproduction that sounds natural to humans, at least 16 bits (2 bytes) must be used to represent each sample. Thus, one second of sound sampled 44,100 times per second requires 88,200 bytes for digital representation. A full minute of sound requires over 5 MB.

Sampling and playback rely on simple devices called analog-to-digital converters and digital-to-analog converters. An **analog-to-digital converter** (**ADC**) accepts a continuous electrical signal representing sound such as microphone input, samples it at regular intervals, and outputs a stream of bits that represent the samples. A **digital-to-analog converter** (**DAC**) performs the reverse transformation, accepting a stream of bits representing sound samples and generating an equivalent continuous electrical signal that can be amplified and routed to a speaker. The conversion hardware, processing power, and communication capacity needed to support sampling and playback have been present in consumer audio devices since the 1980s and in ordinary desktop computers since the early 1990s.

Within a computer system, sound generation and recognition are employed for many purposes, including:

▶ General-purpose sound output, such as warnings, status indicators, and music
▶ General-purpose sound input, such as digital recording for voice email messages
▶ Voice command input
▶ Speech recognition
▶ Speech generation

Limited general-purpose sound output has existed in computer systems for many years. Microcomputers, workstations, and VDTs can generate a single audible frequency at a specified volume for a specified duration of time. This type of sound output is called **monophonic** output because only one frequency (note) can be generated at a time.

As computers have been adapted for multimedia presentation and as general-purpose communication devices, sound generation capability has grown. Most computers now contain **polyphonic**, or multifrequency, sound generation hardware with built-in amplification and high-quality speakers. Modern computers also accept analog or digital sound input from devices such as microphones and CD players.

Speech Recognition

Speech recognition is the process of recognizing and appropriately responding to the meaning embedded within spoken words, phrases, or sentences. Human speech consists of a series of individual sounds called **phonemes**, roughly corresponding to the sounds of each letter of the alphabet. A spoken word is a series of interconnected phonemes. Continuous speech is a series of phonemes interspersed with periods of silence. Recognizing individually voiced (spoken) phonemes is not a difficult computational problem. Sound can be captured in analog form by a microphone, converted to digital form by digital sampling, and the resulting digital pattern compared to a library of patterns corresponding to known phonemes, as shown in Figure 7-18.

Figure 7-18 ►

The process of speech recognition

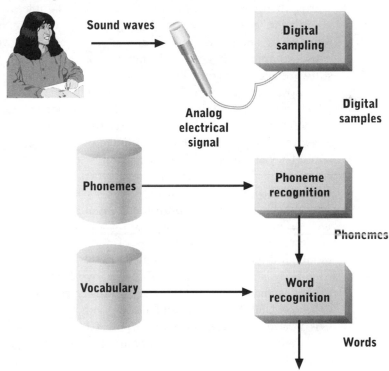

Although the process of speech recognition is conceptually simple, a number of factors complicate it, including:

► Speaker variability
► Phoneme transitions and combinations
► Real-time processing

Speakers vary widely in the exact characteristics of voiced phonemes. Variations arise from physical, geographic, and situational differences. For example, phonemes may differ depending on whether the speaker is male or female, is from Texas or New England, or is making polite conversation as opposed to giving a command. Phoneme matching is inexact because of these variations. Comparison to a library of known phonemes cannot be based on exact match; the closest match must be determined and decisions among multiple possible interpretations must be made. Real-time speech recognition for multiple speakers consumes much of the CPU power available on a typical desktop computer.

Another difficult problem in speech recognition arises from the continuous nature of speech. Individual phonemes sound similar when repetitively voiced by the same speaker. However, when combined with other phonemes in different words, their voicing varies considerably. For example, the letter e is voiced differently in the words "bed" and "neighbor." In addition, a computer must determine where one phoneme ends and another begins, as well as where one word ends and another begins. Performing those tasks in real time requires complex software and powerful CPUs.

Most currently available speech recognition systems are **speaker dependent**, which means they must be "trained" to recognize the sounds of individual speakers. They also are restricted to relatively limited vocabularies, perhaps only a few thousand words. The most limited of these are command recognition systems. They are designed to recognize up to a few hundred spoken words from a single speaker. Such devices are useful when a single user will use the device and when manual input is impractical. These systems have been applied in airplane cockpits, manufacturing control, and input systems for the physically disabled.

A **digital signal processor (DSP)** is a microprocessor specialized to the task of processing continuous streams of audio or graphical data. DSPs are commonly embedded within audio and video hardware such as personal computer sound cards and dedicated audio/video workstations. In some cases, the DSP processes audio input before data is transferred to the CPU. In other cases, the CPU and DSP share processing duties, with each performing the tasks to which it is best suited. Some general-purpose CPUs, such as later generations of the Intel Pentium, now include capabilities that used to be available only in DSPs.

Speech Generation

A device that generates spoken messages based on textual input is called an **audio response unit**. Typical applications include delivering limited amounts of information over conventional telephones, such as automated telephone bank tellers and automated call routing. Audio response units enable a telephone to act as an information system's output device.

Simple audio response units store and play back words or word sequences. Words or messages are digitally recorded and stored. To output a message, the stored message is retrieved and sent to a device that converts the digitized voice into analog audio signals. All possible outputs must be recorded in advance. Phrases and sentences can be generated from individual stored words, though pauses and transitions among the words tend to sound unnatural.

A more general and complex approach to speech generation is **speech synthesis,** in which individual vocal sounds, or phonemes, are stored within the system. Character outputs are sent to a processor within the output unit, which assembles corresponding groups of phonemes to generate synthetic speech. The quality of speech output from such units varies considerably. Creating natural-sounding transitions between phonemes within and between words requires sophisticated processing. However, the difficulties of speech generation are far less formidable than those of speech recognition.

General-purpose audio hardware can be used for speech generation. Digital representations of phoneme waveforms can be combined mathematically to produce a continuous stream of digitized speech, which then is sent to the digital-to-analog converter of the audio hardware. Once converted to an analog waveform, the speech is amplified and routed to a speaker or telephone.

General-Purpose Audio Hardware

General-purpose audio hardware is typically packaged as an expansion card that connects to the system bus of a workstation. Common names for such hardware include **sound card** and **multimedia controller.** The latter term often is used when a compact disc controller and drive are part of the purchased package.

At minimum, sound cards include an ADC, DAC, a low-power amplifier, and connectors (jacks) for a microphone and a speaker or headphones (see Figure 7-19). More elaborate cards may provide:

► Two pairs of converters for stereo input and output

► A general-purpose Musical Instrument Digital Interface (MIDI) synthesizer

► MIDI input and output jacks

► A more powerful amplifier to drive larger speakers and generate more volume

Other differences among sound cards include degree of polyphony (the number of simultaneous sounds that can be generated), supported sampling rates, and the power and speed of embedded DSPs.

As stated earlier, sampling rates of 40,000 Hz or greater are required to capture and reproduce waveforms that sound natural to the human ear. Reduced sampling rates may be adequate when a full range of sound isn't required, such as in speech generation, or when processing power or disk space is limited. Halving the sampling rate also halves processing, data communication, and storage requirements.

Figure 7-19 ►

Components and
connections of a
typical sound
card

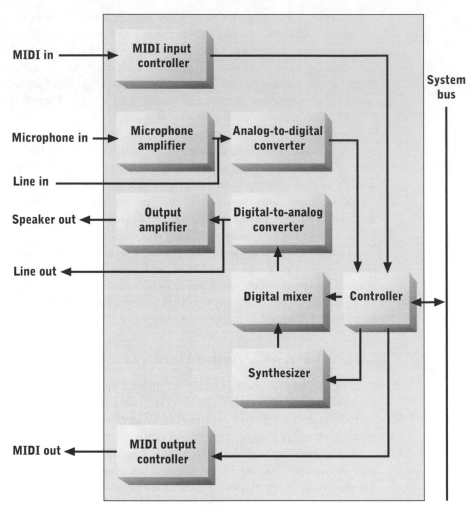

Musical Instrument Digital Interface (MIDI) is a standard for storing and transporting control information among computers and electronic musical instruments. MIDI's original purpose was to enable electronic keyboards to control remotely connected synthesizers. The MIDI standard defines a serial communication channel and a standard set of 1- to 3-byte commands that can:

► Turn individual notes on and off

► Modulate pitch

► Mimic various musical instrument controls, such as the sustain pedal of a piano

► Select specific instruments or artificial sounds

Up to 16 channels of MIDI data can be sent over the same serial transmission line, so 16 separate instruments or instrument groups can be played at once.

Most sound cards include some MIDI capability. At minimum they provide a simple synthesizer capable of simulating the sound of a few dozen instruments. Files containing MIDI control data are transferred to the sound card over the system bus, and then the synthesizer processes the commands and generates digital waveforms for each instrument. The digital waveforms are combined (mixed) and sent through the DAC, amplifier, and speakers.

More complex sound cards provide more synthesized instruments, better-quality waveforms, and the ability to accept or transmit MIDI data through external MIDI connections. The quality of synthesized sound varies widely from one synthesizer and/or sound card to another. The poorest-quality sounds are generated by inexpensive DSPs executing simple instrument simulation programs and operating at low sampling rates. The most realistic sounds are achieved with sample-based synthesis, which uses stored digitized recordings of musical instruments.

The primary advantage of MIDI is its compact storage format. A command to sound a single note on a piano for one second requires 6 bytes: 3 bytes each for the note on and note off control messages. A 16-bit digital recording of that same sound sampled at 44,100 Hz requires 88,200 bytes of storage. The disadvantage of MIDI is a lack of control over the exact nature of the generated sound. A MIDI command selects instruments and notes, but doesn't control how instrument sounds are generated. Sound quality can vary widely among sound cards and synthesizers depending on sample rate, synthesis technique, and hardware capability. The sound for a given instrument may not even exist on the synthesizer receiving the MIDI commands.

SUMMARY

► Display surfaces can be divided into rows and columns similar to a large table or matrix. Each cell, or pixel, in that table represents one simple component of an image. The number of pixels per inch is called the resolution of the display. A stored set of numeric pixel descriptions is called a bitmap. Monochrome display uses a single bit to indicate whether each pixel is on or off. Multiple bits can represent varying levels of intensity, such as a grayscale.

► Any color in the visible spectrum can be represented as a mixture of varying intensities of the additive (RGB) colors red, green, and blue or the subtractive (CMY) colors cyan, magenta, and yellow. Three separate numbers represent a pixel in full color. Significant drawbacks to bitmaps include their large size and device dependence. Compression can be used to reduce size. Image description languages can compactly describe images using vectors, display objects, or both.

► Video display terminals (VDTs) consist of an integrated keyboard and video display surface. VDTs with limited processing capabilities are called network computers or thin clients. Modern video displays are separate from keyboards and are called monitors. A computer system communicates with a monitor via a video controller. Monitors may be implemented as cathode ray tubes (CRTs), liquid crystal displays (LCDs), or plasma displays.

► Printer types include dot matrix, inkjet, and laser printers. Dot matrix printers are inexpensive but are also slow, noisy, and produce relatively poor-quality output. Inkjet printers produce output of excellent quality but are relatively slow. Laser printers are relatively fast and produce excellent-quality output. Modern plotters are similar to inkjet printers except that they print on wide-format paper.

► Manual input devices include keyboards, mice, and input pads. Keyboards are used to enter text and commands. Mice and other pointing devices are used to point to and select buttons or menu items, for drawing, and for moving the position of a cursor. Touch pads can perform many of the same functions as a mouse. Artists and drafters use large-format touch pads called digitizers. Touch pads are also used for devices such as signature pads and touch screens.

► Optical input devices include mark sensors, bar-code readers, optical scanners, and digital cameras. All detect light reflected off a printed surface or object into a photosensor. A mark sensor scans for light or dark marks at specific locations on a page. A bar code is a series of vertical bars of varying thickness and spacing. Optical scanners generate bitmap representations of printed images. Optical character recognition (OCR) devices combine optical scanning technology with intelligent interpretation of bitmap content. Digital cameras capture single or still images and store them as raw compressed bitmaps.

► Sound is an analog waveform that can be sampled and stored as digital data. Computer sound recognition and generation hardware and software can be used to play or record speech and music, synthesize speech from text inputs, or recognize spoken commands or text. Sound cards include an analog-to-digital converter (ADC), a digital-to-analog converter (DAC), a low-power amplifier, and connectors for a microphone, speaker, or headphones. MIDI inputs, MIDI outputs, and a music synthesizer can also be provided. Digital signal processors are commonly used for complex sound processing such as music synthesis and speech recognition.

This chapter discussed how I/O technology supports human/computer interaction. The next two chapters focus on another important class of I/O technology— data communications and computer networks. These technologies enable computers to interact and allow users to interact simultaneously with multiple computer systems. Modern computer systems rely heavily on these technologies to deliver powerful information retrieval and processing capabilities to end users.

Key Terms

24-bit color
active matrix display
additive colors
analog-to-digital converter (ADC)
audio response unit
bar code
bar-code scanner
bitmap
cathode ray tube (CRT)
chromatic depth
chromatic resolution
CMY
CMYK
cursor
digital signal processor (DSP)
digital-to-analog converter (DAC)
digitizer
dithering
dot matrix printer
dots per inch (dpi)
dual porting
flat panel display
font
graphics accelerator
Graphics Interchange Format (GIF)
grayscale

half-toning
image description language (IDL)
inkjet printer
input pad
Joint Photographic Experts Group (JPEG)
keyboard controller
large format printer
laser printer
liquid crystal display (LCD)
mark sensor
monitor
monochrome
monophonic
mouse
Moving Picture Experts Group (MPEG)
multimedia controller
Musical Instrument Digital Interface (MIDI)
network computer
optical character recognition
optical scanner
palette
passive matrix display
phoneme
photosensor

pixel
plasma display
plotter
point
polyphonic
Portable Document Format (PDF)
PostScript
refresh cycle
refresh rate
resolution
RGB
sampling
scan code
scanning laser
sound card
speaker dependent
speech recognition
speech synthesis
subtractive colors
terminal
thin client
thin film transistor (TFT)
vector
vector list
video controller
video display terminal (VDT)
video RAM (VRAM)

Vocabulary Exercises

1. Color display can be achieved by using separate, but closely spaced, elements colored _____, _____, and _____.

2. The display device used in most computer monitors and televisions is a(n) _____.

3. A color display system that uses eight bits to represent each primary video display color intensity is called _____.

4. The printing industry generally uses inks based on the _____ colors, which are _____, _____, and _____. A(n) _____ ink also can be used as a fourth color.

---

5. A(n) _____ is another name for a wide format printer.

6. _____ and _____ are commonly used image description languages.

7. The print head of a(n) _____ forces liquid ink onto a page by heating it or by forcing it out with a piezoelectric membrane.

8. The _____ format is commonly used to compress graphic images.

9. The _____ format is commonly used to compress motion video.

10. A(n) _____ sound card can generate multiple notes simultaneously. A(n) _____ sound card can generate only one note at a time.

11. The _____ of a display device is the number of colors that can be displayed simultaneously.

12. Within a computer, an image is stored as a(n) _____, where each pixel is represented by one or more numbers.

13. A(n) _____ is a primitive component of voiced human speech.

14. A(n) _____ is a small device found within optical scanners and digital cameras that converts light into an electrical signal.

15. When a user presses one or more keyboard keys, the keyboard controller sends a(n) _____ to the computer system.

16. A(n) _____ recognizes input in the form of alternating black and white vertical lines of varying width.

17. A(n) _____ converts analog sound waves to a digital representation.

18. One output cell of a video display surface is called a(n) _____.

19. A(n) _____ illuminates a pixel using excited gas and a colored phosphor.

20. _____ speech recognition programs must be trained to recognize one person's voice.

21. The output resolution of a printer is described by the unit(s) _____.

22. The _____ of a monitor describes the number of times per second that the screen contents are redrawn.

23. A(n) _____ is one seventy-second of an inch.

24. A(n) _____ accepts data or commands over the system bus and generates output for a video monitor.

25. A group of symbols of a similar style is called a(n) _____.

26. A(n) _____ compresses an image by replacing bitmaps, simple lines, and shapes with equivalent drawing commands.

27. _____ is dual-ported, which enables cache contents to be written by a video processor while they're simultaneously being read by display generator circuitry.

28. A(n) _____ is an LCD with one or more transistors to control each display pixel.

29. _____ creates a pattern of colored pixels that fools the eye into thinking a composite color is being displayed.

30. _____ is a standardized method of encoding notes and instruments for communication with synthesizers and sound cards.

31. A modern VDT with limited internal computing power is called a(n) _____ or a(n) _____.

Review Questions

1. Describe the process by which keystrokes are recognized by software.

2. What is a font? What is point size?

3. What are the additive colors? What are the subtractive colors? What types of I/O devices use each kind of color?

4. What is a bitmap? How does the chromatic resolution of a bitmap affect the size of a bitmap image?

5. What is an image description language? What are the advantages of representing images using image description languages?

6. What is JPEG encoding? What is MPEG encoding?

7. Why haven't CRTs been completely replaced by more modern display technologies?

8. Describe the various technologies used to implement flat panel displays. What are their relative advantages and disadvantages?

9. How does the operation of a laser printer differ from that of a dot matrix printer?

10. Describe the various types of optical input devices. For what type of input is each device intended?

11. Describe the process of automated speech recognition. What types of interpretation errors are inherent to this process?

12. Describe the components and functions of a typical sound card. How is sound input captured? How is speech output generated? How is musical output generated?

Research Problems

1. Many organizations have begun to store documents as graphic images rather than encoded textual and numeric data. Examples of applications that use such storage methods include medical records, engineering data, and patents. Identify one or more commercial software products that support this form of data storage. Examine the differences between such products and more traditional data storage approaches. Concentrate on efficient use of storage space, internal data representation, and methods of data search and retrieval. What unique capabilities are provided by image-based storage and retrieval? What additional costs and hardware capabilities are required? Examine sites such as http://www.imagemax.com and http://www.virage.com.

2. Many computer scientists are researching the problem of textual data understanding. The problem of recognizing characters and converting them to ASCII or Unicode is largely solved. However, basic understanding and classification of the textual data still is poorly developed. Examples of applications that could be built with such capabilities include automatic indexing, abstracting, and summarization of news articles or other large textual inputs. Several products that address these needs are emerging from research labs and can be seen at http://ciir.cs.umass.edu and at commercial and governmental domains such as http://www.islanddata.com and http://thomas.loc.gov. Investigate one or more of these products. What limitations must be placed on text input to ensure reasonable success in automated recognition? What methods are used to perform classification and summarization? What is the relative performance, or speed and error rate, of such systems compared to humans performing the same tasks? What traditional information processing applications might use this technology?

3. The field of flat panel display technology changes so rapidly that many of the comparative statements made in this chapter may be out of date before the textbook is printed. Investigate current CRT, LCD, and plasma display product offerings and answer the following questions:

    a. Do CRT-based monitors still hold a price performance edge over LCDs and plasma displays?

    b. Have plasma display operational lifetimes advanced sufficiently to make them viable monitors for most desktop computer systems?

    c. Are any new flat panel display technologies coming into the marketplace?

Chapter **8**

Data and Network Communication Technology

Chapter Goals

- ► Explain communication protocols

- ► Describe signals and the media used to transmit digital signals

- ► Compare and contrast methods of encoding and transmitting data using analog or digital signals

- ► Describe methods for efficiently using communication channels

- ► Describe methods for detecting and correcting data transmission errors

Successful and efficient communication is a complex endeavor. People have long labored with the intricacies of communicating concepts and data. The method of expression (such as words, pictures, and gestures), language syntax and semantics, and communication rules and conventions vary widely across cultures and contexts. This makes successful communication among people a long learning process covering all of these areas.

Understanding computer and network communication can be daunting as well, because these topics also cover a broad range of interdependent concepts and technologies. Some of those topics were addressed in Chapter 7. This chapter extends and deepens that coverage, as shown in Figure 8-1, and lays the foundation for a detailed discussion of computer networks in Chapter 9.

Figure 8-1 ►

Topics covered in this chapter

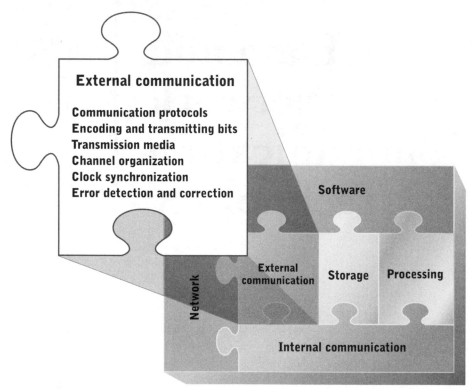

COMMUNICATION PROTOCOLS

A **message** is a unit of data or information transmitted from a sender to one or more recipients. Messages can be loosely categorized into two types—data messages and command messages. Data messages vary in format and content and can include any of the primitive CPU data types or more complex data types (see Chapter 3). For purposes of computer communication, data message content is not important. That is, the message is simply moved from one place to another with no attempt to understand its content.

A command message contains one or more instructions that control some aspect of the communication process. Examples of command messages include many of the ASCII device control characters, addressing and routing instructions, and error detection and correction information. Command messages can also be used to transmit information about data messages such as format, content, length, and other information the receiver needs to interpret the data message correctly.

At the most primitive level, a message is simply a sequence of bits. Sender and receiver must agree on a common method of encoding, transmitting, and interpreting those bits. A **communication protocol** is a set of rules and conventions for communication. Although this definition sounds simple, it encompasses many complex details.

Consider the communication protocol that applies during a classroom lecture or presentation. First, all participants must agree on a common language. The grammatical and semantic rules of the language are part of the classroom communication protocol. For spoken languages, the air in the room serves as the transmission medium. Large classrooms might require auxiliary means to support adequate transmission, such as sound insulation and voice amplification equipment.

Command and response sequences ensure efficient and accurate information flow. The instructor coordinates access to the shared transmission medium. Students use gestures such as raising hands and pointing to ask for permission to speak (transmit messages). Message content is restricted; messages that are off-topic or too long are ignored or explicitly cut off.

Some rules and conventions ensure accurate transmission and reception. For example, participants must speak loudly enough to be heard by all. The instructor might pause and look at the students' faces to determine whether the message is being received and interpreted correctly. The very act of entering the classroom signals participant acceptance of the communication protocol.

Communication among computers also relies on complex communication protocols. Figure 8-2 presents a hierarchical organization of the components of a computer communication protocol. Each node of this hierarchy has many possible technical implementations. A complete communication protocol is a complex combination of subsidiary protocols and the technologies used to implement them. You might find it helpful to return to this hierarchy often while reading the remainder of this chapter so that topics can be placed in their proper context.

Figure 8-2 ▶

Components of a
communication
protocol

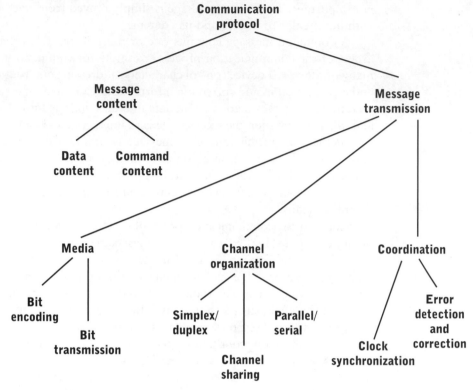

ENCODING AND TRANSMITTING BITS

Chapter 3 described how data items such as integers and characters are encoded within a bit string. This section describes how bits are represented and transported among computer systems and computer hardware components.

Carrier Waves

Light, radio frequencies, and electricity travel through space or cables as a **sine wave** (see Figure 8-3). The energy content of a sine wave varies continuously between positive and negative states. Three characteristics of a sine wave can be manipulated to represent data:

▶ Amplitude
▶ Phase
▶ Frequency

Figure 8-3 ▶

Characteristics
of a sine wave

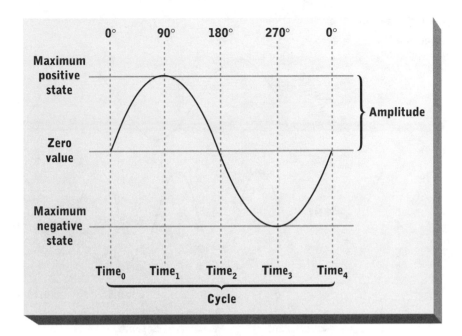

Amplitude is a measure of wave height or power—the maximum distance between a wave peak and its zero value. Amplitude is the same whether measured from zero to a positive peak or zero to a negative peak. A complete **cycle** (also called a period) of a sine wave follows its full range from zero to positive peak, back to zero, to its negative peak, and then back to zero again.

Phase is a specific time point within a wave's cycle. Phase is measured in degrees, with 0° representing the beginning of the cycle and 360° representing the end. The point of positive peak is 90°, the zero point between a positive and negative peak is 180°, and the point of negative peak is 270°.

Frequency is the number of cycles that occur in one second. Frequency is measured in hertz (Hz). Figure 8-4 illustrates two sine waves with frequencies of 2 Hz and 4 Hz.

Waves are important in communications because:

▶ Waves travel, or propagate, through space, wires, and fibers.
▶ Patterns can be encoded within waves.

Figure 8-4 ▶

Sine waves with
frequencies of
2 Hz and 4 Hz

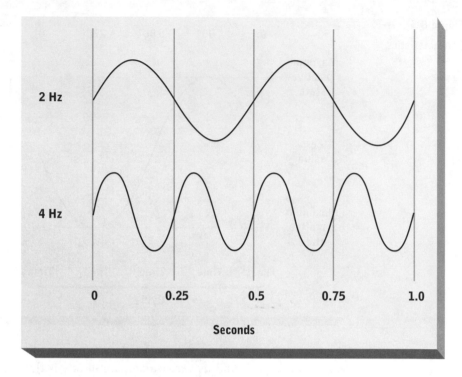

Bits are encoded within a wave by precisely manipulating, or **modulating**, amplitude, frequency, or phase. A wave with encoded bits is called a **carrier wave** because it transports, or carries, bits from one place to another. The receiver observes a modulated carrier wave characteristic and interprets the modulations as bit values. A **signal** is a specific data transmission event or group of events that represents a bit or group of bits.

As a simple example, consider ripples in a pond as carrier waves. Using a stick, the sender can generate waves of varying height that travel across the pond to a receiver on the other side. Strong strikes make large waves and weak strikes make smaller waves. A large wave can be interpreted as a one bit and a small wave as a zero bit, or vice versa. It doesn't matter which interpretation is used as long as the sender and receiver use the same interpretation.

Data can be encoded as bits (large and small waves) by any shared coding method. For example, text messages could be encoded with Morse code (see Table 8-1) with one bits representing dashes and zero bits representing dots (see Figure 8-5). Data could also be encoded using ASCII, Unicode, or any of the other coding methods described in Chapter 3.

Table 8-1 ▶

Morse code

A	•—	N	—•	0	—————		
B	—•••	O	———	1	•————		
C	—•—•	P	•——•	2	••———		
D	—••	Q	——•—	3	•••——		
E	•	R	•—•	4	••••—		
F	••—•	S	•••	5	•••••		
G	——•	T	—	6	—••••		
H	••••	U	••—	7	——•••		
I	••	V	•••—	8	———••		
J	•———	W	•——	9	————•		
K	—•—	X	—••—	?	••——••		
L	•—••	Y	—•——	.	•—•—•—		
M	——	Z	——••	,	——••——		

Figure 8-5 ▶

Wave amplitude
representing
Morse code dots
and dashes

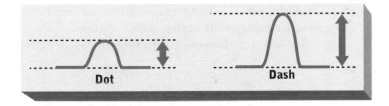

Modulation Methods

Amplitude modulation (AM) represents bit values as specific wave amplitudes. Amplitude modulation is sometimes called **amplitude shift keying (ASK)**. Figure 8-6 shows an amplitude modulation scheme using electricity where 1 volt represents a zero bit value and 10 volts represents a one bit value. Amplitude modulation holds frequency constant while varying amplitude to represent data. The amplitude must be maintained for at least one full wave cycle to be correctly interpreted by the receiver.

Figure 8-6 ▶

The bit string
11010001
encoded in a
carrier wave
using amplitude
modulation

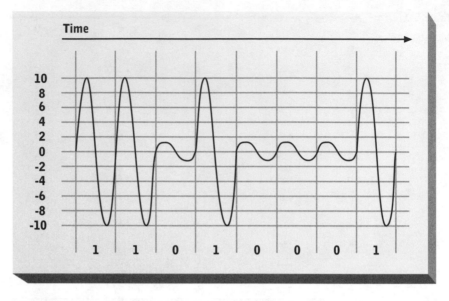

Frequency modulation (FM) represents bit values by varying carrier wave frequency while holding amplitude constant. Frequency modulation is also called **frequency shift keying (FSK)**. Figure 8-7 shows a frequency modulation scheme where 2 Hz represents zero and 4 Hz represents one.

Figure 8-7 ▶

The bit string
11010001
encoded in a
carrier wave
using frequency
modulation

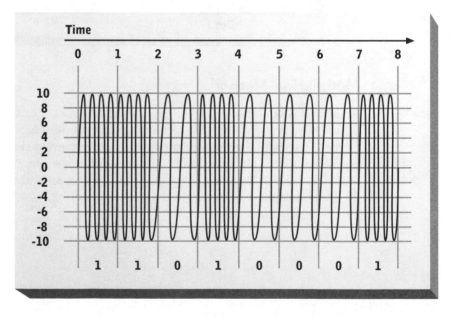

Phase is a wave characteristic fundamentally different from amplitude or frequency. The nature of a sine wave makes it impossible to hold phase to a constant value. However, phase can be used to represent data by making an instantaneous shift in the phase of a signal or by switching quickly between two signals of different phases. The sudden shift in signal phase can be detected and interpreted as data, as shown in Figure 8-8. This method of data encoding is called **phase shift modulation** or **phase shift keying (PSK)**.

Figure 8-8 ▶

The bit string 11010001 encoded in a carrier wave using phase shift modulation

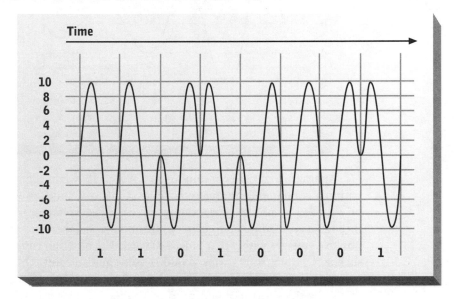

Multilevel coding is a technique for embedding multiple bit values within a single wave characteristic, such as frequency or amplitude. Groups of bits are treated as a single unit for the purpose of signal encoding. For example, two bits can be combined into a single amplitude level if four different levels of the modulated wave characteristic are defined. Figure 8-9 illustrates a four-level amplitude modulation coding scheme. Bit pair values of 11, 10, 01, and 00 are represented by amplitude values of 8, 6, 4, and 2, respectively.

Analog Signals

An **analog signal** uses the full range of a carrier wave characteristic to encode continuous data values. The measured value of a wave characteristic is either equivalent to a data value or can be converted to a data value by a simple mathematical function. For example, assume that data is to be encoded as an analog signal using sound frequency. The numeric value 100 could be encoded by setting the wave frequency to 100 Hz. The number 9999.9 could be encoded by setting the wave frequency to 9999.9 Hz. Data values anywhere within the range of frequencies that can be generated and detected could be encoded within the signal.

Figure 8-9   ▶

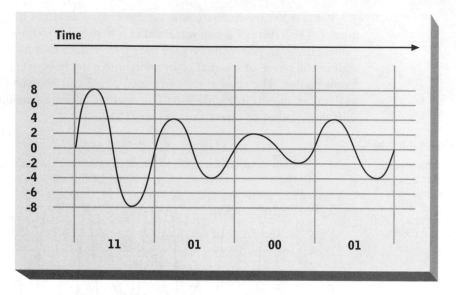

The bit string
11010001
encoded in a
carrier wave
using 2-bit
multilevel
amplitude
modulation

Analog signals are continuous in nature. They can represent any data value within a continuum, or range, of values. In theory, the number of values that can be encoded is infinite. For example, even if the voltage range that can be sent through a wire is limited to between 0 and 10 volts, there are an infinite number of possible voltages within that range—0 volts, 10 volts, and values between 0 and 10, such as 4.1 volts, 4.001 volts, 4.00001 volts, and so forth. The number of different voltages within any range is limited only by the ability of the sender to generate them, the transport mechanism to carry them accurately, and the receiver to distinguish among them.

Digital Signals

A **digital signal** can contain one of a finite number of possible values. A more precise term is **discrete signal**, where discrete means a countable number of possible values. The modulation schemes in Figure 8-6 and Figure 8-7 can encode one of two different values in each signal event and are called **binary signals**. Other possible digital signals include trinary (three values), quadrary (four values), and so forth. The modulation scheme in Figure 8-9 generates a quadrary signal.

Digital signals can also be generated by using a square wave instead of a sine wave as the carrier wave. A square wave contains abrupt amplitude shifts between two different values. Figure 8-10 shows data transmission using an electrical square wave with one bits represented by 5 volts and zero bits represented by no voltage. Square waves are generated by rapidly switching, or pulsing, an electrical or optical power source. Binary data transmission via square waves is sometimes called **pulse code modulation (PCM)** or **on-off keying (OOK)**.

Figure 8-10 ▶

The bit string 11010001 encoded in square waves (digital signals)

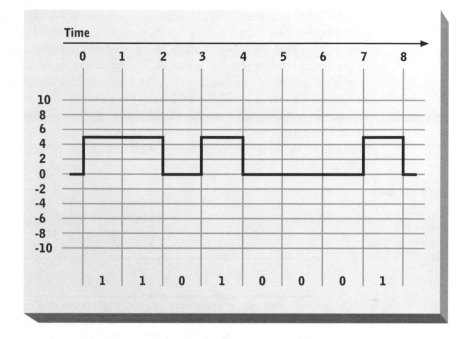

Square waves are the preferred method of digital data transmission over relatively short distances, for example, on a computer system bus. However, electrical square waves cannot be reliably transmitted over long distances, such as those greater than a kilometer.[1] Power loss, electromagnetic interference, and noise generated within the wiring all combine to "round off" the sharp edges of a square wave. Pulses also tend to spread out due to interactions with the transmission medium. Without abrupt voltage level changes, receivers are unable to decode data values reliably. Optical square waves are subject to similar problems, though to a lesser degree. They can be reliably transmitted over distances up to a few dozen kilometers through high-quality optical fibers.

Wave characteristics such as frequency and amplitude are not inherently discrete. For example, electrical voltage can be 0, 0.1, 0.001, 5, 10, 100, and many other values. To send a digital signal with electrical voltage, sender and receiver choose two different values, such as 0 and 5, to represent two different bit values. However, the voltage that reaches the receiver might not be exactly 0 or 5 for a number of reasons that will be discussed shortly. How should the receiver interpret voltage that doesn't precisely match the assigned values for zero and one bits? For example, should 3.2 volts be interpreted as a 0, 1, or nothing?

---

[1] In most communication and network standards, distance is stated in metric units. One meter is approximately 39 inches, and one kilometer is approximately 0.625, or ⅝, of a mile.

A digital signaling scheme defines a range of wave characteristic values to represent each bit value. For example, Figure 8-11 shows two ranges of voltage. Any value within the lower range is interpreted as a zero bit. Any value within the upper range is interpreted as a one bit. The dividing line between the two ranges is called a threshold. The sender encodes a bit value by sending a specific voltage, such as 0 volts for a zero bit and 5 volts for a one bit. The receiver interprets the voltage by comparing it to the threshold. If the received voltage is above the threshold, then the signal is assumed to contain a one bit. If the voltage is below the threshold, then the signal is assumed to contain a zero bit. In Figure 8-11, 0, 0.1, 1.8, and 2.4999 volts would each be interpreted as a zero bit.

Figure 8-11 ▶

A binary signaling method using voltage ranges

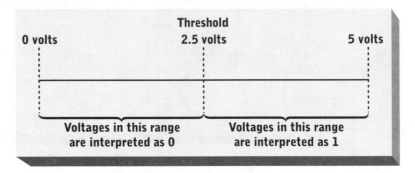

Signal Capacity and Errors

Two important differences between analog and digital signals are their message-carrying capacity and susceptibility to error. Analog signals can carry more information than digital signals within a fixed time interval. Higher data carrying capacity results from the large number of possible messages that can be encoded within an analog signal during one time period.

For example, assume the use of an analog signaling method using electrical voltage with a signal duration of one second. Assume further that the range of voltage that can be reliably transmitted on the wire is 0 to 127 volts, and that sender and receiver can accurately distinguish 1 volt differences. Under these assumptions, there are 128 different signals that can be transmitted during one second. This number would be much larger, theoretically approaching infinity, if sender and receiver were capable of distinguishing smaller voltage differences.

Now consider a binary electrical signal under the previous assumptions, using 64 volts as the threshold. During any single transmission event the number of possible data values sent is only two (either a zero or a one). In this example, the message-carrying capacity of the analog signal is 64 times greater than the binary signal: 128 possible values with analog divided by two possible values with binary.

To transmit values larger than one with binary signals, sender and receiver combine adjacent signals to form a single data value. For example, the sender could transmit seven different binary signals in succession, and the receiver could combine them to form one numeric value. By combining seven consecutive binary signals, it is possible to transmit 128 different messages ($2^7 = 128$ possible combinations of 7 binary signals). But an analog signal could have communicated any of 128 different values in *each* signal.

Although analog signals have higher message-carrying capacity than digital signals, they are much more susceptible to transmission error. If the mechanisms for encoding, transmitting, and decoding electrical analog signals were perfect, this would not be an issue. But errors are always possible. Electrical signals and devices are subject to noise and disruption because of electrical and magnetic disturbances. The noise that one hears on a radio during a thunderstorm is an example of such a disturbance. Voltage, amperage, and other electrical wave characteristics can easily be altered by such interference.

Consider the following hypothetical scenario. Assume that your bank's computer communicates with its automated teller machines via analog electrical signals. You are in the process of making a $100 withdrawal. The value $100 is sent from the automated teller to the bank's computer by a signal of 100 millivolts. The computer checks your balance and decides that you have enough money, so its sends an analog signal of 100 millivolts back to the automated teller as a confirmation. During the transmission of this signal, a bolt of lightning strikes, inducing a 2-volt (2,000 millivolt) surge in the wire carrying the signal. When the signal reaches the automated teller, it dutifully dispenses $2,000 dollars to you in clean, crisp $10 and $20 bills.

The susceptibility of electrical signals to noise and interference can never be completely eliminated. Computer equipment is well shielded to prevent noise from interfering with internal signals, but external communications are more difficult to protect. Also, errors can arise from resistance within wires or from the magnetic fields generated within a device, such as by a transformer or fan.

A digital electrical signal is not nearly as susceptible to noise and interference. Consider the previous example using digital encoding instead of analog encoding. If we use a threshold of 2.5 volts, 0 volts to represent zero, and 5 volts to represent one, a voltage surge less than 2.5 volts can be tolerated without misinterpretation of a signal (see Figure 8-12). If a zero was being sent at the instant the lightning struck, the automated teller would still interpret the signal as a zero because the 2 volts induced by the lightning are below the threshold value of 2.5. If a one was being sent, then the lightning would raise the voltage from 5 to 7 volts, still above the threshold. Resistance in the wire or other factors degrading voltage would not be a problem as long as the voltage did not drop more than 2.5 volts.

Figure 8-12 ►

Margin of
transmission
error (voltage
drop or surge)
before the data
value encoded
within a binary
signal is altered

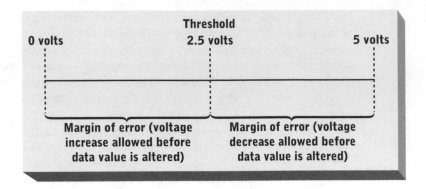

TRANSMISSION MEDIA

The communication path that transports signals is called a **transmission medium**. Copper wire and optical fiber are two types of transmission media. The atmosphere or space can also be a transmission medium for radio or microwave transmissions. Messages must be encoded in signals that can be conducted or propagated through the transmission medium. In copper wires, signals are streams of electrons. In optical fiber, signals are pulses of light. Microwave and satellite transmissions are broadcast as **radio frequency (RF)** radiation through space.

A **communication channel** (see Figure 8-13) consists of a sending device, receiving device, and the transmission medium that connects them. In a less physical sense, it also includes the communication protocol(s) used in the channel. Figure 8-13 implies a relatively simple channel design, but most channels are of complex construction. A complex channel has multiple media segments using different signal types and communication protocols. As a message moves from one media segment to another, it must be altered to conform to the next segment's signal characteristics and protocol.

Figure 8-13 ►

Elements of a
communication
channel

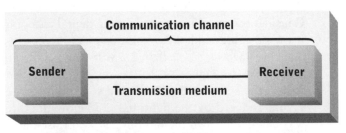

Consider an analog modem connection from a home computer to an Internet service provider (ISP) (see Figure 8-14). Digital electrical signals travel across the system bus to an internal modem. The modem converts the digital signal to an analog signal of limited bandwidth, which then travels through twisted-pair copper wiring to the house telephone interface. From there it travels via lower-gauge (thicker) copper wiring to the nearest telephone pole, and then to the local telephone switching center. The switching center converts the signal to a digital signal and routes it via fiber optic cable to the ISP. Messages sent from the ISP to the home computer travel the same path in reverse.

Figure 8-14 ►

Analog modem connection to an Internet service provider

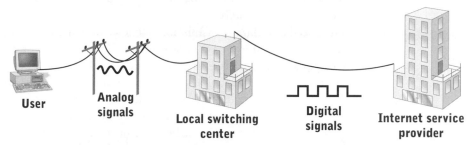

Characteristics of transmission media that affect their ability to transmit messages successfully and efficiently include:

► Speed and capacity
► Bandwidth
► Noise, distortion, and susceptibility to external interference

In general, transmission media are better when their speed, capacity, and bandwidth are high and their noise, distortion, and external interference are low. However, media with all of these desirable characteristics are costly. Reliability, efficiency, and cost also depend on the signaling method and communication protocol. Different combinations of transmission medium, signaling method, and protocol are cost-effective for different types of communication links. Those differences account for the variety of media, signaling methods, and protocols that comprise the communication channel between a modem and an ISP, or between any sender and receiver.

Speed and Capacity

The fundamental transmission speed limit for any medium is the rate at which a carrier wave propagates through the medium. Electrical signals travel through wires at close to the speed of light. Optical signals transmitted through fibers and radio frequencies transmitted through the atmosphere or space also travel at close to the speed of light. Raw speed varies little among commonly used transmission media. What does vary is length of the media, the ways in which multiple media segments are interconnected, and the rate at which bits are encoded in signals and recognized by the receiver. It is these factors that account for transmission speed differences among different media.

Speed and capacity are interdependent. That is, a faster communication channel has a higher transmission capacity per time unit than a slower channel. Capacity is not solely a function of transmission speed. It also is affected by the relative efficiency with which the channel is used. Different communication protocols and different methods of dividing a channel to carry multiple signals alter transmission capacity.

The speed and data transmission capacity of a channel are jointly described as a data transfer rate. A **raw data transfer rate** is the maximum number of bits or bytes per second that the channel can carry. A raw data transfer rate ignores the communication protocol and assumes error-free transmission.

A net or **effective data transfer rate** describes the transmission capacity actually achieved with a particular communication protocol. It is always less than the raw data transfer rate, because no medium is error-free, and because most communication protocols use part of the channel capacity to do things other than transmit raw data. Examples include sending command messages and retransmitting data when errors are detected.

Frequency and Bandwidth

Carrier wave frequency is a fundamental measure of data-carrying capacity. Carrier wave frequency limits data-carrying capacity because a change in amplitude, frequency, or phase must be held constant for at least one wave cycle to be detected reliably by a receiver. If a single bit is encoded in each wave cycle, then the raw data transfer rate is equivalent to the carrier wave frequency. If multiple bits are encoded in each wave cycle via multilevel coding, then the raw data transfer rate is an integer multiple of the carrier wave frequency.

Figure 8-15 shows the range of electromagnetic frequencies and the terms that describe various subsets of that range. Electromagnetic energy can be transmitted through the atmosphere or free space. Wireless networks and cellular telephones use shortwave radio frequency energy transmitted through the atmosphere. Ships sometimes use light to transmit Morse code messages over short distances.

Signals can also be encoded in electromagnetic frequencies transmitted through metal, plastic, or glass wires. Radio frequency energy is typically transmitted through copper wires and light energy is transmitted through plastic or glass fibers. Copper wire can propagate frequencies up to approximately 1 GHz. Optical fibers can propagate frequencies up to 10,000 THz.

Fiber optics have higher potential data transmission capacity because light is a higher-frequency carrier wave. However, note the word "potential" in the previous sentence. Achieving the bit transmission rates implied by carrier wave frequency requires encoding and decoding equipment that can operate at the carrier wave frequency. If the source or destination of the transmission is an electrical device such as a conventional telephone or computer CPU, then signals

Data and Network Communication Technology

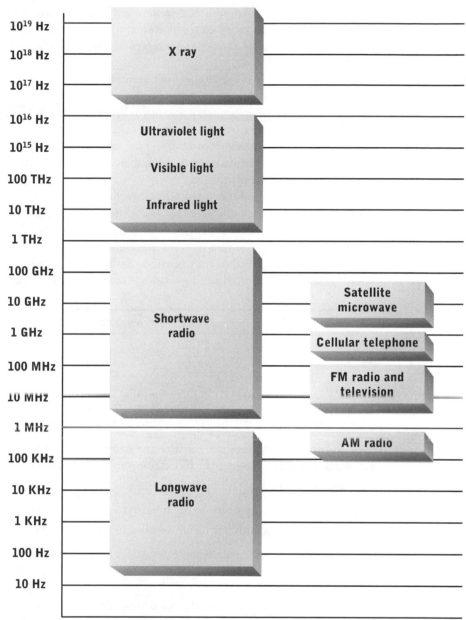

must be converted between electrical and optical formats. Such conversions limit the effective data transmission rate of optical channels.

The difference between the maximum and minimum frequencies of a signal is called the signal **bandwidth**. The difference between the maximum and minimum frequencies that can be propagated through a transmission medium is called the medium bandwidth. Maximum frequency and bandwidth limit the raw transmission capacity of a communication channel. For example, consider the frequency range of an analog voice-grade telephone channel, which is 300 Hz to 3,400 Hz.

The bandwidth of the channel is the difference between the highest and lowest frequencies that can be carried (3400 – 300 = 3100 Hz). In contrast, the frequency range of human speech is approximately 200 Hz to 5000 Hz, a bandwidth of 4800 Hz; the frequency range of the human ear is approximately 20 Hz to 20,000 Hz, a bandwidth of 19,980 Hz. Analog telephone channels can transmit only an approximation of human speech because many high and low frequencies cannot be transmitted.

Modulator-demodulator (modem) technology sends digital signals over voice-grade telephone channels by encoding them in an analog carrier wave. Early modulation techniques used four separate frequencies to carry bit values (see Table 8-2). The lowest of these frequencies (1070 Hz) in effect created an upper limit on the speed of transmission. Assuming that a single bit is transmitted per signal cycle, the maximum transmission rate is 1070 bps.

Table 8-2 ►

Frequency assignments for data transmission over 300-bps analog telephone lines

Mode	Signal Frequency (Hz)	Binary Value
Transmit	1070	0
Transmit	1270	1
Receive	2025	0
Receive	2225	1

Subsequent modem standards have exceeded this theoretical capacity, but they have had to employ ever more complex schemes to do so, including:

► Hardware-based data compression

► Multilevel coding of up to 32 bits in a single carrier wave cycle

► An elaborate combination of amplitude and phase modulation

Current modem transmission rates are as high as 56,000 bps, but the 3100-Hz bandwidth of the voice-grade telephone channel is an effective barrier against higher transmission rates.

Digital signals transmitted over ordinary telephone wire can achieve transmission rates of up to 100 Mbps, which contradicts earlier statements about the relative capacity of analog and digital signals. The contradiction arises because analog telephone signaling standards don't use all of the bandwidth available in a modern telephone circuit. Modern telephone wiring can transmit frequencies much higher than 3400 Hz, but older analog telephone switching equipment was designed to use only the frequencies between 300 and 3400 Hz. Most of this

Data and Network Communication Technology

equipment has been replaced with modern digital switching equipment that converts the electrical voice signals to digital form. Bandwidth above 3400 Hz can be used to transmit digital signals as long as analog voice signals below 3400 Hz are filtered out. Modern digital subscriber lines (DSL) use this method to transmit analog voice and digital data signals over the same telephone line.

Signal-to-Noise Ratio

Within a communication channel, **noise** refers to unwanted signal components added to the original data signal that might be incorrectly interpreted as data. Noise can be heard in many common household devices. Turn on a radio and tune it to an AM frequency (channel) on which no local broadcasting station is transmitting. The hissing sound is amplified background radio frequency noise. Internally generated noise can also be heard on any home stereo system. Set the amplifier or receiver input to a device that isn't turned on, such as a tape or CD player, and turn the volume up relatively high (be sure to turn it back down when you're finished!). The hissing sound is the amplified noise in the signal transmission and amplification circuitry. Any sound other than hissing represents externally generated noise picked up by the amplifier, the speakers, or the cabling among them.

Noise can be introduced into copper, aluminum, and other wire types by **electromagnetic interference (EMI)**. EMI can be produced by a variety of sources, including electric motors, radio equipment, and nearby power transmission or communication lines. In an area that is dense with wires and cables, EMI problems are compounded because each wire both generates and receives EMI.

Attenuation is a reduction in the strength (amplitude) of a signal caused by interactions between the signal's energy and the transmission medium. Attenuation occurs with all signal and transmission medium combinations, though different combinations experience different amounts of attenuation. Electrical resistance is the primary cause of attenuation with electrical signals and transmission media. Both optical and radio signals experience significant attenuation when transmitted through the atmosphere. Optical signals experience low attenuation when transmitted through fiber optic cable. For all signal types, attenuation is proportional to the length of the medium. Doubling the length of the medium doubles the total attenuation.

Another source of errors in communication is **distortion**—changes to the original data signal caused by interaction with the communication channel. The transmission medium itself is a primary source of distortion, though auxiliary equipment such as amplifiers, repeaters, and switches may also distort the signal. Resonances within a transmission medium can amplify certain portions of a complex signal. Attenuation within any medium varies with signal frequency, distorting complex signals with many frequency components.

For a receiving device to interpret encoded data correctly, it must be able to distinguish the encoded bits from noise and distortion. Distinguishing valid signals from extraneous noise becomes increasingly difficult as transmission speed increases. The effective speed limit of a channel is determined by the power of the message-carrying signal in relation to the power of the noise in the channel. This relationship is called the **signal-to-noise (S/N) ratio** of the channel. The S/N ratio is measured at the receiving end of the channel, usually in units of signal power called decibels (dB).

For example, consider listening to a speech. The difficulty or ease of understanding the speech is directly related to the speed at which it is delivered and to the volume (amplitude) of the speech in relation to background noise. Accurate reception of speech is impaired by sources of noise such as other people talking, an air conditioning fan, or a nearby construction project. The speaker can compensate for noise by increasing speech volume, thus increasing the signal-to-noise ratio. Accurate reception is also impaired if the speaker speaks too quickly, because the listener has too little time to interpret one word or speech fragment before the next is received.

The S/N ratio incorporates the effects of signal attenuation and distortion. In Figure 8-16, the S/N ratio is positive for distances up to 5 kilometers. Successful transmission is theoretically impossible beyond 5 kilometers because the S/N ratio is negative.

Each transmitted bit is a period of time during which a signal representing a zero or one is present on the channel. As transmission speed is increased, the duration of each bit in the signal, known as the **bit time**, decreases. If signal-generating equipment could generate a full-amplitude signal instantaneously, then this would not be a problem. But no device, including the human voice, can go from zero amplitude to full power instantaneously. Short bit times don't give the signal-generating device sufficient time to "ramp up" to full power before the next bit must be transmitted. The S/N ratio decreases because the amplitude of each individual signal is decreased. Eventually, a limit is reached at which the bit time is so short that the signal is no louder than background noise. This is the point at which the S/N ratio becomes zero.

As higher speeds are attempted, the error rate increases. A transmission error represents a wasted opportunity to send a message, reducing the effective data transfer rate. A further difficulty is that the noise level may not be constant. In electronic signals, noise usually occurs in short intermittent bursts, for example, due to a nearby lightning strike or the startup of an electric motor. As discussed later in this chapter, the receiver can request retransmission of a message if errors are detected. If noise bursts are infrequent, retransmissions might not diminish the overall data transfer rate significantly, and a relatively high raw transmission speed could be used.

Figure 8-16 ▶

S/N ratio as a
function of
distance for a
hypothetical
channel

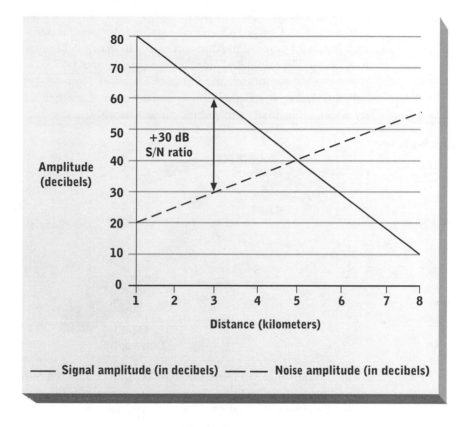

Electrical and Optical Cabling

Electrical signals are usually transmitted through copper wiring. Copper is used because it is relatively inexpensive, highly conductive, and easily formed into wire. Copper wiring varies widely in construction details including gauge (diameter), insulation, purity and consistency, and number and configuration of conductors per cable. The two most common forms of copper wiring used for data communication are twisted pair and coaxial cable.

Twisted pair wire is the most common transmission medium for telephone and local area network connections. It contains two copper wires that are twisted around one another. The wires are usually encased, or shielded, by nonconductive material such as plastic. The primary advantages to twisted pair are its low cost and ease of installation. Its primary disadvantages are a high susceptibility to noise and limited transmission capacity due to low bandwidth, generally less than 250 MHz, and a relatively low amplitude (voltage) limit.

The Electronics Industries Association and the Telecommunications Industries Association have defined the most common standards for twisted pair wire and connectors. The most widely used LAN wiring standard is called Category 5. A **Category 5** cable contains four twisted pairs and can transmit at speeds up to 1 Gbps (250 Mbps on each wire pair). All four wire pairs are bundled in a single thin cable, and standardized modular RJ-45 jacks similar to modular telephone connectors are used at both ends (see Figure 8-17). A newer standard, called **Category 6,** has the same construction as Category 5 but uses higher quality wire and insulators to reduce transmission errors.

Figure 8-17 ►

Common cable connectors

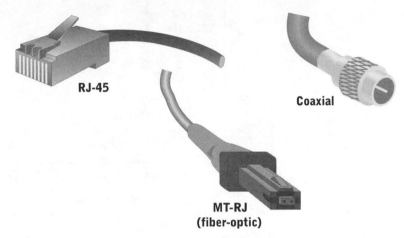

RJ-45

Coaxial

MT-RJ
(fiber-optic)

Coaxial cable contains a single copper conductor surrounded by a thick plastic insulator, a metallic shield, and a tough outer plastic wrapping. Due to its heavy shielding, coaxial cable is very resistant to EMI. It has a relatively high bandwidth of up to 500 MHz and relatively high data transmission capacity. Coaxial cable is widely used for cable television because of its EMI resistance and its ability to carry many channels of analog video signals.

Its primary disadvantages are that it is more expensive and harder to install than Category 5 twisted pair wire. Coaxial modular connectors (see Figure 8-17) are not as easily installed as RJ-45 connectors. Coaxial cable is also very stiff, making it more difficult to route through conduits and around corners than Category 5 wire. Coaxial cable is infrequently used in modern computer networks due to improving twisted pair quality and declining costs of both twisted pair and fiber optic cable and devices.

Twin-axial cable is similar to coaxial cable except that it contains two internal conductors and is thinner. Twin-axial cables support higher data transmission rates and are less susceptible to EMI because both the signal and return wire are heavily shielded. Bundles of twin-axial cables are sometimes used for high-speed network connections over short distances.

Fiber optic cable contains one or more strands of light-conducting filaments made of plastic or glass. Electrical shielding is not used because light waves neither generate nor respond to EMI. A tough, thin plastic coating protects the inner fibers from physical damage. Older fiber optic cables, called **multimode cables,**

surround the glass fibers with a reflective coating called cladding, which acts as a mirror to bounce light waves back into the fiber core. Multimode cables split light waves into multiple signals, each of which travels a different path through the cable. Modern optical cables use fiber cores of varying density so that the core itself contains light waves. Maximum transmission speed is limited because light pulses spread out as they travel through the cable.

The highest quality optical cables are called **single-mode cables**. The fiber core is very narrow and constructed so that all light waves travel a straight path down the center of the fiber. Single-mode cables can transmit data at much higher speeds than multimode cables, but they are also much more expensive. Single-mode cable can transmit data at up to 10 Gbps, and higher speeds will be achieved as signal encoding and decoding devices improve.

The primary disadvantage of fiber optic cable is its high cost. Optical cable is much more expensive than copper wiring. Until recently, optical cable was much more expensive to install than other cable types due to a lack of standardized connectors and lack of personnel trained in splicing techniques. Standardized connectors have been developed (see Figure 8-17), and installation cost now approximates that of twisted pair cable.

No cable type can carry signals more than a few dozen kilometers at high transmission rates with low error rates. Transmitting data over longer distances requires devices such as amplifiers and repeaters to increase signal strength and to remove unwanted noise and distortion.

An **amplifier** increases the strength, or amplitude, of a signal. An amplifier can extend the range of an electrical signal by boosting signal power to overcome attenuation. However, the effective length over which the signal travels cannot be increased indefinitely; it is limited by two factors. The first is noise and interference introduced during transmission. An amplifier amplifies whatever signal is present at its input. If that signal includes noise along with the intended message, the noise is amplified as well as the message. In addition, amplifiers are never perfect. Some distortion or noise is generally introduced in the amplification process. A signal that is amplified many times will contain noise from multiple transmission line segments as well as distortion introduced by each stage of amplification.

A **repeater** performs much the same function as an amplifier but operates on a different principle. It can be used with electrical or optical cable carrying analog or digital signals. Rather than amplifying whatever is sent to it, a repeater interprets the message it receives and retransmits it. A repeater does not retransmit noise or distortion. As long as the noise introduced since the last transmitting or repeating device does not cause a misinterpretation of the message, the retransmitted message will be the same as the original. Repeaters are typically required every 2 to 5 kilometers for coaxial cable and every 40 or 50 kilometers for single-mode fiber optic cable.

Wireless Data Transmission

Wireless transmission uses shortwave radio or infrared light waves to transmit data through the atmosphere or space. The shortwave radio spectrum covers frequency bands commonly used for FM radio, VHF and UHF broadcast television, cellular telephones, land-based microwave transmission, and satellite relay microwave transmission (Figure 8-15). Infrared transmissions occupy higher-frequency bands that can carry more data but are more susceptible to atmospheric interference, limiting transmission distance to a few hundred meters.

The primary advantage of wireless transmission is its relatively high bandwidth and the fact that it avoids wired infrastructure. Another advantage in some situations is broadcast transmission, because many receivers can broadcast and receive simultaneously. The primary weaknesses of wireless transmission are susceptibility to many forms of external interference, the cost of transmitting and receiving stations, high demand for unused radio frequencies, and the ability of anyone with appropriate receiving equipment to listen to the transmission.

Wireless transmission frequencies are regulated heavily, and regulations vary in different parts of the world. The United States, Europe, and Japan each have their own regulatory agencies. Different regulations apply to different transmission power levels. In general, low-power transmission doesn't require a license, but the transmission equipment must be certified to meet appropriate regulations. Current wireless local area network technology uses unlicensed but regulated radio frequency (RF) bands. High-power transmissions, such as long-distance microwave transmission, require certified transmitters and a license to use a specific frequency band. License costs are high because available bandwidth is a limited commodity.

Long-distance wireless networks are seldom implemented by a single user or organization because of the relatively high cost of the necessary transmission equipment and licenses. Instead, companies are formed to purchase and maintain the required licenses and infrastructure. Users purchase capacity from these companies on a fee-per-use or bandwidth-per-time-interval basis.

The Technology Focus on wireless data communication that appears later in this chapter discusses wireless standards and encoding techniques.

CHANNEL ORGANIZATION

A communication channel's configuration and organization have a direct impact on its cost and efficiency. Configuration and organization issues include the number of transmission wires or bandwidth assigned to each channel, the assignment of those wires or frequencies to carry specific signals, and sharing, or lack thereof, of channels among multiple senders and receivers.

Simplex, Half Duplex, and Full Duplex

The most basic form of channel organization for electrical transmission through wires is shown in Figure 8-18. A single communication channel requires two wires: a **signal wire**, which carries data, and a **return wire**, which completes an electrical circuit between the sending and receiving devices. Optical transmission requires only one optical fiber per channel because a complete circuit is not required. In some transmission modes, it may be necessary to provide two or more communication paths between sender and receiver, as shown in Figure 8-19. Multiple electrical transmission lines can share a single return wire to complete all electrical circuits.

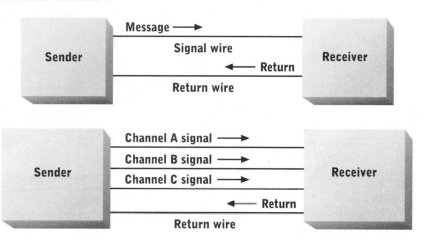

Figure 8-18 ►

Configuration of a two-conductor electrical communication channel

Figure 8-19 ►

Multichannel communication with two or more signal wires and one common return wire

A single communication line can transmit messages in either simplex or half-duplex mode. In **simplex** mode, messages flow in only one direction, as shown in Figure 8-20(a). This is a useful communication mode when data needs to flow in only one direction and when the chance of transmission error is small or irrelevant, such as in ordinary radio broadcasts. However, if transmission errors are common or if error correction is important, then simplex mode is inadequate. If the receiver detects transmission errors, there is no way to notify the sender or to request retransmission. A typical use of simplex mode is to send file updates or system status messages from a host processor to distributed data storage devices within a network. In such cases, the same message is transmitted to all devices on the network simultaneously, or in **broadcast mode**.

Half-duplex mode uses a single shared channel. Each node takes turns using the transmission line to transmit and receive. The nodes must agree which node is going to transmit first. After sending a message, the first node signals its intent to cease transmission by sending a special control message called a **line turnaround**. The receiver recognizes this message and subsequently assumes the role of transmitter. When its transmission is finished, it sends a line turnaround message and the original transmitter once again becomes the sender.

Half-duplex mode transmission allows the receiver to request retransmission of a message if it detects errors, as shown in Figure 8-20(b). If ASCII control characters are used, a **negative acknowledge (NAK)** character is sent if errors are detected, and an **acknowledge (ACK)** control character is sent if no errors are detected. In half-duplex mode, receipt of NAK causes the sender to retransmit the preceding message after turning the line around again.

Figure 8-20 ▶

Configurations for simplex (a), half-duplex (b), and full-duplex (c) transmission modes

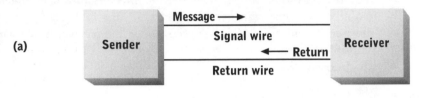

(a)

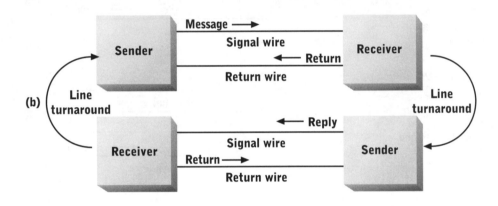

(b)

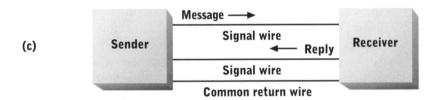

(c)

Communication line cost is the same in simplex and half-duplex modes. However, the added reliability of half-duplex mode is achieved at a sacrifice in overall data transfer rate. When the receiver detects an error, the receiver must wait until the sender transmits a line turnaround before it can send a NAK. If the error occurs near the beginning of a message, the transmission time used to send the remainder of the message is wasted. Also, because errors typically occur in bursts, many retransmissions of the same message might be required before the message is correctly received.

Data and Network Communication Technology

The inefficiencies of half-duplex mode may be avoided by using two transmission lines, as shown in Figure 8-20(c). This two-channel organization permits **full-duplex** communication. In full-duplex mode, the receiver can communicate with the sender at any time over the second transmission line, while data is still being transmitted on the first line. If an error is sensed, the receiver can immediately notify the sender, which can halt the data transmission and retransmit. The speed of full-duplex transmissions is relatively high even if the channel is noisy. Errors are corrected promptly, with minimal disruption to message flow. Speed is increased for the cost of a second transmission line.

Parallel and Serial Transmission

Parallel transmission uses a separate transmission line for each bit position (see Figure 8-21). The width, or number of lines, of a parallel channel is typically one byte or one word, plus a common return line. Parallel communication is relatively expensive due to the cost of multiple transmission lines. Its comparative advantage is the higher data transfer rate that results from combining the capacity of multiple transmission lines.

Figure 8-21 ▶

Parallel transmission of a data byte (8 bits)

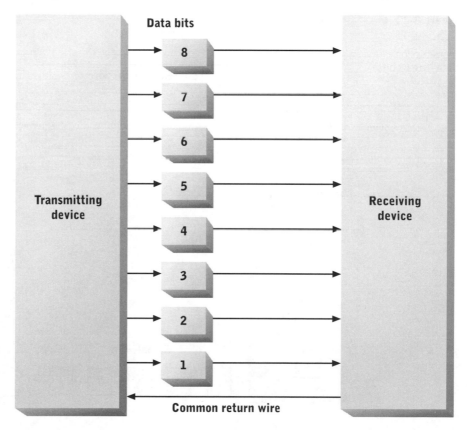

The maximum distance over which data can be sent reliably via parallel transmission is limited. Because of slight differences among parallel transmission lines, data bits might arrive at the receiver at slightly different times. The timing difference among bits, called **skew**, increases with distance and transmission rate. Excessive skew causes bit values to be misinterpreted. To limit skew errors, high-speed parallel communication channels must be very short, typically less than one meter.

Another problem with parallel channels is **crosstalk**, which is noise added to the signal in the wire from EMI generated by adjacent wires. Because of their limitations, parallel channels are used only over short distances where high data transfer rates are required; for example, the system bus and connections among memory caches and the CPU.

Serial transmission uses only a single transmission line and return wire. Bits are sent sequentially through the transmission line and are reassembled by the receiver into larger groups (see Figure 8-22). Digital communication over distances greater than a few meters usually employs serial transmission to avoid skew and crosstalk and minimize wiring and cable cost. LAN, WAN, and long-distance telecommunication lines all use serial transmission.

Figure 8-22 ▶

Serial transmission of a data byte (8 bits)

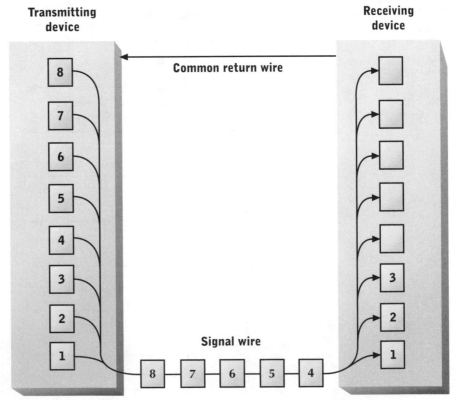

Data and Network Communication Technology

Technology
Focus

Serial and Parallel Storage Connections

For several decades, most computer systems have used parallel cables to connect storage devices and device controllers. The most common examples include cables that follow the ATA and SCSI standards. Until recently, parallel connections have been the only way to provide sufficient data transfer capacity to meet the needs of typical computer systems. However, a few important trends have combined to increase the desirability of serial storage buses.

First, typical desktop and small server computers are much smaller and they continue to shrink. As a result, routing cables among the motherboard, device controllers, storage devices, and I/O devices is problematic. Because they contain many wires, parallel cables are relatively bulky (see Figure 8-23) and difficult to route within a typical small computer cabinet. Their size also restricts airflow, which can cause heat dissipation problems. In contrast, serial cables are thin and flexible. They occupy less space and are easier to route around computer system components.

Figure 8-23 ▶

SCSI (top) and
ATA (bottom)
ribbon cables

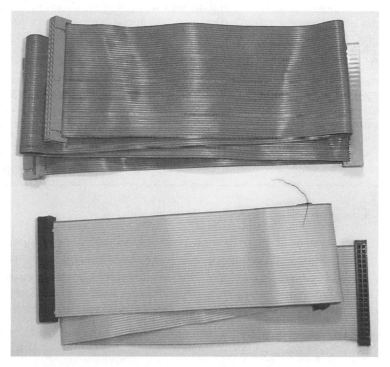

Second, while desktop and small server computer systems have grown smaller, the storage subsystems used by large-scale computers have grown much larger. Many transaction processing and data-mining applications require terabytes of secondary storage. Fulfilling that storage need requires hundreds of disk drives connected to dozens of storage controllers spread across multiple storage cabinets. Connecting all of those devices with parallel cables is problematic because skew and crosstalk increase with cable length.

When electronic switching devices were relatively slow, operating many of them in parallel was the only feasible way to provide high data transfer rates. But advances in semiconductor fabrication and electronic switching technology have yielded ever smaller and faster devices, which are the basic building blocks of computer system components such as network interface cards and storage device controllers. As a result, it is now possible to achieve data transfer speeds of hundreds or thousands of megabits per second across a single cable pair.

Recognizing these trends, standard-setting organizations for storage devices and cables have developed modern serial versions of older parallel standards. In the personal computer arena, the **serial ATA (SATA)** standard has been widely adopted by storage device and computer manufacturers. SATA uses a software interface standard that is compatible with older parallel ATA standards. But the SATA substitutes a seven-wire cable for parallel ATA's 40-wire cable. The SATA cable uses two wire pairs for serial data transmission in each direction and three wires for shielding and ground. Personal computer manufacturers are rapidly switching to SATA storage devices for higher performance and simpler low-cost assembly. SATA can also be used in servers due to improved features such as command queuing, hot-swapping, and improved error detection.

In the server arena, SCSI standards are also moving toward serial communication with the adoption of a new standard called **serial-attached SCSI (SAS)**. Like SATA, SAS retains the command set and software interface of older standards while abandoning parallel cables and connectors in favor of a serial approach. This enables much higher data transfer rates over substantially longer cables. In addition, SAS and SATA share a common physical connector standard which enables a single device controller to control disk drives of both types.

Channel Sharing

Few users require high data transmission capacity on a continuous basis. Rather, transmission capacity is typically needed for short periods, or bursts. Channel sharing techniques efficiently use available capacity by combining the traffic of multiple users. As long as many users don't need high capacity at the same time, the combined load on the channel averages to an acceptable level.

Most local telephone service is based on a channel-sharing strategy called **circuit switching**. When a user makes a telephone call, the full capacity of a channel between the user and the nearest telephone switching center is devoted to that call. The user has a continuous supply of data transfer capacity whether it is needed or not. The channel is unavailable to other users until the call is terminated.

Time division multiplexing (TDM) describes any technique by which data transfer capacity is split into time slices and allocated to multiple users. Packet switching is the most common type of TDM. Packet switching divides messages from all users/applications into relatively small pieces called packets (see Figure 8-24). Each packet contains a header that identifies the sender, recipient, sequence number, and other information about its contents. Packets are sent to their destination as channel capacity becomes available. Packets can be held temporarily in a buffer, pending channel availability. If multiple channels are available, multiple packets can be sent simultaneously.

Figure 8-24 ►

Packet switching—the most common form of time division multiplexing (TDM)

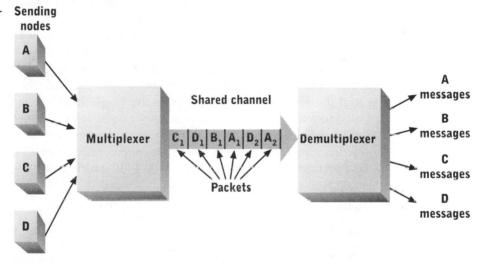

The primary advantage of packet switching is that the telecommunication service provider determines how to use available data transfer capacity and channels most effectively. In most situations, the provider can make rapid and automatic decisions that efficiently allocate available capacity among users. The result is substantially reduced total cost for telecommunication service.

The primary disadvantages of packet switching are varying delays in message transmission and the complexity of creating and routing packets. Because a user does not have a dedicated channel, the time required to send a message is unpredictable; it rises and falls with the total demand for the available data transfer capacity. Delays can be substantial when many users simultaneously send messages and insufficient capacity is available.

Messages must be broken into packet-sized chunks and appropriate headers must be created. The packets must be sent through the network to their intended destination. The receiver must reassemble packets into the original message in their intended sequence even if they arrive in a different order. Transmission errors must be localized to a specific packet in order to request retransmission of the correct packet.

Despite its complexities, packet switching is the dominant form of intercomputer communication. Its cost and complexity are more than offset by more efficient utilization of available communication channels. Circuit switching is used only in situations where data transfer delay and available data transfer capacity must be within precise and predictable limits.

Another method for sharing communication channels is **frequency division multiplexing (FDM)**, shown in Figure 8-25. Under FDM, a single **broadband** (high bandwidth) channel is partitioned into multiple **baseband** (low bandwidth) subchannels. Each subchannel represents a different frequency range, or **band**, within the broadband channel. Signals are transmitted within each subchannel at a fixed frequency or narrow frequency range. Cable television (CATV) is the most common example of FDM. Long-distance telecommunication providers sometimes use FDM to multiplex single-mode optical fibers. That application of FDM is commonly called **wavelength-division multiplexing (WDM)**.

Figure 8-25 ►

Channel sharing using frequency division multiplexing (FDM)

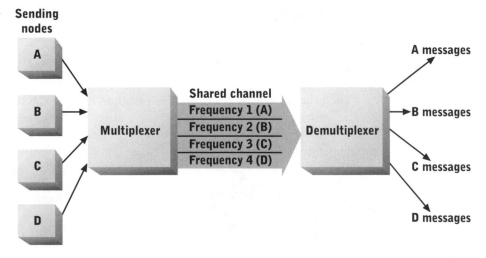

Data and Network Communication Technology

FDM might require moving signals originally intended for one frequency band into another frequency band. For example, in most cable TV systems, channels 2 through 13 are carried across a coaxial cable at their original broadcast frequencies. Channels above 13 are remapped to frequencies other than their assigned broadcast frequencies so that they occupy contiguous subchannels within the coaxial broadband channel.

Baseband channels within a single broadband channel can use different signaling methods, communication protocols, and transmission speeds. Baseband channels can be shared using packet switching or other time division multiplexing methods. Multiple baseband channels can be combined into a parallel transmission channel.

Technology
Focus

Infiniband

Infiniband is a data interconnection standard developed by the Infiniband Trade Association, a consortium founded by Dell, Hewlett-Packard, IBM, Intel, Microsoft, and Sun Microsystems. Infiniband is a high-speed communication architecture intended to interconnect devices such as servers, secondary storage appliances, and network switches. The goal of the standard is to replace the current jumble of competing interconnection standards with a unified standard and architecture that provide significantly increased data transfer rates.

Infiniband is based on an interconnection architecture known as **switched fabric**, which interconnects multiple devices with multiple data transmission pathways and a mesh of switches that visually resembles the interwoven threads of fabric or cloth (see Figure 8-26). A fabric switch connects any sender directly to any receiver and can support many simultaneous connections. Connections are created on request and deleted when they are no longer needed, freeing data communication capacity to support other connections. Switched fabric architecture is not new, but the digital switching technology that supports it has only recently become cost-effective.

Figure 8-26 ▶

A 3x3 switched
fabric

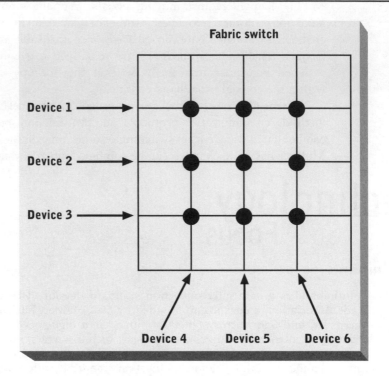

Each device is connected to an Infiniband switch by a **host channel adapter (HCA)** or a **target channel adapter (TCA)**. HCAs are used by devices such as general-purpose servers that can initiate and respond to data transfer requests. An HCA has a direct connection to the host's primary storage via a device controller attached to the system bus or via a special-purpose memory interface. TCAs are used by simpler devices such as network switches and storage appliances.

Infiniband devices are connected with copper wires or fiber optic cables. The Infiniband standard specifies cable connectors and operational characteristics, but not physical cable construction. Conventional twisted pair or coaxial cables cannot meet the Infiniband requirements. Some vendors employ a bundle of twin-axial cables. Data transmission rates range from 2.5 to 10 Gbps depending on the number of conductors. Copper cables can be up to 25 meters long and fiber optic connections can extend up to 10 kilometers.

Figure 8-27 shows a typical architecture for a medium- to large-scale server farm used by large e-commerce Web sites. The architecture includes multiple general-purpose servers configured for specific tasks, network storage servers, and network switches. Specializing the functions of each device makes it easier to enlarge or reduce overall capacity, and device redundancy increases reliability and fault tolerance. However, many high-speed transmission lines and switches are necessary to interconnect all system components. Those interconnections are Infiniband's current target market.

Figure 8-27 ▶

A server farm with network interconnections

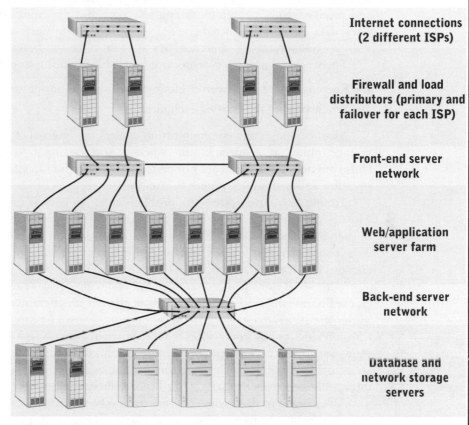

Internet connections (2 different ISPs)

Firewall and load distributors (primary and failover for each ISP)

Front-end server network

Web/application server farm

Back-end server network

Database and network storage servers

Most current Infiniband products are switches and interconnection bridges for existing components that use other data communication standards such as the PCI bus, Gigabit Ethernet, and FibreChannnel. Most current devices are manufactured by smaller companies and incorporated by customer request into computer systems marketed by larger companies such as Dell and IBM. Time will tell whether customer demand for Infiniband products will result in greater support by larger computer and peripheral manufacturers and a new generation of switches and peripherals that abandon earlier standards such as PCI and FibreChannel.

CLOCK SYNCHRONIZATION

An important part of any communication protocol is a common transmission rate. Senders place bits onto a transmission line at precise intervals. Receivers examine the signal at or during specific time intervals to extract encoded bits. Unless sender and receiver share a common timing reference and use equivalent transmission rates, data cannot be extracted correctly from the signal. When sender and receiver share a common timing reference they are said to be synchronized.

There are two primary synchronization problems during message transmission:

▶ Keeping sender and receiver clocks synchronized during transmission

▶ Synchronizing the start of each message

Most computer and communication devices use internal clocks that generate voltage pulses at a specific rate when a specific input voltage is applied. Unfortunately, such clocks are not perfect. Timing pulse rates can vary with temperature and with minor power fluctuations.

Timing fluctuations are not a problem when all devices use the same clock. Most parallel transmission standards assign a separate transmission line to carry the sender's clock pulse. The receiver continuously monitors this clock line and uses it to ensure that it reads and interprets bits at the same rate that the sender encodes and transmits them.

With serial transmission, sending and receiving devices each have their own clock. Because the clocks are independent, there is no guarantee that their timing pulses are generated at exactly the same time. Most communication protocols transmit clock pulses from sender to receiver when a connection is first established. However, even if perfect synchronization is achieved at that time, clocks can subsequently drift out of synchronization. As clocks drift out of synchronization, bit interpretation errors by the receiver become more likely.

An example of the errors resulting from unsynchronized clocks is illustrated in Figure 8-28. The receiving device clock pulses are being generated at different times than the sender clock pulses, resulting in misalignment of bit time boundaries. In this example, the receiver is unable to interpret several bits correctly because they contain two different signal levels.

Sender and receiver must occasionally transmit timing signals over the data transmission line to ensure that their clocks are synchronized. They also must agree on the boundaries of each message, and in order to do so they must agree on message length. Two common approaches to synchronizing clocks and coordinating message boundaries are synchronous transmission and asynchronous transmission. These methods are sometimes referred to as **character framing** methods when messages consist of ASCII or Unicode characters.

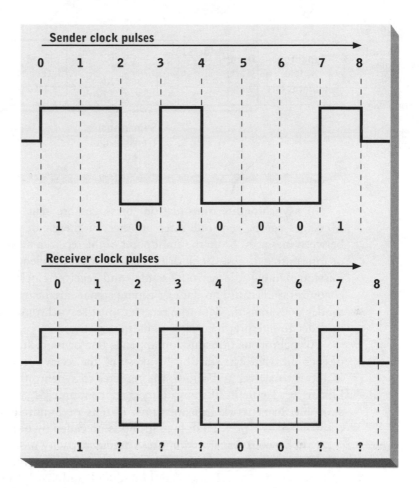

Synchronous transmission ensures that sender and receiver clocks are always synchronized by sending continuous data streams. Messages are transmitted in fixed-size byte groups called blocks (see Figure 8-29). Because block size is always the same, the receiver always knows where one block ends and another begins. If necessary, transmission blocks are separated by a continuous stream of **synchronous idle messages**. A synchronous idle message has a predetermined pattern of signal transitions designed for easy clock synchronization. The data start flag is an easily recognizable bit pattern that differs from the synchronous idle message and data stop flag bit patterns. Those differences enable the receiver to detect the beginning of a new transmission block.

Figure 8-29 ▶

Typical format
for messages
transmitted
using
synchronous
character
framing
techniques

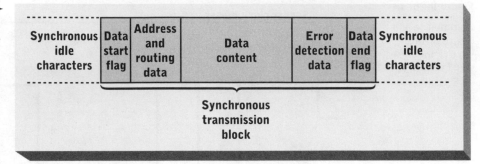

| Synchronous idle characters | Data start flag | Address and routing data | Data content | Error detection data | Data end flag | Synchronous idle characters |

Synchronous transmission block

In **asynchronous transmission**, messages are sent on an as-needed basis. Messages can be sent one after another, or there can be periods of inactivity between messages. From the standpoint of the receiver, messages arrive at random or intermittent times. Clock drift is a significant problem in asynchronous transmission. During idle periods, sender and receiver clocks can drift out of synchronization because no data or timing signals are being transmitted. When the sender does transmit data, the receiver must resynchronize its clock immediately in order to interpret incoming signals correctly.

Asynchronous transmission appends one or more **start bits** to the beginning of each message. The start bit "wakes up" the receiver, informs it that a message is about to arrive, and allows the receiver to synchronize its clock before data bits arrive. Figure 8-30 shows a message containing a single byte preceded by a start bit. In network transmissions, dozens or hundreds of bytes are usually transmitted as a unit. Each byte group is preceded by one or more start bits.

Synchronous transmission uses channel capacity more efficiently than asynchronous transmission because data is transmitted in large blocks with relatively few bits used to mark message boundaries and synchronize clocks. However, because transmission is continuous, synchronous transmission can't be used when multiple senders and receivers share the same channel. Asynchronous transmission is used in most computer networks, modem communication, and communication between computer systems and peripheral devices, such as keyboards and printers. Synchronous transmission is used only for high-speed communication between a dedicated sender and receiver, such as between two mirrored or clustered mainframe computer systems.

Figure 8-30 ▶

Asynchronous character framing for serial transmission, including a start bit

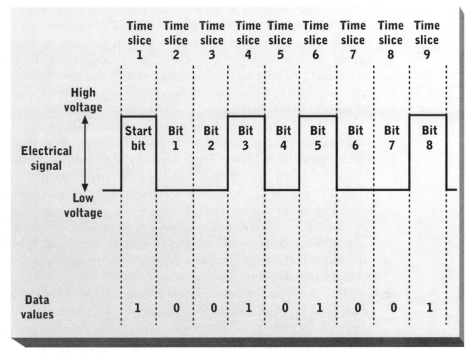

ERROR DETECTION AND CORRECTION

A crucial component of any communication protocol is a method or methods for detecting and correcting errors in data transmission, reception, or interpretation. All widely used error detection methods are based on some form of redundant transmission. A redundant message or message component is transmitted with or immediately after the original message. The receiver compares the original message with the redundant transmission and if the two match, the original message is assumed to have been transmitted, received, and interpreted correctly. If the two don't match, then a transmission error is assumed to have occurred, and the receiver asks the sender to retransmit the message.

Error detection and correction methods vary by the:

▶ Size and content of the redundant transmission
▶ Efficient use of the communication channel
▶ Probability that an error will be detected
▶ Probability that an error-free message will be identified as an error
▶ Complexity of the error detection method

Size and content of the redundant transmission are inversely related to efficient use of the channel. For example, one possible error detection message is to send three copies of each message and verify that they are identical. This method is simple to implement, but it uses the channel at only one-third of its capacity, or less if many errors are correctly or incorrectly detected. Changing the method to send only two copies of each method increases maximum channel utilization to 50 percent of raw capacity.

For most methods and channels, the probability of detecting errors can be mathematically or statistically computed. The probability of *not* detecting a real error is called **Type I error**. The probability of incorrectly identifying good data as an error is called **Type II error**. For any given error detection method, Type I and Type II errors are inversely related. That is, a decrease in Type I error is accompanied by an increase in Type II error.

Type II errors result in needless retransmission of data that was correctly received but incorrectly assumed to be in error. Increasing Type II error decreases the channel efficiency because a greater proportion of channel capacity is used to retransmit data needlessly.

In some types of communication channels, for example a system bus or a short-haul fiber optic cable, the probability of a transmission or reception error is extremely remote. In other types of channels, for example high-speed analog modem channels, errors are common. Different methods of error detection are appropriate to different channels and different purposes. For example, error detection is not normally employed for digital voice transmissions because users are relatively insensitive to occasional small errors. At the other extreme, communication between bank computers and automated teller machines over copper wires normally employs extensive error checking because of the importance of the data and because of the relatively high error rate in long-distance electrical transmission.

Commonly used methods of error detection include:

▶ Parity checking (vertical redundancy checking)
▶ Block checking (longitudinal redundancy checking)
▶ Cyclical redundancy checking

Each of these is described in detail in the sections that follow.

Parity Checking

Character data is often checked for errors using **parity checking,** also called **vertical redundancy checking**. In a character-oriented transmission, one parity bit is appended to each character. The **parity bit** value is a count of other bit values within the character. Parity checking can be based on even or odd bit counts.

With **odd parity**, the sender sets the parity bit to zero if the count of 1-valued data bits within the character is odd. If the count of 1-valued data bits is even, the parity bit is set to one (see Figure 8-31). Under **even parity**, the sender sets the parity bit to zero if the count of 1-valued data bits is even or to one if the count of 1-valued data bits is odd. The receiver counts the 1-valued data bits in each character as they arrive and then compares the count against the parity bit. If the count doesn't match the parity bit, then the receiver asks the sender to retransmit the character.

Figure 8-31 ►

Sample parity bits

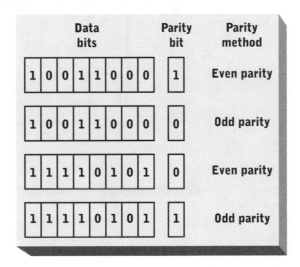

Parity checking has a high Type I error rate. For example, a transmission error that flipped the values of 2, 4, or 6 bits within an ASCII-7 character would not be detected. Parity checking is unreliable in channels that are subject to error bursts affecting many adjacent bits. Parity checking is more reliable in channels with rare errors that are usually confined to widely spaced bits.

Block Checking

Parity checking can be expanded to groups of characters or bytes using **block checking**, also called **longitudinal redundancy checking (LRC)**. To implement LRC, the sending device counts the number of 1-valued bits at each bit position within a block. The sender combines parity bits for each position into a **block check character (BCC)** and appends it to the end of the block (see Figure 8-32). The receiver counts the 1-valued bits in each position and derives its own BCC to compare to the BCC transmitted by the sender. If the BCCs are not the same, the receiver requests retransmission of the entire block. In Figure 8-32, an even parity bit is computed for each bit position of a block of 8 bytes. The set of parity bits forms a block check character that is appended to the block for error detection.

Figure 8-32 ►

Block checking (longitudinal redundancy checking)

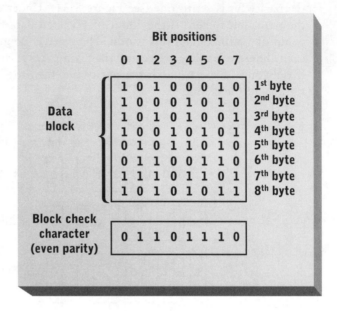

LRC is vulnerable to the same types of errors as parity checking. Type I error rates can be reduced by combining parity checking and LRC. However, even with this approach, some compensating errors might go undetected.

Cyclic Redundancy Checking

Cyclic redundancy checking (CRC) is currently the most widely used error detection technique. Like LRC, CRC produces a block check character for a group of characters or bytes. The CRC character is generated by a complex mathematical algorithm. The CRC is usually more than 8 bits long and can be as large as 128 bits. CRC bit strings can be computed by software or by special-purpose microprocessors incorporated directly into data communication and network hardware.

CRC has much lower Type I and Type II error rates than parity checking and LRC checking. Both error rates depend on the size of the transmitted data block and CRC bit string. CRC bit strings of 64 or 128 bits are commonly used within network packets and for magnetic tape data backups.

Technology
Focus

802.11 Wireless Network Standards

The Institute of Electrical and Electronics Engineers (IEEE) creates many network and telecommunication standards. (These standards will be discussed more fully in Chapter 9.) The 802.11 committee was formed in the early 1990s to develop standards for metropolitan and local area wireless networks. In 1997 the committee published the first version of the 802.11 standard. A second version (802.11b) followed in 1999, a third (802.11a) in 2001, and a fourth (802.11g) in 2003.

Some of the goals of the IEEE 802.11 standards are to:

► Meet the requirements of national and international regulatory bodies and pursue adoption by those bodies

► Support communication by radio waves and infrared or visible light

► Support stationary stations and mobile stations moving at vehicular speeds

To expedite its work, the committee initially focused on RF transmission in the 2.4 GHz band because RF waves can travel several miles and penetrate walls and other obstacles. The 2.4 GHz band was chosen because most of that band is regulated but unallocated in the United States, Europe, and Japan.

The first **802.11** standard defined two different RF transmission methods with a maximum raw data transfer rate of 2 Mbps. One of the transmission methods, frequency hopping spread spectrum (FHSS), was abandoned in later standards because its transmission speed could not be increased within FCC-mandated transmitter power limits. The other method, direct sequence spread spectrum (DSSS), was enhanced in the 802.11b standard to provide raw data transmission at 5.5 and 11 Mbps. The 802.11a standard employs a transmission method called orthogonal frequency division multiplexing (OFDM), which transmits across several channels simultaneously.

Error detection is a major problem in any RF data transmission scheme. RF transmissions are subject to many types of interference, and receivers encounter varying degrees of interference when in motion. Devices such as computers, televisions, microwave ovens, and cordless telephones can interfere with wireless network communication. Another significant problem is multipath distortion, which occurs when an RF signal bounces off stationary objects (see Figure 8-33), causing the receiving antenna to receive multiple copies of the same signal at different times. Multipath distortion blurs or smears the signal content, causing bit detection errors. The problem is especially severe indoors, where there are many hard, flat surfaces that reflect RF waves.

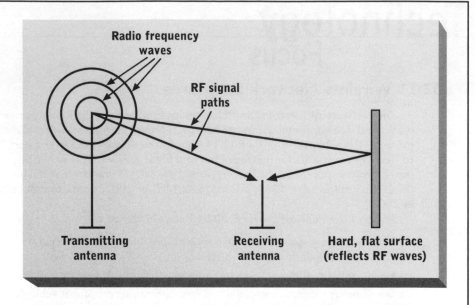

Figure 8-33 ►

Multipath distortion caused by receiving multiple signal copies at slightly different times

Radio frequency waves

RF signal paths

Transmitting antenna

Receiving antenna

Hard, flat surface (reflects RF waves)

   The **802.11b** standard divides the 2.4 GHz band into 14 channels, each with 22 MHz of bandwidth. Transmissions within each band encode bits using two-level, four-level, or eight-level phase shift modulation, which yields raw data transfer rates of 22, 44, and 88 Mbps, respectively. A substantial portion of the raw data transfer rate is used for redundant bit transmission to improve error detection. The four standard transmission speeds (1, 2, 5.5, and 11 Mbps) use different bit encoding and signal modulation methods, each representing a different trade-off among reliability, speed, and transmission distance. The fastest method uses eight-level modulation and encodes bits with minimal redundancy. Slower transmission uses two-level modulation with greater redundancy.

   Transmitters continuously monitor signal quality and error rates. As noise, interference, and errors increase, or as the sender and receiver move further apart, transmitters change modulation and bit encoding methods to compensate. Transmitters step down through the 5.5, 2, and 1 Mbps transmission methods until reliable transmission is achieved and then move upward again as conditions improve.

   The **802.11a** standard divides frequency bands in the 5.2, 5.7, and 5.8 GHz ranges into 12 channels. The standard transmission speeds are 6, 9, 12, 18, 24, 36, 48, and 54 Mbps. Bit transmission rates in each channel can be varied to account for differing interference levels and error rates.

   The **802.11g** standard combines the channel and bit encoding methods of 802.11b with the OFDM transmission method of 802.11a. Standard raw transmission speeds are the same as in 802.11a. 802.11a transmission is generally more reliable over short distances than 802.11b and 802.11g because there are fewer sources of noise and interference. However, RF signals in the 5 GHz band attenuate more quickly than 2.4 GHz signals and are more easily absorbed by walls and other obstacles. Thus, effective 802.11a transmission distance is shorter than for 802.11b and 802.11g.

Business
Focus

Upgrading Storage and Network Capacity (Part I)

The Bradley Advertising Agency (BAA) specializes in designing and developing printed and video advertisements. Most of its work is performed under contract to other advertising agencies. BAA employs 30 people, most of whom are actively engaged in design and production work.

Printed ad copy is designed and produced almost entirely by computer. BAA uses Adobe Illustrator and other software on 15 networked Dell and Macintosh desktop computers. BAA uses many peripheral devices including several scanners, various types of color printers, and a high-speed color copier. A Compaq server stores most data and software files on five 72-GB disks. The file server also has an 80-GB DLT tape drive and a DVD-RW drive.

Until last year, BAA used analog video cameras and videotape editing equipment for video production projects. BAA purchased two Sun Microsystems workstations, an expensive ($125,000) video editing software package, and several digital video cameras and recorders. BAA expects to add two more Sun workstations this year and two or three more the following year.

The files used to store print ads are normally 200 to 500 MB per project, but can range as high as 5 GB. BAA often has several dozen print projects in process at a time. Video production files are much larger than print files. Files for a typical 30-second television ad require 10 GB of storage.

File sharing performance was already slow when the first Sun workstations were purchased. A consultant determined that the file server was being used at only about 20 percent of its capacity and that the network itself was the performance bottleneck. The existing network is an Ethernet LAN using twisted pair wiring that was installed in 1993. The LAN originally operated at 10 Mbps, but the network hub was upgraded to a 10/100 Mbps switch. The desktop computers all operate at 100 Mbps. The Sun workstations and the server have network interfaces that can support 100 Mbps or 1 Gbps speed, though they're limited to 100 Mbps by the current switch.

Network bottlenecks have become more common since the two Sun workstations were acquired. Access to print ad files that took a few seconds a year or two ago now can take up to 5 minutes during busy periods. Recently, two 72–GB disks were added to each Sun workstation so video files can be stored locally during editing, but downloading and uploading a project to or from the server can take 15 minutes or more. Project files are sometimes moved between Sun workstations several times per day if both machines are being used for the same project. Because BAA is growing rapidly, it expects its disk storage requirements to double every year for the next three years.

BAA is studying alternatives to increase data storage, data communication, and network capacity. Options under consideration include:

► Upgrading the existing network to Gigabit Ethernet using existing cabling

---

- Installing a parallel 1- or 10-Gbps network for the server and Sun workstations using new copper or fiber optic cable
- Replacing the disks in the existing server with larger disks
- Buying a network-attached storage server with at least 500 GB of disk capacity
- Purchasing one or more Infiniband storage devices and interfacing them to the existing file server and video editing workstations via an Infiniband switch

BAA is unsure whether its cabling will work with Gigabit Ethernet because it isn't certified as Category 5. It's also concerned that even if the existing cable will work, the network will soon be saturated, and the cable will be unable to support a future move to 10 Gbps Ethernet. BAA is also concerned about investing in relatively new and expensive Infiniband technology.

Questions

1. If BAA's current cable is tested and determined to be reliable with Gigabit Ethernet, should the company use its existing wiring or install new wiring?

2. Assume that cabling costs (including installation labor) for Catgory 5, Category 6, and fiber optic are $20,000, $25,000, and $40,000, respectively. Which cable should BAA choose if the existing network must be rewired for Gigabit Ethernet?

3. What are the advantages of operating a separate high-speed network to support video editing applications in addition to the existing network? What are the disadvantages? What cable type is most appropriate?

4. Should BAA seriously consider moving to Infiniband and stand-alone storage servers? Why or why not? If they should adopt Infiniband, how quickly should they do so?

SUMMARY

- A communication protocol is a set of rules and conventions covering many aspects of communication, including message content and format, bit encoding, signal transmission, transmission medium, and channel organization. It also includes procedures for coordinating the flow of data, including media access, clock synchronization, and error detection and correction.

- Data bits can be encoded into analog or digital signals. Signals can be electrical, optical, or radio frequency. Bits are encoded in sine waves by modulating one or more wave parameters. Possible modulation techniques include frequency, amplitude, phase shift, and pulse code modulation.

- Important characteristics of transmission media include raw data transfer rate, bandwidth, and susceptibility to noise, distortion, external interference, and attenuation. Bandwidth is the difference between the highest and lowest frequencies that can be propagated through a transmission medium. High-bandwidth channels can carry multiple messages simultaneously.

- The effective data transfer rate can be much lower than the raw data transfer rate due to attenuation, distortion, and noise. Attenuation is loss of signal power as it travels through the transmission medium. Distortion is caused by interactions between the signal and the transmission medium. Noise can be generated internally or added through external interference. The signal-to-noise (S/N) ratio is a measure of the difference between noise power and signal power.

- Electrical cables are of two primary types—twisted pair and axial. Both use copper but differ in construction and shielding. Twisted pair is relatively cheap but limited in bandwidth, signal-to-noise ratio, and transmission speed due to limited shielding. Coaxial and twin-axial cables are more expensive but offer higher bandwidth, greater signal-to-noise ratio, and lower distortion.

- Optical cables are of two types—multimode and single mode. Single-mode cable provides higher transmission rates than multimode cable but at much greater cost. Optical cables provide very high bandwidth, little internally generated noise and distortion, and immunity to electromagnetic interference.

- Data can be transmitted without wires via radio waves and infrared light. Short-distance wireless transmission is increasingly used for local area networks. Long-distance radio frequency channels are in short supply due to their limited number and intense competition for licenses. Radio channels are generally leased from telecommunication companies on an as-needed basis. Infrared transmission has very limited transmission distance.

- Channel organization describes the number of lines dedicated to a channel and the assignment of specific signals to those channels. A simplex channel uses one optical fiber or copper wire pair to transmit data in one direction only. A half-duplex channel is identical to a simplex channel but sends a control signal to reverse transmission direction. Full duplex channels use two fibers or wire pairs to support simultaneous transmission in both directions.

- Parallel transmission uses multiple lines to send several bits simultaneously. Serial transmission uses a single line to send one bit at a time. Parallel transmission provides higher channel throughput, but it is unreliable over distances greater than a few meters due to skew and crosstalk. Serial transmission is reliable over much longer distances. Serial channels are cheaper to implement because fewer wires or wireless channels are used.

► Channels are often shared when no single user or application needs a continuous supply of data transfer capacity. In circuit switching, an entire channel is allocated to a single user for the duration of one data transfer operation. Packet switching allocates time on the channel by dividing many message streams into smaller units called packets and intermixing them during transmission. Frequency division multiplexing divides a broadband channel into several baseband channels.

► Sender and receiver must synchronize clocks to ensure that they use the same time periods and boundaries to encode and decode bit values. A single shared clock is the most reliable synchronization method, but it requires that clock pulses be sent continuously from sender to receiver. Asynchronous transmission relies on specific start and stop signals (usually a single bit) to indicate the beginning and end of a message unit. Synchronous transmission maintains a continuous flow of signals from sender to receiver to provide continual opportunities for clock synchronization.

► Error detection is always based on some form of redundant transmission. The receiver compares redundant copies of messages and requests retransmission if they don't match. Increasing the level of redundancy increases the chance of detecting errors at the expense of reducing channel throughput. Common error detection schemes include parity checking (vertical redundancy checking), block checking (longitudinal redundancy checking), and cyclical redundancy checking.

At this point, you have a thorough understanding of how data is encoded and transmitted among computer systems. Data communication technology is the foundation of computer networks, but other software and hardware technologies are also required. In Chapter 9, we'll discuss computer network hardware, protocols, and architecture. In Chapter 13, we'll discuss system software that supports distributed resources and applications.

Key Terms

802.11
802.11a
802.11b
802.11g
acknowledge (ACK)
amplifier
amplitude
amplitude modulation (AM)
amplitude shift keying (ASK)

analog signal
asynchronous transmission
attenuation
band
bandwidth
baseband
binary signal
bit time
block check character (BCC)

block checking
broadband
broadcast mode
carrier wave
Category 5
Category 6
character framing
circuit switching
coaxial cable

communication channel
communication protocol
crosstalk
cycle
cyclic redundancy checking
   (CRC)
digital signal
discrete signal
distortion
effective data transfer rate
electromagnetic interference
   (EMI)
even parity
fiber optic cable
frequency
frequency division multiplexing (FDM)
frequency modulation (FM)
frequency shift keying (FSK)
full-duplex
half-duplex
host channel adapter (HCA)
Infiniband
line turnaround
longitudinal redundancy
   checking (LRC)

message
modulation
modulator-demodulator
   (modem)
multilevel coding
multimode cable
negative acknowledge (NAK)
noise
odd parity
on-off keying (OOK)
packet
packet switching
parallel transmission
parity bit
parity checking
phase
phase shift keying (PSK)
phase shift modulation
pulse code modulation (PCM)
quality of service
radio frequency (RF)
raw data transfer rate
repeater
return wire
serial ATA (SATA)
serial-attached SCSI (SAS)

serial transmission
signal
signal wire
signal-to-noise ratio (S/N
   ratio)
simplex
sine wave
single-mode cable
skew
start bits
switched fabric
synchronous idle message
synchronous transmission
target channel adapter (TCA)
time division multiplexing
   (TDM)
transmission medium
twin-axial cable
twisted pair wire
Type I error
Type II error
vertical redundancy checking
wavelength-division multiplexing (WDM)

Vocabulary Exercises

1. _____ transmission sends bits one at a time using a single transmission line.

2. During half-duplex transmission, sender and receiver exchange roles after a(n) _____ is transmitted.

3. _____ encodes data by varying the distance between wave peaks within an analog signal.

4. A(n) _____ converts a digital signal to an analog signal so that digital data can be transmitted over analog telephone lines.

5. In synchronous data transmission, a(n) _____ signal is transmitted continuously during periods when no data is being transmitted.

6. Serial data transmission standards including _____ and _____ are replacing older parallel transmission standards for connecting secondary storage devices and controllers.

7. _____ cabling meets an industry standard that specifies four twisted wire pairs and can support data transmission rates up to 1 Gbps.

8. The _____ of a sine wave is measured in Hertz.

9. A local telephone grid uses _____ switching to route messages from a wired home telephone to the local telephone switching center.

10. Most networks use _____ switching to send messages from sender to receiver.

11. In _____, a bit is appended to each character or byte and the bit value is determined by counting the number of 1 bits.

12. A(n) _____ signal is a discrete signal that can encode only two possible values.

13. A(n) _____ wave carries encoded data through a transmission medium.

14. With parity checking, sender and receiver must agree whether error detection will be based on _____ or _____.

15. The _____ of a channel describes the mathematical relationship between noise power and signal power.

16. _____ is any change in a signal characteristic caused by components of the communication channel.

17. Methods of error detection and correction represent specific trade-offs among effective data transfer rate, processing complexity, _____, and _____.

18. Wireless transmission uses the _____ or _____ frequency bands.

19. _____ cannot affect optical signals but can affect electrical or RF signals.

20. A communication channel that uses electrical signals must have at least two wires—a(n) _____ and a(n) _____ to form a complete electrical circuit.

21. _____ measures the theoretical capacity of a channel. _____ measures the actual capacity of a channel when a specific communication protocol is employed.

22. Simultaneous or interleaved transmission of multiple messages on a single transmission line can be accomplished by _____ or _____ multiplexing.

23. A(n) _____ signal can encode an infinite number of possible numeric values.

24. _____ is a measure of peak signal strength.

25. In asynchronous transmission, at least one _____ is appended to the beginning of each character or block.

26. The term _____ describes the encoding of data as variations in one or more physical parameters of a signal.

27. In _____ transmission, blocks or characters arrive at unpredictable times, and no signal is transmitted during idle periods.

28. The _____ of a communication channel is the difference between the highest and lowest frequencies that can be transmitted.

29. _____ transmission implements two-way transmission with two separate communication channels; _____ transmission implements two-way transmission with only one communication channel.

30. _____ encodes data by varying the magnitude of wave peaks within an analog signal.

31. _____ transmission uses multiple lines to send multiple bits simultaneously.

32. A(n) _____ extends the range of data transmission by retransmitting a signal.

33. _____ generates a(n) _____ consisting of a single parity bit for each bit position within the bytes of a block.

34. In _____ transmission, signals are transmitted continuously, even when there is no data to send, to ensure clock synchronization.

35. _____ uses more than two signal parameter levels to encode multiple bits within a single signal event.

36. _____ encodes bit values in brief bursts of electrical or optical power.

37. _____ is the interference among adjacent transmission lines in a parallel communication channel.

38. Frequency-division multiplexing of optical channels is sometimes called _____.

39. The length of a parallel communication channel is limited by _____, which can cause bits to arrive at slightly different times.

40. A receiver informs a sender that data was received correctly by sending a(n) _____ message. It informs the sender of a transmission or reception error by sending a(n) _____ message.

41. _____ is used in some networks (for example, cable television networks) as a transmission medium.

42. _____ is a term that describes loss of signal strength as it travels through a transmission medium.

43. Messages to be transmitted by time division multiplexing are divided into _____ prior to physical transmission.

44. _____ cable is an improved version of Category 5 cable that reduces transmission errors.

Review Questions

1. What are the components of a communication channel?

2. What are the components of a communication protocol?

3. Describe frequency, amplitude, phase shift, and pulse code modulation.

4. How does multilevel coding increase the effective data transfer rate of a communication channel?

5. Describe the relationship among bandwidth, data transfer rate, and signal frequency.

6. How can noise and distortion be introduced into a transmission medium? How does a channel's signal-to-noise ratio affect the reliability of data transmission?

7. Compare and contrast twisted pair, coaxial, twin-axial, multimode fiber optic, and single-mode fiber optic cable in terms of raw data transfer rate, cost, and maximum cable segment length.

8. What are the advantages of wireless transmission using RF waves as compared to infrared waves?

9. Describe simplex, half-duplex, and full-duplex transmission. Compare and contrast them in terms of cost and effective data transfer rate.

10. Why is the actual data transfer rate of a channel usually less than the theoretical maximum of the technology used to implement the channel?

11. Compare and contrast serial and parallel transmission in terms of channel cost, data transfer rate, and suitability for long-distance data communication. Why are standards for connecting secondary storage devices migrating from parallel to serial transmission?

12. What are the differences between synchronous and asynchronous data transmission?

13. What is character framing? Why is it generally not an issue in parallel data transmission?

14. Describe the differences between even and odd parity checking.

15. What is a block check character? How is it derived and used?

16. Compare and contrast frequency and time division multiplexing. What physical characteristics of the communication channel are required by each type? Which provides greater data transmission capacity?

17. What is the difference between an amplifier and a repeater?

Data and Network Communication Technology

Problems and Exercises

1. Assume that data is transmitted on a dedicated channel in blocks of 48 bytes at 100 Mbps. Also, assume that an 8-bit LRC check character is used for error detection and that an error is detected, on average, once every 1,000 transmitted blocks. What is the effective data transfer rate of the channel?

2. Assume that data is transmitted over twisted pair cable. The power of the signal when placed on the channel is 75 dB. Signal attenuation is 5 dB per hundred meters. Average noise power on the cable is 0.1 dB per meter. What is the signal-to-noise ratio of a 50-meter channel? What is the signal-to-noise ratio of a 1-kilometer channel? What is the maximum channel length?

Research Problems

1. Parity checking often is used in computer memory to detect errors. Investigate currently available memory types, including parity and error correcting circuits (ECC) memory. How is parity checking implemented within the memory module? How are errors detected, and what happens when an error is detected? What types of computer systems should use parity or ECC memory and why?

2. Research InfiniBand® Architecture (*www.infinibandta.org*) to investigate some issues not addressed in the chapter. What signal encoding, cables, and connectors are used? Is the channel serial or parallel? Is transmission synchronous or asynchronous? How are errors detected and corrected?

3. Investigate the IEEE Firewire data communication standard. How is data encoded, and what is the raw transmission speed? Describe how the standard enables quality-of-service guarantees when transmitting multimedia and other time-sensitive data.

4. Investigate the IEEE 806.16 (WiMax) wireless networking standard. How do transmission speeds and distances compare to the IEEE 802.11 standards? What method(s) is/are used to detect and correct transmission errors? Will 802.16 networks eventually replace 802.11 networks? Why or why not?

Chapter 9

Computer Networks

Chapter Goals

- ► Compare and contrast bus, ring, and star network topologies

- ► Describe packet routing across local and wide area networks

- ► Describe the CSMA/CD media access control protocol

- ► Describe network hardware devices, including network interface units, routers, and switches

- ► Describe the OSI network model, the TCP/IP protocol suite, and IEEE network standards

In the previous chapter, we discussed data communication between a single sender and a single receiver, including data transfer among hardware devices within a computer system, between a computer system and a peripheral device such as a printer, and between two directly connected computer systems. Communication among computer systems in a network is much more complex. Figure 9-1 shows the additional complexities that are addressed in this chapter.

Figure 9-1 ▶

Topics covered in this chapter

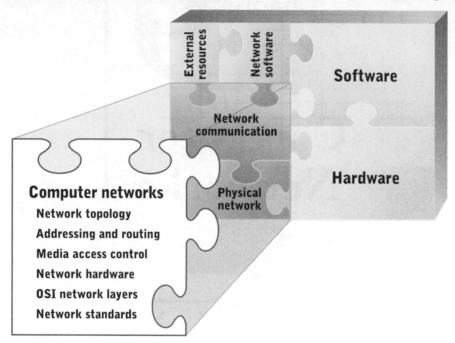

NETWORK TOPOLOGY

The term **network topology** refers to the spatial organization of network devices, physical routing of network cabling, and flow of messages from one network node to another. An end node is a device such as a workstation, server, or printer that can be the source or destination of a message. Physically linking two nearby end nodes is usually straightforward. A point-to-point transmission line is laid over the shortest path and connected directly to both end nodes.

Point-to-point network topology requires many transmission lines if the number of end nodes is large. To connect four end nodes, six transmission lines and three connections per node are required. As the number of end nodes (n) increases, the number of point-to-point transmission lines rises quickly (see Figure 9-2(a)), as described by the formula:

$$(n-1)! = 1 + 2 + 3 + \ldots + (n-1)$$

Figure 9-2 ▶

End nodes
connected by
(a) point-to-
point and
(b) shared
connections

(a)

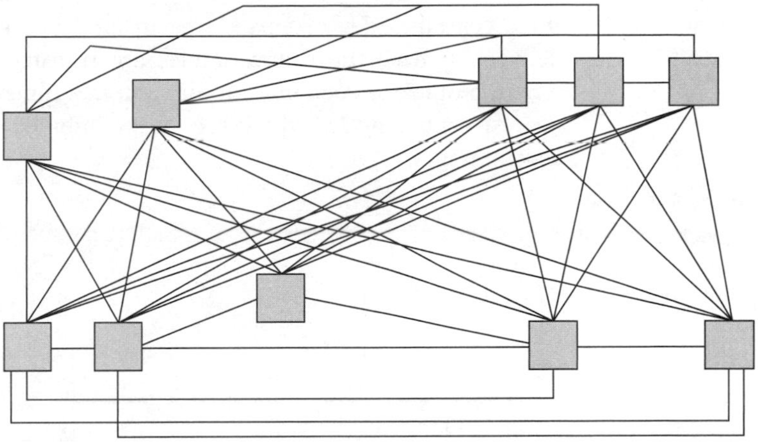

(b)

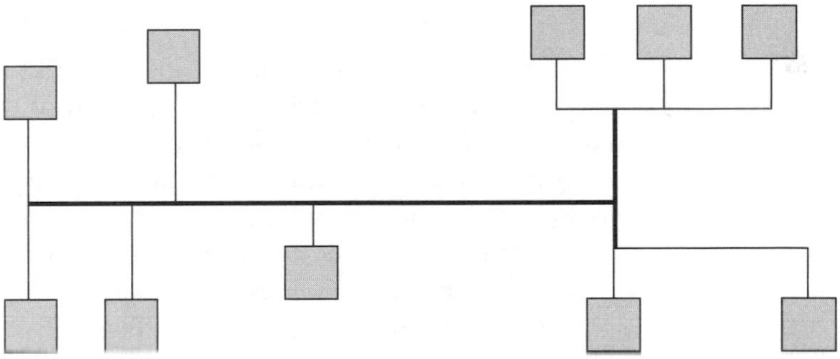

Point-to-point network topology is impractical for all but very small networks. For larger networks, a segmented approach that shares links among end nodes is more efficient. Note the similarity of Figure 9-2(b) to a modern roadway system. Individual end nodes are served by low-capacity links that connect to higher-capacity shared links, just as driveways connect houses to residential streets. Smaller shared links are connected to larger shared links in the same manner that residential streets connect to higher-capacity roads and highways.

If the end nodes in Figure 9-2(b) were parts factories and assembly plants, and the links among them were roads, how would you design a delivery service to move goods among them? You could buy a large number of trucks and use each truck to deliver goods between two specific locations. In the worst case of goods needing to flow between every pair of end nodes, you would need (n–1)! trucks. How could you efficiently move the goods with fewer trucks?

Most large-scale delivery services use a **store and forward** system to connect source and destination nodes. Figure 9-3 shows the network in Figure 9-2(b) with three central nodes of small capacity and one of large capacity. In a physical delivery system, the central nodes are transfer points located at or near the junction of major roadways or air routes. Shipments move from end nodes to the nearest central node, where they are combined into larger shipments to other central nodes.

Figure 9-3 ▶

Shared connections and central nodes

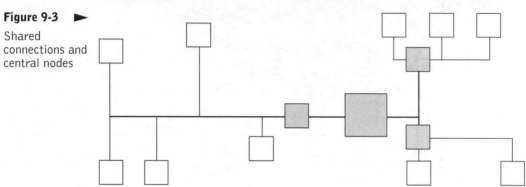

Computer networks and shipping networks are similar in that both connect large numbers of end nodes using a system of interconnect transmission (shipment) routes and central nodes. Efficiency is paramount for both network types to ensure reliable and rapid movement of shipments or messages among end nodes with minimal investment in network infrastructure. Both network types achieve reliability and cost-efficiency by carefully organizing end nodes, shipping or transmission routes, and central nodes.

Network topology is an essential factor in computer network efficiency and reliability. Network topology can be referred to in physical or logical terms. **Physical topology** refers to the physical placement of cables and device connections to those cables. **Logical topology** refers to the path that messages traverse as they travel among end and central nodes. Logical topology can differ from the underlying physical topology, as later examples will illustrate.

There are three common network topologies:

▶ Star

▶ Bus

▶ Ring

The characteristics that differentiate these topologies include the length and routing of network cable, type of node connections, data transfer performance, and susceptibility of the network to failure.

The **star topology** uses a central node to which all end nodes are connected (see Figure 9-4). The central node can be a transfer point, in which case it relays messages among end nodes. The central node can also be an end node in its own right, such as a mainframe computer, in which case it accepts messages addressed to itself and relays messages addressed to other nodes.

Figure 9-4 ▶

Star topology

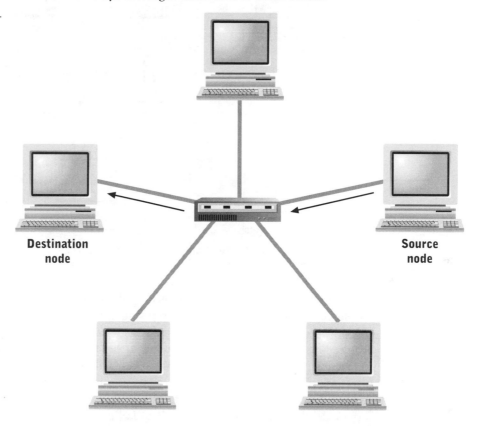

Destination node

Source node

The primary advantage of the star topology is relatively simple wiring. A transmission line is routed from each end node to the central node, which is typically located in a central location on the floor where the network is located or somewhere within the building if the network spans more than one floor. The primary disadvantage of the star topology is that failure of the central node disables the entire network.

The **bus topology** connects each end node to a common transmission line (see Figure 9-5). Transmitted messages travel from an end node across the line to all other attached nodes, each of which listens for incoming messages and ignores those not addressed to it. Terminating resistors are located at each end of the bus to absorb signals and prevent them from bouncing back down the common transmission line.

Figure 9-5 ▶

Bus topology

Destination node

Source node

The primary advantages of the bus topology are relatively simple wiring and low susceptibility to failure. The bus transmission line is installed from one end of a floor or building to the other end in an easily accessible place, such as above the drop ceiling in a hallway. Each end node attaches to the bus with its own cable and connection. There is no central node that can fail and disable the entire network. Also, failure of any one end node does not prevent other nodes from transmitting and receiving messages. However, the bus itself is a central point of failure. Also, it is possible, though unusual, for a node to fail in such a way that it floods the bus with noise or spurious packets, thus disabling the entire network.

The **ring topology** connects each end node to two other end nodes. The entire network forms a closed loop (see Figure 9-6). Messages are passed around the ring in one direction. Each node monitors the incoming link and removes messages addressed to it. Messages addressed to other nodes are repeated on the outgoing link. Some networks use two rings carrying messages in opposite directions.

The primary advantages of the ring topology are long maximum network length and low susceptibility to noise and distortion. Because each node repeats outgoing messages, there is little degradation in signal quality as a message traverses the ring. The disadvantages of the ring topology are susceptibility to failure and difficulty in adding, removing, or moving nodes. Rings are susceptible to failure because any node failure breaks the ring. Reconfiguring the network is relatively difficult compared to other topologies because each node is connected to two others. Two links must be changed when adding, deleting, or moving a node.

Figure 9-6 ▶

Ring topology

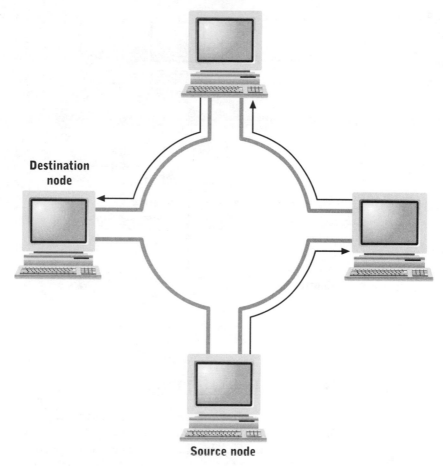

The strengths of two different topologies can be combined by using one topology for physical layout and another for message routing. Star topology is the most common physical topology because it is the simplest to wire and the easiest to reconfigure. A physical star network can be a logical bus or logical ring network. The logical topology of the network is implemented within the central node. For a physical star/logical bus network, each central node connection actually connects to an internal bus (see Figure 9-7).

Figure 9-7 ▶

Physical star/
logical bus
topology

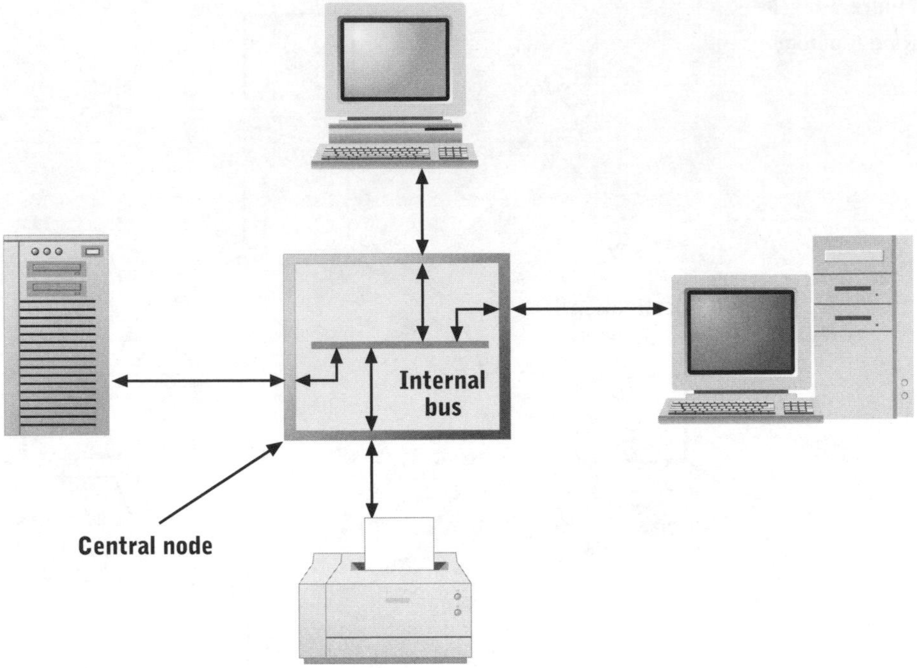

Internal
bus

Central node

ADDRESSING AND ROUTING

This section discusses how messages sent by end nodes find their way through transmission lines and central nodes to their ultimate destination. There are many types of central nodes that can be used in a network including hubs, switches, and routers. Although they have significant internal differences, they share similar methods for routing messages from sender to receiver. To simplify the discussion of routing, we'll ignore their differences and continue to use the generic term *central node* to describe any device that forwards messages. Later in the chapter, we'll describe the various types of central nodes in more detail.

Consider a typical college or corporate campus network (Figure 9-8). Entire buildings, such as Jones Hall, or parts of buildings, such as the first three floors of Smith Hall, are organized into small networks wired as physical stars. Each network covering a floor or building is called a **local area network (LAN)**.

A zone network connects all LAN central nodes in a group of buildings. A zone central node connects each zone network to the campus backbone network. A campus central node connects the campus backbone network to the Internet. The entire campus network is a **wide area network (WAN)**, including end nodes, LANs, zone networks, the campus backbone network, and central nodes.

Messages can travel through many transmission lines and central nodes before reaching their destination. Each message includes a destination address that each central node examines to determine where to send the message. But how does a central node "know" the addresses of other nodes in its network and how does it forward a message to a distant address?

Local Area Network Routing

Each central node maintains a table of node addresses and transmission lines or connection ports, called a **routing table,** and uses that table to make routing decisions. Each time an end or central node is powered on, it sends a message announcing its presence and its address to all nearby nodes. Central nodes listen for such messages and update their tables appropriately.

The exact procedure for routing a message between two end nodes in the same LAN depends on the logical network topology. For a logical bus, the central node examines the destination address to see whether it is local. If the address is local, the central node sends the message to all other local nodes. If the address is not local, the central node forwards the message to the next central node. The central node needs to distinguish only between local and nonlocal addresses to determine whether to forward the message to another central node.

Figure 9-8 ▶

A typical large campus network

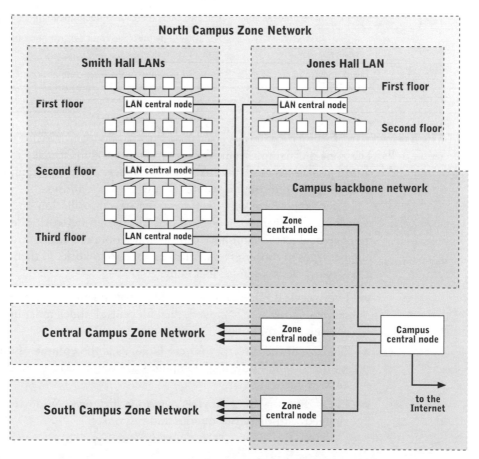

Routing of local messages within a logical star network is more complex. Because each node is attached to its own transmission line, the central node must keep track of which transmission line is connected to which address. A local message is routed from the source node's transmission line to the destination node's transmission line. The central node connects local sending and receiving nodes as needed for each incoming message.

Messages coming from outside the local network to a local node are handled in a manner similar to messages between local nodes. For logical bus networks, messages arriving from another central node are copied onto the local bus. For a logical star network, the switch looks up the target address and copies the message to the target node's transmission line. Table 9-1 summarizes the routing decisions made by LAN central nodes.

Table 9-1 ▶

LAN central node routing decisions

Source and Destination	Routing Action
Local node to local node	For bus: Propagate message through local medium
	For star: Copy message from source node's transmission line to target node's transmission line
Nonlocal node to local node	For bus: Copy message to local medium
	For star: Copy message from central node to target node's transmission line
Local node to nonlocal node	Forward message to the next central node

Wide Area Network Routing

The zone and campus central nodes in Figure 9-8 must make more sophisticated routing decisions than the LAN central nodes. Zone and campus central nodes need to know the route or routes to any network address.

There are several ways to store and access information on all addresses. One way is to store a master directory of all known network addresses on one or more central servers and have central nodes obtain routing information by sending messages to those servers. Such an approach works in theory, but in practice it has many shortcomings, including:

▶ Every central node must know the physical route to one or more directory servers. If a server is moved, then all central nodes must update their server routing information.

▶ The size of the directory is very large, as is the volume of directory service requests. Many powerful computers are required to store the directory and answer queries.

▶ Every directory query travels over long distances. Wide area networks are clogged with directory queries and responses.

► Directory updates must be copied to each server, requiring either multiple update messages (one for each server) or regular synchronization of directory content among servers. Either approach further clogs wide area networks.

Most wide area network routing protocols use a distributed approach to maintaining directories. Each central node knows the addresses and physical locations of other nodes on its network and knows other nearby central nodes and the groups of addresses that they control. Each central node also has a default destination for addresses that it doesn't know. Central nodes periodically exchange information, so they all are kept abreast of changes to network nodes and topology.

For example, consider a message sent from an end node in Jones Hall to an end node in the south campus zone network in Figure 9-8. The Jones Hall LAN central node knows the addresses of all Jones Hall end nodes and the address of the north campus zone central node. When the central node receives a message to any node outside its LAN, it forwards the message to the north campus zone central node.

The north campus zone central node examines the destination address and compares it to addresses in its routing table. The LAN central nodes in Jones Hall and Smith Hall periodically exchange routing information with the north campus central node so it knows the range of addresses serviced by each LAN central node. Because the destination address doesn't match any of its known address ranges, the north campus central node forwards the message to the campus central node.

The campus central node knows that the destination address falls within a range of addresses controlled by the south campus central node, so it forwards the message to that central node. The south campus central node knows which addresses are controlled by which LAN central nodes and forwards the message to the appropriate central node, which in turn forwards the message to its final destination.

This description of distributed wide area routing is generic and omits many details. There are many ways to implement distributed wide area routing, but the method just described approximates the standard Internet routing method used in most modern networks.

MEDIA ACCESS CONTROL

If multiple nodes attempt to transmit across the same medium at the same time, their messages will mix, producing noise or interference, which is called a **collision**. Methods of dealing with collisions fall into two broad categories—those that allow collisions but detect and recover from them, such as CSMA/CD, and those that avoid collisions altogether, such as token passing. Nodes follow a **media access control (MAC)** protocol to determine when they may access the shared transmission medium.

Carrier Sense Multiple Access/Collision Detection (CSMA/CD) is a MAC protocol commonly used in bus network topologies. The basic strategy is not to avoid collisions, but to detect and recover from them. The protocol details are as follows:

► A node that wants to transmit listens (carrier sense) until no traffic is detected.

► The node then transmits its message.

► The node listens during and immediately after its transmission. If abnormally high signal levels are heard (a collision is detected), then the node ceases transmission.

► If a collision is detected, the node waits for a random time interval and then retransmits its message.

The **token passing** MAC protocol is commonly used in ring network topologies. A control message called a **token** is passed from node to node and only the node that "possesses" the token is allowed to transmit messages. All other nodes can only receive and repeat messages. The token is passed from node to node in a predetermined order that includes all nodes on the network. A node must transmit the token to another node after a specified time interval or as soon as it has no more messages to transmit. Token passing is no longer used in LANs and rarely employed in WANs.

The primary advantage of CSMA/CD is its simplicity. There are no tokens to be passed between nodes and no inherent ordering of nodes in the network. Nodes can be added or deleted without updating the token passing order. The hardware and software that implement the protocol are simpler, faster, and less expensive than hardware and software that implement more complex token-passing protocols.

The primary disadvantage of CSMA/CD is its potentially inefficient use of data transfer capacity. Network transmission capacity is wasted each time a collision occurs. As network traffic increases, collisions also become more frequent. At high traffic levels, network throughput decreases due to excessive numbers of collisions and message retransmissions. Figure 9-9 shows the effect of this phenomenon on effective communication capacity. In networks that use CSMA/CD, network throughput is typically no more than one-half the theoretical data transfer rate.

Figure 9-9 ►

Effect of CSMA/CD on network throughput

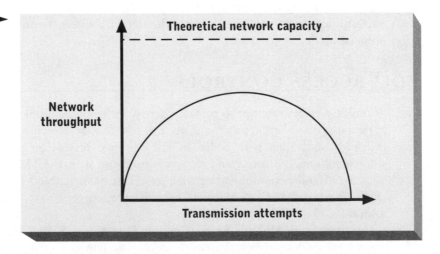

NETWORK HARDWARE

Network topology, addressing and routing, and MAC functions are carried out by network hardware devices that include:

► Network interface units (NIUs) or network interface cards (NICs)
► Hubs
► Bridges
► Routers
► Switches

Table 9-2 summarizes the functions of each device type. Network hardware is constantly evolving, and device functions are frequently combined within a single device, particularly in later generations of hardware. When selecting devices, it is important to understand their functional capabilities, because device names might be misleading.

Table 9-2 ►

Network hardware devices

Device	Function
NIU	• Connects an end node to the network • Performs appropriate MAC and message forwarding functions • Acts as a bridge between computer system bus and a LAN
Hub	• Acts as a central connection point for LAN wiring • Implements the logical network topology • Usually bridges a LAN to a WAN
Bridge	• Connects two separate networks to form a single virtual network • Copies packets between networks as needed
Router	• Connects two or more networks • Forwards packets between networks as needed • Makes intelligent choices among alternate routes
Switch	• Connects networks or network nodes to form virtual networks on a per-packet basis • Quickly routes packets using hardware-based switching

Network Interface Units

A device that connects a network node such as a computer or a network printer to a network transmission cable is called a **network interface unit** (**NIU**) or **network interface card** (**NIC**). An NIU for a single computer is usually a printed circuit board, or card, attached directly to or inserted in a slot in the system bus. An operating system device driver controls the NIU and directs the hardware actions that move packets between the NIU and primary storage. An NIU for a peripheral device such as a printer is more complex because it can't rely on the processing and storage resources of a complete computer system.

In a bus network, the NIU scans the destination address of all packets and ignores all those not addressed to it. In ring networks, the NIU scans the destination address of all packets and retransmits all packets not addressed to it. When a correctly addressed packet is received, the NIU stores the packet in a buffer and generates an interrupt on the system bus. The NIU also implements media access control functions including listening for transmission activity, detecting collisions, and retransmitting messages in CSMA/CD networks.

Hubs

A **hub** is a central connection point for nodes in a LAN. Most hubs are Ethernet devices that perform the central node function depicted in Figure 9-7. They combine separate point-to-point connections between nodes and the hub into a single shared transmission medium by repeating all incoming packets to every connection point. As technology has progressed, the role filled by hubs has been assumed by more complex devices such as routers and switches. But hubs are still available as low-cost alternatives for home and small office networks.

Bridges

A **bridge** copies or repeats packets from one network to another. A bridge's complexity and exact functions depend on the differences between the connected networks. The simplest bridges connect networks that use identical transmission speeds, packet types, and protocols. More complex bridges connect dissimilar networks and translate packet formats and network protocols.

A bridge scans packets on each network for destination addresses of nodes on the other network and copies those packets onto the other network. As the bridge scans network packets it also looks at source addresses and updates internal tables of node addresses on each network segment.

Bridges are most commonly used to:

► Construct a virtual LAN from two separate LANs
► Divide a network into segments in order to minimize congestion

A network designer sometimes needs to build a LAN that is larger than allowed by design standards. For example, a 100-Mbps Ethernet LAN cannot be longer than 210 meters. If a 300-meter LAN is required, then two shorter LANs can be joined with a bridge. Bridged LANs are sometimes called **virtual LANs**.

If a network is routinely overloaded with traffic, throughput can be improved by dividing the network into two or more segments and joining the segments with bridges. Nodes that have high communication volume with one another are attached to the same network segment, which minimizes the number of packets that need to cross the bridge(s).

Routers

A **router** intelligently routes and forwards packets among two or more networks. A router constantly scans the network to monitor traffic patterns and network node additions, modifications, and deletions. Routers use this information to build an internal "map" of the network. Routers periodically exchange information in their internal tables with other routers to gain knowledge of networks beyond those to which they are directly connected. They use this information to forward messages from local nodes to distant recipients and to choose among multiple possible routes to a recipient.

Unlike hubs and bridges, routers can be connected to more than two networks and can forward packets based on information other than the destination address. For example, an organization might have two internal networks—one dedicated to "ordinary" data traffic and another dedicated to audio or video streams. A router can be configured to examine packets arriving from an external network and forward them to one of the two internal networks based on their content. A router might also be connected to two Internet service providers (ISPs) and route outgoing packets to a specific ISP based on criteria such as shortest path to recipient or maximal use of the least expensive ISP. Routers can also examine incoming and outgoing packets and discard those with errors.

A stand-alone router is essentially a special-purpose computer with its own processor and storage. Routing functions can be embedded within other devices such as LAN hubs or general-purpose computers. Any computer system with multiple NIUs connected to different segments or networks can be a router if the proper software is installed. Routing software is usually a standard network operating system component, and might or might not be enabled by the server administrator.

Routing is not a computationally complex task, but it does require fairly extensive I/O capabilities. Every network packet must be examined and forwarded. In a busy network, packet volume can consume most or all of the bus capacity of a general-purpose computer. Such a heavy load can leave insufficient bus or network I/O capacity to perform server functions such as file transfer and printer sharing.

Switches

Like a hub, a **switch** generally has a dozen or more input connections for computers and other network nodes. Unlike a hub, each input connection is treated as a separate LAN. A switch examines the destination address of each incoming packet and temporarily connects the transmission line of the sender to the transmission line of the receiver. The switch creates a new virtual LAN for each packet and destroys the virtual LAN as soon as the packet has reached its destination.

Switches dramatically increase network performance because:

► Connection decisions are made by hardware based only on the destination address

► Each virtual LAN has only one sending and one receiving node, thereby eliminating congestion

In a router, packets are temporarily stored in a buffer, examined by a CPU, and forwarded according to the destination decision. The packet is then copied out of the buffer to the appropriate network segment. The entire process is complex and there is a significant delay between the packet receipt and retransmission. In a switch, only the packet destination address is examined. Specialized hardware quickly connects the appropriate network segments and the packet proceeds through the connection with very little delay. In essence, switches trade intelligent decision-making for connection speed.

Switching is particularly advantageous for LANs that use CSMA/CD. Switches internally segment the LAN, reducing or eliminating collisions and retransmissions. Switches can also be used in place of bridges to join LANs with multiple segments. As with bridging, the network designer must group network nodes into LANs based on shared traffic in order to minimize the number of packets that must be replicated across LANs.

OSI NETWORK LAYERS

In the late 1970s, the International Standards Organization (ISO) developed a conceptual model for network hardware and software called the **Open Systems Interconnection (OSI) model**. The OSI model is useful as a general model of networks, a framework for comparing networks, and an architectural roadmap that enhances interoperability among various network architectures and products. The OSI model organizes network functions into seven layers (see Figure 9-10). Each layer uses the services of the layer immediately below it and is unaware of the internal functions of other layers.

Application Layer

The **application layer** contains programs that make and respond to high-level requests for network services. Programs in the application layer include end-user network utilities such as Web browsers and e-mail clients, network services embedded in the operating system (such as service routines that access remote files), and network service providers such as FTP and Web server programs.

The OSI application layer of one network node generates a service request and forwards it to the OSI application layer of another network node. The application layers of the two network nodes "talk" to each other (shown by the topmost dashed arrow in Figure 9-10) using all of the lower-level OSI layers on both nodes as a communication channel (shown by solid arrows in Figure 9-10). Other layers also communicate with their counterparts in this way and use the lower-level layers to transmit messages.

Figure 9-10 ▶

Layers of the OSI network model

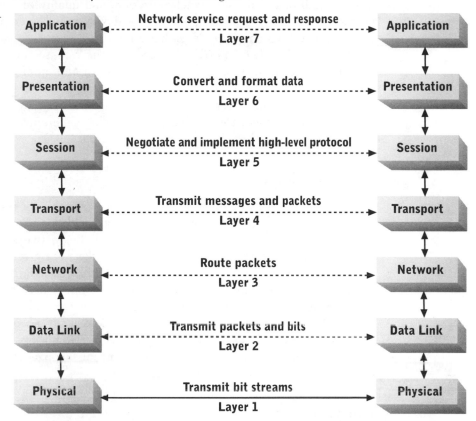

Presentation Layer

The **presentation layer** ensures that data transmitted by one network node is correctly interpreted by the other network node. Specific tasks that the presentation layer might perform include encryption and decryption, compression and decompression, converting data between EBCDIC and ASCII, and font substitution. The presentation layer is primarily used by applications that format data for user display. For example, the presentation layer of a Web browser formats compressed JPEG images for visual display and decides which font(s) to use for text display.

Session Layer

The **session layer** establishes and manages communication sessions. When a session is first established, the session layers negotiate protocol parameters such as timeout, half or full duplex, synchronization, and quality of service. Once the protocol has been established, the session layers monitor communication to detect and deal with any problems that arise.

For example, when an order is placed over the Web, session layers negotiate session parameters such as encrypted transmission and monitor the session for disruptions. If the communication channel is broken before the transaction is completed, the session layer on the vendor's node ensures that the transaction is completely abandoned, for instance, by canceling an unfinished order so that charges aren't applied to the customer's credit card.

Transport Layer

The **transport layer** formats messages into packets suitable for transmission over the network. The transport layer places messages within a packet data area and adds required header and trailer information including network addresses, error detection data, and packet sequencing data. Packets are given to the network layer for delivery to the recipient. When packets are received, the transport layer examines them for errors and requests retransmission as necessary.

If a message doesn't fit within a single packet, the sending transport layer divides the message content among several packets, adds a sequence number to each packet, and adds special coding within the first and last packets. The receiving transport layer requests retransmission of any missing packets and reassembles data from multiple packets in the proper order.

Network Layer

The **network layer** routes packets to their proper destination. As described earlier in this chapter, the network layers of sending nodes typically route packets to the nearest central node. Network layers within the central node interact with one another to exchange routing information and update internal routing tables. Network layers within end nodes announce their presence to other end nodes and central nodes when they are initialized.

Data Link Layer

The **data link layer** is the interface between network software and hardware. In an end node such as a computer system or network printer, the NIU and device drivers implement the data link layer. Device drivers manage packet transfer from secondary or primary storage to the NIU, which transmits packets as bit streams across the physical network link. The media access protocol is also implemented within the data link layer.

Physical Layer

The **physical layer** is the layer at which communication between devices actually takes place. The physical layer includes hardware devices that encode and decode bit streams and the transmission lines that transport them.

TCP/IP

The **Transmission Control Protocol (TCP)** and **Internet Protocol (IP)**, jointly called **TCP/IP,** were originally developed for the U.S. Department of Defense (DOD) Advanced Research Projects Agency Network (ARPANET) in the late 1960s and early 1970s. Many DOD researchers worked at universities, which soon adopted ARPANET technology in their own networks. Networks based on TCP/IP eventually evolved into the Internet.

Most of the services that we normally associate with the Internet are delivered via TCP/IP. These services include file transfer via the File Transfer Protocol (FTP), remote login via the Telnet protocol, electronic mail distribution via the Simple Mail Transfer Protocol (SMTP), and access to Web pages via the Hypertext Transfer Protocol (HTTP). TCP/IP is the glue that binds private networks together to form the Internet and the World Wide Web.

TCP/IP predates the OSI model by almost a decade. Thus, it is understandable that TCP/IP protocol layers correspond poorly to the OSI model. IP is roughly equivalent to the OSI data link, network, and transport layers. TCP is roughly equivalent to the OSI session layer.

IP provides packet routing and forwarding services to higher network layers. IP is independent of the underlying physical network layer and effectively hides it from layers above. An IP layer is implemented for virtually every known physical network layer. IP accepts packets called **datagrams** from TCP and other session layer protocols.

IP translates datagrams into a format suitable for transport by the physical network. If a datagram is larger than the physical layer data transfer units—for example, an Ethernet packet—then the IP layer divides the datagram into appropriately sized units and transmits them individually. IP attaches header information to each unit, including its sequence within the datagram. The IP layer at the receiving end reassembles units in the proper order and passes the datagram to TCP.

IP assumes that a datagram will traverse multiple networks via nodes called gateways. It determines the transmission route via a number of related protocols including the **Internet Control Message Protocol (ICMP)** and the **Routing Information Protocol (RIP)**. A **gateway** is any node that connects two or more networks or network segments. A gateway might be physically implemented as a workstation, server, hub, bridge, router, or switch. Figure 9-11 shows the TCP layers involved in a route between two nodes that includes two gateways. Note that only the IP layer is implemented within the gateways.

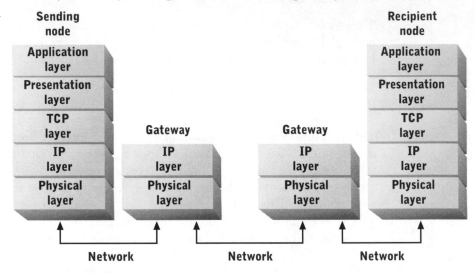

All IP nodes are identified by a unique 32-bit address of the form *nnn.nnn.nnn.nnn*, where *nnn* is a decimal number between 0 and 255. Each gateway maintains a routing table of node IP addresses and physical network addresses such as Ethernet addresses. Each node and gateway knows the physical network address of at least one other gateway. If a received message contains an IP address that doesn't appear within a gateway's internal tables, the packet is forwarded to the default gateway.

IP tables can include entries for partial IP addresses. For example, most node addresses within the University of New Mexico begin with 129.24. A node or gateway might not know the physical address for a specific node, such as 129.24.8.1, but its routing tables might contain the physical address of a gateway to any node with 129.24 in the first 16 address bits. IP nodes periodically exchange routing information to keep their tables current. Nodes and gateways announce their presence to nearby devices whenever they are started.

When TCP/IP was first developed, 32 bits seemed large for a network address. Today, there aren't enough unique 32-bit IP addresses for all of the world's Internet-attached devices. In the late 1990s, an IP standard called **IPv6** was developed with 128-bit addresses. To date, the standard hasn't been widely adopted due to the high cost of updating the IP layers of all devices currently connected to the Internet.

IP is an example of a **connectionless protocol**, in which the sender does not attempt to verify the existence of a recipient or ask its permission before sending data. IP sends a datagram to the appropriate gateway and assumes that the datagram will find its way to the intended recipient. IP cannot implement some forms of error control because it is connectionless. Because datagrams are "launched" toward a recipient IP layer without telling it that data is coming, there is no way for the recipient to know whether a datagram is "lost" in transit. The recipient node cannot send a negative acknowledgment or request retransmission.

TCP is a **connection-oriented protocol**, which provides the required framework to check for lost messages by explicitly establishing a connection with an intended recipient before transmitting messages. TCP can perform several connection management functions, including verifying receipt, verifying data integrity, controlling message flow, and securing message content. The sender and recipient TCP layers maintain information about one another, including message routes, errors encountered, transmission delays, and the status of ongoing data transfers.

TCP uses a positive acknowledgment protocol to ensure data delivery. A recipient TCP layer sends an ACK signal to the sender TCP layer to acknowledge data receipt. The sender TCP layer waits for ACK signals for a specific time interval that is initially set based on the time required to establish a connection. It can be extended or shortened during the life of the connection to adapt to changing network conditions. If an ACK is not received, the sending TCP layer assumes that the message has been lost and retransmits it. If several consecutive messages are lost, the connection "times out" and the sender assumes that the connection itself has been lost.

TCP connections are established through a port and socket. A port is a TCP connection with a unique integer number. A **socket** is the combination of an IP address and a port number. Many ports are standardized to specific Internet services. For example, port 21 is the default TCP connection for FTP, and port 53 is the default TCP connection for name service queries. The socket 129.24.8.1:53 is the TCP connection for Internet name services at the University of New Mexico.

Technology
Focus

Voice over IP

Voice over IP (VoIP) is a family of technologies and standards that enables voice messages and data to be carried over a single packet-switched network. As discussed in Chapter 8, transmitting messages via packet-switched networks generally uses available transmission capacity more efficiently than circuit-switching. With VoIP, consumers of telephone services can place phone calls over the Internet or other networks using their computers, VoIP cell phones, or other VoIP-enabled devices, at a much lower cost than over a traditional public switched telephone network (PSTN).

Because of its advantages, many organizations are considering changing to VoIP. However, there are many challenges to be addressed by any organization that wants to use VoIP including:

- ► Choosing appropriate VoIP standards
- ► Acquiring and configuring compatible VoIP equipment and software
- ► Guaranteeing service quality
- ► Providing emergency telephone connectivity

One of the challenges of VoIP deployment is that there are many complex and competing standards. As with data networks, VoIP networks rely on a protocol suite—an integrated set of protocols, each of which performs a specific function such as creating connections, packetizing the audio signal, routing packets, ensuring quality of service, and ensuring call security. The oldest and most widely deployed VoIP protocol suite is **H.323**, which also addresses video and data conferencing. H.323 is an umbrella for many component protocols (see Figure 9-12). Many of the component protocols have multiple versions, which can cause compatibility problems for VoIP equipment and software that support different mixes of standards and versions.

Figure 9-12 ►

H.323 VoIP protocol suite

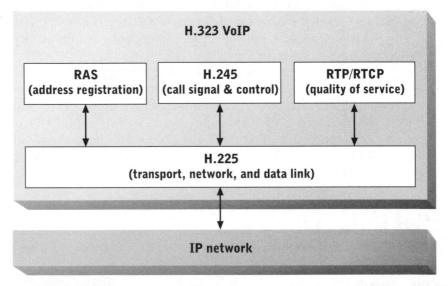

Further complicating VoIP deployment is the fact that H.323 has several competitors including Session Initiation protocol (SIP), H.248, and Media Gateway Control Protocol (MGCP). Each protocol suite offers a different mix of capabilities, limitations, and interoperability with competing standards. Different vendors support different protocol suites, and equipment capability varies widely among vendors and often among products from the same vendor. As a result, full interoperability is the exception rather than the norm.

VoIP equipment and software can be loosely divided into two groups: end nodes and intermediate nodes. An end node is a device that a user uses to participate in a telephone conversation. It may be a traditional telephone with an external IP interface, a telephone with an integrated IP interface, a computer, or a portable device that combines a telephone and computer capabilities. An intermediate node performs many of the same functions as routers, bridges, and switches in a data-oriented network including routing and protocol conversion. Intermediate nodes also perform call and connection management functions that are unique to VoIP.

In the PSTN, a caller opens a connection to the local switching center, receives a dial tone, and then enters a series of numbers to identify the intended recipient. The local switching center uses a protocol suite that describes how a telephone number is used to establish a two-way connection to the receiver across multiple networks and telephone switches. VoIP must perform similar call management functions, including:

► Establishing connections among VoIP end nodes on IP networks

► Establishing connections among VoIP end nodes and PSTN and other telephones

► Performing protocol conversions as needed

► Managing available network capacity to ensure quality of service

Different VoIP protocol suites allocate these tasks to different hardware and software components. At one extreme, most call management functions can be embedded in end nodes. At the other extreme, end nodes can be relatively simple devices and all call management tasks can be performed by VoIP intermediate nodes including specialized bridges, routers, and switches. Again, the plethora of available standards yields a large number of choices in end and intermediate nodes with widely varying capabilities and interoperability. Some vendors, such as Cisco, market general-purpose equipment that can be customized to particular protocols and functions by adding software and hardware options.

Guaranteeing service quality is not a simple matter in most IP networks. The IP protocol cannot guarantee minimal levels of throughput between a specific sender and receiver nor guarantee that packets will be delivered in a timely fashion, in the proper order, or at all. Transmission quality problems include:

► *Packet loss*—VoIP packets may be lost during transmission due to network congestion, router overload, and intermittent device failures. Call participants perceive lost packets as *dropouts*, which are missing audio segments of short or long duration.

► *Latency*—VoIP packets may be delayed in transit due to network congestion or a large number of routers in the chain of devices between end nodes. Call participants perceive periods of silence before the other participant's speech is heard. If the delays are too long, one party may assume that other is silent and start speaking, only to then hear both parties' overlapped speech.

► *Jitter*—VoIP packets may be subject to latency that fluctuates over short time periods. The user perceives jitter as a combination of periods of silence intermixed with overlapped voice signals.

If the same organization controls both ends of a call and the network in between, then intermediate nodes can detect VoIP packets and take actions to ensure their timely delivery. Those actions might include routing the packets over dedicated communication channels or assigning them higher priority during periods of heavy traffic. Little can be done to ensure service quality across a public IP network or when the same organization doesn't control both ends of the call.

Providing emergency telephone service is problematic with VoIP for two primary reasons. First, ordinary telephones are designed to operate during power outages but most VoIP nodes are not. An ordinary telephone receives all of the electrical power it requires through the same wires that carry the analog voice signal. If that signal is carried over a separate circuit switched network then telephones can usually function during power blackouts. Because most VoIP end and intermediate nodes require external power, they don't function during power outages.

The second problem with emergency VoIP telephone service is that few communities have implemented VoIP within their emergency (911) call centers. Traditional telephones and telephone networks are the only means of reaching most of those call centers. To provide reliable emergency telephone service, most organizations maintain at least some traditional telephone equipment and connections.

NETWORK STANDARDS

The Institute of Electrical and Electronics Engineers (IEEE) has drafted a number of telecommunication and network standards, collectively referred to as the **IEEE 802 standards**. These standards describe network hardware, transmission media, transmission methods, and protocols. Current standards and standards under development are listed in Table 9-3.

Table 9-3

IEEE 802
network
standards

IEEE Standard	Functional Description
802.1	Media access control (MAC)
802.2	Logical link control (LLC)
802.3	CSMA/CD and Ethernet
802.4	Token bus
802.5	Token ring
802.6	Metropolitan area networks (MANs)
802.7	Broadband LANs
802.8	Fiber optic LANs
802.9	Integrated data and voice networks
802.10	LAN/MAN security
802.11	Wireless LANs
802.12	Demand priority
802.14	Cable TV broadband networking
802.15	Wireless personal area networks
802.16	Broadband wireless access

The 802.1 and 802.2 standards roughly correspond to the data link layer of the OSI model. The 802.1 standard addresses a number of issues including media access and bridging. It also describes an architectural framework into which the other standards fit. The 802.2 standard addresses issues such as routing, error control, and flow control. The methods defined in these standards are incorporated into most of the other 802 standards.

The most common standards for local area networks are 802.3 (CSMA/CD in bus networks) and 802.11 (wireless LANs). Each standard has a number of subsidiary standards, indicated by letters added to the standard numbers, such as 802.11g, that cover additional implementation parameters.

Many commercial networking products are based on the IEEE standards. For example, Ethernet is based on the 802.3 standard, and Switched Multimegabit Data Services (SMDS) is based on 802.6.

The IEEE standards development process is heavily influenced by companies and organizations in the telecommunications and networking industries. Standards are developed by committees whose membership is drawn from industry, government, and academia. Standards committees work closely with other standards and regulatory agencies such as the American National Standards

Institute (ANSI), the ISO, and the U.S. Federal Communications Commission and its international counterparts. The development of standards can be influenced as much by politics as by technological considerations.

The committees set criteria for evaluation and then invite proposals for adoption as standards. Proposals are usually supplied by vendors based on products and technologies already under development. Occasionally the process backfires. Adopted standards sometimes represent significant compromises among very different proposals. Companies might decide not to adjust their developmental products to meet the compromise standard, thus generating an "orphan standard," such as 802.4. A company might also choose to release products that don't adhere to any published standard and hope that its technology becomes a de facto standard.

The marketplace ultimately decides which technologies and products will succeed. The standard-setting process usually provides an accepted target for implementation and, in the case of a widely implemented standard, a degree of compatibility among competing products.

Technology
Focus

Ethernet

Ethernet is a local area network technology closely related to the IEEE 802.3 standard. Ethernet was developed by Xerox in the early 1970s. Digital Equipment Corporation (DEC), which is now part of Hewlett-Packard, was the driving force in developing commercial Ethernet products in the late 1970s. DEC, Intel, and Xerox jointly published a specification for Ethernet networks in 1980. The 802.3 standard was published a few years later and incorporated most aspects of the 1980 Ethernet standard. The 802.3 standard has been extended many times.

Slower-speed Ethernet standards use a bus topology, Category 5 twisted pair cable, and CSMA/CD. The 802.3 packet format is shown in Figure 9-13. There is no provision for packet priorities or guarantees of quality of service, though incorporating those capabilities is a high priority for vendors and standard-setting organizations.

Figure 9-13 ►
Ethernet packet format

Synch (56 bits)	Start flag (8 bits)	Recipient address (48 bits)	Sender address (48 bits)	Length (16 bits)	Data (12,208 bits maximum)	CRC (32 bits)

The original Ethernet standard transmits data at 10 Mbps. The standard was updated (802.3u) in the late 1990s to increase transmission speed to 100 Mbps. There were several competing and incompatible proposals for 100-Mbps Ethernet, including the never-completed 802.13 and the dormant 802.12 standards.

Gigabit Ethernet is based on the 802.3z standard published in 1998 and the 802.3ab standard published in 1999. Several physical implementations are defined, each representing a trade-off between cost and maximum cable length.

10 Gigabit Ethernet is based on the 802.3ae standard ratified in 2002 and the 802.3ak standard ratified in 2004 (see Table 9-4). The standards support 10 Gbps transmission speeds over multiple fiber-optic cable types at LAN distances of 100–300 meters and WAN distances of 2–40 kilometers. Copper-based transmission is only supported over short distances though another copper-based standard is under development. 10 Gigabit Ethernet is the first Ethernet to completely abandon CSMA/CD in favor of a full-duplex point-to-point switched architecture.

Table 9-4 ▶

10 Gigabit Ethernet specifications

IEEE Standard	Configuration Name	Cable and Laser Type	Maximum Length
802.3ae	10GBASE-SR	Short wavelength laser over single or multimode fiber	300 meters
802.3ae	10GBASE-LX4	Wavelength-division multiplexed laser over single or multimode fiber	10 kilometers
802.3ae	10GBASE-LR	Long wavelength laser over single mode fiber	10 kilometers
802.3ae	10GBASE-ER	Long wavelength laser over single mode fiber	40 kilometers
802.3ak	10BASE-CX4	Infiniband (four twinaxial cables)	15 meters

Ethernet has been a popular networking technology for over two decades. Many competitors, such as the IBM token ring, have fallen by the wayside; others, such as Asynchronous Transmission Mode, have been relegated to niche markets. If ongoing efforts to update the technology are successful, then Ethernet should continue to dominate other network standards well into the current decade.

Business
Focus

Upgrading Network Capacity (Part II)

Note: Part I of this Business Focus was presented in Chapter 8.

The Bradley Advertising Agency (BAA) has decided to implement a Gigabit Ethernet network to support general data communications and file access. They will purchase an additional server dedicated to high-speed, high-capacity storage for video editing. The new server will have 420 GB of storage, expandable to 1.5 TB. Their reasons are:

► The existing twisted pair cable was tested and found to meet Category 5 requirements

► Gigabit Ethernet will meet current and near-term demands for video file sharing among video editing workstations

► Gigabit Ethernet equipment is significantly cheaper than Infiniband equipment

BAA is not sure how to integrate its microcomputers, video editing workstations, and servers into the Gigabit Ethernet network. They are considering three options:

1. Connect all microcomputers to the existing 10/100-Mbps Ethernet switch; connect the video editing workstations to a new 100/1000-Mbps switch; and connect both servers and both switches to a new Gigabit Ethernet switch.

2. Connect all microcomputers to the existing 10/100-Mbps Ethernet switch and connect the video editing workstations, the servers, and the 10/100-Mbps switch to a Gigabit Ethernet switch.

3. Create an entirely separate network to support video editing by installing a Gigabit Ethernet switch and connecting the video editing workstations and new server only to that switch.

Questions

1. What are the costs of each option? Which option provides the greatest network performance? Which option provides the greatest operational flexibility?

2. Which option should BAA choose to implement now? Why?

3. BAA anticipates that they will need to transmit high-quality video to and from client locations within the next three years. Video transfer modes will include transfer of large files and real-time video conferencing at HDTV resolution. Which current option will better enable BAA to adapt to that longer-term need?

SUMMARY

▶ Network topology refers to the spatial organization of network devices, the physical routing of network cabling, and the flow of messages from one network node to another. Topology can be physical (arrangement of node connections) or logical (message flow). Star topology connects every end node to a central node. Bus topology connects every end node to a shared transmission medium. Ring topology connects each end node to two neighbors, and all connections form a closed loop.

▶ LANs are interconnected to form WANs. Specialized network hardware devices route packets from source to destination. Packet routing within a LAN is usually handled by the LAN hub or switch. Packet routing across WANs uses a store and forward approach that is like the delivery and warehouse network of a package delivery service. Forwarding stations can be implemented using bridges, routers, or switches.

▶ A media access control (MAC) protocol specifies rules for accessing a shared transmission medium. CSMA/CD is commonly used in bus networks. Each node listens for an idle state, transmits a packet, listens for a collision, and retransmits after a random waiting period, if necessary.

▶ Network hardware devices include NIUs, hubs, bridges, routers, and switches. An NIU is the interface between a network node and the network transmission medium. A hub connects nodes to form a LAN. A bridge is a device that connects two networks or network segments and copies packets between them. Routers can connect to more than two networks and exchange information to improve routing decisions. Switches are high-speed devices that create virtual LANs on a per-packet basis.

▶ The Open Systems Interconnection (OSI) model is an ISO conceptual model that divides network architecture into seven layers: application, presentation, session, transport, network, data link, and physical. Each layer uses the services of the layer below and is unaware of other layers' implementations.

▶ TCP/IP is the core Internet protocol suite. IP provides connectionless packet transport across LANs and WANs. Internet addresses are 32-bit values, though a 128-bit address standard has been approved but not implemented. TCP provides connection-oriented packet transport to higher-level Internet service protocols including HTTP, FTP, and Telnet.

▶ The IEEE 802 standards cover many types of networks. The 802 standards are developed by committees composed of interested parties from industry, government, and academia. IEEE standards help ensure compatibility among products from competing vendors. Some standards, such as the Ethernet standard (802.3) are very successful, and others are never implemented.

In this chapter and the previous chapter, we described all of the hardware and only some of the software technology that underlies modern computer networks. In Chapter 10, we'll examine software tools used to build application software. In Chapters 11 and 12, we'll look at operating system technology, including resource allocation and file management systems. Then we'll return to computer networks in Chapter 13 and discuss distributed resources and applications.

--------------------------------- **Key Terms** ---------------------------------

10 Gigabit Ethernet
application layer
bridge
bus topology
Carrier Sense Multiple
  Access/Collision Detection
  (CSMA/CD)
collision
connection-oriented protocol
connectionless protocol
data link layer
datagram
Ethernet
gateway
Gigabit Ethernet
H.323
hub
IEEE 802 standards

Internet Control Message
  Protocol (ICMP)
Internet Protocol (IP)
IPv6
local area network (LAN)
logical topology
media access control (MAC)
network interface card (NIC)
network interface unit (NIU)
network layer
network topology
Open Systems Interconnection
  (OSI) model
physical layer
physical topology
port
presentation layer
ring topology

router
Routing Information Protocol
  (RIP)
routing table
session layer
socket
star topology
store and forward
switch
TCP/IP
token
token passing
Transmission Control Protocol
  (TCP)
transport layer
virtual LAN
Voice over IP (VoIP)
wide area network (WAN)

--------------------------------- **Vocabulary Exercises** ---------------------------------

1. The _____ standards define many aspects of computer networking.

2. The _____ layer establishes connections between clients and servers.

3. A(n) _____ is a device that connects two separate networks to form one virtual network.

4. A(n) _____ is a special-purpose packet representing the right to originate network messages.

5. The _____ layer determines the routing and addressing of packets.

6. A TCP/IP _____ is the combination of an Internet address and a port number.

7. The _____ layer refers to programs that generate requests for network services.

8. A network using physical _____ topology connects all end nodes to a central node.

9. A physical connection between two different networks is implemented using a(n) _____, _____, _____, or _____.

10. Packet loss can't be detected by a receiver if a(n) _____ protocol is in use.

11. Under TCP, a(n) _____ is the basic data transfer unit.

12. _____ transmits at 10 Mbps over twisted pair cabling.

13. The _____ defines a generic set of software and hardware layers for computer networks.

14. In a network that uses _____, a packet is passed from one network node to the next until it reaches the destination node.

15. When two messages are transmitted at the same time on a shared medium, a(n) _____ has occurred.

16. Under the _____ media access strategy, collisions can occur, but they are detected and corrected.

17. A(n) _____ protocol defines the rules that govern a network node's access to a transmission medium.

18. A microcomputer or workstation hardware interface to a network transmission medium is called a(n) _____

19. The _____ layer is responsible for data conversion functions such as compression and encryption.

20. The oldest and most widely deployed VoIP protocol suite is _____.

Review Questions

1. Describe the function of each layer of the OSI network model.

2. Describe the movement of packets in physical bus, ring, and star networks.

3. What are the comparative advantages and disadvantages of physical ring, bus, and star networks?

4. How does a packet from one LAN node find its way to a recipient on the same LAN? How does a packet find its way to a recipient on another LAN?

5. Describe the CSMA/CD media access protocol including the IEEE standard(s) on which it is based, how collisions are detected or avoided, potential impacts on network throughput, and its inclusion in current Ethernet networks.

6. What is a LAN hub? What functions does it normally perform?

7. What is the function of a bridge? How does it differ from a router? How does it differ from a switch?

8. As compared to a hub, how does a switch alleviate potential network throughput problems associated with CSMA/CD?

9. What is a connectionless protocol? What is a connection-oriented protocol? List one example of each protocol type.

10. How many bits are in an IP address? What is an IP port? What is an IP socket?

11. Describe various aspects of Ethernet networks including transmission speed, logical topology, supported transmission media, and relevant IEEE standards.

12. What protocols are commonly used to implement VoIP? Are all VoIP protocols compatible with one another?

13. Describe the service quality problems that can occur in VoIP. Why are these problems so difficult to solve?

Research Problems

1. Investigate the VoIP equipment and software offerings of a major networking or telecommunication vendor such as Cisco Systems (www.cisco.com), 3Com (www.3com.com), Tandberg (www.tandberg.net), Radvision (www.radvision.com), and Polycom (www.polycom.com). What VoIP and related standards are supported in each company's product line? What are the complexities of configuring a functional VoIP network using products from more than one company?

2. Investigate Multiprotocol Label Switching (MPLS), an emerging family of standards for routing and Internet traffic control. What current problems do the standards address? How do the standards solve those problems? Have the standards been fully ratified? Are there current products that adhere to the standards?

3. Investigate the Ethernet connectivity devices offered by a major vendor such as 3Com or Cisco. What is the range of features available in LAN switches? What is the range of features available in WAN switches and routers? What devices does the vendor offer that don't clearly fall into the categories described in this chapter? What is the cost per port of 100/1000-Mbps switches and 10-Gbps switches? What options are provided to link groups of hubs or switches?

Chapter 10

Application Development

Chapter Goals

- Describe the application development process and the role of methodologies, models, and tools

- Compare and contrast programming language generations

- Explain the function and operation of program translation software, including assemblers, compilers, and interpreters

- Describe link editing, and contrast static and dynamic linking

- Describe integrated application development software, including programmer's workbenches and CASE tools

Application software development is a complex process that follows a series of steps to translate users' needs into executable programs. There are automated tools to support each step, from user requirement statements through system models and program source code to executable code. Understanding the role and function of application development tools will make you a more efficient and effective analyst, designer, and programmer. Figure 10-1 shows the topics covered in this chapter.

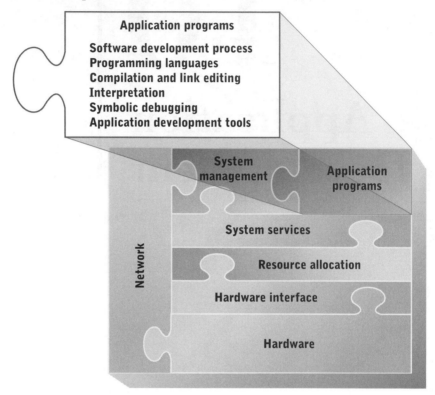

THE APPLICATION DEVELOPMENT PROCESS

The process of designing and constructing software translates users' information processing needs into CPU instructions that, when executed, address those needs. User needs are stated abstractly in a natural language, for example, "I need to process accounts receivable and payable." Software programs are detailed, precise statements of formal logic written as sequences of CPU instructions. Software development involves two translations—from an abstract need statement to a detailed implementation that satisfies the need, and from natural language to CPU instructions (see Figure 10-2).

Figure 10-2 ▶

Application development translates user needs to CPU instructions

Developing software requires significant effort and resources. A user need stated in a single sentence might require millions or billions of CPU instructions to address. Developing and testing CPU instructions requires software development methods and tools, highly trained specialists, and months or years of effort. Software has surpassed hardware to become the most costly component of most information systems.

Information system owners expect their investment in software development to produce software that meets their needs. Unfortunately, the complexity of software development creates many possibilities for error resulting in software that doesn't work or fails to meet user needs. The economic cost of software development errors can be much greater than the cost of developing the software. Reduced productivity, dissatisfied customers, and poor managerial decisions are just a few of the indirect costs of software that don't completely or correctly address user needs.

Systems Development Life Cycle

In this chapter we return to the systems development life cycle (SDLC) introduced in Chapter 1, but with a specific focus on software development.

Figure 10-3 shows the disciplines of the Unified Process (UP) and their organization into iterations of a typical software development project. The UP is a way of breaking the complex software development process into smaller and more manageable pieces. In this chapter, we're primarily concerned with the requirements, design, and implementation disciplines.

Figure 10-3 ►

Disciplines and
iterations of the
Unified Process

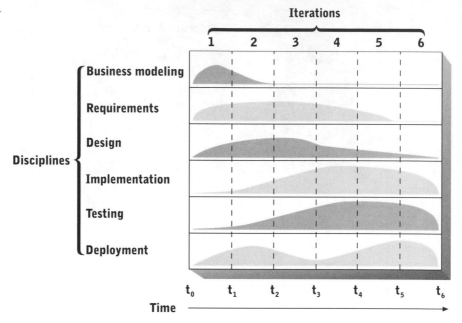

As part of the requirements discipline, members of the software development team interact with users to understand and document users' needs. Much of the translation from abstract concepts to specific and detailed facts occurs during requirements discipline activities. The components of a general statement such as, "I need to process accounts receivable and payable" are examined and documented in minute detail. What is an account payable or account receivable? What specific tasks are implied by the term "process"? What is the data content of system inflows and outflows? What data must be stored internally? What tasks are performed when, and by which people or programs? The answers to these questions are summarized in a variety of forms including text descriptions and diagrams that together compose a system requirements model.

System requirements models provide the detail needed to develop a specific system to meet user needs. The detailed functions and architecture of computer and related hardware, system software, and application software are determined by completing design activities, which produce **design models** that provide an architectural blueprint for system implementation. The blueprint specifies required features of system software such as operating systems and network server software, and required functions of application software subroutines, procedures, methods, objects, and programs.

Implementation activities acquire or construct software and hardware components that match the design models. Typically, computer hardware and system software are purchased, and some or all applications software is constructed, sometimes using purchased or previously developed software components. As acquisition and construction proceed, hardware and software are tested to ensure that they function correctly and meet user needs as described in the analysis model. The end result of the UP is a fully tested system that is ready to be put to productive use.

Methodologies and Models

Developers attempt to minimize errors by using proven development methodologies. A methodology is an integrated collection of models, tools, and techniques. The Unified Process is a methodology that employs object-oriented analysis, design, and deployment models. Class diagrams (see Figure 10-4) and other diagram types document user and system requirements. Design and deployment discipline activities also employ specialized model types. Many graphical models are supplemented with detailed text descriptions.

Tools

Many automated tools support software development. Some tools, such as word processors and general-purpose drawing programs, can be used to build many types of models. Other tools are customized to specific models and development methodologies. Related tools are often integrated into a single suite that supports an entire system development methodology, in much the same fashion that word processing, graphics, database, and spreadsheet programs can be integrated into a single "office" suite.

Software development tools vary with the target system deployment environment. Tools can be specific to a particular operating system, programming language, database management system, or type of computer hardware. Wide variation in underlying methodology, supported models and techniques, and target deployment environment makes proper tool selection a critical and difficult undertaking. The task is further complicated by a dizzying array of descriptive terminology and varying levels of support for different parts of the development process.

The declining cost of computer hardware and the increasing cost of application development labor have spurred the rapid development and widespread adoption of automated system and software development tools. Substituting automated tools for manual methods is a classic economic tradeoff between labor (such as analysts, designers, and programmers) and capital resources (sophisticated application development tools and the computer hardware required to run them).

In the early days of computers, hardware was so much more expensive than labor that it made economic sense to use labor-intensive processes for application development. As hardware costs decreased, this economic balance shifted, leading to the introduction of automated tools to support software development.

Figure 10-4 ▶

A class diagram

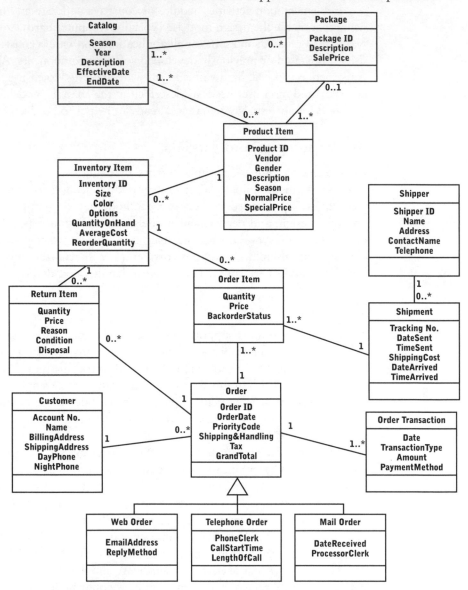

Programming languages, program translators, and operating system service layers are the earliest examples of this shift. They are all methods by which computing hardware can support software development directly. The modern proliferation of application development tools is a continuation of this economic shift. Modern analysts, designers, and programmers use a wide array of tools to support or completely automate software development tasks.

PROGRAMMING LANGUAGES

A **programming language** is used to instruct a computer to perform a task. Program instructions are sometimes called **code** and the person who writes the instructions is called a **programmer**. Developers of programming languages continually strive to make software easier to develop by:

▶ Making the programming language easier for people to understand

▶ Developing languages and program development approaches that require people to write fewer instructions to accomplish a given task

Programming languages have evolved through several types or "generations." Figure 10-5 summarizes the programming language evolution. Table 10-1 summarizes the characteristics of each programming language type, and details are provided in the following sections.

Figure 10-5 ▶

Programming language evolution

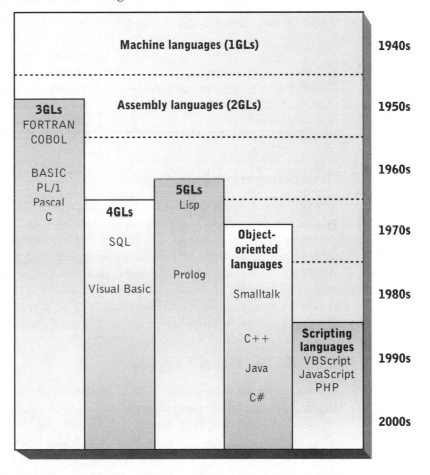

Table 10-1 ►

Programming
language
characteristics

Generation	Name or sample languages	Description or characteristics
1st	Machine language	Binary instructions for a specific CPU
2nd	Assembly language	Mnemonic instructions for a specific CPU
3rd	FORTRAN, COBOL, BASIC, PL/1, Pascal, and C	Instruction explosion, machine independence, and (usually) standardization
4th	Visual Basic, SQL	Higher instruction explosion, interactive and graphical I/O support, database support, limited nonprocedural programming, and proprietary standards
5th	Lisp, Prolog	High instruction explosion, nonprocedural programming, expert systems, and artificial intelligence applications
Object-oriented	Smalltalk, C++, Java, C#	High instruction explosion, support for modern code development, and code reuse methods such as inheritance and message passing
Scripting	VBScript, JavaScript, PHP	Very high instruction explosion, interpreted execution, primarily used for Web-based applications

First-Generation Languages

Binary CPU instructions, called **machine languages**, or **first-generation languages** (**1GLs**) are the earliest programming languages. It is difficult for people to remember and manipulate long strings of binary digits, yet that is what early machine language programmers did. Programmers had to remember the binary codes that represented each CPU instruction and had to specify all operands as binary numbers. Programming with binary numbers is tedious and error-prone. As primary storage capacity and program size grew, it became less and less feasible for people to develop error-free programs using only binary numbers.

Second-Generation Languages

A **second-generation language** (**2GL**) is more commonly known as an **assembly language**. The instructions and programming examples defined in Chapter 4 used a simplified assembly language. Assembly languages use a short character sequence called a mnemonic to represent each CPU instruction. Assembly language programmers can define their own mnemonics to represent the memory addresses of instructions and data items. Today, we usually use the term **variable** to describe a mnemonic that represents a data item memory address and the term **label** to describe a mnemonic that represents a program instruction memory address.

Short names are easier for people to manipulate than binary numbers. As a result, 2GLs became the most commonly used programming languages during the early and mid-1950s. 2GLS are still used today in some types of system programming, such as writing device drivers.

Because the CPU processes only binary digits, any language other than a 1GL must be translated into a 1GL before a CPU can execute it. An **assembler** is a program that translates an assembly language program into binary CPU instructions. An assembler reads the assembly language program file one instruction at a time. It translates each mnemonic into its corresponding binary digit sequence and writes those digits to an output file ready to be loaded into memory and executed by the CPU. Assemblers are the earliest example of automated program development tools.

Although 2GLs made programming easier, they did nothing to minimize the number of instructions that a programmer had to write to implement a task. Each assembler instruction is translated into a CPU instruction, so a 2GL program still directs every individual CPU action. As memory capacity and program size grew, it became more difficult for a programmer to specify every CPU action. To increase programmer productivity, the 1:1 correspondence between program instructions and CPU actions had to be broken.

Third-Generation Languages

Programming languages beyond the second generation allow programmers to specify many CPU actions with a single program instruction or statement. The one-to-many (1:N) relationship between later generation programming language statements and the CPU actions that implement them is called **instruction explosion**. For example, if 1000 machine instructions result from translating a 20-statement program, then the degree of instruction explosion can be stated as a 50:1 ratio.

Different programming languages have different degrees of instruction explosion. Within the same programming language, different types of statements have different degrees of instruction explosion. In general, statements that describe mathematical computation typically have low instruction explosion (10:1 or less), and statements that describe I/O operations typically have high instruction explosion (100:1 or greater).

A **third-generation language** (3GL) is a programming language that uses mnemonics to represent instructions, variables, and labels, and has a degree of instruction explosion greater than 1:1. The first 3GL was FORTRAN. Later 3GLs include COBOL, BASIC, PL/1, Pascal, and C. Some 3GLs are still used today, though their popularity fades with each passing year.

Like 2GL programs, 3GL programs must be translated into binary CPU instructions before the program is executed. Compilers, link editors, and interpreters are automated tools used to translate 3GL programs; they are covered in more detail later in this chapter. 3GL program translation is much more complex than 2GL program translation, so 3GL translation programs are much larger and consume more computer resources than assemblers.

A single 3GL program can be translated to execute on many different CPUs, a characteristic that is called **machine independence**. Compilers, link editors, and interpreters translate machine-independent instructions into machine language instructions for a specific CPU. Each CPU uses a different 3GL translation program that is customized to that CPU's instruction set.

Most 3GLs were developed before graphical user interfaces, database management systems, and the Internet. These limitations account for much of the decline in popularity of languages such as FORTRAN, COBOL, and C.

Fourth-Generation Languages

In the 1970s and 1980s, **fourth-generation languages** (**4GLs**) were designed to address 3GL limitations. They shared many features, including:

▶ Significantly higher instruction explosion than 3GLs

▶ Instructions or prewritten functions to implement interactive and graphical user interfaces

▶ Instructions to interact directly with an internal or external relational database

▶ Ability to describe processing requirements without precisely specifying solution procedures

The majority of 4GLs were proprietary, and many were optional components of database management systems. Few of the original 4GLs survive today, though some modern programming languages are their direct descendants. Two notable exceptions are Visual Basic and Structured Query Language (SQL). Visual Basic evolved from the 3GL BASIC and is adapted to modern software requirements. Visual Basic has a rich variety of graphical user interface capabilities and can interact directly with Microsoft Access databases. Visual Basic is widely used, but it lacks some features needed by "industrial strength" software.

SQL is specialized for adding, modifying, and retrieving data from relational databases, and is usually used in combination with database management systems such as Oracle or Microsoft SQL Server. SQL is not a full-fledged programming language because it lacks many features of a general-purpose programming language, including many control structures and user interface capabilities. SQL programs are usually embedded within other programs, such as Web-based applications written in scripting languages. Figure 10-6 shows two programs that perform similar functions, one in C and the other in SQL. The number of program statements is much larger in the C program, which indicates a lower level of instruction explosion.

Figure 10-6 ▶

Equivalent programs in SQL and C

```
                  Structured Query Language Example

open database banking;
select customer.acct_num, customer.name,balance+sum(transaction.
 amount)
from      customer,transaction
where     customer.acct_num=transaction.acct_num
group by  customer.acct_num,customer.name;



                          C Example

balance_report ()  {
    FILE  *cust_file, *trans_file;
    int     status,acct_num,a_num,balance,amount;
    char name [256];

    cust_file=fopen("customer","r");
    trans_file=fopen("transaction","r");
    status=scanf(cust_file, "%d%s%d\n", &acct_num,name, &balance);
    while (status != EOF {
        status=scanf (trans_file, "%d%d\n," &a_num, &amount);
        while (status != EOF){
            if (acct_num == a_num) {
                balance+-amount;
            }
            status=scanf(trans_file,"%d%d\n", &a_num,&amount);
        }
        printf("%d  %s  %d",acct_num,name,balance);
        trans_file=freopen("transaction","r");
        status=scanf(cust_file,"%d%s%d\n", &acct_num,name,&balance);
    }
    close(trans_file);
    close(cust_file);
    exit(0);
} /* end balance_report */
```

The programs also differ in their approach to solving a specific data retrieval problem. The C program describes a specific procedure for extracting and displaying data that includes control structures and many database management functions such as opening and closing files, and testing for end of file while reading records. In contrast, the SQL program contains no control structures, file manipulation, or output formatting commands. The SQL program states what fields should be extracted, what conditions extracted data should satisfy, and how data should be grouped for display. SQL is a **nonprocedural language** because it describes a processing requirement without describing a specific procedure for satisfying the requirement. The compiler or interpreter determines the most appropriate procedure for retrieving the data and generates or executes whatever CPU instructions are needed to do so.

4GLs are sometimes called nonprocedural languages, though the description is not completely accurate. Most 4GLs support a mixture of procedural and nonprocedural instructions. A 4GL typically provides nonprocedural instructions for database manipulation and report generation. Some 4GLs also provide nonprocedural instructions for interactively updating database contents. Other types of information processing such as complex computations require procedural instructions.

Fifth-Generation Languages

A **fifth-generation language** (5GL) is a nonprocedural language suitable for developing software that mimics human intelligence. 5GLs first appeared in the late 1960s with Lisp but weren't widely used until the 1980s. Lisp and Prolog are the most common general-purpose 5GLs, but there are many proprietary 5GLs used to build applications in areas such as medical diagnosis and credit application scoring.

A 5GL program contains a set of nonprocedural rules that mimic rules people use to solve problems. A rule processor accepts a starting state as input and iteratively applies rules to achieve a solution. Figure 10-7 shows a sample Prolog program for solving the problem of moving a farmer, fox, chicken, and corn across a river on a boat that only can hold two. The rules describe valid moves and unsafe states; for example, the chicken and corn can't be left without the farmer because the chicken would eat the corn. The program prints trace messages as it applies rules and prints the set of moves that solve the problem.

Figure 10-7 ▶

Sample Prolog
program

```
go (S,G) :- write ('start at state'),write (S),nl,path(S,G,[S]).

path (G,G,L) :- write ('SUCCESS, the path is:'),nl,ppt(L).

path (s(B,F,C,F,Cn),G,L) :-
move(s(B,F,C,F,Cn),s(Boat,Farmer,Chicken,Fox,Corn)),
   not(member(s(Boat,Farmer,Chicken,Fox,Corn),L)),
   not(unsafe(Boat,Farmer,Chicken,Fox,corn)),
   write('use this move to get to s('),
   write(Boat), write(','), write(Farmer), write(','), write(Chicken),
   write(','), write(Fox), write(','), write(Corn), write(')'), nl,
   path(s(Boat,Farmer,Chicken,Fox,Corn),
        G,
        [s(Boat,Farmer,Chicken,Fox,Corn) |L]), !.
/* s(Boat,Farmer,Chicken,Fox,Corn)   */

Move(s(e,e,e,Fox,corn),s(w,w,w,Fox,Corn)) :-
   Write('move farmer and chicken from east to west'), nl.

Move(s(w,w,w,Fox,Corn),s(e,e,e,Fox,Corn)) :-
   Write('move farmer and fox from west to east'), nl.

Move(s(e,e,Chicken,e,Corn),s(w,w,Chicken,e,Corn)) :-
   Write('move farmer and fox from east to west'), nl.

Move(s(w,w,Chicken,e,Corn),s(e,e,Chicken,e,Corn)) :-
   Write('move farmer and fox from west to east'), nl.

Move(s(e,e,Chicken,Fox,e),s(w,w,Chicken,Fox,w)) :-
   Write('move farmer and corn from east to west'), nl.

Move(s(w,w,Chicken,Fox,w),s(e,e,Chicken,Fox,e)) :-
   Write('move farmer and corn from west to east'), nl.

Move(s(e,e,Chicken,Fox,Corn),s(w,w,Chicken,Fox,Corn)) :-
   Write('move farmer only from east to west'), nl.

Move(s(w,w,Chicken,Fox,Corn),s(e,e,Chicken,Fox,Corn)) :-
   Write('move farmer only from west to east'), nl.

unsafe(X,X,Y,Y,_).
unsafe(X,X,Y,_,Y).

number(X,[X|T]).
number(X,[Y|T]) :- member(X,T).

ppt([]).
ppt([H|T]) :- ppt(T), write(H), nl.
```

Object-Oriented Programming Languages

3GLs and 4GLs have a similar underlying programming paradigm under which data and functions are distinct. Program instructions manipulate data and move it from source to destination. Data is processed by passing it among code modules, functions, or subroutines, each of which performs one well-defined task.

Software researchers in the late 1970s and early 1980s began to question the efficacy of this program/data dichotomy. They developed a new programming paradigm called **object-oriented programming** (**OOP**) that better addressed program reuse and long-term software maintenance. The fundamental difference between the object-oriented and traditional programming paradigms is that OOP views data and programs as two parts of an integrated whole called an **object**. Objects contain data and programs or procedures, called **methods**, that manipulate the data.

Objects reside in a specific location and wait for messages to arrive. A **message** is a request to execute a specific method and return a response. The response can include data, but the data included in a response is only a copy. The original data can be accessed and manipulated only by object methods. Figure 10-8 shows the relationships among data, methods, messages, and responses.

Figure 10-8 ▶

Objects, data, methods, messages, and responses in OOP

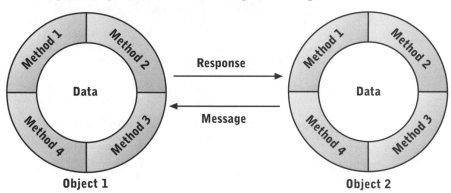

Smalltalk was the first commercial OOP language. C++ (an extension of C) and Java are more widely used OOP languages. Other languages, such as Visual Basic, have been extended with OOP concepts, and many proprietary OOP languages such as Oracle Power Objects have been developed.

Object-oriented programming and design is uniquely suited to developing real-time programs such as operating systems and interactive user interfaces. Most current operating systems and windowing environments such as Microsoft Windows, UNIX, and X Window were constructed with OOP languages. OOP languages and good object-oriented software design promote reusability and portability of source code. New objects can incorporate data and methods from existing objects through a feature called inheritance.

The fact that OOP languages are described by a name rather than a generation number is evidence that the traditional concept of language generations is breaking down. OOP languages are not a clear successor to 5GLs because they are neither nonprocedural nor exclusively used for developing artificial intelligence and expert system applications. Some OOP languages such as C++ and C# are direct extensions of 3GLs, and others such as Smalltalk and Java are more closely related to 4GLs.

Scripting Languages

Scripting languages evolved from 4GLs, though most now incorporate object-oriented programming concepts. A **scripting language** enables programmers to develop applications that do most of their work by calling other applications and system software. Scripting languages provide the normal set of control structures, mathematical operators, and data manipulation commands, but they extend these tools with the ability to call external software programs such as Web browsers, Web servers, and database management systems. Scripting languages enable programmers to assemble application software rapidly by "gluing" together the capabilities of many other programs.

VBScript, JavaScript, and PHP are widely used languages for developing applications that use a Web browser as their primary user interface. Many proprietary scripting languages are also provided with Web server and database server software.

Programming Language Standards

The American National Standards Institute (ANSI) and the International Organization for Standardization (ISO) set standards for some programming languages. Examples of ANSI standard languages include FORTRAN, COBOL, C, and C++. A programming language standard defines:

► Language syntax and grammar
► Machine behavior for each instruction or statement
► Test programs with expected warnings, errors, and execution behavior

Compilers, link editors, and interpreters are submitted to the standard-setting body for certification. Test programs are translated, and the behavior of the translator program and the executable code is examined. The behavior of the program is compared to the behavior defined by the standard to determine whether the translation program complies with the standard.

Standard programming languages guarantee program portability among operating systems and application programs. A program can be moved to a new operating system or CPU by recompiling and relinking or reinterpreting it with certified translation programs in the new environment. Portability is important because operating systems and hardware change more frequently than most program source code.

Setting standards for programming languages has become much less relevant over the last two decades. There are many reasons for this, including a significant movement of basic software research from academia into industry, reluctance by software companies to share proprietary technology, variability in applications and run-time environments, and increasing interdependence between application and system software. Modern software standards are frequently more concerned with system software and interfaces between programs than with the languages used to build programs.

COMPILATION

Figure 10-9 shows application development using a program editor, compiler, and link editor. Input to the program editor comes from a program template, CASE tool, programmer, or any combination thereof. The output of the program editor is a partial or complete program, called **source code**, in a programming language such as C++ or Visual Basic. Source code is normally stored in a file that is named to indicate both its function and programming language; for example, ComputePayroll.cpp, where cpp is an abbreviation for C++.

Figure 10-9 ▶

Program development with a program editor, compiler, and link editor

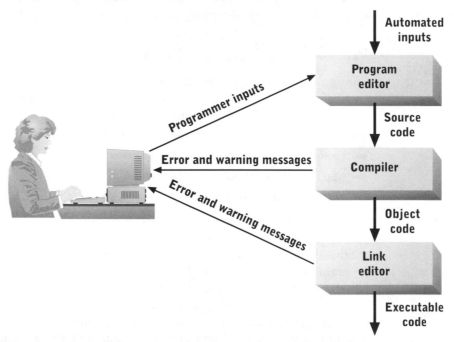

The compiler and link editor translate source code into executable code. **Executable code** is stored in a file such as an .exe file in Windows, and contains CPU instructions that are ready to be loaded and executed by an operating system.

The compiler and link editor each perform a different part of the translation process. The **compiler** translates some source code instructions directly into executable code and other source code instructions into library calls, which are further processed by the link editor. Library calls and link editing are described in detail in the next section. Compiler output is called **object code** and contains a mixture of CPU instructions, library calls, and other information needed by the link editor. Object code is usually stored in a file with an .o or .obj extension, for example, ComputePayroll.obj.

The compiler reads and translates source code instructions. While reading the source code, the compiler:

▶ Checks for syntax and other errors and issues appropriate warning or error messages

▶ Updates internal tables that store information about data items and program components such as functions and subroutines

▶ Generates either CPU instructions or library calls to carry out the source code instruction

The exact actions taken by the compiler vary depending on the source code instruction type. For most programming languages, there are four source code instruction types:

▶ Data declarations

▶ Data operations

▶ Control structures

▶ Function, procedure, or subroutine calls

Data Declarations

A **data declaration** defines the name and data type of one or more program variables. For example, consider the following C++ source code instructions:

```
float f_temperature;
float c_temperature;
```

These instructions declare two floating point variables named f_temperature and c_temperature. When a compiler reads a data declaration, it allocates memory to store the data item. The amount of memory the compiler allocates depends on the data type and the number of bytes used to represent that data type in the target CPU.

The compiler updates an internal table, called a **symbol table,** to keep track of the data names, types, and assigned memory addresses. Information stored in the table includes the variable name, what the name refers to, the data type, and memory location. Sample entries are shown in Table 10-2.

Table 10-2 ►

Name	Type	Length	Address
f_temperature	Single precision floating point	4	1000
c_temperature	Single precision floating point	4	1004

Data Operations

A **data operation** is any instruction that updates or computes a data value. Examples of data operations include assignment statements and computations. The compiler translates data operation instructions into an equivalent sequence of data movement and data transformation instructions for the target CPU. The compiler refers to entries in the symbol table to determine source and destination memory addresses for data movement instructions.

For example, consider a source code instruction that assigns a constant value to a program variable such as:

```
f_temperature = 212;
```

The compiler looks up f_temperature in the symbol table to find its memory address (1000) and generates a CPU instruction such as:

```
MOV 1000 <212>
```

where <212> is the internal floating point representation of the value 212. The compiler translates an instruction that copies data from one variable to another such as:

```
f_temperature = c_temperature;
```

into a CPU instruction such as:

```
MOV 1000 1004
```

Now consider a more complex example such as:

```
c_temperature = (f_temperature - 32) ÷ 1.8;
```

that combines computation and data movement. The compiler translates this statement into a sequence of CPU instructions such as:

```
MOV R1 1000        ; move f_temperature to register 1
MOV R2 <32>        ; store constant 32 to register 2
FSUB R1 R2 R3      ; subtract R2 from R1 and
                   ; store the result in register 3
MOV R2 <1.8>       ; store constant 1.8 in register 2
FDIV R3 R2 R3      ; divide register 3 by register 2 and
                   ; store the result in register 3
MOV R3 1004        ; copy register 3 to c_temperature
```

More complex formulas or larger numbers of input variables and constants require longer CPU instruction sequences.

Control Structures

A **control structure** is a source code instruction that controls the execution of other source code instructions. Control structures include unconditional branches such as a Goto statement, conditional branches such as an If-Then-Else statement, and loops such as While-Do and Repeat-Until.

The common thread among all control structures is the transfer of control among CPU instructions. A CPU branch instruction requires a single operand containing the address of another instruction. Because all control structures require a CPU branch instruction, the compiler must keep track of where CPU instructions for each source code instruction are located in memory.

For example, consider the following source code instructions:

```
       f_temperature = 0;
loop: f_temperature = f_temperature + 1;
       goto loop;
```

and the equivalent CPU instruction sequence:

```
2000 MOV 1000 <0>    ; copy zero to f_temperature
2004 MOV R2 <1>      ; copy 1 to register 2
2008 MOV R1 1000     ; copy f_temperature to register 1
200C RADD R1 R2 R1   ; add R1 and R2 store result in R1
2010 MOV 1000 R1     ; copy result to f_temperature
2014 JMP 200C        ; loop back to add
```

The numbers to the left of each CPU instruction are the hexadecimal memory addresses of those instructions. The name *loop* in the source code is a label. When the compiler reads a label, it stores the label in the symbol table along with the address of the first CPU instruction generated for the corresponding source code instruction. When the label is used as the target of a Goto instruction, the compiler retrieves the corresponding memory address from the symbol table and uses it as the operand of the corresponding CPU branch or jump instruction.

The If-Then-Else, While-Do, and Repeat-Until control structures are based on a conditional branch. For example, consider the source code instruction sequence:

```
f_temperature = 0;
while (f_temperature < 10) {
    f_temperature = f_temperature + 1;
}
```

and the equivalent CPU instruction sequence:

```
2000 MOV 1000 <0>      ; copy zero to f_temperature
2004 MOV R2 <1>        ; copy 1 to register 2
2008 MOV R3 <10>       ; copy 10 to register 3
200C MOV R1 1000       ; copy f_temperature to register 1
2010 RADD R1 R2 R1     ; add R1 and R2 store result in R1
2014 MOV 1000 R1       ; copy result to f_temperature
2018 FLT R1 R3         ; floating point less than comparison
201C CJMP 2010         ; loop back to add if less-than is true
```

As before, each CPU instruction is preceded by its memory address. When the compiler reads the While statement, it temporarily stores the contents of the condition and marks the memory address of the first instruction within the While loop (2010). The compiler then translates each statement within the loop. When there are no more statements within the loop, as indicated by the closing brace, the compiler generates a CPU instruction to evaluate the condition (2018) followed by a conditional branch statement to return to the first CPU instruction in the loop.

An If-Then-Else statement selects one of two instruction sequences using both a conditional and unconditional branch. For example, the source code instruction sequence:

```
if (c_temperature >= -273.16)
    f_temperature = (c_temperature * 1.8) + 32;
else
    f_temperature = -999;
```

causes the compiler to generate the following CPU instruction sequence:

```
2000 MOV R1 1004       ; copy c_temperature to register 1
2004 MOV R2 <-273.16>  ; store constant -273.16 to register 2
2008 FGE R1 R2         ; floating point > or =
200C CJMP 2018         ; branch to if block if true
                       ; begin else block
2010 MOV 1000 <-999>   ; copy constant -999 to f_temperature
2014 JMP 202C          ; branch past if block
                       ; begin if block (calculate f_temp)
2018 MOV R2 <1.8>      ; store constant 1.8 to register 2
201C FMUL R1 R2 R3     ; multiply registers 1 and 2 and
                       ; store the result in register 3
2020 MOV R2 <32>       ; store constant 32 in register 2
2024 FADD R2 R3 R3     ; add registers 2 and 3 and
                       ; store the result in register 3
2028 MOV 1000 R3       ; copy register 3 to f_temperature
202C                   ; next instruction after If-Then-Else
```

Function Calls

In most languages, a programmer can define a named instruction sequence called a **function, subroutine,** or **procedure**[1] that is executed by a **call** instruction. A call instruction transfers control to the first instruction in the function, and the function transfers control to the instruction following the call by executing a **return** instruction.

Functions are declared in a manner similar to program variables, and the compiler adds descriptive information to the symbol table when it reads the declaration. For example, consider the following function:

```
float fahrenheit_to_celsius(float F) {
    /* convert F to a celsius temperature */
    float C;
    C = (F - 32) / 1.8;
    return(C);
}
```

When the compiler reads the function declaration, it adds the name fahrenheit_to_celsius to the symbol table and also records the memory address of the function's first CPU instruction, as well as other information such as the function type (float) and the number and type of input parameters. When the compiler reads a later source code line that refers to the function, such as:

```
c_temperature = fahrenheit_to_celsius(f_temperature);
```

it retrieves the corresponding symbol table entry. The compiler first verifies that the types of the input variable (f_temperature) and result assignment variable (c_temperature) match the types of the function and function arguments.

To implement the call and return instructions, the compiler generates CPU instructions to:

▶ Pass input parameters to the function

▶ Transfer control to the function

▶ Execute CPU instructions within the function

▶ Pass output parameters back to the calling module

▶ Transfer control back to the calling module at the statement immediately following the function call statement

The flow of control to and from functions is similar to the flow of control to and from an operating system interrupt handler, which is covered in Chapter 6.

---

[1] The terms function, subroutine, and procedure all describe a named instruction sequence that receives control via a call instruction, receives and possibly modifies parameters, and returns control to the instruction after the call. For the remainder of the chapter, we will use the term function exclusively, although everything said about functions also applies to procedures and subroutines.

The calling function must be suspended, the called function must be executed, and the calling function must then be restored to its original state so it can resume execution. As with interrupt handlers, register values are pushed on the stack just before the called function begins executing and popped from the stack just before the calling function resumes execution. The stack serves as a temporary holding area for the calling function so it can be restored to its original state after the called function completes execution.

Table 10-3 shows symbol table entries for the function fahrenheit_to_celsius, its internal parameters, and the variables within the function call:

```
c_temperature = fahrenheit_to_celsius(f_temperature);
```

Table 10-3 ▶

Symbol table entries

Name	Context	Type	Address
fahrenheit_to_celsius	Global	Single precision floating point, executable	3000
F	fahrenheit_to_celsius	Single precision floating point	2000
C	fahrenheit_to_celsius	Single precision floating point	2004
f_temperature	Global	Single precision floating point	1000
c_temperature	Global	Single precision floating point	1004

The compiler reads data names used in the function call and return assignment and looks up each name in the symbol table to extract the corresponding address. It also looks up the name of the function parameter and extracts its addresses and lengths. The compiler then creates a MOV4 (move 4 bytes) instruction using this address and places it immediately before the PUSH instruction:

```
MOV4 2000 1000    ; copy fahrenheit temperature
PUSH              ; save calling process registers
JMP 3000          ; transfer control to function
```

The return from the function call is handled in a similar manner:

```
MOV4 1004 2004    ; copy function value to caller
POP               ; restore caller register values
```

The compiler generates code to copy the function result to a variable in the calling program. Control is then returned to the caller by executing a POP instruction.

LINK EDITING

The examples in the previous section showed how a compiler directly translates assignment, computation, and control structure source code instructions into CPU instructions. The compiler translates other types of source code instructions such as file, network, and interactive I/O statements into external function calls. The compiler also generates external function calls when it encounters call statements without corresponding function declarations. An **external function call**, sometimes called an **unresolved reference**, is a placeholder for missing executable code. An external function call contains the name and type of the called function as well as the memory addresses and types of function parameters.

A **link editor** searches an object code file for external function calls. When one is found, the link editor searches other object code files or compiler libraries to find executable code that implements the function. A **compiler library** is a file that contains a related set of executable functions and an index of the library contents. When the link editor finds executable code in a library or another object code file, it inserts the executable code into the original object code file and generates instructions to copy parameters and the function return value. The result is a single file containing only executable code. If the link editor cannot find executable code for an external function call, it generates an error message and produces no executable file.

At this point, you might be wondering why the compiler couldn't have generated all of the executable code, eliminating the need for a link editor. Link editors provide two key benefits in program translation:

► A single executable program can be constructed from multiple object code files compiled at different times.

► A single compiler can generate executable programs that run under multiple operating systems.

Consider how a program to convert temperatures from Fahrenheit to Celsius and vice versa might be compiled and linked if each function were written by a different programmer. Each programmer would independently develop a function and store the source code in a file. Each file would be compiled separately, and none of the object code files would contain a complete program. The link editor patches together the call and return instructions in each object code file to create a single executable file (see Figure 10-10).

Figure 10-10 ▶

A link editor combines separately compiled functions into a single executable program

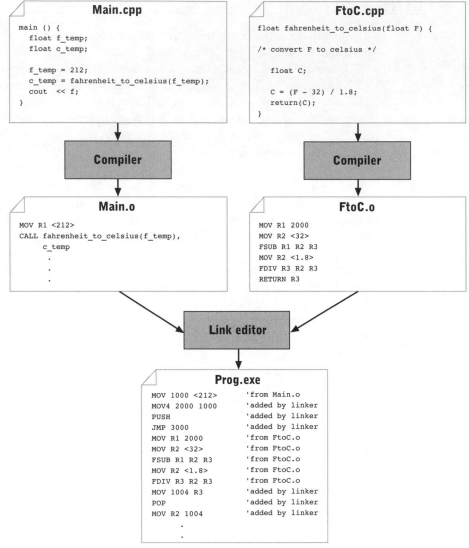

The link editor also makes it possible to use a single compiler for multiple operating systems. For example, consider the C++ temperature conversion program in Figure 10-11 to read temperature input from the keyboard (lines beginning with // are comments). Because the CPU instructions for interactive I/O vary by operating system, if the compiler were responsible for generating CPU instructions for I/O operations, then different compilers would be required to implement this program for each operating system.

Figure 10-11 ►

A C++ program
that reads input
from the
keyboard

```
main() {
    float f;
    float c;

    // print prompt
    cout << "Enter a fahrenheit temperature: ";
    // get keyboard input
    cin >> f;
    // compute celsius equivalent
    c = fahrenheit_to_celsius (f);
    // print celsius temperature
    cout << f << " degrees fahrenheit is ";
    cout << c << " degrees celsius." << endl;

    return(0);
}
```

Using external function calls and link editing allows a single compiler to serve multiple operating systems. CPU instructions that implement I/O functions for each operating system are stored in a compiler library. When object code files are linked, the programmer instructs the link editor to use the library for the target operating system. The link editor replaces external function calls for interactive I/O with the appropriate CPU instructions for a specific operating system.

Link editing with alternate compiler libraries can provide flexibility in other ways. For example, encryption algorithms vary in complexity and maximum length of the encryption key. Different versions of encryption algorithms with the same name and parameters can be stored in different libraries. An application programmer can choose the appropriate level of encryption to embed in an executable program by choosing which encryption library is read by the link editor.

Dynamic and Static Linking

The linking process described above is often called **static linking** or **early binding**. The term "static" is used because library and other subroutines cannot be changed once they are inserted into the executable code. **Dynamic linking**, or **late binding**, is performed during program loading or execution.

Calls to operating system service routines sometimes are implemented with dynamic linking. For example, most Windows operating system service routines are stored in **dynamic-link libraries** (DLLs). During program development, the link editor inserts code to call a special DLL interrupt handler and passes pointers to the service routine name and parameters in general-purpose registers. During program execution, the DLL interrupt handler accepts DLL service requests, locates the service routine by name, loads it into memory (if it isn't already there), copies input parameters, and then executes the service function. When the service routine finishes, the DLL interrupt handler copies output parameters, terminates the service routine, and passes control back to the calling program.

Dynamic linking provides two key advantages over static linking. The first advantage is smaller application program executable files. Static linking incorporates copies of service layer functions directly into executable programs. Depending on the nature and number of functions included, this can add dozens of megabytes to each executable program. Much of this additional storage is redundant, because commonly used subroutines (such as those to open and close files or windows) are used by many application programs. Dynamic linking avoids redundant storage of service functions within application program executable files. Instead, each program contains only the code needed to call the DLL interrupt handler, which is typically only a few hundred kilobytes.

The second advantage of dynamic linking is flexibility. Portions of the operating system can be updated by installing new DLLs. If application programs use dynamic linking, then updating the operating system updates all application programs with the new service layer subroutines. The new subroutines are loaded and executed the next time an application program is executed after an upgrade. In contrast, static linking locks an application program to a particular version of the service layer functions. The functions can be updated only by providing an updated compiler library and relinking the application object code.

The primary advantage of static linking is execution speed. When a dynamically linked program calls a service layer function, the operating system must find the requested function, copy parameters, and transfer control. If the subroutine is not in memory, the application program is delayed until the service function is located and loaded into memory. All of these steps are avoided if the application program is statically linked, resulting in significantly faster execution.

Static linking also improves the reliability and predictability of executable programs. When an application program is developed, it is usually tested with a specific set, or version, of operating system service functions. Service function updates are designed to be backward-compatible with older versions, but there is always the risk that new functions dynamically linked into an older application program will result in unexpected behavior. Static linking ensures that this cannot happen. However, the reverse situation can also occur. That is, a new service function version might fix bugs that were present in an older version. Static linking prevents these fixes from automatically being incorporated into existing programs.

INTERPRETERS

Source and object code files are compiled and linked as a whole. In contrast, **interpretation** interleaves source code translation, link editing, and execution. An **interpreter** reads a single source code instruction, translates it into CPU instructions or a DLL call, and immediately executes the instructions or DLL call before the next program statement is read. In essence, one program (the interpreter) translates and executes another (the source code) one statement at a time.

The primary advantage of interpretation over compilation is that it provides flexibility to incorporate new or updated code into an application program. This flexibility is a direct result of dynamic linking. Program behavior can be easily updated by installing new versions of dynamically linked code. It also is possible for programs to modify themselves directly—a feature required for certain types of applications such as expert systems.

The primary disadvantage of interpretation as compared to compilation/linking is increased memory and CPU requirements during program execution (see Table 10-4). With compilation and link editing, each program used to develop executable code resides in memory for only a short time. When a program is loaded and executed, all of the software components that helped create it (the program editor, compiler, and link editor) are no longer in memory. The memory requirements of a compiled and linked application program consist only of the memory required to store its executable code.

Table 10-4 ▶

Memory and CPU resources used during application execution

Resource	Interpretation	Compilation
Memory contents (during execution):		
Interpreter or compiler	Yes	No
Source code	Partial	No
Executable code	Yes	Yes
CPU instructions (during execution):		
Translation operations	Yes	No
Library linking	Yes	No
Application program	Yes	Yes

An interpreter is a large and complex program that must reside in memory during the entire execution of a source code program. Essentially, there are two programs in memory at run time—the interpreter itself and the application program that it is translating and executing. Also, there are two programs consuming CPU and other systems resources—a critical performance difference for many CPU-intensive applications such as numerical modeling and data mining.

Technology
Focus

Java

Java is an object-oriented programming language and program execution environment. It was developed by Sun Microsystems during the early and mid-1990s. It first debuted in the Netscape Navigator Web browser version 3.0 and, as a result, has been widely described as a Web or network programming language. Although it does satisfy that description, Java is a full-featured programming language that supports virtually any combination of hardware platform and operating system (OS).

Java is similar in syntax and capability to C++, so programmers familiar with C++ can be trained quickly in Java. Java incorporates support for object-oriented features, including encapsulation, inheritance, and polymorphism (multiple inheritance is not supported). Java is designed to maximize reliability of applications and reusability of existing code.

What makes Java unique is a standardized target machine language for Java interpreters and compilers. Compilers and interpreters for other programming languages translate source code into executable CPU instructions and service function calls to a specific operating system. A Java compiler or interpreter translates Java source code into machine instructions and service function calls for a hypothetical computer system and operating system called the **Java Virtual Machine (JVM)**. Instructions and library calls to the JVM are called Java byte codes.

The term JVM also describes an interpreter that translates JVM instructions into actual CPU instructions and operating system service function calls. The JVM simulates a Java computer system with the CPU and operating system of a conventional computer system (see Figure 10-12). The JVM byte code interpreter translates JVM machine instructions into CPU instructions for a real CPU such as an Intel Pentium or Sun UltraSPARC. Calls to JVM service functions are translated into equivalent calls to a real operating system such as Microsoft Windows or UNIX.

Java programs come in two forms: applets and stand-alone applications. Stand-alone Java applications execute within the JVM. A Java **applet** executes within another program such as a Web browser. Applets run within a protected area called the **sandbox**. The sandbox provides extensive security controls to prevent applets from accessing unauthorized resources or damaging the hardware, operating system, or file system.

In 1997, Sun announced that it had developed a microprocessor that directly executed Java byte codes. Sun hoped that the microprocessor would be widely applied in embedded applications such as Web-enabled televisions. Sun apparently has abandoned that hardware approach, because its latest chip offerings don't directly execute Java byte codes, though Sun claims that the chips are optimized to execute Java programs.

Figure 10-12 ►

How Java and
the JVM relate
to the CPU
and OS

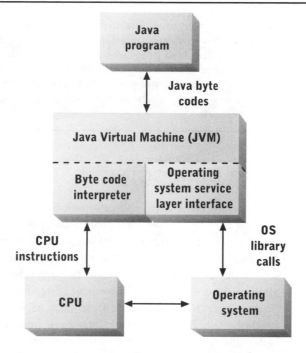

Native support for Java OS calls has been added to the Sun Solaris operating system, which is Sun's version of UNIX. Other operating system vendors have not added native support for Java, though it is rare to find a general-purpose computer system without an installed JVM. Incorporating the JVM into the native operating system reduces some of the inefficiency inherent in translating OS service calls. It also reduces the need for Java-enabled application software to install and use private and sometimes incompatible JVMs.

The popularity of Java has far exceeded most initial expectations due to a number of factors including:

► Sun's strategy of providing Java compilers and virtual machines at little or no cost

► Incorporation of JVMs into Web browsers

► Increasing development of application software that uses a Web browser as the primary I/O device

► The ability of Java programs to execute on any combination of computer hardware and operating system

The most important drawback of Java is reduced execution speed due to the use of interpreted byte codes and operating system translation. Emulation processes are inherently inefficient users of hardware resources. **Native applications**—those compiled and linked for a particular CPU and operating system—typically

> execute 10 times faster than interpreted Java byte code programs. Some JVMs employ a just-in-time compiler to reduce the speed difference to approximately 3:1, but Java programs always will be slower than native programs when executed on anything other than a true Java machine. This is an important consideration for programs that make maximal demands of available hardware resources, such as simulation applications and database management systems.

SYMBOLIC DEBUGGING

Compilers, link editors, and interpreters produce error messages to warn programmers of actual or potential mistakes in source code. However, even if a program is compiled and linked successfully, it might still produce errors when executed. Diagnosing run-time errors based on a typical run-time error message is difficult, because the error message refers to the executable code generated by the compiler and link editor instead of the source code written by the programmer. Variable names have been replaced by memory addresses, and symbolic source code instructions have been replaced by CPU instructions. Determining the source code instructions and variables that correspond to these memory addresses is a difficult task.

A compiler can be instructed to incorporate its symbol table into the object code file, and a link editor can be instructed to produce a **memory map,** or **link map,** based on the symbol table contents. A memory map lists the memory location of every function and program variable. A programmer can use a memory map to trace error messages containing memory addresses to corresponding program statements and variables. Tracing those addresses to particular source code statements or data items requires a very detailed memory map and a thorough understanding of machine code and the compiling and linking processes.

For example, consider the partial memory map in Figure 10-13 (the complete map is several hundred text lines) and the run-time error message in Figure 10-14. Based on the instruction address in the error message and the starting addresses of the modules listed in the memory map, a programmer can determine that the error occurred somewhere within the function _main, but the programmer can't tell exactly which instruction caused the error or determine the contents of program variables at the time the error occurred.

Figure 10-13 ▶

Small portion of a memory map

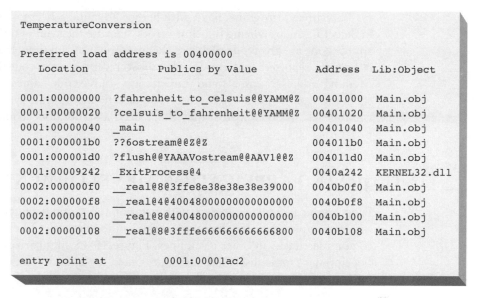

```
TemperatureConversion

Preferred load address is 00400000
    Location              Publics by Value          Address    Lib:Object

  0001:00000000     ?fahrenheit_to_celsuis@@YAMM@Z    00401000    Main.obj
  0001:00000020     ?celsuis_to_fahrenheit@@YAMM@Z    00401020    Main.obj
  0001:00000040     _main                             00401040    Main.obj
  0001:000001b0     ??6ostream@@Z@Z                   004011b0    Main.obj
  0001:000001d0     ?flush@@YAAAVostream@@AAV1@@Z     004011d0    Main.obj
  0001:00009242     _ExitProcess@4                    0040a242    KERNEL32.dll
  0002:000000f0     __real@8@3ffe8e38e38e38e39000     0040b0f0    Main.obj
  0002:000000f8     __real@4@40048000000000000000     0040b0f8    Main.obj
  0002:00000100     __real@8@40048000000000000000     0040b100    Main.obj
  0002:00000108     __real@8@3fffe666666666666800     0040b108    Main.obj

  entry point at            0001:00001ac2
```

Figure 10-14 ▶

Typical run-time error message

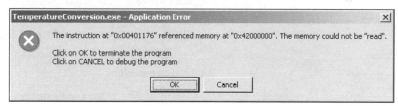

TemperatureConversion.exe - Application Error   ✕

❌ The instruction at "0x00401176" referenced memory at "0x42000000". The memory could not be "read".

Click on OK to terminate the program
Click on CANCEL to debug the program

[OK]   [Cancel]

A symbolic debugger is an automated tool for testing executable programs. It provides a number of features, including the ability to:

▶ Trace calls to specific source code statements or subroutines

▶ Trace changes to variable contents

▶ Execute source code instructions one at a time

▶ Detect run-time errors and report them to the programmer in terms of specific source code instructions and variables

A symbolic debugger uses the symbol table, memory map, and source code files to trace memory addresses to specific source code statements and variables. It also inserts debugging checkpoints after each source code instruction so program execution can be paused. An executable program that contains symbol table entries and debugging checkpoints is sometimes called a **debugging version**. In contrast, the **production version**, or **distribution version**, of a program omits the symbol table and debugging checkpoints to reduce program size and increase execution speed.

Interpreted programs have an inherent advantage over compiled programs in detecting and reporting run-time errors, because the symbol table and program source code are always available to the interpreter at run time. If an error occurs, it can be reported in terms of the most recent source code line translated. Memory addresses can also be converted to source code names by looking them up in the symbol table. Most interpreters directly incorporate symbolic debugging capabilities. In other words, they don't need a stand-alone symbolic debugging program.

INTEGRATED APPLICATION DEVELOPMENT TOOLS

Programming languages, program translation tools, and debugging tools address only the implementation discipline of the Unified Process. Other tools are needed to complete tasks in other disciplines. Figure 10-15 summarizes the application development tools discussed, introduces new tools, and ties each tool to a specific part of the software development process.

Figure 10-15 ▶

Application development tools

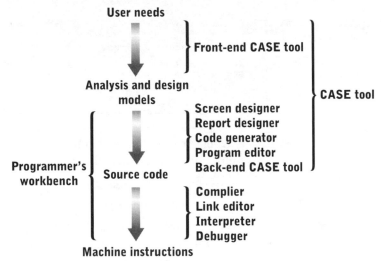

Programmer's Workbenches

A **programmer's workbench** is an integrated set of automated support tools to speed development and testing. Programmer's workbenches generally provide the following components:

▶ Smart program editor

▶ Compiler and/or interpreter

► Link editor and a large library of classes or subroutines

► Interactive tool for prototyping and designing user interfaces

► Symbolic debugger

► Integrated window-oriented graphical user interface

The key feature of a programmer's workbench is the level of integration among tools, not the specific tools provided. A programmer's workbench leverages programmer effort and skill so programs can be written and tested in minutes, hours, or days. The workbench automates tedious and time-consuming tasks, streamlines other tasks, and maximizes reuse of existing source code. Using a programmer's workbench, a programmer can rapidly develop and test many program versions.

Editors included in the workbench assist the programmer in writing syntactically correct code by performing rudimentary syntax checking as the code is being typed and visually identifying errors. Compiler and link editor errors displayed in one window are linked directly to specific source code instructions in another window. Some compilers provide suggested corrections or context-sensitive help in addition to error messages.

Workbenches also can provide program templates and skeletons that can be stored in a library or generated based on input from the programmer or from other tools. Workbenches usually contain large libraries of predefined subroutines or classes that address user interface, database manipulation, network resource access, and other common functions. Library routines can be imported directly into program source code or added via function or method calls and linked into the executable program. Libraries allow programmers to reuse source and executable code, thus speeding up application development.

Workbenches include extensive symbolic debugging support. Method calls, function calls, and changes to program variables or object data can be traced during program execution. Programs can be started and stopped at programmer-specified locations so testing can be performed on specific code segments. Debugging and tracing can also be performed on machine code and operating system service calls. Testing and debugging occur in a simulated run-time environment so program crashes can't disable the workbench or the entire workstation.

Workbenches need powerful hardware and operating software. Complex tools and simulated execution environments require substantial CPU power and memory. Large libraries, complex tools, and extensive documentation require significant amounts of secondary storage. Programmers need large monitors to display all of the windows used in the development process. Workbenches typically require two to three times the CPU power, memory, and disk storage of a typical business workstation.

Technology
Focus

Oracle JDeveloper

Oracle JDeveloper is a programmer's workbench for developing object-oriented Java software using Oracle and other database management systems. A JDeveloper workspace is a container for application components including program files, system development models, user interface windows, and database connections. A programmer can interact with different views of a project, including software development models such as class and package diagrams, files (see the upper-left subwindow in Figure 10-16), and classes/methods (see the middle-left subwindow in Figure 10-16).

Figure 10-16 ►

Oracle
JDeveloper

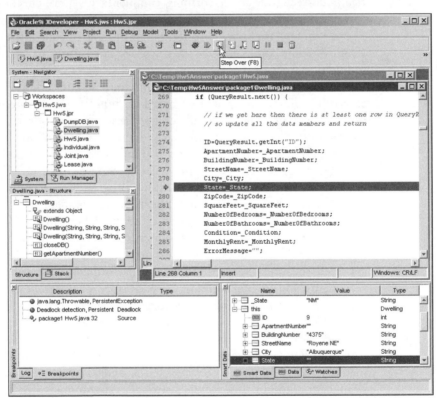

JDeveloper contains many software development tools, including compilers, link editors, program text editors, symbolic debuggers, and interactive tools for designing user interface elements such as menus, toolbars, icons, and forms.

JDeveloper keeps track of which tools are used to create and modify each programmer-defined object and opens the appropriate tool automatically when the programmer selects an application component. Overlapping menus, tool-bars, and icons provide access to various tool features and commands.

Development tools are automatically integrated whenever possible. For example, a command to build a complete program automatically executes user interface resource compilers, program compilers for each component programming language, and the link editor, in the appropriate order. JDeveloper keeps track of file dependencies, so a change to one source code file causes all dependent files to be rebuilt automatically. JDeveloper manages data movement among tools and displays progress and results within an output window.

JDeveloper also provides extensive symbolic debugging capabilities. A programmer can pause program execution at specified source code lines (in Figure 10-16, execution is paused before line 279). Further execution can be directed on a line-by-line or function-by-function basis. A programmer can examine variable content while the program is paused and trace variable changes during execution (see lower right window in Figure 10-16).

JDeveloper provides extensive documentation that can be accessed from a hard disk or the Web. A sophisticated documentation browser provides extensive search and indexing capabilities. The documentation contains many cross-references and hyperlinks to glossary terms, examples, and tutorials.

CASE Tools

The term **computer-assisted software engineering (CASE) tool** usually refers to a tool that supports the UP requirements and design disciplines. The term "CASE tool" is a bit of a misnomer, because any tool that assists system or program development fits within the literal definition of a CASE tool. The singular form of the term also is a misnomer because CASE tools are normally highly integrated tool suites.

The key feature of a CASE tool is support for a broad range of system development activities with particular emphasis on model development. CASE tools support most or all system development activities up to, and sometimes including, program translation. CASE tools usually support a specific system development methodology such as structured system development or the UP. A tool suite that primarily supports model development is sometimes called a **front-end CASE tool**. A tool suite that primarily supports program development based on specific analysis and design models sometimes is called a **back-end CASE tool** or **code generator**.

CASE tools can perform very specialized model development and analysis tasks due to high integration levels and specialization to particular methodologies and models. CASE tools can perform consistency and other types of error checking on models. For example, a CASE tool that supports structured analysis can verify that all data flows on a data flow diagram have corresponding definitions in the data dictionary. Such a tool can also verify that the data outputs of each data flow diagram process can be produced from the data inflows based on the process logic specification.

CASE tools with extensive back-end capabilities can automate much of the software development process. For example, a CASE tool that supports object-oriented development can generate class declarations in Java or C++ source code files. Such a tool can also generate database schema descriptions and component registration entries for specific database and component management systems.

The most comprehensive CASE tools automate the entire process of generating working systems from fully specified analysis and design models. Such tools generate program and other source code from models, compile and link the programs, create databases, and create, register, and install all components. CASE tools can also provide comprehensive facilities for planning and managing development projects and for maintaining and modifying existing systems. Comprehensive CASE tools are usually expensive, require weeks or months of training, and support very specific target deployment environments.

Business
Focus

Building the Next Generation of Application Software

Southwestern Gifts, Incorporated (SGI) is a small catalog retailer of southwestern art, jewelry, and gifts. SGI distributes two catalogs per year, in May and October. Most orders are received by telephone and the rest by mail. SGI employs up to 50 people, including 15 warehouse staff and 30 order-processing clerks during the Christmas shopping season, 4 accountants and bookkeepers to handle payroll and accounts payable, 3 buyers, a catalog designer, 3 IS staff, a warehouse manager, and the owner-operator.

SGI currently has a Compaq midrange computer running the Linux operating system, a small Dell server running Windows Server, 6 microcomputers, and 12 Wyse video display terminals. The Wyse terminals are connected to the Compaq computer by serial cables and are used only for order entry. The Dell server provides file and printer sharing for the PCs and printer sharing for the Compaq computer. The PCs are used by the accountants, warehouse manager, buyers, catalog designer, and IS staff. A 100-Mbps Ethernet LAN connects the PCs, Dell server, and Compaq midrange computer.

Application software on the personal computers includes Quicken (an accounting package), various productivity tools such as Microsoft Office, and graphical design software. SGI purchased an order entry and inventory control package eight years ago from an out-of-state vendor. The package stores data in ordinary Linux files and uses a proprietary application development and operating environment called DERQS (Data Entry, Retrieval, and Query System).

DERQS provides forms-based data entry, a forms-based data query facility, a file definition tool, and a simple report generator. Screen forms are defined using an interactive layout tool that compiles and stores screen layout and content in a screen definition file (SDF). Files also are defined using an interactive tool, and descriptions of file format and content are stored in a data description file (DDF). Data content is stored in ordinary UNIX files.

Application programs are written in a proprietary interpreted scripting language to display and manipulate screens, manipulate file content, and generate simple reports. Most of the existing application programs are DERQS scripts, but some are written in C and C++. DERQS provides a compiler library with functions that allow C programs to display DERQS screens and interact with DERQS files. The library also works with C++ programs because C++ is a superset of C.

One IS staff member specializes in DERQS data entry functions and script-based applications. A second IS staff member specializes in the DERQS query language and develops and maintains C and C++ applications. A third IS staff member manages the LAN and personal computers.

SGI would like to make several improvements to its information systems, including:

- ▶ A Web site to provide catalog information and order entry
- ▶ A direct interface to financial information in DERQS data files so data can be uploaded to Quicken and Excel
- ▶ An automated system to support the buyers that interacts directly with accounts payable
- ▶ Modern graphical user interfaces for existing and newly developed application programs using modern thin client terminals and full-featured microcomputers

SGI recently learned that the company that developed and supported DERQS has filed for bankruptcy. No company has shown an interest in purchasing the rights to DERQS, so it appears that no further upgrades or technical support will be available.

Questions

1. Should SGI develop any new software using DERQS? If not, what tools should they acquire for new system development?
2. Should SGI re-implement the functions of its existing DERQS-based applications using more up-to-date development tools?
3. What risks and hidden costs might SGI face if it acquires new tools to implement old software functions and/or develop new software?

SUMMARY

► Application systems are developed by following the disciplines and iterations of the Unified Process. Application development follows a development methodology, develops a set of models, and uses automated tools. Most software development cycles include two sets of models: a requirements model and a design model.

► Executable software consists entirely of CPU instructions. Modern programmers write programs in higher-level programming languages that are translated into CPU instructions. Programming languages have evolved through many generations, the most recent of which are object-oriented and scripting programming languages. Later generations have incorporated improvements and capabilities such as instruction explosion, database access, support for graphical user interfaces, and nonprocedural programming.

► All programming language generations other than the first must be translated into CPU instructions prior to execution. A compiler translates an entire source code file, building a symbol table and object code file as output. An interpreter translates, links, and executes source code programs iteratively, one source code instruction at a time.

► Compiled and interpreted programs must be linked to libraries of executable functions or methods. A link editor statically links external reference calls in object code to library functions and combines them into a single file containing executable code. A link editor can also dynamically link external calls by statically linking them to an operating system service function that loads and executes dynamic-link library (DLL) functions at run time. Interpreters always use dynamic linking.

► Integrated suites of automated tools support the entire application development process. A programmer's workbench integrates tools such as a program editor, screen and report designers, an intelligent compiler and link editor, a symbolic debugger, and extensive documentation. A CASE tool provides a tool suite to support a wider range of development tasks. A front-end CASE tool supports development of requirements and design models. A back-end CASE tool generates program source code from models.

In the next two chapters, we'll look at operating systems in detail. Chapter 11 describes the internal architecture of an operating system and discusses how it manages processes and memory. Chapter 12 concentrates on file and secondary storage device management.

applet
assembler
assembly language
back-end CASE tool
call
code
code generator
compiler
compiler library
computer assisted software
    engineering (CASE) tool
control structure
data declaration
data operation
debugging version
design model
distribution version
dynamic linking
dynamic-link library
    (DLL)
early binding
executable code
external function call

fifth-generation language (5GL)
first-generation language (1GL)
fourth-generation language
    (4GL)
front-end CASE tool
function
instruction explosion
interpretation
interpreter
Java
Java Virtual Machine (JVM)
label
late binding
link editor
link map
machine independence
machine language
memory map
message
method
native application
nonprocedural language
object

object code
object-oriented programming
    (OOP)
procedure
production version
programmer
programmer's workbench
programming language
return
sandbox
scripting language
second-generation language
    (2GL)
source code
static linking
subroutine
symbol table
symbolic debugger
system requirements model
third-generation language (3GL)
unresolved reference
variable

Vocabulary Exercises

1. A compiler allocates storage space and makes an entry in the symbol table when it encounters a(n) _____ in source code.

2. A(n) _____ is produced as output during activities of the UP requirements discipline.

3. A link editor searches object code for _____.

4. A(n) _____ is produced as output during activities of the UP design discipline.

5. A 4GL has a greater degree of _____ than a 3GL.

6. _____ contains CPU instructions and external function calls.

7. A(n) _____ produces a(n) _____ to show the location of functions or methods in executable code.

8. A(n) _____ is a model or system component that combines a set of related data and the methods that manipulate that data.

9. The compiler adds the names of data items and program functions to the _____ as they are encountered in source code.

10. A 2GL is translated into executable code by a(n) _____.

11. A(n) _____ translates an entire program before linking and execution. A(n) _____ interleaves translation and execution.

12. A Java _____ runs within the _____ of a Web browser.

13. _____ in a source code program are translated into machine instructions to evaluate conditions and/or transfer control from one program module to another.

14. A(n) _____ uses the content of the symbol table to aid the programmer in tracing memory locations to program variables and instructions.

15. FORTRAN, COBOL, and C are examples of _____.

16. A(n) _____ tool supports system model development. A(n) _____ tool generates program source code from systems models.

17. A link editor performs _____ linking. An interpreter performs _____ linking.

18. Java programs are compiled into object code for a hypothetical hardware and system software environment called the _____.

19. Widely used scripting languages include, _____, _____, and _____.

Review Questions

1. Describe the relationships between application development methodologies, models, and tools.

2. Compare and contrast the various generations and types of programming languages.

3. What is instruction explosion? What types of programming languages have the most instruction explosion? What types of programming languages have the least instruction explosion?

4. What are the differences between source code, object code, and executable code?

5. Compare and contrast assemblers, compilers, and interpreters.

6. What does a compiler do when it encounters data declarations in a source code file? Data (manipulation) operations? Control structures?

7. Compare and contrast the execution of compiled programs to interpreted programs in terms of CPU and memory utilization.

8. What is a link editor? What is a library? How and why are they useful in program development?

9. What types of programming language statements are likely to be translated into machine instructions by a compiler? What types are likely to be translated into library calls?

10. How does interpretation differ from compilation?

11. Compare and contrast the error detection and correction facilities of interpreters and compilers.

12. Compare and contrast static and dynamic linking.

13. With respect to the requirements of modern applications, what are the shortcomings of 3GLs?

14. What are the primary differences between object-oriented programming languages and more traditional programming languages?

15. What components are normally part of a programmer's workbench? In what ways does a workbench improve programmer productivity?

16. What is a CASE tool? What is the relationship between a CASE tool and a system development methodology?

17. What is the difference between a front-end CASE tool and a back-end CASE tool?

Problems and Exercises

1. Develop a set of machine instructions to implement the following source code fragment:

```
a = 0;
i = 0;
while (i < 10) do
    a = a+i;
    i = i+1;
endwhile
```

2. Some early RISC processors did not provide PUSH and POP instructions. The rationale is that they are complex instructions that can be built from simpler components. Write a short machine language program to implement the PUSH and POP operations using only data movement and computation instructions.

Research Problems

1. Investigate a modern CASE tool such as those offered by Computer Associates (http://ca.com) or Oracle (www.oracle.com). On what system development methodology or methodologies is the tool based? What types of system models can be built with the tool? How is an analysis model translated into an implementation model? What programming languages, operating systems, and database management systems are supported by the back-end CASE tool?

2. Investigate a modern programmer's workbench such as Microsoft Visual Studio .NET (www.microsoft.com/vstudio), or Borland Delphi (www.borland.com). Are application programs interpreted, compiled, or both? What program editing tools are provided? What tools are available to support run-time debugging? What DBMSs can be accessed by application programs?

3. Investigate the latest Java-oriented microprocessors offered by Sun (www.sun.com). In what ways are they optimized to execute Java programs? What architectural features are similar to microprocessors of competitors such as Intel? Is the Java orientation of Sun microprocessors a fact or just marketing hype?

4. Investigate the graphical user interface (GUI) features of different operating systems such as Windows XP and Mac OS and the features of a later 3GL such as C or an object-oriented language such as Smalltalk, C++, or Java. How would you go about developing a suite of interactive test programs to ensure compliance with a programming language standard? What GUI features couldn't be tested because they're not embedded within the programming language? Could a language standard be constructed that would use equivalent instructions or statements to implement an interactive interface under both Windows XP and Mac OS? Could the behavior and appearance of the GUI be exactly the same when a single program is translated and executed under both operating systems?

Chapter 11

Operating Systems

Chapter Goals

- ► Describe the functions and layers of an operating system

- ► List the resources allocated by the operating system and describe the complexities of the allocation process

- ► Explain how an operating system manages processes and threads

- ► Compare and contrast alternative CPU scheduling methods

- ► Explain how the operating system manages memory

Operating systems are critical components of all modern information systems. Users and application programs rely on operating systems to provide services and to manage software, data, and hardware resources. As an information systems professional, you will often be asked to select, install, configure, and upgrade operating systems. Successfully performing those tasks requires a detailed understanding of how operating systems work. Figure 11-1 shows some of the topics that you will learn in this chapter.

Figure 11-1 ▶

Topics covered in this chapter

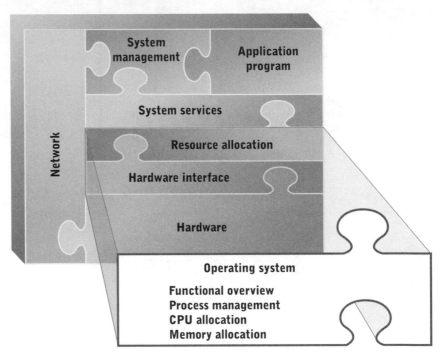

OPERATING SYSTEM OVERVIEW

The operating system is an important part of all information systems. It plays a dual role as a high-level manager and a tireless low-level worker. As a manager, the operating system oversees all hardware resources and allocates them to users and applications as needed. It also shifts resources to meet rapidly changing demands while enforcing resource security and sharing policies. With respect to resources, the operating system is similar to "Big Brother"—an omniscient government with unquestioned authority.

At the same time, the operating system performs many low-level tasks on behalf of users and application programs such as accessing files and directories, creating and moving display windows, and accessing resources over a network. The operating system provides these services for two reasons. First, it is efficient.

If many application programs perform similar tasks, it makes sense to create services accessible to all applications. That way, application programmers are free to address tasks unique to their application area. Operating system services are a form of code reuse.

Second, providing so many low-level functions enables the operating system to maintain absolute control over hardware resources. For example, the operating system provides file access services so it can maintain complete control over secondary storage devices. With complete control, the operating system can enforce system-wide security policies, prevent programs from accidentally or maliciously destroying one another's data, and efficiently allocate disk and I/O capacity to meet the needs of all users and programs.

Operating System Management Functions

Operating system management functions (see Figure 11-2) can be loosely divided between those oriented to hardware resources (on the left) and those oriented to users and their programs (on the right). Each hardware resource type (CPU, memory, secondary storage, and I/O devices) has an associated operating system management function, as do users and programs. This chapter primarily explores CPU, memory, I/O, and process management. Secondary storage management is described in Chapter 12 and some aspects of user management are described in Chapter 14.

Figure 11-2 ►

Operating system management functions

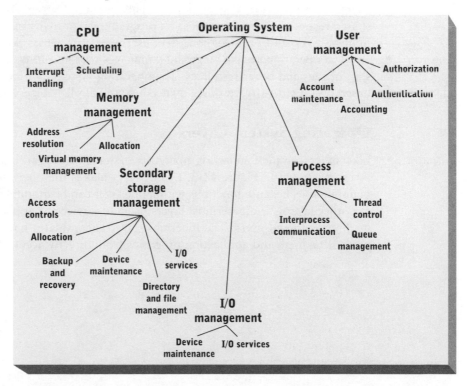

All programs require CPU cycles and memory to execute. Managing programs, CPUs, and memory is complex because a modern computer system can have hundreds or thousands of programs executing at the same time. The operating system tracks each program from start to finish and processes all requests for resource allocation and communication. All executing programs compete for a limited memory pool and access to a small number of CPUs. The operating system controls which programs access which resources and for how long they control those resources.

Most computer systems have gigabytes of secondary storage organized into tens of thousands or hundreds of thousands of files. Keeping track of the locations of files on disk requires a complicated set of directories and allocation tables. The operating system updates directories and allocation tables as files are created, modified, and deleted. Because it performs all low-level file access functions on behalf of users and programs, it can maintain tight control over directories and allocation tables. The operating system also enforces access controls and provides utility functions such as backup and recovery. Secondary storage functions are described in detail in the next chapter.

The operating system manages all I/O devices and accesses those devices on behalf of users and programs. It also manages data movement among I/O devices, memory buffers, and programs. As the user moves from program to program, the operating system reconnects interactive I/O devices such as keyboards and sound cards. The operating system provides many I/O services, including output to printers and display devices. The operating system efficiently manages shared resources, relieves application programmers of many user interface development tasks, and provides consistent user interfaces across programs.

The operating system creates and manages user accounts and tracks ownership of files and other resources. The operating system enforces access controls based on user identity, resource ownership, and explicit access authorizations.

Operating System Layers

Like other complex software, operating systems are organized internally into layers as shown in Figure 11-3. Using layers makes the operating system more maintainable, because functions within one layer can be modified without affecting other layers. The outermost layers provide services to application programs or directly to end-users. The innermost layer encapsulates hardware resources, preventing users and application programs from directly accessing them.

Figure 11-3 ►

Operating system
layers (shaded)

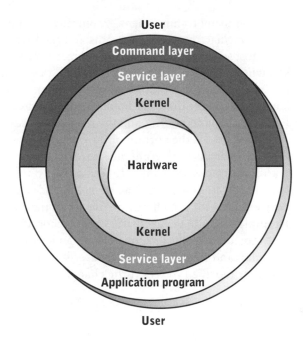

The **command layer,** sometimes called the **shell,** is the user's interface to the operating system. Through the command layer, a user or system administrator can execute programs (including operating system utility programs) and directly manage files and directories.

The command layer can be implemented with a textual or a graphical user interface. Textual interfaces accept user input from the keyboard via a command prompt. The set of commands and their syntax requirements are called a **command language,** or sometimes a **job control language (JCL).** Command languages tend to be difficult to learn and use. The user or programmer must know the syntax and semantics of the language in order to issue commands. Examples of command languages and command sets are MS-DOS, IBM MVS JCL, and UNIX Bourne Shell.

Modern operating systems provide a graphical user interface (GUI), which reduces or eliminates the need for users to learn a command language or JCL. With GUI-based command layers, users can execute many operating system commands by manipulating graphical images on a video display using a pointing device such as a mouse. For example, programs can be executed by double-clicking their icon. The visual representation of files, programs, and commands is easier to understand and manipulate for most users.

The **service layer** contains a set of functions that are executed by application programs and the command layer. A request to execute a service layer function is called a **service call.** Most operating systems provide thousands of service calls, including many to implement graphical user interfaces. The service layer is also an intermediary between programs and the kernel.

Most service calls are actually indirect requests for system resources. For example, when an application program asks the operating system to retrieve data from a file, it is indirectly asking for access to memory buffers, the system bus, the disk controller, and the disk that holds the file. An I/O service call is an implicit request for access to an I/O device and the resources needed to communicate with it.

The **kernel** is the portion of the operating system that manages resources and directly interacts with computer hardware. Because the kernel directly interacts with hardware, service layer functions rely on the kernel to allocate and interact with hardware resources. The kernel includes a set of interface programs called device drivers for each hardware device in the computer system (see Figure 11-4). Examples include device drivers for keyboards, video display devices, disk drives, tape drives, and printers. Using device drivers makes the operating system modular and flexible. As hardware is added or upgraded, device drivers are added or modified to reflect the changes.

Figure 11-4 ►

Components of the kernel

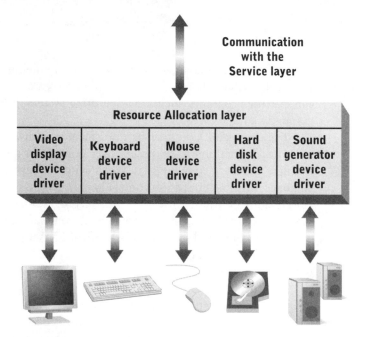

RESOURCE ALLOCATION

Early computer systems were relatively simple devices with few hardware resources. Because hardware resources were limited, complex resource allocation methods were unnecessary. An application program "took control" of the entire computer system when it executed. Because only one application program executed at a time, there was no need for operating system software to resolve competing demands for hardware resources.

Over time, computers became more powerful and their hardware resources became more numerous and diverse. Extra processing power and storage capacity made it possible for multiple users and programs to use a computer at the same time. Operating system support for multiple executing programs is called **multitasking**. Multitasking operating systems employ elaborate procedures to share resources among processes and users and to prevent them from interfering with one another.

Single-Tasking Resource Allocation

Resource allocation in a single-tasking operating system involves only two executing programs—an application and the operating system. Single-tasking operating systems are small and efficient because they do not need complex resource allocation procedures. There is rarely any contention between the operating system and the application program for resources. The operating system reserves whatever resources it requires when it first boots up. All resources not allocated to the operating system are available to an application program.

In a single-tasking environment, when an application program begins executing, the operating system grants it control of all unused hardware resources. The program controls those resources until it terminates, an error occurs, or it needs service from the operating system. Errors and service calls are normally processed through interrupts (interrupts were discussed in Chapter 6). When an interrupt is detected, control of the CPU is passed back to the operating system. If an error has occurred, the operating system can attempt to correct it so the program may continue. If this is not possible, the application program is terminated and the operating system reclaims control of all hardware resources. In a single-tasking operating system, no elaborate mechanisms are needed to check for competing resource allocation because there is only one active program.

MS DOS is the most common single-tasking operating system. MS-DOS was developed for the original IBM Personal Computer and was widely used until the early 1990s. It is still sometimes employed in specialized applications that do not require multitasking capabilities such as building security systems and control of hardware devices such as elevators and irrigation systems.

Multitasking Resource Allocation

Multitasking operating systems are the norm for modern general-purpose computers. Under multitasking, application and system software can be more flexible, because large software programs can be built from smaller independent modules or processes. Small modules are easier to develop and update, and they can be loaded, executed, suspended, and resumed as needed.

For example, a typical word processing program often needs to format output and send it to a printer. With a multitasking operating system, the word processor's printer formatting and interface functions can be a separate module. That module can be loaded and executed when printing is required and suspended when not in use. The printing module can execute independently of other word processing modules, so the user can perform other tasks while a document is being printed.

System software also benefits from multitasking capabilities. Network interface units, for example, typically have a dedicated operating system module that detects and responds to incoming data. With multitasking, this module is loaded into memory when the operating system is booted and placed in a suspended state until data arrives. Other I/O devices such as printers and terminals also are managed by independent modules. Dividing operating system tasks among multiple independent modules makes the operating system easier to build and upgrade. It also results in more efficient use of hardware resources.

Resource Allocation Goals

A multitasking operating system manages hardware resources to achieve the following goals:

► Meet the resource needs of each program
► Prevent programs from interfering with one another
► Efficiently use hardware and other resources

Related goals include maximizing the amount of work performed per time unit, ensuring reliable program execution, protecting resource security, and minimizing the resources consumed by the operating system itself.

Resource Allocation Tasks

Consider the resource allocation tasks performed by the audiovisual support service for a school. The service maintains a pool of equipment, accepts equipment requests from instructors and faculty, allocates equipment to satisfy requests, and moves equipment from place to place as needed. Managing the equipment inventory and responding to user requests is a complex task that requires detailed knowledge of what equipment is available and when it is available. The service maintains detailed records of equipment capabilities, operational status, commitments, and movement schedules. The records are input data for future scheduling decisions and are modified as a result of those decisions. The service follows well-defined policies and procedures to prioritize conflicting demands for service.

An operating system's resource allocation functions are similar to the audio-visual service scenario. The operating system keeps detailed records of available resources and knows which resources can satisfy which requests. It schedules resources based on specific allocation policies to meet present and anticipated demand, and it continually updates records to reflect resource commitment and release by programs and users.

Resource allocation goals and policies are defined as a set of procedures or algorithms implemented in software. The records that support the procedures are implemented as a set of data structures. The resource allocation functions of an operating system are similar to any other program; they are a set of algorithms and data structures organized to accomplish a particular purpose.

Computer scientists have long studied data structures and the most efficient ways to manipulate them. They have created a large body of knowledge that describes which algorithms and data structures are most efficient for particular types of processing tasks.

System programmers apply that body of knowledge to designing and implementing operating system resource allocation procedures. They take great care to implement procedures that consume as few machine resources as possible. The reason is simple—every resource consumed by operating system tasks is unavailable to application programs. Think of the resource allocation function as similar to middle management functions in a business. Every business has a limited set of resources. Every resource devoted to a management function is unavailable for other business functions such as producing and selling products.

The resources consumed by the resource allocation procedures sometimes are referred to as **system overhead**. An important design goal for most operating systems is to minimize system overhead while achieving acceptable levels of throughput and reliability. Note, however, that these goals are in direct conflict. Allocation decisions that provide higher levels of resource availability to application programs typically require more elaborate and complex allocation procedures. Ensuring higher levels of reliability also requires more extensive oversight procedures. Yet, increases in the complexity of allocation and oversight procedures consume more hardware resources when those procedures are implemented in software. Operating system designers must determine the right balance among these competing objectives.

Real and Virtual Resources

A computer system's physical devices and associated system software are called **real resources**. As allocated by the operating system, the resources that are apparent to a process or user are called **virtual resources**. Modern operating systems make the set of virtual resources available to programs and users appear to be equal to or greater than the available set of real resources. For example, a typical personal computer usually has only one printer. But if a user runs multiple programs such as a word processor and spreadsheet application at the same time, each program "thinks" it has a printer all to itself. The single physical printer is a real resource. The printers that each program "thinks" it exclusively controls are virtual resources.

Application design and programming is much simpler if the application is not concerned with resource availability. Programs can be written under the assumption that whatever resources are requested will be provided. The programmer doesn't need to implement schemes to lock and unlock resources for exclusive use. For example, a word processing program doesn't need to check whether the physical printer is currently being used by another program. It simply prints to its own virtual printer and leaves it to the operating system to figure out how to print output from multiple programs without interference.

Providing virtual resources that meet or exceed real resources is accomplished by:

► Rapidly shifting resources unused by one program to other programs that need them

► Substituting one type of resource for another when possible and necessary

Although each program "thinks" it has control of all hardware resources in the computer system, it is rare for any one program to need them all at the same time. The operating system shifts resources among programs as demand rises and falls. In this way, the sum of virtual resources for all active programs can substantially exceed the real resources of the computer system. Interrupt processing is an example of resource shifting. When a program makes an I/O request that can't be immediately satisfied, the operating system suspends it and stores its register contents in memory (a stack PUSH). Other programs receive control of the CPU while the I/O request is being processed. Later, the register contents are copied back from memory (a stack POP) and the program resumes execution and control of the CPU.

Certain types of resources can be substituted for each other. For example, memory is temporarily substituted for CPU registers while a program is suspended pending completion of an I/O request. In the printing example above, an application program "thinks" that it is interacting directly with the printer. Instead, the operating system temporarily stores output in a file. In essence, the operating system substitutes secondary storage resources for the printer and related I/O resources. When the printer is available, the operating system copies the data to the printer and deletes the file, thus reversing the temporary substitution.

PROCESS MANAGEMENT

A **process** is a unit of executing software that is managed independently by the operating system. A process can request and receive hardware resources and operating system services. Processes can be stand-alone entities or part of a group of processes that cooperate to achieve a common purpose. Processes can communicate with other processes executing on the same computer system or with processes executing on other computer systems.

Process Control Data Structures

The operating system keeps track of each process by creating and updating a data structure, called a **process control block (PCB)**, for each active process. The operating system creates a PCB when a process is created, updates the PCB as process status changes, and deletes the PCB when the process terminates. Using information stored in the PCB, the operating system can perform a number of functions, including resource allocation, secure resource access, and protecting active processes from interference by other active processes.

PCB content varies among operating systems and also can vary within a single operating system, depending on whether certain functions such as resource accounting and auditing are enabled. The following data items are typically included:

► A unique process identification number

► The current state of the process; for example, executing or suspended

► Events for which the process is waiting

► Resources allocated exclusively to the process, including memory, files, and I/O devices

► Machine resources consumed; for example, CPU seconds consumed and bytes of data transferred to/from disk

► Process ownership and access privileges

► Scheduling priority or data from which scheduling priority can be determined

PCBs are normally organized into a larger data structure such as a linked list. The list of PCBs is sometimes called a **process queue** or **process list**. The process queue is often searched by operating system components. For example, a user control process might search the process queue for all processes owned by a certain user so they can be terminated when that user logs off. Search and update speed is an important characteristic of PCBs and the process list. Process list and PCB contents are changed frequently as processes are created and terminated and as resources are allocated and released. Using efficient data structures and processing algorithms minimizes system overhead.

Processes can create, or **spawn**, other processes and communicate with them. The original process is called the **parent process** and the newly created process is called the **child process**. Parent processes can spawn multiple child processes, and the child processes of a single parent are collectively called **sibling processes**. Child processes can, in turn, spawn children of their own. A group of processes descended from a common ancestor, including the common ancestor itself, is called a **process family**.

Application programs often spawn a child process to execute general-purpose utility programs that are not part of the operating system service layer. For example, a Web browser can spawn a process to view specialized content such as a spreadsheet or Acrobat document. The process can be another program within the same application, a utility program supplied with an application suite, or a utility program supplied with the operating system. Subdividing application programs into multiple cooperating processes can improve execution speed, especially if the computer system has multiple CPUs, because multiple processes can execute simultaneously.

Even if only one CPU is available, subdividing application programs into cooperating processes increases execution speed due to more efficient resource allocation. For example, consider a word processor user who wants to edit one document while generating an Acrobat version of another. If the word processor spawns a separate Acrobat formatting process, then both tasks can execute concurrently. Because the two tasks use different mixes of computer resources (heavy I/O and light CPU for the editing process and the reverse for the Acrobat formatting process), the operating system can allocate resources efficiently so both processes move quickly toward completion.

Threads

The benefits of subdividing large processes into smaller ones is subject to the law of diminishing returns. As the number of active processes increases, the operating system overhead required to track and manage them also increases. The process list grows and operating system processes that search and update the list become less efficient. System overhead increases and the resources available to execute application processes are reduced.

In many operating systems, processes can subdivide themselves into more easily managed subunits called threads. A **thread** is a portion of a process that can be scheduled and executed independently. Process threads can execute concurrently on a single processor or simultaneously on multiple processors. Threads share all resources allocated to their parent process, including primary storage, files, and I/O devices.

The advantage of organizing an application program into a family of threads instead of multiple processes is that operating system overhead for resource allocation and process management is reduced. Because all threads of a process share storage and I/O resources, the operating system can track allocation of those resources on a per-process basis. This reduces the number of PCBs and speeds up searching and updating the process queue.

The operating system keeps track of thread-specific information in a **thread control block (TCB)**. Each PCB contains a set of pointers to its related TCBs. All active TCBs are organized into a data structure called a **run queue** or **thread list**. As with the process list, the run queue is implemented with efficient data structures that can be searched or updated quickly.

A process or program that divides itself into multiple threads is said to be **multithreaded**. An operating system that supports threads automatically creates a thread for each process, but subdividing processes into multiple threads is not automatic. The programmer or compiler must specifically identify code segments that can execute independently and create a thread for each segment. Operating systems that support multithreaded processes include Windows XP and later versions of UNIX.

CPU ALLOCATION

Threads[1] progress toward completion only when they have CPU cycles with which to execute their instructions. A multitasking operating system can execute dozens, hundreds, or thousands of threads within the same time frame. Most computer systems have only one or two CPUs, so threads must share CPUs.

The operating system makes rapid decisions about which threads receive CPU control and for how long that control is retained. Typically, a thread controls the CPU for no more than a few milliseconds before it relinquishes control and the operating system gives another thread a turn. Figure 11-5 shows three threads sharing a single CPU using small time slices. This method of CPU sharing is called **concurrent execution** or **interleaved execution**.

---

[1] We will use the term "thread" throughout this chapter, though all references to threads also apply to processes in operating systems that don't support threads.

Figure 11-5 ▶

Concurrent
(interleaved)
thread execution
on a single CPU

	Time slice 1	Time slice 2	Time slice 3	Time slice 4	Time slice 5	Time slice 6	Time slice 7
Thread 1	Running	Idle	Idle	Idle	Idle	Running	Idle
Thread 2	Idle	Running	Idle	Running	Idle	Idle	Idle
Thread 3	Idle	Idle	Running	Idle	Running	Idle	Running

Thread States

An active thread can be in only one of the following states:

► Ready
► Running
► Blocked

Figure 11-6 shows the three thread states and the events that move a thread from one state to another. Threads in the **ready state** are waiting for access to a CPU. There can be many threads in the ready state at any point in time. When a CPU becomes available, the operating system chooses a ready thread to execute on that CPU.

Figure 11-6 ▶

Thread
movement
among states

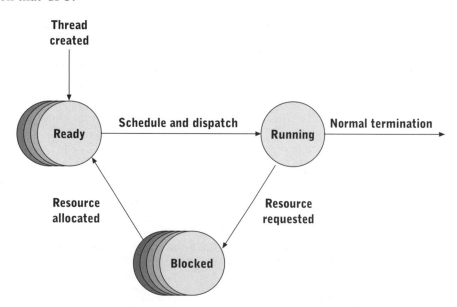

442   Operating Systems

The act of giving control of a CPU to a ready thread is called **dispatching**. The operating system dispatches a thread by loading the instruction pointer with an instruction address that belongs to the dispatched thread. During the next fetch and execute cycle, the thread takes control of the CPU and its directly related resources, such as general-purpose registers.

Once dispatched, a thread has entered the **running state** and retains control of the CPU until one of the following events:

► The thread or its parent process terminates (halts) normally

► An interrupt occurs

When a process or thread terminates its own execution, control returns to the operating system. The exact way in which the operating system regains CPU control varies across operating systems and CPUs. The most common method is for the process or thread to execute an exit service call. The exit call triggers a software interrupt.

Interrupts can occur for a variety of reasons, including:

► Executing a service call, such as a file I/O request

► A hardware-generated interrupt indicating an error such as overflow or a critical condition such as a power failure alarm from an uninterruptible power supply

► An interrupt generated by a peripheral device such as a network interface unit when a packet arrives

Recall from Chapter 6 that when any interrupt is received, the CPU automatically suspends the currently executing thread, pushes current register values onto the stack, and transfers control to the operating system master interrupt handler. The suspended thread remains on the stack until interrupt processing is completed. During this period of time the thread is in a **blocked state**. Once the interrupt has been processed, the operating system can leave the suspended thread in the blocked state, move it to the ready state, or return it to the running state.

Interrupt Processing

A blocked thread is waiting for an event to occur, such as allocation of a requested resource or correction of an error condition. Error conditions might or might not be correctable. If they are not correctable—for example, overflow or memory protection faults—the thread is halted. If the error can be corrected, the thread remains in the blocked state until the error condition is resolved.

For example, if a thread attempts to access a file on a removable disk and the disk is not in the drive, the disk drive controller generates an interrupt to signal the error. The CPU automatically suspends the thread and passes CPU control to an operating system interrupt handler. In an operating system such as Windows, the interrupt handler displays an error message asking that a disk be inserted. If the user complies, the error condition is resolved and the thread leaves the blocked state. If the user does not comply, the error is not corrected and the requesting thread is terminated.

Most service calls directly or indirectly require resource allocation. For example, a service call that requests additional memory for a thread is a direct request for resource allocation. A request to open a file is an indirect request for memory buffers to hold file data as it is transferred to or from secondary storage. If the request is for read/write access, it also indirectly requests exclusive access to the file. A subsequent request to read data from the file is an indirect request for access to the secondary storage device, its controller, and the I/O channel(s) that connect the secondary storage device to its memory buffers.

Some service calls, such as requests for additional memory, can usually be performed immediately. But many service calls, such as requests for input from a file or device, require a period of time to complete. In that case, the thread remains in the blocked state until the request is satisfied.

The following event sequence is typical when a process or thread reads from a file:

1. The thread requests file input by issuing an appropriate software interrupt.

2. The thread is automatically pushed onto the stack and the appropriate interrupt handler is called.

3. The interrupt handler executes. It checks to see whether the requested data is already in a memory buffer. If it is, the interrupt handler transfers the data from the memory buffer to the data area of the requesting process. The interrupt handler exits, and the operating system places the suspended thread in the ready or running state.

4. If the requested data isn't already in a memory buffer, the interrupt handler generates an appropriate read command and sends it to the secondary storage controller. The interrupt handler then exits without waiting for the data to be returned. The operating system places the suspended thread in a blocked state pending arrival of the requested data.

5. When the data is read from disk, the secondary storage controller sends an interrupt to the CPU. The interrupt handler transfers the data from the secondary storage controller to a memory buffer over the system bus and then copies it to the data area of the requesting process. The interrupt handler exits, and the operating system places the blocked thread in the ready or running state.

Scheduling

The decision-making process used by the operating system to determine which ready thread moves to the running state is called **scheduling**. The portion of the operating system that makes scheduling decisions is called the **scheduler**. Operating systems vary widely in their scheduling methods, though the following methods are typical:

▶ Preemptive scheduling

▶ Priority-based scheduling

▶ Real-time scheduling

Preemptive Scheduling In **preemptive scheduling**, a thread can be removed involuntarily from the running state. A running process or thread controls the CPU by controlling the content of the instruction pointer. CPU control is lost whenever an interrupt is received and the CPU pushes the current thread onto the stack and transfers control to the operating system. The portion of the operating system that receives control is called the **supervisor**.

The supervisor performs two important functions:

▶ Calls the appropriate interrupt handler

▶ Transfers control to the scheduler

The supervisor uses the value in the interrupt register as an index into the interrupt table. It extracts the corresponding address, pushes itself on the stack, and transfers control to the interrupt handler by placing its address in the instruction pointer. When the interrupt handler finishes, it passes control back to the supervisor by popping the stack. The supervisor then passes control to the scheduler by executing an unconditional branch instruction.

The scheduler performs four tasks:

▶ Update the status of any process or thread affected by the last interrupt

▶ Decide which thread to dispatch to the CPU

▶ Update thread control information and the stack to reflect the scheduling decision

▶ Dispatch the selected thread

Processing an interrupt usually changes the state of at least one thread. If the interrupt resulted from a request for a resource that couldn't be provided immediately, the thread currently on top of the stack must be moved from the running state to the blocked state. If processing the interrupt cleared an error condition or supplied a resource for which some thread was waiting, the thread must be moved from the blocked state to the ready state. The scheduler updates the appropriate TCBs to reflect these state changes. Figure 11-7 illustrates the processing sequence for both a service request and a delayed satisfaction of that request. The processing steps on the left occur after Thread 1 makes an I/O service call. The processing steps on the right occur after the I/O device completes the I/O operation.

Figure 11-7 ►

Interrupt
processing

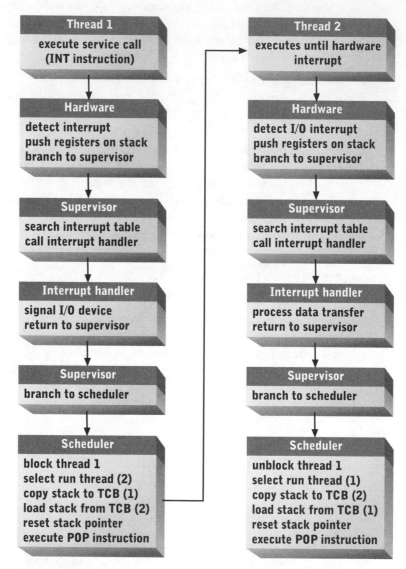

Once the scheduler decides which thread will run next, it updates the thread's TCB and manipulates the stack contents, if necessary. If the suspended thread is to become the running thread, then the scheduler pops the stack and the thread resumes execution. If another thread is selected as the running thread, then the scheduler must save the entire stack contents to the appropriate TCB, because elements below the topmost element represent suspended subroutines or functions of the interrupted thread.

For example, assume that a thread called main begins execution, calls one subroutine (A), that subroutine calls another subroutine (B), and an interrupt is received before any of the subroutines completes execution. After the interrupt is received and control is passed to the supervisor, there are three thread states on the stack (see Figure 11-8). The stack contents are main, which was pushed when subroutine A was called; A, which was pushed when subroutine B was called; and B, which was pushed when the interrupt was detected. The entire stack must be saved to the thread's TCB so it can be restored when the thread is dispatched at some future point.

Figure 11-8 ►

Subroutine calls (left) and associated stack content (right)

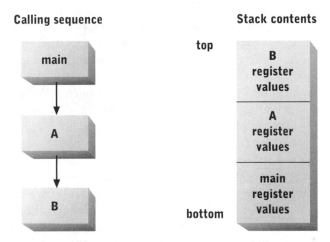

Calling sequence

main

A

B

Stack contents

top

B register values

A register values

main register values

bottom

After the suspended thread's stack is saved to its TCB, the operating system copies the saved stack of the next running thread from its TCB to the stack. The scheduler then transfers control to the thread by issuing a POP instruction.

Timer Interrupts Because interrupt arrival is unpredictable, the time that a thread remains in the ready state is also unpredictable. Theoretically, it is possible for a thread to control the CPU indefinitely if it makes no service calls, generates no error conditions, and receives no external interrupts. Under these conditions a thread stuck in an infinite loop would execute forever.

Most CPUs periodically generate an interrupt to give the scheduler an opportunity to suspend the currently executing thread. A **timer interrupt** is generated at regular intervals of between several dozen and several thousand CPU cycles. As with other interrupts, the currently executing process is pushed onto the stack and control is transferred to the supervisor. Because a timer interrupt isn't a "real" interrupt, there is no interrupt handler to call, so the supervisor passes control to the scheduler.

Timer interrupts are an important CPU hardware feature for multitasking operating systems. They guarantee that no thread can hold the CPU for long periods, because the scheduler will have regular opportunities to place another thread in the running state.

Priority-Based Scheduling Methods   Priority-based scheduling determines which ready thread should be dispatched to the CPU according to one or more of the following methods:

▶ First come first served

▶ Explicit priority

▶ Shortest time remaining

Under **first come first served (FCFS)** scheduling, the scheduler always dispatches the ready thread that has been waiting the longest. The simplest way to determine which thread has been waiting the longest is to order the run queue by waiting time. Each time a thread is moved into the ready state, the scheduler places its TCB at the back of the run queue. The scheduler then searches the run queue from front to back until it finds a ready thread, and that thread is the next to run. Implementing the run queue in this fashion is relatively simple, so FCFS scheduling has minimal system overhead.

Explicit priority scheduling uses a set of priority levels and assigns a level to each process or thread. A priority level can be assigned to a newly created thread based on the default priority of the process owner, a priority stated explicitly in an operating system command, or a number of other methods. Many operating systems automatically assign high priority levels to their own internal threads under the assumption that operating system threads must be completed quickly to maintain total system throughput.

The scheduling method can use priority levels in two ways:

▶ Always dispatch the highest-priority ready thread

▶ Assign larger time slices to high-priority threads

The first strategy ensures that high-priority processes are always dispatched before lower-priority processes. However, scheduling decisions based solely on initial priority level can result in extremely long or infinite idle time for long-running threads with low priority. Such threads are always moved to the back of the run queue and are unable to obtain sufficient CPU time to complete. To prevent this, many operating systems automatically increase the priority level of "old" threads. Priority-based scheduling using variable-length time slices also addresses this problem, though that method is seldom implemented.

Shortest time remaining (STR) scheduling chooses the next process to be dispatched based on the expected amount of CPU time needed to complete the process. This can be implemented directly by tracking time to completion for each thread and ordering the run queue with shortest time remaining threads first. STR scheduling can be accomplished indirectly by increasing the explicit priority of a thread as it nears completion.

In either case, the scheduler must know how much CPU time is required for each thread to execute and how much time already has been used. This implies that the required CPU time was provided to the scheduler when the thread was created and that this information was stored in the TCB. It also implies that the amount of CPU time used by a thread is stored in its TCB and is updated each time the thread leaves the ready state. Creating, updating, and using time-to-completion information increases system overhead, so STR scheduling is rarely implemented.

Real-Time Scheduling   **Real-time scheduling** guarantees a minimum amount of CPU time to a thread if the thread makes an explicit real-time scheduling request when it is created. Real-time scheduling guarantees a thread enough resources to complete its function within a specified time. Real-time scheduling is often used in transaction processing, data acquisition, and automated process control. For example, in a transaction processing application, an automated bank teller machine typically expects a response to an account balance inquiry within a specified amount of time.

Data acquisition and process control applications use data that arrives at a constant rate from one or more hardware devices. Data acquisition programs, such as those used to record radio astronomy observations, copy incoming data to a storage device such as tape or disk. Data analysis is performed later by a separate process. Process control programs actively process the data as it arrives. For example, a process control program in a chemical manufacturing plant processes data from sensors within pipelines and reaction vessels.

Data inputs flow from data collection devices or process sensors into memory buffers. A data acquisition or process control program must extract data from the buffers for analysis quickly enough to keep the buffers from overflowing. The program must be allocated sufficient resources to ensure that it can process or store incoming data at least as quickly as it arrives.

When a thread with a real-time scheduling requirement is created, it informs the operating system of the maximum CPU time needed to complete one thread cycle and the frequency of the cycles. A **thread cycle** can execute instructions to process a single transaction, retrieve and store data from an I/O device, or retrieve and analyze one set of process variables. For example, data acquisition equipment might send data every second, and the thread might need a maximum of 150 milliseconds to complete processing the data.

The thread signals the operating system at the start of each thread cycle, and the scheduler starts a timer. The scheduler tracks thread progress by updating CPU time used in the TCB each time it enters and leaves the ready state. The scheduler checks the timer, CPU time used, and maximum thread cycle time each time it makes a scheduling decision to ensure that the thread completes its cycle before the timer expires. The scheduler can manage multiple real-time threads, though real-time scheduling overhead can quickly consume available CPU resources.

Technology
Focus

Windows Scheduling

Current Windows desktop and server operating systems support a number of advanced features including multitasking, multithreaded processes, and pre-emptive priority-based scheduling and dispatching. Scheduling, dispatching, and interrupt handling are performed by an operating system component called the microkernel, which acts as both supervisor and scheduler.

Each process managed by the microkernel has a base priority level described by an integer value from 0 to 31. Higher values represent higher scheduling priority. Priority levels are grouped into four categories called base priority classes: Idle (priority levels 0-6), Normal (priority levels 6-10), High (priority levels 11-15), and Real-time (priority levels 16-31). A process is assigned a base priority class when it is first created.

Most user processes start in the Normal base priority class. In desktop computers and workstations, the microkernel raises the base priority class to High when a process is moved from the background to the foreground; in other words, when the process is receiving input from the keyboard and mouse. The Real-time base priority class does not meet the definition of real-time scheduling given earlier in the chapter. It simply describes the highest group of priority levels.

Threads inherit the base priority class of their parent process. Thread priority classes are named Idle, Lowest, Below Normal, Normal, Above Normal, Highest, or Time Critical. The precise meaning of thread priority class names within process base priority classes is summarized in Table 11-1.

The current priority level of a thread is called its dynamic priority. The initial value of a thread's dynamic priority level is the same as the base priority class of the thread and parent process. The microkernel can alter dynamic priority to improve system performance and to reflect high or low demand for system resources by a thread.

Table 11-1 ▶

Thread priority
classes and
levels

Base priority class	Thread priority class	Priority level
Real-time	Time Critical	31
	Highest	26
	Above Normal	25
	Normal	24
	Below Normal	23
	Lowest	22
	Idle	16
High	Time Critical	15
	Highest	15
	Above Normal	14
	Normal	13
	Below Normal	12
	Lowest	11
	Idle	1
Normal	Time Critical	15
	Highest	10
	Above Normal	9
	Normal	8
	Below Normal	7
	Lowest	6
	Idle	1
Idle	Time Critical	15
	Highest	6
	Above Normal	5
	Normal	4
	Below Normal	3
	Lowest	2
	Idle	1

Figure 11-9 shows a Windows utility called Process Viewer. Process Viewer shows information about active processes, including each process's base priority level and the priority class and dynamic priority of its threads. This example shows information for a process named *explorer* and its five threads. The base priority class of *explorer* is Normal. The initial base priority class for the highlighted thread (0) is Above Normal. The initial dynamic priority for this thread was 9. Note that the current value shown at the bottom of the display is 14. Some time after the thread was created, the microkernel adjusted its dynamic priority upward.

Figure 11-9 ▶

Windows
Process Viewer

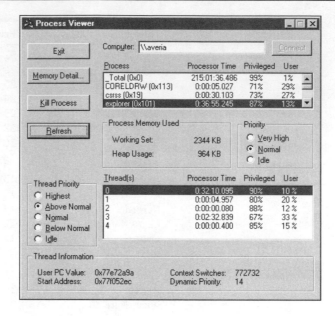

The microkernel implements a strict priority-based scheduling algorithm. A higher-priority thread in the ready state is always executed before a lower-priority thread. When a new thread is created with a higher priority level than the currently executing thread, the microkernel moves the currently executing thread to the ready state, where it remains until all higher-priority threads terminate or become blocked.

Business
Focus

Choosing a Server Operating System

Microsoft has steadily gained market share in server operating systems starting with Windows NT and continuing with various versions of Windows Server. Windows Server faces competition from several operating systems including OS/390 (IBM) and OpenVMS (Compaq/Hewlett-Packard). UNIX is also a significant competitor in the server arena with Linux and products from hardware vendors such as Solaris (Sun Microsystems), AIX (IBM), and Tru64 UNIX (Compaq/Hewlett-Packard).

Early Windows versions such as Windows 3.1 and Windows 95 lacked many features needed for a server or multiuser operating system. Beginning with Windows NT, Microsoft has aggressively added server-oriented features including:

▶ A reliable and robust file management system (NTFS)

▶ Access controls for key system resources including files and directories

▶ Preemptive priority-based scheduling and multithreading

▶ Memory protection and virtual memory management

▶ Interprocess communication and Internet-oriented networking support

▶ Server clustering with load-sharing and failover

Windows Server includes many server-oriented services and functions over and above the base operating system capabilities, including:

▶ Core Internet services such as DNS, DHCP, and routing

▶ User- and business-oriented Internet services such as e-mail, FTP, HTTP, and e-commerce

▶ File and printer sharing

▶ Support for component-based software with the COM+ and .NET component frameworks

By providing such functions with the "base" operating system, Microsoft has forced other vendors to bundle software that they used to sell separately with their "base" server operating systems.

UNIX has a long history beginning in the early 1970s as an experimental operating system developed at AT&T Bell Laboratories. AT&T didn't initially commercialize UNIX and allowed others to use both the operating system and its source code for free. Many universities experimented with UNIX and made changes and enhancements to it. Eventually, commercial vendors began to offer UNIX operating systems with proprietary extensions such as graphical user interfaces, advanced networking support, and database management systems.

At the same time, various nonprofit organizations also enhanced UNIX and developed useful applications and services that ran under UNIX. Those enhancements and extensions were always distributed freely and with complete source code (called open source) so that others could make improvements or modify the software as they saw fit. Currently, the driving force behind UNIX development and standardization is the Open Group (www.opengroup.org), which has owned the UNIX trademark since 1993. In early 2003, the International Organization for Standardization adopted the Single UNIX Specification. A few vendors have adopted the 2003 version of the standard and many more have adopted earlier versions.

Linux is a UNIX variant developed by Linus Torvalds in the early 1990s. Linux combines many early open-source enhancements with new ones to create a unified package that is generally easier to install and administer than other UNIX variants. Many Linux applications and services are based on non-proprietary standards such as MIT's X Window GUI and the CORBA component standard (CORBA is discussed in detail in Chapter 13). Organizations can deploy Linux and many Linux-based applications and services without paying the per-user or per-server licensing fees that are required with Windows and proprietary UNIX variants.

Linux, UNIX, and all of the current UNIX variants provide essentially the same base operating system functions as Windows Server. In addition, there is a rich variety of available applications and services ranging from personal productivity software to industrial-strength database management systems and computer-aided design and manufacturing software. The primary differences from Windows Server are:

- ► Many commonly used services and server software (e.g., Web services) are not provided as part of the base operating system. A wide variety of for-profit and nonprofit organizations develop and distribute such software for Linux and UNIX.

- ► Most Linux/UNIX application and server software is provided with source code, which enables users to make modifications for their own purposes.

- ► Application and server software can't be distributed in binary (executable) form because Linux and UNIX run on so many different CPUs and computer systems. Software is normally distributed in source code format with installation scripts that recompile and relink it for a specific hardware platform.

- ► Linux/UNIX application and server software is usually based on open standards, controlled by nonprofit organizations with wide representation from computer hardware and software vendors, academia, and user communities.

The differences in standards, bundling, binary compatibility, and the number of software sources yield a complex set of trade-offs that must be considered when deciding which operating system to acquire. For example, the open-source nature of Linux and UNIX is seen as a security and reliability advantage by many people, under the assumption that open source enables rigorous testing by a large user and developer community, as well as faster incorporation of security patches and improvements. On the other hand, many people consider Windows Server to be much easier to use and configure due to binary software compatibility, an extensive set of similar configuration tools, and the fact that the most commonly used server software is provided and supported by a single vendor.

MEMORY ALLOCATION

Threads need memory to hold instructions and data during execution. The operating system allocates memory when threads are created and responds to requests for additional memory during a thread's lifetime. The operating system also allocates memory to itself and for other needs such as buffers and caches.

Note that most of what is said about memory allocation in this section also applies to managing secondary storage devices. Keep this in mind as you read this section, because many of the terms and concepts will be used again in the next chapter.

Single-Tasking Memory Allocation

Memory allocation in a single-tasking operating system is fairly simple. A memory map for a single-tasking operating system is shown in Figure 11-10. The bulk of the operating system normally occupies lower memory addresses, and the application program is loaded immediately above it. For the application program in Figure 11-10, memory allocation is **contiguous**; that is, all portions of the program and operating system are loaded into sequential locations within memory.

Figure 11-10 ►

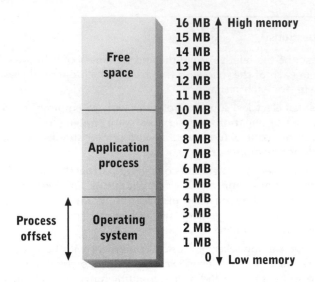

To conserve physical memory, some operating system components are loaded into memory only when needed. When loaded, they are usually placed in upper memory, as shown in Figure 11-11. The memory allocation for the operating system shown in Figure 11-11 is **noncontiguous,** or **fragmented.** Although most of the operating system is stored contiguously in low memory, a portion is stored in high memory, with intervening space allocated to an application process. MS-DOS employs contiguous memory allocation for application programs and noncontiguous memory allocation for the operating system itself.

Figure 11-11 ►

Noncontiguous
memory
allocation

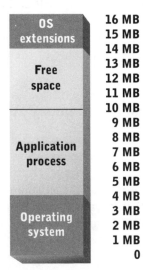

As discussed in Chapter 5, most CPUs use some form of offset addressing. With offset addressing, programs can be located in physical memory locations other than those assumed at the time of compilation and linking. Normally, programs are compiled and linked assuming that the first instruction will be located at the first physical memory address (zero).

When a CPU instruction uses a memory address as an operand, the operand is called a memory reference. Examples of CPU instructions that contain memory references include load, store, and branch. The process of determining the physical memory address that corresponds to a memory reference is called **address mapping** or **address resolution**. Address resolution is a simple process when memory is allocated contiguously. Each memory reference is mapped to its equivalent physical memory address by adding the offset value. For the application program shown in Figure 11-11, all memory references are mapped to physical memory addresses by adding the value 4,194,304 (4 MB) to the reference.

Programs can grow or shrink during execution by calling operating system service functions to allocate or deallocate memory. To service such requests, the operating system keeps track of allocated and free memory. Each time memory is allocated to a program or the operating system, the allocation is recorded so that the same memory region is never allocated more than once.

Multitasking Memory Allocation

Memory allocation is much more complex when the operating system supports multitasking. The operating system finds free memory regions in which to load new processes and threads and reclaims memory when processes or threads terminate. The goals of multitasking memory allocation are:

► Allow as many active processes as possible

► Respond quickly to changing memory demands of processes

► Prevent unauthorized changes to a process's memory region(s)

► Implement memory allocation and addressing as efficiently as possible

Multitasking operating systems use partitioned memory, which divides memory into equally sized regions, each of which can hold all or part of a process or thread. Figure 11-12 shows 16 MB of physical memory divided into 1 MB partitions, each of which can hold an operating system component, a process, or nothing at all (free space).

Figure 11-12 ►

Partitioning
main memory
into fixed-size
regions

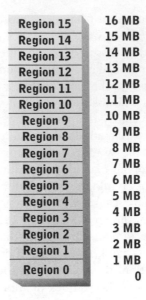

Region 15	16 MB
Region 14	15 MB
Region 13	14 MB
Region 12	13 MB
Region 11	12 MB
Region 10	11 MB
Region 9	10 MB
Region 8	9 MB
Region 7	8 MB
Region 6	7 MB
Region 5	6 MB
Region 4	5 MB
Region 3	4 MB
Region 2	3 MB
Region 1	2 MB
Region 0	1 MB
	0

Figure 11-13 shows three processes and the operating system loaded into fixed-size memory partitions. Process 1 occupies three memory regions and Process 2 occupies two memory regions. Each process uses all memory within its allocated regions. Process 3 occupies four complete regions and part of a fifth. Partitions 14 and 15 are unallocated.

Figure 11-13 ►

Several
processes and
the operating
system loaded
into 1MB fixed-
size memory
partitions

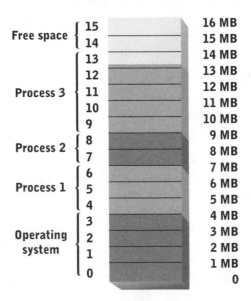

The operating system maintains a memory partition table and updates it each time a partition is allocated or freed (see Table 11-2). When a process is ready to be loaded, the operating system searches the table to find a sufficient number of contiguous free partitions. If they are found, the table is updated and the process is loaded into the free partition(s). When a process terminates, the table is updated to free the memory partition(s).

Table 11-2 ▶

Fixed-size memory partitions maintained by the operating system

Partition	Starting address	Status	Allocated to
0	0 MB	Allocated	OS
1	1 MB	Allocated	OS
2	2 MB	Allocated	OS
3	3 MB	Allocated	OS
4	4 MB	Allocated	Process 1
5	5 MB	Allocated	Process 1
6	6 MB	Allocated	Process 1
7	7 MB	Allocated	Process 2
8	8 MB	Allocated	Process 2
9	9 MB	Allocated	Process 3
10	10 MB	Allocated	Process 3
11	11 MB	Allocated	Process 3
12	12 MB	Allocated	Process 3
13	13 MB	Allocated	Process 3
14	14 MB	Free	
15	15 MB	Free	

Contiguous program loading, coupled with fixed-size memory partitions, usually results in wasted memory space. For example, Process 3 uses only part of its last partition. The remainder is wasted because the operating system can't allocate less than a full partition. Wasted space can be reduced by reducing partition size. In general, the smaller the size of the partitions, the less wasted space. However, smaller partitions require a larger partition table, which increases operating system memory requirements and the time needed to search and update the table.

Memory Fragmentation

As processes are created, executed, and terminated, memory allocation changes accordingly. Over time, memory partition allocation and deallocation leads to an increasing number of small free partitions separated by allocated partitions. Figure 11-14 shows changes in memory allocation over time as processes are created and terminated. Figure 11-14(a) shows a starting point with four processes in memory. Figure 11-14(b) shows memory allocation after Process 3 has terminated and Process 5 has been loaded into memory. A new free space area has been created because Process 5 is smaller than Process 3. In Figure 11-14(c), Process 2 has terminated and its former partition is now free.

Figure 11-14 ▶

Changes in memory allocation as processes execute and terminate

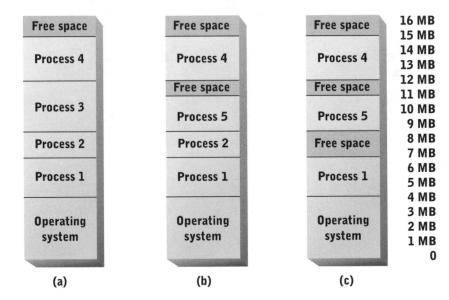

(a)   (b)   (c)

There is now 4 MB of free memory but only two contiguous free partitions. Although there might be other processes waiting to execute that require 4 MB or less, only those that require 2 MB or less can be loaded due to the fragmentation of available free space.

Over time, memory will become more and more fragmented, as shown in Figure 11-15(a), and larger processes will have increasing difficulty finding enough contiguous partitions. One way to address the problem of fragmented free space is to periodically relocate all programs in memory in a process called **compaction**. After compaction, all free partitions form a contiguous block in upper memory, as shown in Figure 11-15(b). Compaction is a time-consuming process because entire programs are moved within memory and many partition table entries are updated. The overhead required for compaction is generally larger than the overhead required to implement a more common strategy—noncontiguous memory allocation.

Noncontiguous Memory Allocation

In an operating system that supports noncontiguous memory allocation, portions of a process can be allocated to free partitions anywhere in memory. Noncontiguous memory allocation uses small fixed-size partitions, usually no larger than 64 KB, though we will continue to assume 1 MB partitions to simplify the current discussion and figures.

Examine Figure 11-15(a) and assume that a new process (Process 6) awaits loading and needs 4 MB of memory. Under contiguous memory allocation Process 6 can't be loaded unless memory first is compacted. Under noncontiguous memory allocation, Process 6 can be loaded into the four free partitions, as shown in Figure 11-16. The process is divided into four 1 MB partitions that are allocated to available free memory partitions.

<table>
<tr><td>

Figure 11-15 ►

Compaction combines multiple free space fragments (a) into a single contiguous region (b)
</td><td>

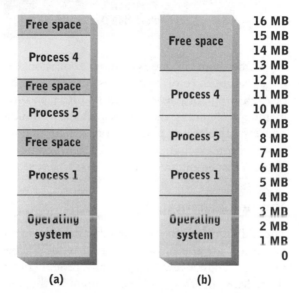

(a)   (b)
</td></tr>
</table>

Noncontiguous memory allocation is much more flexible than contiguous memory allocation, but flexibility comes at a price. Process 6 was compiled and linked under the assumption that it would occupy contiguous memory locations starting at address 0. But memory references within Process 6 must now cross noncontiguous partitions. For example, consider a branch instruction in partition 7 that references an instruction in partition 15. What offset value should be used to resolve the memory reference to a physical address?

Figure 11-16 ▶

Process 6 is
allocated to
noncontiguous
memory
partitions

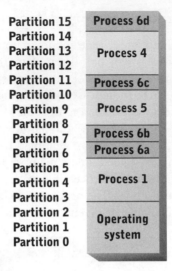

Partition 15	Process 6d
Partition 14	
Partition 13	Process 4
Partition 12	
Partition 11	Process 6c
Partition 10	
Partition 9	Process 5
Partition 8	
Partition 7	Process 6b
Partition 6	Process 6a
Partition 5	
Partition 4	Process 1
Partition 3	
Partition 2	Operating
Partition 1	system
Partition 0	

Under noncontiguous memory allocation, each process partition has its own offset value. To resolve addresses correctly, the operating system must keep track of each program partition and its offset and make appropriate adjustments to memory references that cross partition boundaries. The partition tables and address calculations that support noncontiguous memory allocation are much more complex than for contiguous memory allocation. Table content and address resolution are described in the next section.

Virtual Memory Management

The only portions of a process that must be in memory at any given time during execution are the next instruction to be fetched and any operands stored in memory. Only a few bytes of any process must reside in memory at any one time. Most operating systems minimize the amount of process code and data stored in memory at any one time, which frees large quantities of memory for use by other processes and greatly increases the number of processes that can execute concurrently.

Virtual memory management divides a program into partitions called **pages**. Each page is a small fixed-size portion of a program, normally between 1 and 4 KB. Memory is also divided into pages of the same size. Each memory page is called a **page frame**. During program execution, one or more pages are allocated to page frames and the rest are held in secondary storage. As pages in secondary storage are needed for current processing, the operating system copies them into page frames. If necessary, pages currently in memory are written to secondary storage to make room for pages being loaded.

Each memory reference is checked to see whether the page it refers to is currently in memory. A reference to a page held in memory is called a **page hit** and a reference to a page held in secondary storage is called a **page fault**. **Page tables** store information about page locations, allocated page frames, and secondary storage space. Each active process has a page table or portion of a page table dedicated to it. Table contents include page numbers, a status field indicating whether the page currently is held in memory, and the page frame number in main memory or in secondary storage. A sample page table is shown in Table 11-3.

Page number	Memory status	Frame number	Modification status
1	In memory	214	no
2	On disk	101	n/a
3	On disk	44	n/a
4	In memory	110	yes
5	On disk	252	n/a

Because page size is fixed, memory references can be easily converted to the corresponding page number and offset within the page. The page number can be determined by dividing the memory address by the page size. The whole portion of the result is the page number and the remainder is the offset into that page. For example, if page size is 1 KB, a reference to address 1500 is equivalent to an offset of 476 (1500 - 1024) into page number 2 (1500/1024 + 1). If the table is stored sequentially, the corresponding entry in the process's page table can be computed as an offset into the table.

If the reference is to an address in a page held in memory, the corresponding memory address is an offset into the memory page indicated in the table. Using Table 11-3 as an example, a reference to address 700 (offset of 700 into page 1) would resolve to an offset of 700 into memory frame 214. Once again, fixed page size is used to calculate the corresponding memory address as 219836 (214 × 1024 + 700). Note that these calculations are similar to those used for addressing array contents, as described in Chapter 3.

A secondary storage region, called the **swap space**, **swap file**, or **page file** is reserved exclusively for the task of holding pages not held in memory. The swap space is divided into page frames in the same manner as memory. A memory reference to a page held in the swap space results in that page being loaded into a page frame in memory. As with address resolution, page location within the swap space can be computed by multiplying the page number by the page size.

If all page frames are currently allocated, a page currently in memory, called the **victim,** must be written to the swap space before the reference page is loaded into a page frame. Some commonly used methods for selecting the victim are:

► Least recently used
► Least frequently used

Both methods require the operating system to maintain information about accesses to pages in memory. Searching and updating this information is a part of the system overhead associated with virtual memory management.

When a victim has been selected, it might or might not be copied back to the swap space. The sample data in Table 11-3 shows an entry that indicates whether a page has been modified since it was swapped into memory. If a page has not been modified, the copy held in the swap space is identical to the copy in memory, which means that the page does not have to be copied to the swap space if it has been selected as the victim.

Memory Protection

Memory protection refers to the protection of memory allocated to one program from unauthorized access by another program. This can apply to interference between programs in a multitasking environment or interference between a program and the operating system in either a single or multitasking environment. If memory is not protected, errors in one program can generate errors in another. If the program being interfered with is the operating system, this might result in a system crash.

In the simplest form of memory protection, the operating system checks each write to a memory location to ensure that the address being written is allocated to the program performing the write operation. Complicating factors include various forms of indirect addressing, virtual memory management, and cooperating processes, which are two or more processes that "want" to share a memory region. Memory protection adds overhead to each write operation.

Memory Management Hardware

Early types of multitasking, protected memory access, and virtual memory management were implemented exclusively by the operating system, which imposed severe performance penalties. Consider, for example, the overhead required to map program memory references using virtual memory management. Each reference requires the operating system to search one or more tables to locate the appropriate page and to determine the corresponding memory or disk location of that page. A memory reference that should consume only one or a few CPU cycles consumes many additional cycles for paging, swapping, and address mapping functions.

The benefits of advanced memory addressing and allocation schemes are offset by reduced performance when they are implemented in software. Because of this, modern CPUs and computer systems incorporate advanced memory allocation and address resolution functions in hardware. For example, all Intel microprocessors since the 80386 have included hardware support for virtual and protected memory management.

Technology
Focus

Intel Pentium Memory Management

Intel Pentium family CPUs dedicate six registers to hold data structures called segment descriptors. The registers are named CS, DS, ES, FS, GS, and SS. Segment descriptors contain various information about a segment, including:

▶ The physical memory address of the first byte of the segment, that is, its base address

▶ The size of the segment

▶ The segment type

▶ Access restrictions

▶ Privilege level

A memory address used in an operand consists of a reference to one of the segment registers and an offset value. For example, the operand DS:018C refers to byte 018C (hexadecimal) within the segment pointed to by the DS segment descriptor. The CPU converts this to a physical memory address by adding the segment base address and the offset.

Segments range in size from 1 byte to 4 GB. Segment types include data segments and code segments (executable instructions). The data type is further subdivided into "true" data and stack data. The CS register always contains a segment descriptor for a code segment. The DS, ES, FS, and GS registers always contain segment selectors for a data segment, or a null descriptor. The SS registers always contain the segment descriptor of a stack.

Pentium CPUs protect process memory by checking instruction operands that reference segment descriptors. The simplest check is a limit check on the offset value that prevents processes from reading or writing data outside their own segments. For each memory reference, the CPU compares the offset value to the size parameter stored in the segment descriptor. If the offset is larger than the segment size, an error interrupt is generated.

Access types must be specified for each segment. Data segments can be marked as read-only (RO) or read-write (RW). Code segments can be marked as execute-only (EO) or execute-read (ER). A segment's access type restricts the types of instructions that can execute using memory references within the segment. The CPU generates an error interrupt if an instruction attempts to write to a segment marked RO or read from a segment marked EO. Write operations to code segments always generate an error interrupt. The CPU also checks segment types when loading segment descriptors into a register. Segment descriptors for data segments cannot be loaded into the CS register. A segment descriptor for a code segment cannot be loaded into the DS, ES, FS, GS, or SS registers.

Pentium CPUs use two different methods to prevent access to code segments not allocated to a process. The first method specifies a privilege level for each process and segment. Privilege levels are numbered 0 through 3, with 0 being the most privileged. Every process is assigned a privilege level and cannot access a segment with a privilege level numbered lower than its own. This method protects operating system components from interference by application programs. Segments allocated to the kernel typically have a privilege level of 0. Segments allocated to other operating system components typically have a privilege level of 1 or 2. Application processes typically have a privilege level of 3.

Memory segments can be completely hidden from a process. The processor maintains a global descriptor table (GDT) containing descriptors for all segments. If an application process uses the GDT, then it can "see" all segments. The operating system can execute an instruction that allocates a local descriptor table (LDT) to a process. The operating system defines the content of the LDT prior to issuing the instruction. Typically, the operating system populates the LDT only with descriptors allocated to the process. It is impossible for a process using an LDT to "know of" any segments other than its own.

Pentium processors implement virtual memory management tables in hardware. Page size is normally 4 KB. The CPU uses two types of tables—page directories and page tables. A page directory is a table of pointers to page tables. By convention, the entries in the page directory correspond directly to the segment's descriptors contained in a process's LDT (see Figure 11-17). Each entry in a page directory points to one entry in a page table. Each entry in a page table contains descriptive information about one 4-KB page.

A page table entry includes the physical memory address of the page if it is loaded into memory. The entry also contains a number of bit field flags, including a flag that indicates whether the page is in memory, a flag that indicates whether or not the page has been accessed since it was loaded, and a flag that indicates whether the content of a page in memory has been written.

Figure 11-17 ▶

Relationship among segment and page (virtual memory) tables

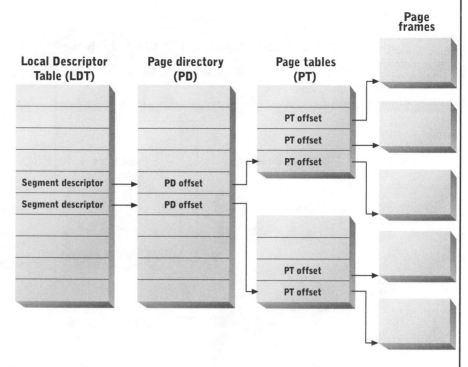

Virtual memory management responsibility is split between the CPU and the operating system. When virtual memory management is enabled, the CPU converts addresses containing a segment descriptor and offset into addresses containing a page directory offset, a page table offset, and an offset into the 4-KB page. The CPU automatically performs a lookup in the page directory and page table based on these offsets to find the base physical address of the page (see Figure 11-18). It then adds the page offset to generate a physical address. If the page table entry indicates that the page is not in memory, then the CPU generates an interrupt. The CPU also sets the access and write flags in the page table as page contents are read or written.

The operating system maintains its own table that maps pages to page frames in the swap space, and clears the access and write flags when a page is loaded into memory. The operating system sets and clears page table "in memory" flags each time pages are loaded or swapped. It also provides an interrupt handler to implement page swaps in response to page fault interrupts.

Figure 11-18►

Address
resolution with
virtual memory
management

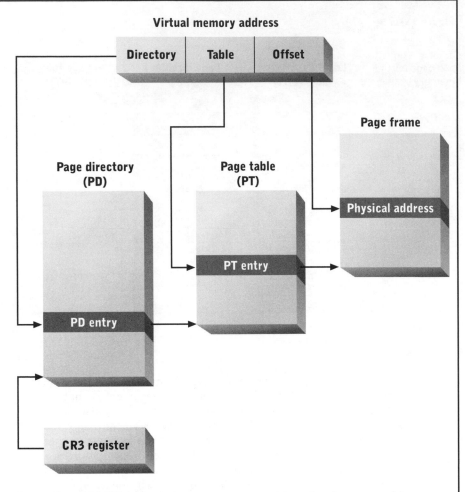

Early Microsoft and UNIX operating systems were unable to implement usable multitasking and virtual memory management on Intel-based personal computers due to the lack of those features in early Intel CPUs. Intel introduced hardware support for memory protection and virtual memory management with the 80386 CPU in 1987 and those features have continued into current Pentium processors. That hardware support enabled significant improvements in Microsoft operating systems and also provided sufficient CPU power to port UNIX to personal computers.

SUMMARY

► The operating system is the most important component of system software. Its primary purpose is to manage hardware resources and to provide support services to users and application programs. In functional terms, the operating system manages the CPU, memory, processes, secondary storage (files), I/O devices, and users. In architectural terms, the operating system consists of the kernel, service layer, and command layer.

► The operating system allocates hardware resources to individual user processes on demand. In multitasking environments, the operating system coordinates access to shared resources by many processes while ensuring that each process receives the resources it needs. The resource allocation function requires extensive recordkeeping and complex procedures to make allocation decisions. An important design goal for any operating system is to minimize system overhead while ensuring that resource allocation goals are met.

► Application software is simpler to develop if programs are unaware of resource allocation functions. Operating systems make a large set of virtual resources appear to be available to each application program. The sum of these apparent virtual resources generally exceeds the real resources that exist within the computer system. The operating system implements virtual resources by rapidly reallocating real resources among programs and by substituting one resource for another.

► The operating system stores information about each process in a process control block (PCB). PCB contents are updated continually as resources are allocated to processes. In some operating systems, processes can create executable subunits called threads, which can be independently scheduled. Threads share all resources with their parent process. Execution speed is increased when multiple threads execute concurrently or simultaneously.

► An active thread is always in one of three states—ready, running, or blocked. In a computer with a single CPU, only one thread may be in the running state at a time. Ready threads are waiting for access to the CPU. Blocked threads are waiting for some event to occur, such as completion of a service request or correction of an error condition. Processes can be scheduled by many methods, including first come first served, explicit prioritization, and real-time scheduling.

► Memory is divided into fixed-size partitions and processes are allocated one or more memory partitions to store instructions and data. The operating system maintains tables to track partition allocations and free space. Memory references are mapped to physical addresses through table lookups and address calculations.

► Modern operating systems implement virtual memory management. Portions of processes, called pages, are allocated to small memory partitions called page frames. Pages are swapped between memory and secondary storage as needed. Complex memory management procedures incur substantial overhead. To reduce operating system overhead, modern CPUs implement many of these functions in hardware.

This chapter presented an overview of internal operating system architecture and described CPU and memory allocation. In Chapter 12, we'll examine secondary storage allocation and file management. You will see that many of the issues addressed in memory allocation must also be addressed in secondary storage allocation. In Chapter 13, we'll turn our attention to external resources and explore how one operating system cooperates with other operating systems to provide local users access to distant resources.

Key Terms

address mapping	page fault	scheduling
address resolution	page file	service call
blocked state	page frame	service layer
child process	page hit	shell
command language	page table	shortest time remaining (STR)
command layer	parent process	sibling process
compaction	preemptive scheduling	spawn
concurrent execution	priority-based scheduling	supervisor
contiguous	process	swap file
dispatching	process control block (PCB)	swap space
explicit priority	process family	system overhead
first come first served (FCFS)	process list	thread
fragmented	process queue	thread control block (TCB)
interleaved execution	ready state	thread cycle
job control language (JCL)	real resource	thread list
kernel	real-time scheduling	timer interrupt
multitasking	run queue	victim
multithreaded	running state	virtual resource
noncontiguous	scheduler	virtual memory management
page		

Vocabulary Exercises

1. A(n) _____ operating system supports multiple active processes or users.

2. Under virtual memory management, the location of a memory page is determined by searching a(n) _____.

3. A(n) _____ occurs when a process or thread references a memory page not currently held in memory.

4. Dispatching a process moves it from the _____ to the _____.

5. The CPU periodically generates a(n) _____ to provide the scheduler an opportunity to allocate the CPU to another ready process.

6. A process in the _____ requires only access to the CPU to continue execution.

7. In the _____ scheduling method, processes are dispatched in order of their arrival.

8. A(n) _____ process contains subunits that can be executed concurrently or simultaneously.

9. _____ scheduling guarantees that a process will receive sufficient resources to complete one _____ within a maximum time interval.

10. Hardware resources consumed by an operating system's resource allocation functions are called _____.

11. _____ scheduling refers to any type of scheduling in which a running process can lose control of the CPU to another process.

12. The act of selecting a running process and loading its register contents is called _____ and is performed by the _____.

13. To achieve efficient use of memory and a large number of concurrently executing processes, most operating systems use _____ memory management.

14. When a process makes an I/O service request, it is placed in the _____ until processing of the request is completed.

15. Memory pages not held in primary storage are held in the _____ of a secondary storage device.

16. On a computer with a single CPU, multitasking is achieved through _____ execution of multiple processes.

17. Under _____, all portions of a process must be loaded into sequential physical memory locations.

18. The _____, _____, and _____ are the primary layers of an operating system.

19. A(n) _____ operating system can support multiple active processes at any one time.

20. A(n) _____ is the unit of memory read or written to the swap space.

21. A(n) _____ resource is apparent to a process or user, although it might not physically exist.

22. Under a(n) _____ memory allocation scheme, portions of a single process might be located physically in scattered segments of main memory.

23. Under virtual memory management, memory references by a process must be converted to an offset within a(n) _____.

24. Information about a process's execution state such as register values and process status are stored in a(n) _____.

25. Under the _____ scheduling method, processes requiring the least CPU time are dispatched first.

26. A(n) _____ causes the currently executing process to be _____ and control passed to the _____.

27. The process of converting an address operand into a physical address within a memory partition or page frame is called _____.

28. A(n) _____ is an executable subunit of a process that is scheduled independently but shares memory and I/O resources.

Review Questions

1. Describe the functions of the kernel, service, and command layers of the operating system.

2. What is the difference between a real resource and a virtual resource?

3. What are the goals of an operating system resource allocation function? Describe the conflicts among them.

4. How and why does a thread move from the ready state to the running state? How and why does a thread move from the running state to the blocked state? How and why does a thread move from the blocked state to the ready state?

5. What is a process control block and for what is it used?

6. What is a thread? What resources does it share with other threads within the same process?

7. Briefly describe the most common methods for making scheduling decisions.

8. What scheduling complexities are introduced by real-time processing requirements?

9. Describe the operation of virtual memory management.

10. What is memory protection and why is it needed? Where is it most efficiently implemented or enforced?

Research Problems

1. The RISC approach to processor design described in Chapter 4 avoids internal CPU complexity to gain raw processor speed. Strict adherence to this design philosophy might be construed to eliminate CPU support for memory protection and virtual memory management. Investigate the memory management techniques of a RISC CPU such as the Sun UltraSPARC (www.sun.com) or IBM POWER5 (www.chips.ibm.com). What support is provided for memory protection and virtual memory management?

2. Investigate the memory protection and multitasking capabilities of Sun Microsystems' Solaris operating system (see the textbook Web site for a set of Web resources). How are application programs isolated from the operating system? How are application programs isolated from each other? Is it possible for one application to cause another application or the operating system to crash? How are CPU cycles allocated to application programs? Can a thread or process lock up the system by never releasing the CPU?

Chapter 12

File and Secondary Storage Management

Chapter Goals

- Describe the components and functions of a file management system

- Compare the logical and physical organization of files and directories

- Explain how secondary storage locations are allocated to files and describe the data structures used to record those allocations

- Describe file manipulation operations, including open, close, read, delete, and undelete operations

- List access controls that can be applied to files and directories

- Describe security, backup, recovery, and fault tolerance methods and procedures

- Compare and contrast storage area networks and network-attached storage

File and secondary storage management are important system software functions because stored programs and data are important individual and organizational resources. The entire collection of system software that performs file and secondary storage management and access functions is known as a **file management system (FMS)**. The FMS usually is part of the operating system, though it is sometimes supplemented by additional software such as database management systems. Figure 12-1 shows the FMS functions described in this chapter.

Figure 12-1 ▶

Topics covered in this chapter

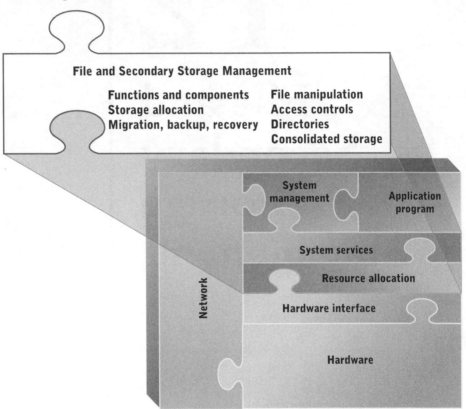

FILE MANAGEMENT SYSTEMS

An FMS is implemented in the following layers, which are similar to those of the operating system as a whole (see Figure 12-2):

▶ Application program or command layer

▶ File control

▶ Storage I/O control

▶ Secondary storage devices

Figure 12-2 ▶

Operating system layers compared to file management system layers

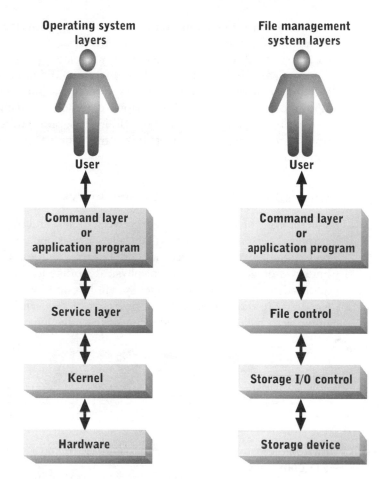

Operating system layers

User

Command layer
or
application program

Service layer

Kernel

Hardware

File management system layers

User

Command layer
or
application program

File control

Storage I/O control

Storage device

Storage devices physically store bits, bytes, and blocks within a storage medium, and storage device controllers interact with the bus and with operating system device drivers to transfer data between storage devices and memory. As discussed in Chapter 6, the device controller presents a logical view of the storage device or media to the device driver. The logical view is a linear sequence of storage locations or linear address space.

The **storage I/O control layer** is the part of the kernel that accesses storage locations and manages data movement between storage devices and memory. Software modules within this layer include:

▶ Device drivers for each storage device or device controller

▶ Interrupt handlers

▶ Buffers and cache managers

The **file control layer** provides a set of service functions for manipulating files and directories. It processes service calls from the application or command layer and issues commands to the storage I/O control layer to interact with hardware. The file control layer maintains the directory and storage allocation data structures used to locate files and their associated physical storage locations.

An FMS provides command layer functions and utility programs for users and system administrators to manage files, directories, and secondary storage devices. With the command layer, users perform common file management functions such as copying, moving, and renaming. Utility programs address more complex functions such as creating text files, formatting storage devices, and creating backup copies of files and directories.

Logical and Physical Storage Views

The file control layer is the bridge between logical and physical views of secondary storage. Users and applications logically view secondary storage as a collection of files organized within directories and storage volumes. The physical view of secondary storage is a collection of physical storage locations organized as a linear address space. A typical computer system has up to a few dozen storage volumes, hundreds to thousands of directories, tens of thousands to millions of files, and millions to billions of physical secondary storage locations (see Figure 12-3).

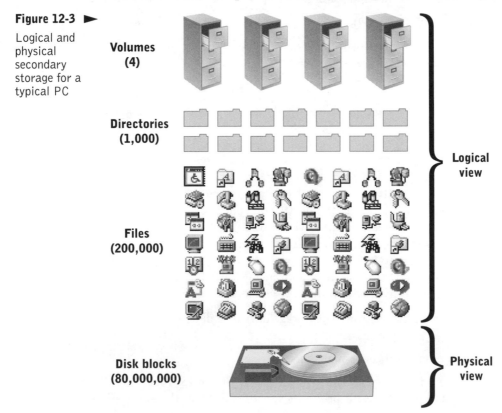

Figure 12-3 ►

Logical and physical secondary storage for a typical PC

Volumes (4)

Directories (1,000)

Files (200,000)

Disk blocks (80,000,000)

Logical view

Physical view

File and Secondary Storage Management

Figure 12-4 shows the logical structure of a typical data file. The file is sub-divided into multiple **records** and each record is composed of multiple **fields**. A record usually contains information about a single person such as a customer or employee, a thing such as a product held in inventory, or an event such as a transaction. A field contains a single data item that describes the record subject.

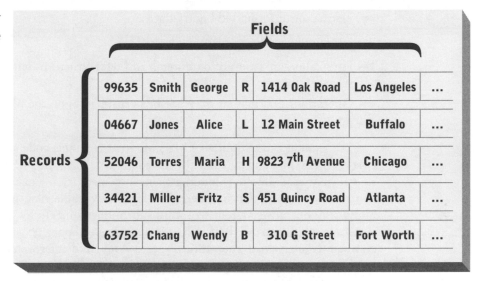

The logical file structure is independent of the physical device on which it is stored. A number of physical structure characteristics are simplified or ignored in the corresponding logical file structure, including:

▶ Physical storage allocation

▶ Data access method(s)

▶ Data encoding method(s)

Physical storage allocation considerations include the placement of fields and records within a file, and the distribution of the file across storage locations, media, or devices. Physical data access considerations include whether file content is accessed sequentially, using an index, or by some other method. Data encoding issues include the data structures and coding methods used to represent individual fields. Related issues include data encryption and data compression.

File Content and Type

A file can store many different data types including text, numbers, complex data structures, and executable instructions. It is difficult to design an FMS that accounts for all possible variations in file content and organization. Such an

FMS would be very complex and have a very large number of file-oriented commands and service routines. To avoid these problems, most FMSs directly support only a limited number of file types, including:

► Executable programs
► Operating system commands
► Textual or unformatted binary data

Among other things, file type determines:

► Physical organization of data items and data structures within secondary storage
► Operations that may or may not be performed upon the file
► Filename restrictions

For example, the Windows FMS supports executable code stored within executable (EXE) files and dynamically linked library (DLL) files. EXE files are stored in a format that simplifies loading them into memory for execution. DLL files contain subroutines that can be called from executable files and an index that allows the operating system to locate subroutines quickly by name. Operating system batch command (BAT) files are stored as ordinary text files, and the text commands are passed automatically to a command interpreter when the user double-clicks a BAT file icon. If the user double-clicks a text (TXT) file icon, the operating system automatically calls the Notepad text editing utility, which opens the file for editing. The relationship between file types and the programs or operating system utilities that manipulate them is called **file association**.

Modern FMSs provide a framework to support additional file types. In this framework, users and application programs can register new file types and install programs to perform common file manipulation operations such as printing, editing, and error checking. Figure 12-5 shows a partial listing of application file types registered on a computer running Windows XP. Figure 12-6 shows registration details for Microsoft Word document files, including the associated programs and commands used to print a document. Once a file type is registered, the operating system loads and executes the registered program each time the user prints, edits, or otherwise manipulates the file.

File type is normally declared when a file is created. In some FMSs, such as UNIX, file type is stored within the directory. In other FMSs, such as Windows, the file type is declared through a filename restriction or convention. For example, in Windows, executable filenames must end in .EXE, dynamically linked library filenames must end in .DLL, and the default text filenames end in .TXT. Windows registered file types also use naming conventions such as .DOC for Microsoft Word document files and .CDR for CorelDraw graphics files (see the Extensions column in Figure 12-5).

Figure 12-5 ►

Registered
Windows XP
file types

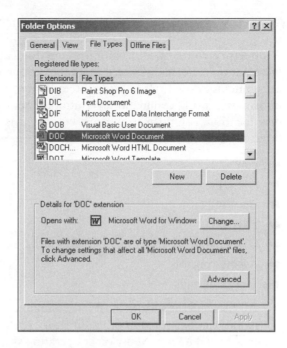

Figure 12-6 ►

Sample
association
details of a
Windows XP
file type

DIRECTORY CONTENT AND STRUCTURE

A **directory** contains information about files and other directories. Directory information is normally stored in a table, though more complex data structures are sometimes used. Users and system administrators can view most directory information using command layer utilities. Some directory information, such as disk location, is hidden from users because it is only used by the FMS. Typical directory contents include:

► Name

► File type

► Location

► Size

► Ownership

► Access controls

► Time stamp(s)

All files and directories must have a unique name within their own directory, but names can be reused in other directories. Operating systems vary in requirements for valid names. Many older operating systems restrict both the length of a name and the characters used. For example, MS-DOS uses a two-part name with a maximum of eight characters in the first part, three characters in the second part, and a mandatory dot (.) symbol separating the parts, for example SAMPLE.DOC (this format is sometimes called 8.3). Embedded space characters are not allowed nor are most nonalphabetic and nonnumeric characters. Modern operating systems are far less restrictive in their requirements for filenames.

File type can be stored implicitly through a file naming convention, such as .EXE for an executable file, or directly in a coded field within the directory. The type field can be displayed in several formats, including icons in a graphical display or special characters appended to the filename in a text-oriented display. Most FMSs store information about directories in the same way as information about files. An FMS uses a special directory code in the file type field to distinguish directories from files.

The location field content varies considerably across FMSs. In simpler FMSs, the location field usually contains the disk address of the file's first disk block. In more complex FMSs, it contains or points to a data structure such as an index that contains the addresses of all disk blocks.

The size field can contain the number of bytes allocated to the file, the actual number of bytes stored in the file, or both. The number of allocated bytes is usually larger than the actual number of bytes because a portion of the last disk block typically is not used. For example, the directory listing of a file with

513 actual bytes might list the file size as 1024 bytes if the allocation unit size is 512 bytes, because the last byte causes the FMS to allocate another entire 512-byte disk block.

The file owner field contains the account name or identifying number of the file creator or the last account to take ownership. Most complex FMSs grant special permissions to the file owner. Access controls list the accounts that have been granted or denied rights to the file. Rights vary among operating systems, but the following are typical:

► List
► Read
► Modify
► Change

With the List right, an account or group can view a file in a particular directory listing or list the contents of a directory. With the Read right, an account or group can view the content of a file—which also implies the right to copy it. With the Modify right, an account or group can add new content, alter or delete existing content, rename the file, move the file to a new directory, or delete the file. With the Change right, which is usually granted to the owner by default, an account or group can alter permissions for other accounts or groups.

Most FMSs store one or more time stamps in the directory for each file. Time stamps can include:

► When the file was created
► When the file was most recently read
► When the file was most recently written
► When the file was last backed up

Hierarchical Directory Structure

In a **hierarchical directory structure**, directories can contain other directories, but a directory cannot be contained within more than one parent. A hierarchical directory structure is sometimes called a **tree directory structure** because directory diagrams resemble upside-down trees. The left pane of Figure 12-7 shows a portion of a hierarchical directory structure as displayed by Windows Explorer. There is a root directory for each secondary storage device (drives A:, C:, F:, and G:, as shown in Figure 12-7). Each root directory can contain other directories, files, or both, as can directories below the root directory. The number of recursively descending directory levels theoretically is unlimited.

Figure 12-7 ▶

A hierarchical
directory
structure

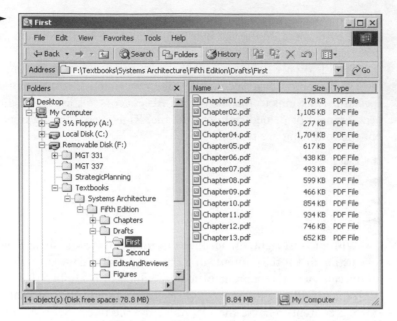

The right pane of Figure 12-7 shows the content of a single directory named First. For each process or user, the operating system maintains a pointer to the directory that is currently being accessed. That directory is called the **current directory** or **working directory**. In Figure 12-7, the current directory is First.

In a multiuser operating system, each user normally has a default working directory, or **home directory**. When a user logs on interactively or runs a batch process, his or her home directory is the default current directory. The user or process can subsequently change the current directory by issuing a command via the command layer or by making the appropriate service call.

Within the hierarchy of directories, names of access paths can be specified in two ways. A **complete path**, also called a **fully qualified reference**, begins at the root directory and proceeds through all directories along a path to the desired file. Directory names are separated by a special character such as "\" in Windows or "/" in UNIX. The fully qualified reference to the current directory is shown in the Address text box near the top of Figure 12-7:

F:\Textbooks\Systems Architecture\Fifth Edition\Drafts\First

The fully qualified reference to the first file in that directory is:

F:\Textbooks\Systems Architecture\Fifth Edition\Drafts\First\Chapter01.pdf

A **relative path** begins at the level of the current directory. For example, the name:

Chapter01.pdf

is a relative path to the file Chapter01.pdf in the current directory displayed in Figure 12-7.

Graph Directory Structure

A **graph directory structure** (see Figure 12-8) is more flexible than a hierarchical directory structure because it relaxes two of the restrictions enforced within a hierarchical directory structure:

► Files and subdirectories can be contained within multiple directories

► Directory links can form a cycle

Figure 12-8 ►

Graph directory structure (links shown as dashed lines)

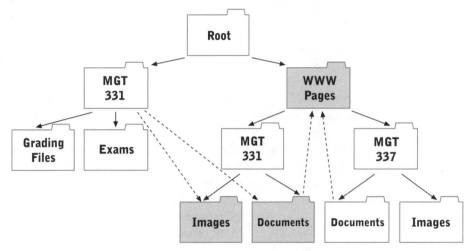

In Figure 12-8, subdirectories contained within multiple directories are shaded. The directories Images and Documents (on the left) are each contained in two directories, and WWW Pages is contained in three directories. The dashed lines in Figure 12-8 are called **links** in UNIX and **shortcuts** in Windows.

Links create the possibility of cycles or loops in the directory structure. There are two cycles in Figure 12-8—one from WWW Pages to MGT 331 to Documents and then back to WWW Pages, and the other from WWW Pages to MGT 337 to Documents and then back to WWW Pages. Cycles create special problems for programs that recursively search directories, because the search program can become stuck in an endless loop as it traverses the cycle over and over again. Special processing procedures are required to avoid this situation. Programs that recursively access directory contents can avoid infinite loops by not traversing links or by keeping a list of directories already "visited" and never rereading a directory on that list.

STORAGE ALLOCATION

The storage I/O control layer allocates secondary storage locations to individual files and directories. Because there are so many directories, files, devices, and secondary storage locations, storage allocation data structures and the functions that maintain their content are complex. The data structures and procedures used to manage secondary storage allocation to files are similar to those used to allocate memory to processes, as described in the previous chapter. The most significant differences are the number of storage locations, which is much larger for secondary storage, and the frequency of allocation changes, which is much lower for secondary storage.

Allocation Units

An **allocation unit** is the smallest number of secondary storage bytes that can be allocated to a file. Allocation units cannot be smaller than the unit of data transfer between the storage device and controller, which is normally called a **block**. Typical secondary storage device block size ranges from 512 bytes to 4 KB in multiples of 512 bytes.

Allocation unit size can be a multiple of block size. For example, a disk might use a 512-byte block size but the system administrator might decide to use 16-KB allocation units, each containing 32 blocks. Allocation unit size usually is set when an operating system or storage device is installed. Some operating systems such as DOS and early Windows versions[1] set unit size automatically based on storage device or media capacity. In other operating systems, such as UNIX and Windows XP, the installer selects the unit size. Once allocation unit size is set it is difficult to change.

Allocation unit size is a tradeoff among:

▶ Efficient use of secondary storage space for files
▶ Size of storage allocation data structures
▶ Efficiency of storage allocation procedures

Smaller allocation units result in more efficient use of storage space. For example, a file containing a single byte must be stored in a full allocation unit. If allocation unit size is 512 bytes, 511 bytes are empty and wasted. If the allocation unit size is 16 KB and only one byte is stored, 16,383 bytes are empty and wasted. If a storage device holds many small files, a large allocation unit can waste a great deal of storage capacity.

The advantage of larger allocation units is that storage allocation data structures can be smaller. As allocation unit size increases, the number of allocation units decreases. For example, consider a 100-GB disk. If unit size is set to 512 bytes, there are 209,715,200 (100 x 1024^3 ÷ 512) allocation units in the device. If unit size is set to 16 KB, there are 6,553,600 (100 x 1024^3 ÷ 16,386) allocation units in the device.

---

[1] All MS-DOS versions and many Windows versions use similar storage allocation and file management methods. For the remainder of this chapter, all references to Windows Me also apply to Windows 98, Windows 95, Windows 3.1, and all MS-DOS versions unless otherwise stated.

Storage Allocation Tables

A **storage allocation table** is a data structure that records which allocation units are free and which belong to files. A storage allocation table contains one entry for each allocation unit. A smaller allocation unit increases the number of entries in the storage allocation table. As the storage allocation table grows larger, the time required to search and update the table increases, which slows down any processing function that creates, deletes, or changes the size of a file.

Storage allocation table format and content vary across FMSs. Operating systems such as Windows Me use relatively simple data structures such as tables and linked lists. More complex operating systems such as UNIX and Windows XP use more complex data structures such as bitmaps and B+ trees. We'll stick to simpler data structure examples in this chapter.

Figure 12-9 shows a hypothetical storage device with 36 allocation units. Allocation units shaded the same color all belong to the same file or to free space. Table 12-1 shows directory entries for the files stored in Figure 12-9. Free allocation units are assigned to a hidden system file called SysFree. A file's allocation units can be stored in any order on the device. The linear address of each file's first allocation unit is stored in the directory. The addresses of other file allocation units are stored in the storage allocation table.

Table 12-2 shows a storage allocation table for the files in Figure 12-9. The format and content of this table are nearly identical to the **file allocation table (FAT)** used in Windows Me. The table contains an entry for each allocation unit. All of a file's allocation units are "chained" together in sequential order by a series of pointers. The table entry for each allocation unit contains a pointer to the next allocated unit in the file. For example, the entry for unit 0 (the first unit allocated to File1) contains a pointer to the table entry for unit 1 (the second unit allocated to File1). The pointers form a linked list that ties together the entries of all allocation units assigned to a specific file. The entry for a file's last allocation unit contains a special code to indicate that it is the last unit. All unallocated units are linked into a single chain, which simplifies finding storage units to allocate to new or expanded files.

Sequential access to a file's allocation units is efficient when the storage allocation table uses linked lists. However, random access is much less efficient, particularly if the file is large. Some FMSs define a separate file type for random access files and store an index or similar data structure in the file's first allocation unit. The contents of this index are redundant with the content of the storage allocation table, but the index allows specific units of a file to be located and accessed efficiently.

Table 12-1 ►

Directory
content for
the files shown
in Figure 12-9

Filename	Owner name	First allocation unit	Length (in allocation units)
File1	Smith	0	14
File2	Jones	3	3
File3	Smith	5	9
SysFree	System	2	10

Table 12-2 ►

A storage
allocation table
matching the
storage alloca-
tions shown in
Figure 12-9

Unit	Pointer	Unit	Pointer	Unit	Pointer	Unit	Pointer
0	01	9	10	18	22	27	32
1	08	10	13	19	20	28	29
2	7	11	12	20	21	29	30
3	04	12	18	21	23	30	34
4	06	13	15	22	28	31	End
5	14	14	16	23	24	32	33
6	End	15	17	24	31	33	35
7	11	16	19	25	26	34	End
8	09	17	25	26	27	35	End

Figure 12-9 ►

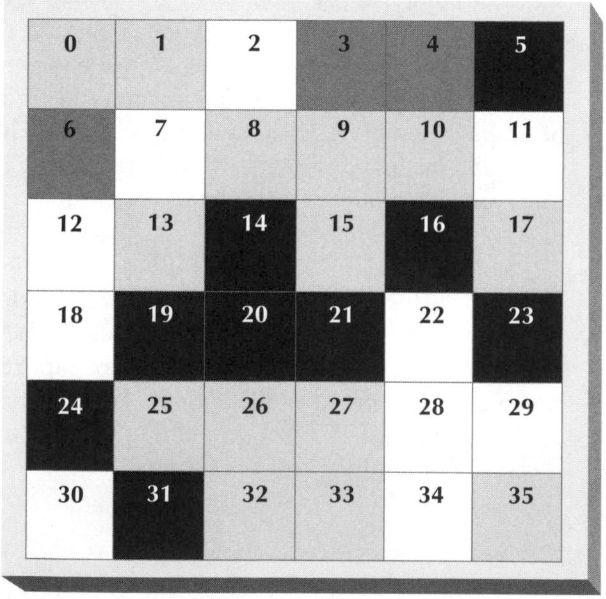

Storage blocks
allocated to
three files

Blocking and Buffering

Some application programs access files by logical records. A **logical record** is a collection of data items, or fields, that is accessed by an application program as a single unit. A **physical record** is the unit of storage transferred between the device controller and memory in a single operation. For disks and other devices that use data transfer units of fixed size, a physical record is equivalent to a block. For storage devices with variably sized data transfer units, such as tape drives, block size might differ from physical record size or might be undefined.

If logical record size is less than physical record size, a single physical record might contain multiple logical records, as shown in Figure 12-10(a). If logical record size is larger than physical record size, multiple physical records are required to hold a single logical record, as shown in Figure 12-10(b). Logical record grouping within physical records is called **blocking**. Blocking is described by a numeric ratio of logical records to physical records called the **blocking factor**. The blocking factor in Figure 12-10(a) is 4:3 and the blocking factor shown in Figure 12-10(b) is 2:3. If a physical record contains just one logical record, then the file is said to be **unblocked**.

Figure 12-10 ▶

Blocking of logical records into physical records

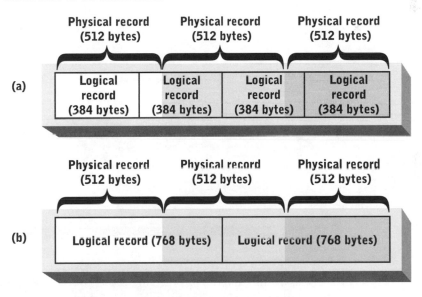

Read and write operations to or from an unblocked file can be implemented with simple and efficient algorithms because of the one-to-one correspondence between logical and physical records. File I/O is much more complex when logical and physical records have different sizes, because the FMS must coordinate physical record I/O and extract logical records on behalf of the requesting program.

The FMS uses buffers in primary storage to store data temporarily as it moves between programs and secondary storage devices. Buffers are allocated automatically when the file is first accessed and are managed by the operating system on behalf of application programs. Each buffer is the size of one allocation unit. As physical records are read from secondary storage, they are stored in the buffer(s). The FMS extracts logical records from the buffers and copies them to the data area of the application program (see Figure 12-11). The procedure is reversed for write operations.

Figure 12-11 ►

Input from secondary storage to an application program using a buffer

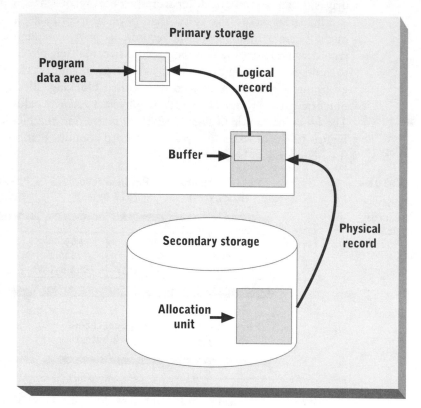

A buffer is a scratchpad for extracting logical records from physical records. As discussed in Chapter 6, buffering also improves I/O performance if a sufficient number of buffers are used. For high blocking factors, a small number of buffers can improve performance dramatically. For example, if each physical record contains 10 logical records, reading a single physical record provides sufficient data for 10 logical read operations. Reading the first logical record results in copying 10 logical records, which is one physical record, into the buffer. The next nine sequential read operations can be satisfied from the buffer without further input from secondary storage.

Low blocking factors, with larger logical records, require more buffers to achieve significant performance improvements. The FMS usually allocates enough buffers to hold at least one physical record. If enough buffers can't be allocated to hold an entire physical record, then a logical record must be moved to the program data area in a series of physical read and buffer copying operations.

An Example

The following example is typical of storage allocation and file I/O procedures in a simple FMS. Assume that the directory entries in Table 12-1 represent actual files stored on a disk drive and that the storage allocation unit size used by the disk drive and FMS is 512 bytes. Allocation units are assigned to files as shown in Figure 12-9, and the file allocation table is shown in Table 12-2. Assume that 55-byte logical records are stored sequentially in File3.

In response to any read operation performed by the application program, the FMS performs the following tasks:

1. Determine which allocation unit contains the requested record.

2. Load that allocation unit into the buffer if it is not already there.

3. Copy the portion of the allocation unit that contains the desired logical record to the application program's data area in memory.

4. Increment a pointer to the current position in the file.

Sequential Access The first allocation unit contains the first logical record or the first part of the logical record if it is larger than an allocation unit. When the first logical record is read, the FMS looks up File3's first allocation unit in the directory (5), issues a read request to the disk controller, and then loads the physical record into the buffer. Once loaded into the buffer, the first logical record's first byte is stored in byte 0. The FMS copies the first 55 bytes from the buffer to the application program. At the conclusion of the operation, the FMS sets the file pointer to 55 to point to the start of the second logical record.

During subsequent read operations, the FMS calculates the allocation unit and offset for each logical record. For the second record, the calculation is as follows:

$$\frac{(Pointer - 1) \times Record.Size}{Block.Size} = \frac{(2 - 1) \times 55}{512} = 0, \text{remainder } 55$$

The second record begins in allocation unit 0 at byte 55. Because that block is already in the buffer, the FMS copies 55 bytes starting at offset 55 and adds 55 to the file pointer.

Direct Access Assume that the first read operation requests the 37th logical record. Using the same formula as before, the calculation is:

$$\frac{(\text{Pointer} - 1) \times \text{Record.Size}}{\text{Block.Size}} = \frac{(37 - 1) \times 55}{512} = 3, \text{remainder } 499$$

The 37th record begins in the fourth allocation unit allocated to File3 at byte 499. Allocation unit 3 is the fourth allocation because allocation unit 0 is the first allocation unit.

To find File3's third allocation unit, the FMS follows the chain of entries in the storage allocation table. The first allocated unit (5) is recorded in the directory. File3's second allocation unit is the pointer field of entry 5 in the file allocation table, which is allocation unit 14. File3's third allocation unit is the pointer field of entry 14 in the file allocation table, which is allocation unit 16. File3's fourth allocation unit is the pointer field of entry 16 in the storage allocation table, which is allocation unit 19. Therefore, the 37th record is stored in allocation unit 19, starting at byte 499.

The FMS loads allocation unit 19 into the buffer and begins transferring 55 bytes to the application program. While copying, the FMS reaches the end of the buffer before it has finished copying 55 bytes. The FMS recognizes this condition, looks in the storage allocation table to determine File3's next allocation unit (20), issues a read request for allocation unit 20, loads it into the buffer, and copies the first 42 bytes to the application program.

FILE MANIPULATION

All FMSs provide service layer functions to enable application programs to create, copy, move, delete, read, and write files. The exact set of service layer functions varies widely among FMSs. Application programs interact directly with the FMS through the operating system service layer, and users interact with the FMS indirectly through the command layer.

File Open and Close Operations

The FMS must perform several tasks, collectively called a **file open operation**, before an application program can read or write a file's contents. The application program executes a file open service call to inform the FMS that it intends to read or write a file. In response to the service call, the FMS performs the following steps:

1. Locates the file within the directory structure and reads its directory entry.

2. Searches an internal table of open files to see whether the file is already open.

3. Ensures that the process has sufficient privileges to access the file.

4. Allocates one or more buffers.

5. Updates an internal table of open files.

The FMS maintains a table of open files to prevent application programs from interfering with each other's file I/O activities. If a file is already open for read-only access and another program tries to open the file, the FMS normally opens the file but allocates a separate set of buffers to each program. If another program attempts to open the file for writing, the request is normally denied, because buffer content is difficult to manage if one program can change content that another program has already read. In some complex FMSs, multiple programs can read or write a file by employing complex schemes for locking physical records.

When an application program finishes reading or writing a file, it executes a file close service call. The FMS completes the **file close operation** by performing the following steps:

1. Flushing the program's file I/O buffers to secondary storage.

2. Deallocating buffer memory.

3. Updating the file's directory entry time stamps.

4. Updating the open file table.

If the program that issued the request is the only program accessing the file, the FMS deletes the file's entry in the open files table. If other programs still are accessing the file, the FMS deletes the program from the list of programs actively using the file.

Delete and Undelete Operations

In most FMSs, files are not immediately removed from secondary storage when they are deleted. Instead, the file's storage allocation units are marked as free and its directory entry is marked as unused. As new files are created or existing files are expanded, the deleted file's allocation units are reassigned to other files and overwritten with new content. Some portion of the deleted file's content remains on secondary storage until all of its allocation units have been reassigned and overwritten. When a new file is created in the deleted file's former directory, the deleted file's directory entry is overwritten.

Implementing file deletion as just described is efficient, but it has two important consequences:

► Files can be undeleted by reconstructing appropriate directory and storage allocation table contents

► File content can be visible to intruders who can bypass the storage allocation table and read allocation units directly

A user might be able to recover a deleted file by performing an **undelete operation**. For example, if File2 in Figure 12-9 and Table 12-1 is deleted, the second row of the directory table is marked as deleted, and allocation units 3, 4, and 6 are added to the chain of allocation units for SysFree. If the user executes a file recovery utility before performing any other file operations, File2 can be recovered based on the information still in the directory entry and the chained contents of the storage allocation table in Table 12-2.

In some environments, such as law enforcement and defense research, users need to know that a deleted file can never be recovered. In some FMSs, users or system administrators can configure the FMS so directory entries and allocation units are immediately overwritten with blanks. This provides additional security but slows the deletion process, particularly for large files.

ACCESS CONTROLS

Because data is an important and valuable organizational resource, an FMS helps prevent loss, corruption, and unauthorized access to files. An FMS relies on the operating system to identify and authenticate users and their processes. In operating systems that enforce access controls, each user has a unique account name or identification number and must authenticate their identity through passwords or other means. Once user identity is authenticated, the user's name or ID is passed to the FMS with every service request.

By default, individual users are the owners of files they create and can grant or deny file access privileges to other users or groups of users. Different FMSs provide different sets of access controls. For example, UNIX defines three access control types:

► Read
► Write
► Execute

With read access, a user or process can view the contents of a file. With write access, a user or process can alter the contents of a file or delete it altogether. With execute access, a user or process can execute a file, assuming that the file contains an executable program or set of operating system commands. A file owner can reserve access privileges to himself or herself, thereby denying those access privileges to all other users except the system administrator. The file owner can grant any access privilege to other members of a user group or to all users. For example, a file owner might grant read and write access to one work-group but only read access to other users. A file owner can revoke his or her own access privileges. For example, a file owner might deny write access to prevent accidental deletion of an important file.

Most FMSs use similar access controls for files and directories. Users are the owners of their home directories and any directories that they create below the home directory in a hierarchical directory structure. Access controls for reading (listing directory contents) and writing (altering directory contents) are defined.

Access controls are enforced automatically in FMS service routines that access and manipulate files and directories. Although access controls are a necessary part of file manipulation, they impose additional processing overhead. In some FMSs, the system administrator might choose among several different levels of enforcement for file access controls to balance overall FMS performance with the relative need for file security.

An FMS restricts access to secondary storage devices, storage allocation tables, and root directories to prevent users and processes from bypassing security controls built into FMS service routines. File and directory accesses also can be logged for later review by the system administrator. Enforcing access controls reduces the speed of many file access operations because of the extra processing required.

Technology
Focus

Windows NTFS

Early Microsoft operating systems (including MS-DOS and Windows versions through Windows Me) use a file system called the file allocation table (FAT). When Microsoft developed Windows NT, which evolved into Windows 2000, XP, and Server, it decided to develop an entirely new file system, the NT File System (NTFS). Windows NT was targeted to high-performance and "mission critical" application software that required several file system features, including:

▶ High-speed directory and file operations

▶ Ability to handle large disks, files, and directories

▶ Secure file and disk content

▶ Reliability and fault tolerance

Windows NTFS organizes secondary storage as a set of volumes. A volume normally corresponds to one partition of a disk drive, but volumes can sometimes span multiple partitions and drives. Volumes contain a collection of storage allocation units called clusters. Cluster size can be 512, 1024, 2048, or 4096 bytes. Each cluster is identified by a 64-bit logical cluster number (LCN) within a linear address space. A volume can be as large as 4096×2^{64} bytes.

A volume's master directory is stored in a data structure called the master file table (MFT). The MFT contains a sequential set of file records, one for each file on the volume. All volume contents are stored as files, including user files, the MFT itself, and other volume management files such as the root directory, storage allocation table, bootstrap program, and bad cluster table. The first 16 MFT entries, numbered 0 through 15, are reserved for the MFT and volume management files. All subsequent MFT entries, numbered 16 and higher, store records about user files.

Conceptually, a file is an object with a collection of attributes including name, global access restrictions such as read only, and a security descriptor that identifies the owner and holds owner-defined access controls. A file's data content is just another attribute, though it is usually much larger than the other attributes. Each attribute type is assigned a numeric code, and file attributes are stored in ascending code order in the file's MFT record, as shown in Figure 12-12(a). MFT record size is 1, 2, or 4 KB and is determined by the operating system when a volume is formatted.

Figure 12-12 ▶

NTFS MFT records for a small file (a), large file (b), and small directory (c)

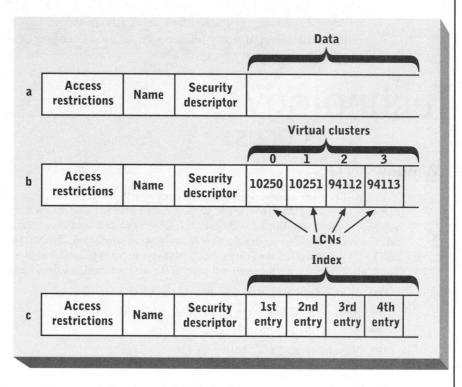

Each file attribute contains a header and a data value. The header contains the attribute name, a resident flag, the length of the header, and the length of the attribute value. Attributes can be either resident or nonresident. A resident attribute is a short value stored within an MFT record immediately following its header such as access restrictions and filename, as shown in Figure 12-12(a).

A nonresident attribute is stored in clusters elsewhere in the storage device, and the cluster addresses are stored in the file's MFT record. Data attributes are the most common nonresident attributes, though normally resident attributes

　　　　　　　　　　　　　　File and Secondary Storage Management

such as the security descriptor can grow too large to fit within an MFT record. The clusters assigned to a nonresident attribute are called virtual cluster numbers (VCNs). The area immediately following the nonresident attribute's header stores a sequence of LCNs, as shown in Figure 12-12(b). The VCN corresponds to the position of an LCN within this sequence. For example, LCN 94112 is VCN 2 in Figure 12-12(b).

The volume root directory and all user-defined directories are stored in the same manner as files. That is, they are stored as sequences of attributes within an MFT record. A directory's data attribute contains an index of files within the directory, as shown in Figure 12-12(c). The index is sorted by filename and also contains the file number (which is equivalent to the MFT record number), time stamp(s), and size. By duplicating this information in the directory index, directories can be listed more quickly. The index of small directories is stored sequentially in the MFT record. Larger directories are stored as B+ trees.

File security is implemented through the object manager facilities of the operating system. Files, I/O devices, and many system services are managed as objects by the Windows NT operating system. Accessing an object automatically invokes a security subsystem that compares the object's security descriptor against the security descriptor of the accessing process or user. An MFT security descriptor has the same structure and content as security descriptors for other object types.

NTFS has several fault tolerance features, including redundant storage of critical volume information, bad cluster mapping, logging of disk changes, and optional RAID (discussed later in this chapter.) The MFT is always stored at the beginning of a volume, but a partial second copy is stored in the middle of the disk in case a block assigned to the primary MFT becomes corrupted or unreadable. The FMS detects unreadable blocks during formatting and subsequent read and write operations. Clusters containing bad blocks are marked as unreadable in a separate bad cluster file.

NTFS uses a delayed, or lazy, write protocol. Disk blocks are cached in memory and write operations are confirmed immediately to the requesting process. A background process performs cache flushing. Write operations that affect volume structure, such as file creation, file deletion, and directory modification are written to the cache and also written immediately to a log file stored on disk. In the event of a system crash, the contents of the log file are always current and can be used to restore the volume structure to a consistent state.

FILE MIGRATION, BACKUP, AND RECOVERY

Most file management systems provide utilities and embedded features to protect files against damage or loss, including:

▶ File migration (version control)
▶ Automatic and manual file backup
▶ File recovery

Smaller scale FMSs such as those within LAN and personal computer operating systems typically do not support file migration but do support backup and recovery.

File Migration

When a user alters a file, the original file version is usually overwritten by the new version. However, there are advantages to maintaining the original file version, including allowing "undo" operations and having the original available as a backup. Many commonly used application programs such as word processors and text editors automatically save original file copies. For example, Microsoft Word can be configured to create a backup of the original version automatically each time it saves a document file. The most recently saved file is stored with a DOC filename extension and the original backup file is stored with a WBK filename extension.

Many transaction processing programs also preserve original versions of input files. For example, the original version of a bank's master account file is usually copied prior to processing daily batch transactions such as checks, interest, and monthly fees. The transaction processing program then reads the master and transaction files and updates the master with the transactions as shown in Figure 12-13. The original account master is commonly called the **father**, and the copy that has been updated to reflect new transactions is called the **son**. After another set of transactions is processed, the father becomes the **grandfather**, the son becomes the father, and the new copy of the account master file becomes the new son.

Figure 12-13 ▶

Batch account transaction update

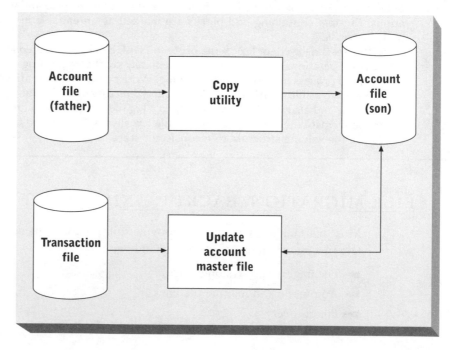

In mainframe file management systems such as IBM's OS/390 and some third-party file management software, the process of naming and storing original versions of altered files is automated. In a process called **versioning**, the original version of a file is archived automatically whenever the file is modified. In such systems, it is common to attach a version number to each filename. When a file is created, it is assigned a version number of 1. The first time it is altered, the altered copy is assigned a version number of 2, and so forth. The file management system automatically generates and stores copies and tracks version numbers. A user or application program can access an older version of a file by explicitly referring to its version number.

As files are modified, older versions accumulate on secondary storage, which can rapidly consume this resource, especially when files are large or frequently altered. To compensate for this, the file management system automatically performs file migration. **File migration** is a file management technique that balances the storage cost of each file version with anticipated user demand for that version. As a file version becomes outdated, the probability of an access decreases. The file management system migrates older file versions from local disks to remote disks and eventually to offline backup storage such as tape.

A simpler file migration method is based on file age rather than multiple file versions. The FMS periodically checks the last access time stamp of each file and migrates older files. The file still appears in directory listings and is reloaded automatically from distant or backup storage if accessed. This file migration method is frequently used in file management systems that manage infrequently accessed files for many users. It often is coupled with robotic devices to access offline tape or removable optical disc volumes.

File Backup

Most FMSs provide general-purpose utility programs to create backup copies of files and directories on removable or remote storage media. Backup utilities can be executed manually or automatically at set periods. Backups protect against data loss due to many causes including storage device or media failure, accidental erasure, or malicious attack. Backup utilities can protect file content, directory content, and storage allocation tables.

Backup copies should be stored on a different storage device to avoid losing both the original and the backup in the event of total device failure. For backups created on removable media, the media should be stored in a different physical location. Large computer centers typically store backup copies in a separate building or site to minimize the probability of a disaster destroying both copies, for example, a fire that destroys an entire building. Backup copies can be transmitted directly to remote storage facilities via a high-speed network, eliminating the cost, delay, and risk inherent in physically transporting storage media.

Types of periodic backups include:

► Full backup
► Incremental backup

When a **full backup** operation is performed, the FMS copies all files and directories for an entire storage volume. A full backup operation can also include storage allocation tables, partition tables, and other important disk management data structures. A full backup is time consuming due to the large number of files copied and the slow write speeds of most backup storage devices. Full backups are usually performed during off-peak hours at long time intervals, such as weekly.

An **incremental backup** operation archives only files that have been modified since the previous incremental or full backup. To make incremental backup operations possible, the FMS must keep track of when backups are performed and when files are modified. The backup utility compares the most recent update and backup time for each file and directory. Only files and directories modified since their last backup operation are copied to backup storage. Incremental backups are usually much faster than full backups because many files are not modified frequently.

Most large-scale FMSs use both full and incremental backups. For example, incremental backups might be created at the end of each business day and full backups created each weekend. Mixing backup types reduces the hardware resources consumed by backup operations but complicates the recovery process. If files must be recovered, the full backup must be restored first, then the incremental backups created since the last full backup must be recovered in the order they were created.

Transaction Logging

Transaction logging, also called **journaling**, is a form of automated file backup. The term transaction in this context should not be confused with the more generic meaning of the term, for example, a business transaction such as a customer purchase. To an FMS, a **transaction** is any single change to file contents or attributes such as a newly added record, modified field, or changed access controls. In an FMS that supports transaction logging, all changes to file content and attributes are recorded automatically in a separate storage area in addition to being written to the file's I/O buffer. Log entries are immediately or frequently written to a physical storage device.

Transaction logging provides a high degree of protection against data loss due to program or hardware failure. When an entire computer system fails, the contents of file I/O buffers are lost. If these buffers were not written to physical storage prior to the failure, file content becomes corrupted and content changes are lost. With transaction logging, the FMS can recover most or all of the lost changes and repair corrupted files. When the system is restarted, the contents of the transaction log are reviewed and compared to the file content on disk. Lost updates are identified and written to the files.

Transaction logging imposes a performance penalty because every file change requires two write operations—one to the file and another to the transaction log. Transaction logging is commonly used only where the costs of data loss are high, such as for large scale e-commerce sites.

File Recovery

Backup procedures and utility programs must be supplemented by a reliable set of recovery procedures to form a complete file protection mechanism. Typically, recovery procedures have both automated and manual components. For example, transaction log replay and subsequent file repair is usually fully automated. Recovery procedures based on full or incremental backups stored on removable media usually rely at least to some degree on manual procedures.

The FMS maintains backup logs to aid in locating backup copies of lost or damaged files. Recovery programs can search these logs for particular files or groups of files. Backup logs record the storage device or medium on which the backup copies are located. At the time of the backup, the backup utility writes an identification number or code to the backup storage medium and sometimes to a label that is manually applied to the medium. With the external label, the system administrator can locate the medium and mount it in the appropriate device. The recovery utility reads the embedded identification number or code to verify that the correct medium has been mounted before beginning recovery operations.

Recovery procedures for a crashed system or physically damaged storage device are usually more sophisticated and highly automated. Damage might have occurred to files, directories, storage allocation tables, and other important disk management data structures. The recovery utility reconstructs as much of the directory and storage allocation data structures as possible and makes a consistency check to ensure that:

► All storage locations appear within the storage allocation table and other data structures

► All files have correct directory entries

► All storage locations of a file can be accessed through the storage allocation table

► All storage locations can be read and/or written

Consistency checking and repair procedures consume a great deal of time— anywhere from a few minutes to several hours. But they mitigate the need to do large amounts of data recovery from backup copies and minimize the amount of current data that is lost. They also can eliminate the need to reinstall system and application software.

Fault Tolerance

As applied to FMSs, **fault tolerance** describes methods of securing file content against hardware failure. Magnetic and optical drives are complex devices with many mechanical parts. To improve performance, manufacturers employ high spin rates and small distances between read/write heads and recording media. The result is devices that provide high performance at the cost of occasional catastrophic failure.

Common causes of disk failure include head crashes, which are contact between a read/write head and a spinning platter, and burned out motors and bearings. Repairing a failed disk drive is prohibitively expensive due to the nature of modern manufacturing methods. The mean time between failures (MTBF) of a modern magnetic disk drive is approximately 10 years of continuous use. But the large number of disks in use guarantees that some failures will occur before 10 years.

File backup, recovery, and transaction logging are forms of protection against disk failure, but they all require time to implement, and the data is unavailable to users during recovery operations. In many processing environments, occasional down time for file recovery is acceptable. In other processing environments such as banking, retail sales, e-commerce, and production monitoring and control, down time is unacceptable or very expensive. In general, any business or organization that performs continuous updates and queries against files and databases is a candidate for advanced methods of fault tolerance such as mirroring or RAID.

Mirroring

Disk mirroring is a fault tolerance technique in which all disk write operations are made simultaneously or concurrently to two different storage devices. In some cases, the two devices might be located physically in different cabinets, rooms, or buildings. If one device fails, the other device contains a duplicate of all data. Data is available continuously because either device can respond to a read request.

Disk mirroring can be implemented through the FMS by configuring it to perform duplicate writes to duplicate storage devices, but that can reduce system performance substantially. Software-based mirroring is required if duplicate disks are not attached to the same disk controller.

When duplicate disks are located within the same cabinet, mirroring is usually implemented within hardware by the device controller, which reduces CPU and system bus overhead. Multiple disk drives are attached to the controller, and write operations are duplicated automatically by the controller to each drive. Read operations are split between the drives to improve performance. Special utility programs are executed to configure the disk controller for mirroring and to initialize a new duplicate drive if a failure occurs.

Disk mirroring provides a high degree of protection against data loss with no performance penalty if implemented in hardware. The primary disadvantages of mirroring are the cost of redundant disk drives and the higher cost of disk controllers that implement mirroring. Mirroring at least doubles the cost of data storage.

Technology
Focus

RAID

Redundant Array of Inexpensive Disks (RAID) is a disk storage technique that improves performance and fault tolerance. The original RAID version, now known as RAID 0, was developed at the University of California, Berkeley in the late 1980s. RAID has evolved considerably since then, and a large number of products are now available commercially. A flurry of RAID development in the early 1990s resulted in many incompatible approaches and products. The RAID Advisory Board (RAB) was formed in 1992 to define standard methods of implementing RAID levels through 5 (see Table 12-3). RAID 1 is disk mirroring, as described in the previous section.

Table 12-3 ►

RAID levels

Level	Description
0	Data striping without redundancy
1	Mirroring
2	Data bit striping with multiple error check sums
3	Data byte striping with parity check data stored on a separate disk
4	Data block striping with parity check data stored on a separate disk
5	Data block striping with parity check data stored on multiple disks

All RAID levels except RAID 1 use some form of data striping. **Data striping** breaks a unit of data into smaller segments and stores those segments on multiple disks. For example, a 16-KB block of data can be divided into four 4-KB segments, each segment written in parallel to a separate disk as shown in

Figure 12-14. A subsequent read of the original 16-KB block accesses all four disks in parallel.

Figure 12-14►

Data striping across four disks

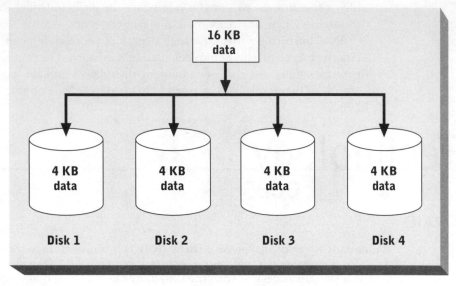

Data striping improves read performance by breaking a large read operation into multiple smaller parallel read operations. The elapsed time to perform the entire read operation is reduced, because multiple disks can perform small parallel read operations faster than a single disk can perform one large read operation. Write performance is also improved because smaller write operations occur more quickly. For both read and write operations, however, overhead is incurred to disassemble and reassemble larger data segments and to issue read or write requests to multiple disks.

RAID levels 1 through 5 achieve fault tolerance by generating and storing redundant data during each write operation. RAID levels 3 through 5 generate parity bits for data bytes or blocks and store those parity bits on one of the disks. If a single drive fails, no data is lost because missing bits can be reconstructed from the data and parity bits on the remaining drives.

For example, assume that even parity is used and that four one-valued bits are stored on four different disks. Also, assume that a fifth disk holds parity data as shown in Figure 12-15. The parity bit for four one-valued data bits also is a one-valued bit because four is an even number of one-valued bits. If the parity disk fails, then the other disks still retain the original data bits. If any of the data disks fails, there is a single missing bit that must have held either a one or zero value. The missing bit value is determined by comparing the parity bit to the remaining bit values. Because there are three one-valued bits and the parity bit is one-valued, the missing bit must be one-valued.

Figure 12-15▶

A RAID write
operation to
multiple disks

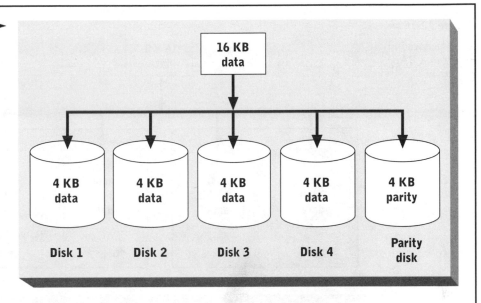

Storing parity data reduces the usable capacity of a disk array. The proportional reduction depends on the number of disks in the array. In Figure 12-15, five disks are used, resulting in 20 percent of the available disk space being used for redundant information. Using a larger number of disks would decrease the portion of space used for parity bits at the expense of a slight increase in the probability of data loss due to multiple drive failures.

Multiple RAID levels can be layered to combine their best features. The most common example is RAID 1+0, also called RAID 10. RAID 10 mirrors individual disks (RAID 1) and then stripes data (RAID 0) across multiple mirrored pairs (see Figure 12-16). Since striping requires at least two disks and each disk is mirrored, RAID 10 requires at least four disks. For RAID 10 with four disks, read and write performance is improved by up to 100% and the system can recover from the failure of any single disk or two disks if they are in different mirror pairs.

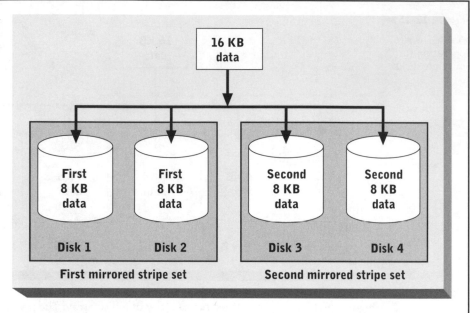

Figure 12-16 ▶

16 KB stored in
a four-disk RAID
10 array

RAID can be implemented in software or hardware. Hardware implementations are common for a number of reasons, including:

▶ The ability to configure all RAID components in a single cabinet

▶ Extended hardware fault tolerance through redundant power supplies and disk controllers

▶ Reduced load on the host CPU

▶ Reduced system software complexity

A RAID storage device looks like a single large disk drive to an operating system. A dedicated controller performs all RAID-related processing including segmenting read and write operations, generating parity data, reconstructing missing data if a drive fails, and repopulating data to a replacement drive. Hardware-based RAID systems for LAN and small WAN servers are typically based on the SCSI bus. RAID systems for larger computer systems usually use other high-capacity communication channels such as Fibre Channel.

STORAGE CONSOLIDATION

The traditional model of storage access by application software relies on an approach commonly called **direct-attached storage (DAS)**. DAS describes any architecture in which software executing on a CPU accesses secondary storage devices within the same computer system. DAS is an efficient approach to storage access when a single computer system interacts with a single storage subsystem. However, DAS can be an expensive and inefficient method of storage access for organizations with dozens or hundreds of servers and terabytes of shared data. In such an environment, storage overlap among servers can be substantial, resulting in high costs and redundant updates of multiple data copies.

Two approaches are commonly employed to overcome the inefficiencies of DAS in multiple-server environments:

▶ Storage area network

▶ Network-attached storage

A **storage area network** (**SAN**) is a high-speed interconnection among general-purpose servers and a separate storage server. Figure 12-17 shows one example of a SAN. Each general-purpose server has a device controller attached to its system bus. The device controller attaches via an external connection to a SAN switch which, in turn, connects to one or more storage servers. A storage server accepts storage access requests from other servers and accesses embedded storage devices on their behalf. As in DAS, SAN storage accesses are at the level of individual disk sectors in a logical address space. Communication within the SAN is based on a high-speed protocol such as Fibre Channel or Infiniband.

Figure 12-17 ▶

A server cluster with a storage area network

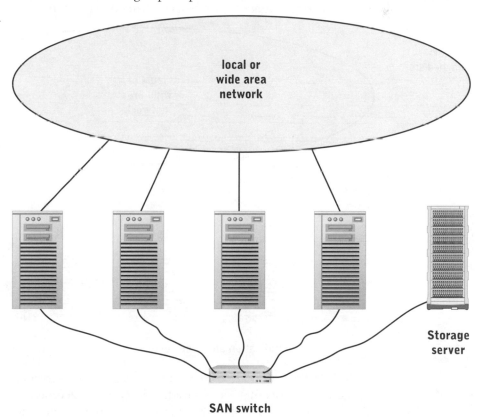

The storage server in a SAN is not a complete general-purpose computer system, though it may have a CPU and a limited-purpose operating system. In essence, the storage server acts as a disk controller for each server, translating accesses in a logical storage address space into physical accesses to one or more disk drives. But the storage server must also handle the complexities associated with shared resource access. For example, if one general-purpose server asks to read a storage location while another server is writing that same location, the storage server must queue the read request until the write request is completed.

The term **network-attached storage (NAS)** describes any architecture in which a dedicated storage server is attached to a general-purpose network to service storage access requests from other servers. Figure 12-18 shows a NAS server with four application and Web servers attached via a local or wide area network. A NAS server can be a general-purpose server customized to storage applications or it may be a limited-purpose server, sometimes called a *server appliance*. In either case, a NAS server has all of the hardware attributes of a complete computer system including a CPU, memory, system bus, storage subsystem, and network I/O devices. It also has an operating system that can manage its hardware resources and respond to storage service requests from other servers.

Figure 12-18 ▶

Network-attached storage

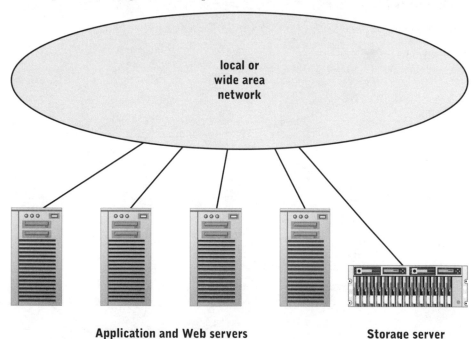

local or wide area network

Application and Web servers   Storage server

A key distinction between SAN devices and NAS servers is the type of storage access requests that are serviced. A SAN device accepts low-level requests to individual locations within the storage subsystems logical address space. This type of access is sometimes called *block-oriented access* or *sector-oriented access*. In contrast, a NAS server accepts access requests to files, which may encompass many storage blocks. This type of access is usually called *file-oriented access*. In NAS, a storage server operating system manages one or more file systems that are shared by other servers or clients. In a SAN, a storage server provides a pool of physical storage locations. Other servers use those locations to store one or more file systems.

SANs and NASs have unique advantages, though many organizations employ a combination of both approaches. SANs are most common in multi-server environments with mainframes or supercomputers and substantial overlap among server storage needs. Clusters commonly employ SANs to share data and system software among many identical computer systems. SANs are expensive to purchase and administer, but they avoid the costs of duplicate storage and storage administration.

NAS is commonly employed when geographically dispersed servers need access to a common file system. One example is a shared file system accessed by servers and clients spread across a university or corporate campus. NAS is much cheaper to acquire and administer than a SAN, but the cost savings come at the price of lower performance. SAN interfaces are high-speed connections to the "back end" of multiple servers. NAS connections are ordinary LAN or WAN connections, which are usually slower and more congested than SAN connections. NAS servers also have the additional task of file system management, which slows their response time compared to storage servers in a SAN.

SUMMARY

► The file management system (FMS), usually a part of the operating system, manages all aspects of user and program access to secondary storage. The FMS presents a logical view of stored data and programs to users as a set of files organized into directories and storage volumes. As users and programs manipulate the logical view, the FMS translates those operations into appropriate commands to physical storage devices. FMSs directly support only a limited number of file types, including executable programs, executable operating system command files, and user data files. In modern FMSs, users and programs can define new file types and install utility programs to manipulate them.

► Directories enable users to organize the thousands of files stored in a typical computer system. Each storage device or volume has a root directory. In hierarchical directory structures, directories can contain other directories, creating a tree structure in which each file belongs to only one directory. In graph directory structures, files can belong to more than one directory, and there is the possibility of loops or cycles in the directory structure. Directories store descriptive information about files and directories, such as name, owner, file type, access controls, and time stamps.

► Secondary storage devices are divided into allocation units, which are typically a few kilobytes in size. The FMS assigns allocation units to files and directories as they are created or expanded and reclaims allocation units as files and directories shrink or are deleted. The FMS uses complex data structures, called storage allocation tables, to track the assignment of allocation units to files and directories. Entries within the table can be linked lists in simpler FMSs or indices or other complex data structures in more complex FMSs.

► The FMS allocates buffers to support program file I/O. A program executes a file open service call before reading or writing a file, which causes the FMS to find the file, verify access privileges, allocate buffers, and update an internal table of open files. Multiple programs can open the same file for reading, but in most FMSs, only one program can open a file for writing. When a program is finished with a file it executes a file close service call, which causes the FMS to flush buffer content to the storage device, release the buffers, update the file time stamps, and update the table of open files.

► The FMS enforces access controls when accessing files on behalf of a user or program. File owners and system administrators can grant or deny access privileges for reading, writing, and executing files. When a user or user program attempts to access a file, the FMS checks the user identification against the access controls stored in the file's directory entry to determine whether access is permitted. Enforcing access controls provides security at the expense of additional FMS overhead.

► FMSs provide utilities to make backup copies of files and directories and to recover them if needed. Backups can be performed manually or may be fully automated. Different types of backups include full and incremental. Some FMSs support automatic storage and backup of old file versions in a process called file migration. As file contents change, the original contents are archived in another file. Older file versions are automatically migrated to slower or offline storage as they become further out of date.

► Organizations that store large amounts of data typically employ some form of consolidated storage. A storage area network (SAN) is a high-speed interconnection among general-purpose servers and one or more storage servers. Network-attached storage (NAS) servers are dedicated to managing one or more file systems and are accessed by other servers and clients over a local or wide area network.

In this chapter and the previous chapter, we covered allocation and management of the CPU, primary storage, and secondary storage. In the next chapter, we'll examine how users and applications interact with external resources including files, I/O devices, and programs.

Key Terms

allocation unit
block
blocking
blocking factor
complete path
current directory
data striping
direct-attached storage (DAS)
directory
disk mirroring
father
fault tolerance
field
file allocation table (FAT)
file association
file close operation

file control layer
file management system (FMS)
file migration
file open operation
full backup
fully qualified reference
grandfather
graph directory structure
hierarchical directory structure
home directory
incremental backup
journaling
link
logical record
network-attached storage (NAS)
physical record

record
Redundant Array of
    Inexpensive Disks (RAID)
relative path
shortcut
son
storage allocation table
storage area network (SAN)
storage I/O control layer
transaction
transaction logging
tree directory structure
unblocked
undelete operation
versioning
working directory

Vocabulary Exercises

1. A(n) _____ is the unit of file I/O to and from an application program.
   A(n) _____ is the unit of file I/O to and from a secondary storage device.

2. The term blocking factor describes the number of _____ contained within a single _____.

3. A file _____ releases allocated buffers and flushes their content to secondary storage.

4. A(n) _____ operation allocates buffers for file I/O and updates a table of files in use.

5. The content of a logically, but not physically, deleted file may be recovered in a(n) _____ operation.

6. _____ describes the tracking of old file versions and their movement to offline and archival storage devices.

7. Under _____, changes to files are written to a log file as they are made.

8. The _____ layer presents a service layer interface to application programs and the command layer. The _____ layer manages the movement of data between secondary and primary storage.

9. A(n) _____ specifies a storage device or volume and all directories leading to a specific file. A(n) _____ specifies file location with respect to the current or working directory.

10. In a(n) _____ directory structure, a file can be located within no more than one directory. This restriction does not apply in a(n) _____ directory structure.

11. A(n) _____ consists of a master (root directory), one or more subdirectories, and a filename.

12. MS-DOS and some Windows versions record storage allocation information in a(n) _____.

13. An FMS can implement _____ through disk mirroring or _____.

14. A(n) _____ records the allocation of storage locations to specific files.

15. When an old version of a master file is saved, the current version can be called the _____, the previous version the _____, and the version before that the _____.

16. RAID 10 combines disk mirroring and _____ to achieve performance improvement and fault tolerance.

17. In a(n) _____, multiple servers share access to the same storage server over a special-purpose network dedicated to low-level storage accesses.

18. Under _____, a specialized server manages one or more file systems and responds to file I/O requests sent across a LAN or WAN.

Review Questions

1. List the layers of a file management system (FMS) and describe their functions.

2. What is the difference between the logical and physical structure of a file? What advantages are realized by *not* having an application program interact directly with physical file structure?

3. What file types are usually supported directly by a file management system?

4. What is an allocation unit? What are the advantages of using small allocation units? What are the disadvantages?

5. Describe the use of buffers in file I/O operations. When are buffers allocated? When are they released?

6. Describe the structure of a hierarchical directory. What are its advantages and disadvantages as compared to graph structure directories?

7. How is file deletion normally accomplished? What security problems might arise from this method?

8. What levels of access privilege can exist for a file?

9. What is transaction logging, or journaling? Describe the performance penalty that it imposes on file update operations.

10. Describe the various levels of RAID. What are their comparative advantages and disadvantages?

11. Compare and contrast storage area networks and network-attached storage. Which is more common in environments where many co-located servers access the same data?

Problems and Exercises

1. Modify the directory in Table 12-1 and the storage allocation table in Table 12-2 to store a new file containing seven allocation units.

2. Assume that the first character of the filename in a deleted file's directory entry is overwritten with an ASCII 0 to mark the file as deleted. Write a step-by-step procedure for undeleting a deleted file, assuming that no file operations have been performed since the deletion.

Research Problems

1. To increase efficiency of application program execution, some FMSs use a large number of file organization and access methods, each optimized to a very specific type of processing. The MVS (IBM mainframe) operating system contains such an FMS. Investigate this FMS to determine the various methods of file organization and access that are supported. For what type(s) of application program file manipulation is each method intended?

2. Investigate the storage area network and network-attached storage products of a major computer system vendor such as IBM, Hewlett-Packard, or Dell. What are the approximate costs of each type of server configured to store 2 terabytes of data and respond to requests from 8 other servers? Which type of device provides higher storage access performance? Which type of device is easier to configure and administer?

Chapter 13

Internet and Distributed Application Services

Chapter Goals

► Describe client/server and multi-tier application architecture and discuss their advantages compared to centralized applications

► Explain how operating systems and network protocol stacks cooperate so users and programs can access remote resources

► Describe low-level protocols for interprocess communication across networks, including sockets, named pipes, RPC, and DCE

► List and describe standard Internet protocols used to access distributed resources

► Discuss component-based application development and describe the protocols and standards that support component-based applications

► Explain the role and function of directory services and the LDAP standard

Users of modern computers and information systems interact with a variety of resources located on computer systems all over the world. In this chapter, we'll examine the complex set of network protocols, infrastructure, and services that make it possible for users to interact with geographically dispersed resources as if they were all located on their own computer. Figure 13-1 shows the topics covered in this chapter.

Figure 13-1 ▶

Topics covered in this chapter

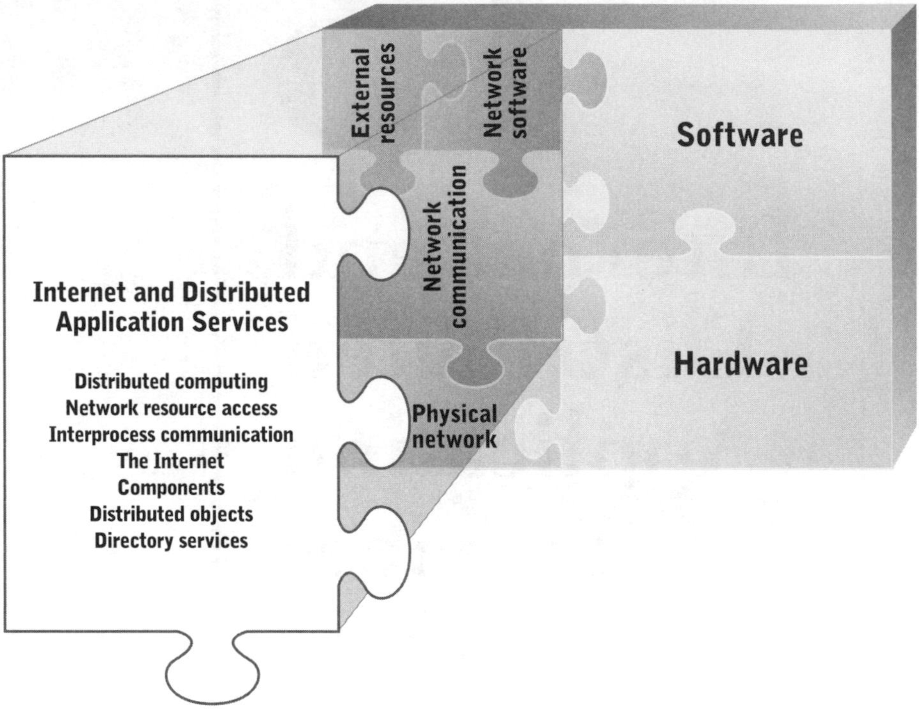

DISTRIBUTED COMPUTING

Modern information systems are often distributed across many computer systems and geographic locations. For example, an organization's corporate financial data might be stored on a mainframe computer in its central office. Midrange computers in regional offices might periodically generate accounting and other reports based on data stored on the mainframe, while microcomputers in branch offices might access and view periodic reports as well as query and update the central database. Similar tasks might be performed with laptop computers or personal digital assistants (PDAs) using wireless networks. Distributing parts of an information system across many computer systems and locations is called **distributed computing** or **distributed processing**.

Client/Server Architecture

Client/server architecture is currently the dominant architectural model for distributing information system resources. **Client/server architecture** divides software into two classes—client and server. A **server** manages one or more system resources and provides access to those resources through a well-defined communication interface. A **client** uses the communication interface to request resources, and the server responds to those requests.

The client/server architectural model can be applied in many different ways. The architecture by which workstations access a shared printer on a LAN, as shown in Figure 13-2, is one approach to client/server architecture. An application program on a workstation sends a document to a server computer that dispatches it to a management process for the specified printer. The server acknowledges the client request and notifies the client when the document is sent to the printer.

Figure 13-2 ►

Network printing services implemented with client/server architecture

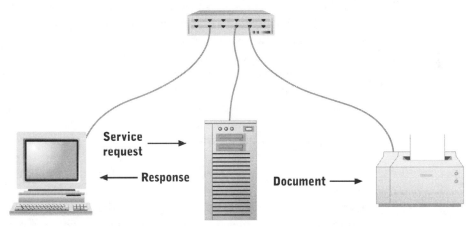

N-Layer Client/Server Architecture

Modern variants of client/server architecture divide application software into the following set of client and server processes called layers or tiers:

► The **data layer** manages stored data, usually in one or more databases
► The **business logic layer** implements the rules and procedures of business processing
► The **view layer** accepts user input and formats and displays processing results

This approach to client/server architecture is sometimes called **three-layer architecture** or **three-tier architecture**. Figure 13-3 illustrates how the three layers interact. The view layer acts as a client of the business logic layer, which in turn acts as a client of the data layer.

Figure 13-3 ▶

Three-layer
architecture

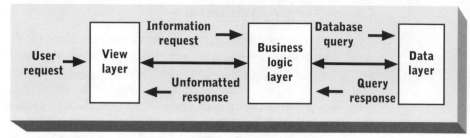

Three-layer architecture simplifies distributing or replicating application software across a network. Interactions between the layers are service requests and responses. Layers can be placed in different processes on the same computer system or on different computer systems. Layers that overload a single computer's capacity can be replicated on multiple machines.

Additional layers can be added when processing requirements or data resources are complex. Architectures that employ more than three layers are called **n-layer architectures** or **n-tiered architectures**. For example, corporate databases are often stored using multiple database management systems. An additional data layer is interposed between the business logic layer and the database management systems to present a unified view of data resources from all corporate databases.

Layers can also be replicated in different forms. A single business logic layer might interact with multiple view layers from different applications. For example, business logic for validating inventory items might be part of an inventory ordering application and part of a customer merchandise return application. A single application can have multiple view layers that interact with a single business logic layer. For example, order entry business logic might interact with a view layer used by telephone sales representatives that runs on simple character-based terminals. The same business logic might interact with a Web-based view layer used by customers.

The connections between layers of client/server and multi-tier application software can be very complex. Every layer can use a unique combination of programming language, operating system, and computer hardware. Well-defined interfaces and communication protocols enable the layers to function as an integrated whole.

The term **middleware** describes software that "glues" together parts of a client/server or multi-tier application. Middleware is a wide-ranging system software category because the connection and communication requirements of applications vary. For example, a simple client/server application might only need to transmit messages between a single client and server in predetermined locations. A more complex multi-tier application might need additional middleware, such as Web servers that support embedded client-side programs, database servers that support stored procedures, and network operating systems that share files among clients on multiple computers.

As operating systems have evolved, they have incorporated more and more middleware functions. For example, software to implement network protocols such as TCP/IP was not embedded in microcomputer operating systems until the mid-1990s. Web server software, once an optional component of most server operating systems, is now standard, and other server operating system functions such as e-mail and document distribution often rely on it.

Other middleware can be obtained as optional operating system utilities or as completely separate packages. The Microsoft BackOffice suite is optional middleware that extends the capabilities of Microsoft Windows Server. Novell GroupWise and BEA Systems Tuxedo are examples of middleware packages that work with many operating systems.

The remaining sections of this chapter explore various forms of middleware and network-related services.

NETWORK RESOURCE ACCESS

The operating system's primary role is to manage hardware, software, and data resources. As part of this role, the operating system accepts and processes resource access requests from users and applications via the service layer. Modern operating systems enable users to interact with resources of the local computer system and of remote computer systems. To provide distributed access, the operating system must be able to distinguish between local and remote resources and interact with distant operating systems. This section covers the operating system components that implement those functions.

Protocol Stacks

An operating system implements network I/O and services as a complex set of software layers. The Open Systems Integration (OSI) network model, covered in Chapter 9, was an early attempt to standardize the number and functions of those software layers. Software that implements the lowest five levels of the OSI model is commonly called a **protocol stack**.

Figure 13-4 shows one possible protocol stack for a workstation. Stacks 1 and 2 have different network, transport, and session layer stacks but share a single physical device driver, Ethernet connection, and physical network connection. The client can access Novell NetWare LAN servers through the Transport Layer Interface (TLI), Sequenced Packet Exchange (SPX), and Internet Packet Exchange (IPX) protocols in Stack 1, and Internet services through the sockets and TCP/IP protocols in Stack 2. With the Open Data Interface (ODI) layer, which is not part of the OSI model, both upper-level protocol stacks share a common physical network interface.

Figure 13-4 ▶

Two upper-level
stacks sharing a
single NIU and
driver

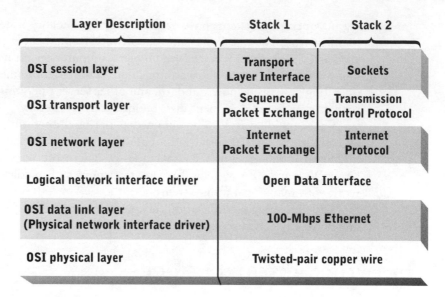

Layer Description	Stack 1	Stack 2
OSI session layer	Transport Layer Interface	Sockets
OSI transport layer	Sequenced Packet Exchange	Transmission Control Protocol
OSI network layer	Internet Packet Exchange	Internet Protocol
Logical network interface driver	Open Data Interface	
OSI data link layer (Physical network interface driver)	100-Mbps Ethernet	
OSI physical layer	Twisted-pair copper wire	

Protocol stacks provide several advantages for implementing network I/O and services:

▶ They divide the task of network interaction into several well-defined pieces that can be separately implemented, installed, and updated

▶ They provide the flexibility needed to keep up with rapid protocol standard evolution

▶ They insulate application programs and many portions of the operating system from details of low-level network communication protocols and physical network implementation, which ensures software portability across a wide range of network protocols and transmission media

Accessing Remote Resources

Connections to remote resources can be either static or dynamic. A **static connection** is initialized by the user or system administrator prior to accessing a remote resource. The remote resource is given a local object, resource, or service name. The window labeled My Computer in Figure 13-5 shows static resource connections to three server directories (drive letters: K:, L:, and M:) under Windows XP. For example, the directory SharedFiles on server Averia is statically connected to the local resource name M:. The figure also shows a dialog box labeled Map Network Drive to create new static connections. Similar static connections can be created for printers and other shared resources.

Figure 13-5 ▶

Displaying and
creating static
directory
connections in
Windows XP

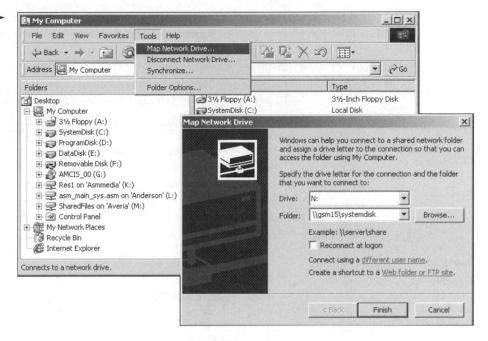

Static connections are inherently difficult to initialize and maintain. The user or system administrator must configure the operating system to establish a connection each time it boots up. If the location or name of a remote resource changes, the configuration of all computers with static connections to that resource must also be changed.

Remote resource access in modern operating systems is based on the following premises:

▶ Operating systems, application programs, and user interfaces are simpler if there is no distinction between local and remote resource access

▶ All resources are potentially shared across a network

▶ Any computer system is potentially both a client and a server

▶ Resources can be moved among computer systems

Software and user interfaces are simplified by providing a common method of accessing both local and remote resources. For example, a word processor executing on a workstation should use the same service call and parameters to access document files stored on local disks and document files stored on server disks. A Web browser should access resources on remote machines in the same manner it accesses resources on the local machine. This characteristic of software and user interfaces is called **location transparency** or **network transparency**.

The second and third premises go hand in hand. If every local resource might be needed by remote users or processes, every computer system is potentially both a client and a server. To provide remote access, all operating systems need to incorporate server-like functions. Operating systems that implement this design feature are said to implement **service-oriented resource access**.

Figure 13-6 illustrates the arrangement of software components that support service-oriented resource access. Two layers, the service provider and resource locator, are interposed between the service layer and device drivers. The service provider is a server interface to a specific resource such as a printer or file system. Service requests from local users or programs are passed down through the operating system service layer. Service requests from remote users or programs are passed through the low-level network protocol stack and directed to the service provider by the resource locator. The service request format is the same regardless of the request's origin.

Figure 13-6 ▶

Software resources used to access local and remote resources

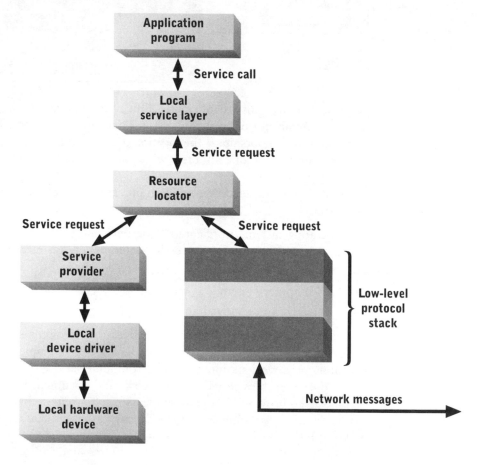

The resource locator has two important tasks:

► Locate resources referred to in service requests from local or remote users and programs
► Forward service requests to the appropriate service provider

The resource locator maintains a **resource registry** containing the names and locations of known resources and services. The format of the names is the same whether the resource or service is local or remote. When local service provider processes are started, they register themselves with the resource locator, which updates the resource registry accordingly. Resource requests from the local service layer are checked against the registry. If the requested resource or service name is not found, the resource locator searches the network for the resource. The exact nature of this search depends on the network service protocols in use.

Most network service protocols use a distributed directory of services and resources. As services and resources are registered with a resource locator, broadcast messages are sent over the network to inform other resource locators of their existence. Some machines on the network can be designated as primary repositories of this information. When a resource locator receives a request for a resource not listed in its registry, it queries the closest primary repository to find the service. That repository can query other repositories until the requested resource is located.

The interaction between a resource locator and a primary resource registration repository is inherently dynamic. Connections established through those interactions are called **dynamic connections**. A local resource locator updates its registry based on query responses from remote registries, which improves the efficiency of future accesses. Resource registry contents must change as local and remote server processes are started, stopped, moved, or reconfigured. The resource locator provides a dynamic mechanism for reconfiguring connections to services and resources.

The resource locator also acts as a router for resource access requests arriving from remote machines. The local physical network layer receives remote service requests and passes them up through the protocol stack to the resource locator. The resource locator then passes the request to the appropriate local service provider process and passes the response back to the requestor via the protocol stack (see Figure 13-6).

It should be clear by now that the traditional distinctions between client and server operating systems have become blurry. All modern operating systems are internally organized as a collection of server processes that can respond to requests from local and remote users and processes. There are still some distinctions between client and server operating systems, such as scalability and security. For example, Windows Server supports up to 32 CPUs and a sophisticated directory-based security system. Windows XP Professional supports one or two CPUs and uses less sophisticated security mechanisms unless it is part of a network managed by Windows Server. But differences between client and server operating systems are primarily a matter of different configuration, not fundamentally different architecture.

INTERPROCESS COMMUNICATION

When an application is split into multiple processes, those processes must communicate with one another to share data and coordinate their activities. But how do processes executing on different computers communicate and coordinate their activities?

A variety of protocols and standards have been developed over the last 20 years to address the problem of process coordination across networks. In this section, we'll concentrate on lower-level **peer-to-peer communication protocols** and standards, which enable processes to communicate synchronously across a network (see Figure 13-7). These protocols often are used by system software such as operating systems and database management systems to exchange data and coordinate activities. Distributed applications usually use higher-level protocols, which are described later in the chapter. However, those higher-level protocols often are layered above the peer-to-peer protocols.

Figure 13-7 ▶

Peer-to-peer
protocol layered
over TCP/IP

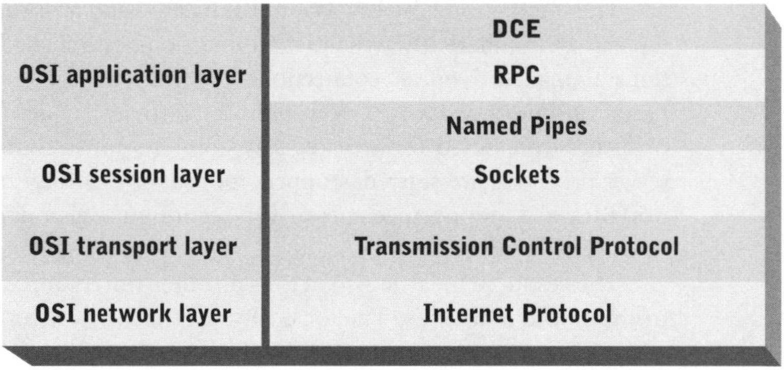

OSI application layer	DCE
	RPC
	Named Pipes
OSI session layer	Sockets
OSI transport layer	Transmission Control Protocol
OSI network layer	Internet Protocol

Sockets

As first described in Chapter 9, a **socket** is a unique combination of an IP number and a port number, separated by a colon. For example, the socket 129.24.8.1:53 is the network listening address for the primary DNS name server at the University of New Mexico. A port number is an unsigned 16-bit integer, so there are 65,536 possible port numbers. Some port numbers are permanently assigned to standard Internet or vendor-specific services, but many port numbers are available for other uses, including client/server or peer-to-peer communication among application programs.

All modern operating systems support sockets and provide system service calls so programs can initialize sockets, receive messages sent to a socket, and send messages to sockets anywhere on the Internet. Figure 13-8 shows a communication scenario between client and server processes on two computer systems. Client processes on Computer A are attached to sockets 129.24.8.212:2 and 129.24.8.212:6 and communicate with two server processes on Computer B attached to sockets 207.46.230.219:1 and 207.46.230.219:6. Each socket uniquely identifies a client or server process on the Internet.

Figure 13-8 ▶

Multiple processes communicating through sockets

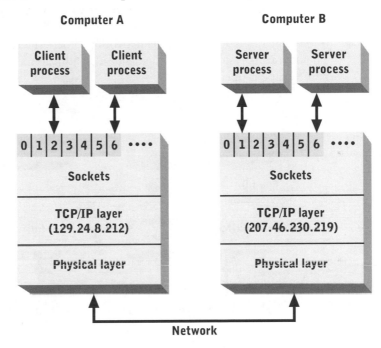

Named Pipes

A **pipe** is a region of shared memory through which multiple processes executing on the same machine can exchange data. Pipes are commonly used for communication among operating system components, for queuing requests to an operating system service such as a Web server, and for exchanging messages among components within a large application program. Processes read or write a pipe as if it were a file and the operating system manages data movement among the processes and the shared memory region.

A **named pipe** is a pipe with two additional features:

▶ A name that is permanently placed within a file system directory

▶ The ability to communicate among processes on different computers

When a named pipe is created, a directory entry is also created in the local file system. Programs can read or write the named pipe as they would read or write an ordinary file. Typically, the server side of a client/server application creates the named pipe. In a peer-to-peer application, either side of the application can create the named pipe. A client or peer opens a named pipe as a network resource and reads or writes it as if it were a file on a shared directory. A server or peer on the machine where the named pipe was created reads and writes the pipe as if it were a local file.

The operating systems at both ends of the pipe manage communication to and from the pipe. Named pipes are actually a high-level interface to sockets, so the operating system assigns a free socket to the named pipe when it is created. The operating system also allocates I/O buffers and routes data flowing in and out of the pipe through the low-level network protocol stack, as shown in Figure 13-9. The operating system on the client side also allocates a socket each time a remote named pipe is opened.

Figure 13-9 ▶

Two processes communicating through a named pipe

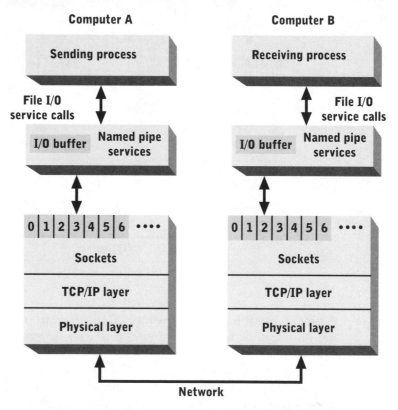

Data can flow in either direction through a named pipe, though bidirectional data flow is usually performed with two one-way pipes. In client/server applications, multiple clients can send messages to a single server by writing to the same named pipe. The server can tell which client sent which messages or data because each client has a unique socket number.

Remote Procedure Calls

With the **remote procedure call** (RPC) protocol, a process on one machine can call a process on another machine. As with function or procedure calls within a single program, the calling process:

1. Passes parameters to the called process
2. Waits for the called process to complete its task
3. Accepts parameters back from the called process
4. Resumes execution with the instruction following the call

Parameter passing among machines is potentially problematic because data representation varies across CPUs and sometimes across operating systems. Common differences include little endian versus big endian memory storage, character coding (ASCII, Unicode, or EBCDIC), and which IEEE floating point format is used for real numbers. If the calling and called process execute on machines with different data representation formats, the parameters must be converted when passed to the called process and when passed back to the calling process.

Technology
Focus

Distributed Computing Environment

Distributed Computing Environment (DCE) is a standard for distributed operating system services defined by the Open Group, formerly known as the Open Software Foundation. DCE is a wide-ranging standard covering network directory services, file sharing services, remote procedure calls, remote thread execution, system security, and distributed resource management. Its primary goal is to promote interoperability of distributed software across operating systems and middleware products. Many operating systems partially comply with DCE, including Windows XP and Server, many versions of UNIX, and OS/390 (MVS). IBM and Hewlett-Packard are principal supporters of and contributors to the standard.

DCE defines a subset of operating system services and a standard interface to those services. DCE functions are incorporated directly into an operating system or supplied as an optional component (see Figure 13-10). In theory, DCE services can replace their counterparts in an existing operating system. In practice, DCE-compliant services usually translate DCE service calls into native operating system service calls.

Figure 13-10 ▶

DCE software
layers

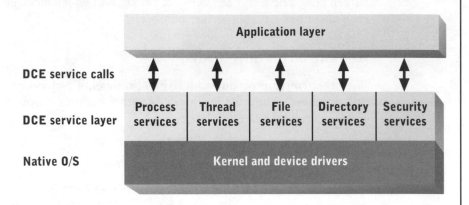

Security is an integral part of every DCE service. DCE security is based on the **Kerberos** security model, which defines interactions between clients, services, and a trusted security service. Clients and servers authenticate one another by asking the security server to authenticate the other party. The security server provides security "tickets" to each party that they exchange to verify their identity.

The security database also maintains an access control list for each service and resource. Once a client has been authenticated, its identity is checked against the access control list for the resource it is attempting to access. If the client is part of the access control list, it is issued a ticket that it presents to the server to gain access. Tickets, passwords, and other security mechanisms are all encrypted during network transmission. Tickets are also time-stamped and expire in minutes or hours.

DCE file services are based on a fully distributed file and directory structure. Every file within a DCE file system has a unique name consisting of its host name, a file system resource name within the host, and its path name within the file system resource. File systems can be distributed or replicated across multiple servers. DCE file services implement transaction logging for all updates to enable rapid recovery in the event of a crash.

Clients, servers, and all DCE services interact through remote procedure calls. RPC messages are formatted according to an Interface Definition Language (IDL). DCE-compliant client and server programs are compiled and linked with a set of DCE library routines that implement the IDL. These routines implement the low-level aspects of message passing, parameter exchange, data format conversions, and interaction with DCE security services.

The DCE standard is implemented in many UNIX operating system variants but it also has penetrated deeply into other operating systems. The standard is widely used, though competing standards such as CORBA, which is discussed later in the chapter, are starting to supplant it. But DCE probably will be around for many years to come, given its widespread operating support and large installed application base.

THE INTERNET

The Internet and World Wide Web are popular frameworks for implementing and delivering information system applications. However, there is a general lack of agreement regarding definitions of the Internet, Web, and related terms. We'll use the following definitions in this chapter:

► The **Internet** is a global collection of networks that are interconnected using TCP/IP.

► The **World Wide Web** (**WWW**), also called the **Web,** is a collection of resources (programs, files, and services) that can be accessed over the Internet by standard protocols such as the File Transfer Protocol (FTP) and Hypertext Transfer Protocol (HTTP).

► An **intranet** is a private network that uses Internet protocols but is accessible only by a limited set of internal users (usually members of the same organization or workgroup). It also describes a set of privately accessible resources that are organized and delivered via one or more Web protocols over a TCP/IP network.

The Internet is the infrastructure upon which the Web is based. In other words, Web resources are delivered to users over the Internet. An intranet uses the same protocols as the Internet and Web but restricts access to a limited set of users. Access can be restricted in many ways, including using privately registered resource names, firewalls, and user/group account names and passwords.

The Web is organized using client/server architecture. Web resources are managed by server processes that can execute on dedicated server computers or on multipurpose computer systems. Clients are programs that send requests using one or more standard Web resource request protocols. Web protocols define valid resource formats and a standard means of requesting resources. Any program, not just a Web browser, can use Web protocols to access Web resources.

Standard Web Protocols and Services

All Web resources are identified by a unique **Uniform Resource Locator** (**URL**) such as http://averia.mgt.unm.edu/default.htm. A URL has four components, as shown in Figure 13-11:

► *Protocol*—an optional header specifying the resource access protocol (http:// is the default value)

► *Host*—the IP number or registered name of an Internet host computer or device

► *Port*—an optional port number that, together with the IP address, specifies a socket as described in Chapter 9 (if omitted, a standard port number for the protocol is assumed)

► *Resource*—the complete path name of a resource on the host (if omitted, the host can return a default resource if appropriately configured)

Figure 13-11 ►

URL
components

IP address or
host name

Optional resource
name on IP host

http:// averia.unm.edu:80/default.htm

Optional protocol
header and separator

Optional port number
and separator

Web standards define a number of protocols for resource format, content, transfer, and manipulation. The number of standard protocols is growing and their content evolves rapidly. Table 13-1 summarizes major Web protocol categories.

Table 13-1 ►

Web protocols

Category	Sample Protocols
Formatted and linked documents	Hypertext Markup Language (HTML) and Extensible Markup Language (XML)
File and document transfer	File Transfer Protocol (FTP) and Hypertext Transfer Protocol (HTTP)
Remote login and process execution	Telnet, tn3270, and Remote Procedure Call (RPC)
Mail and messaging	Simple Mail Transfer Protocol (SMTP), Post Office Protocol (POP), and Internet Message Access Protocol (IMAP)
Executable programs	Java, JavaScript, and Visual Basic Script (VBScript)

The Web as we know it today began with the development of **Hypertext Markup Language (HTML)**. The original version of HTML defined a device-independent document-formatting language in which links to other documents could be embedded. The first Web browser, called Mosaic, displayed HTML documents on any computer and output device to which it was ported. Mosaic spawned the current generation of Web browsers such as Mozilla and Microsoft Internet Explorer. Both HTML and browser software have evolved through several generations to include capabilities such as forms, style sheets, data transfer from client to server, and embedded scripts and programs. HTML will eventually be replaced by **Extensible Markup Language (XML)**, which extends HTML to describe the structure, format, and content of documents.

Hypertext Transport Protocol (HTTP) is a companion protocol to HTML and XML that specifies the language by which clients request documents and how servers respond to those requests. HTTP is an extension of an older Web protocol, called **File Transfer Protocol (FTP)**, that specifies a client/server request and response language for copying files from one Internet host to another. Because HTTP is an extension of FTP, servers that respond to HTTP requests can also respond to FTP requests. **HTTPS** is a secure version of HTTP that encrypts HTTP requests and responses.

With the **Telnet** protocol, users on one Internet host can interact with the operating system command layer of another host, as shown in Figure 13-12. Telnet emulates a character-based VDT and is limited to interacting with command-line-oriented command layers such as Windows COMMAND.COM and the UNIX Bourne shell. Tn3270 is a variant of Telnet that emulates an IBM 3270 VDT to interact with older IBM mainframe operating systems and application programs. **Secure Shell (SSH)** is a modern variant of Telnet that encrypts data flowing between client and server.

Figure 13-12▶

Telnet
connection

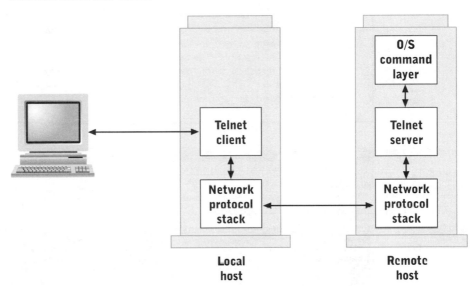

E-mail standards have been part of the Internet since the 1970s. The earliest e-mail protocol is the **Simple Mail Transport Protocol (SMTP)**, which defines how text messages are forwarded and routed among Internet hosts. E-mail client programs on each host interact with a server process to access forwarded messages and to send messages to users on other hosts. SMTP was later extended by the **Multipurpose Internet Mail Extensions (MIME)** protocol to allow non-text files to be included in mail messages. MIME also is used by HTTP and XML.

Modern e-mail server software is based on the **Post Office Protocol 3 (POP3)**, **Internet Message Access Protocol 4 (IMAP4)**, or both. POP3 standardizes the interaction between e-mail clients and servers so the client and server can execute on different Internet hosts. Under POP3, mail messages are temporarily held on the server, downloaded to the client when a connection is established, and deleted from the server as soon as the download is completed. IMAP4 extends

POP3 to permanently store and manage e-mail messages on the server, which enables users to access stored e-mail from any Internet host.

In later HTML versions, program code or scripts can be embedded within HTML documents. Java applets, first described in Chapter 10, can be called from an HTML document with parameters passed in either direction. Figure 13-13 shows sample HTML code to call a Java applet named VocabMan.class. The applet is downloaded to a Web browser along with the surrounding HTML document. Figure 13-14 shows the Web page with the embedded Java applets displayed as

Figure 13-13 ▶

A Java applet call embedded within HTML

```
<li>A compiler allocates storage space and makes an entry in the symbol
    table when a(n)
    <applet width="175"
           height="25"
           code="VocabMan.class"
           codebase="http://asm.unm.edu/sa2e_student/Java/">
           <param name="answer"
                  value="data declaration">
             You must have Java enabled for this feature.
    </applet>
    is encountered in source code.<br><br>
</li>
```

Figure 13-14 ▶

Vocabulary exercise answers displayed or hidden by a Java applet

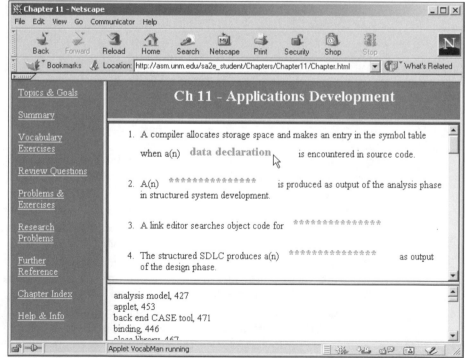

Internet and Distributed Application Services

boxes containing asterisks. The Web browser executes the applet whenever the cursor is placed over a box, and the applet displays answer text when the mouse button is pressed.

As described in Chapter 10, **scripting languages** are slimmed-down programming languages. JavaScript and Visual Basic Script (VBScript) are two widely used scripting languages. Scripts embedded in HTML pages can perform many functions that full-fledged programs can, without the need for compilation or link editing. Typical functions performed by embedded scripts include data validation and customizing page layout or content based on Web browser configuration.

Web servers can execute programs and scripts on the server in response to requests sent by clients as a URL. Internet search engines are a common example of this approach. The Web browser downloads an HTML form in which the user enters search criteria. When the user clicks a Send or Go button, the search criteria ("vacation" and "caribbean", in this case) are transmitted back to the Web server encoded in a URL such as:

http://www.google.com/search?hl=en&ie=UTF-8&oe=UTF-8&q=vacation+caribbean

The URL contains an encoded function call to a search engine program. The text after the question mark contains input parameters for the search engine program. When the program has found the requested data, it encodes processing results in HTML and the Web server sends them back to the Web browser for display.

The Internet as an Application Platform

Internet and Web technologies present an attractive alternative for implementing distributed applications. For example, consider a payroll system for a geographically dispersed computer consulting firm. Such an organization might have dozens or hundreds of offices and many employees traveling or on extended assignment at client locations. Employees need to update withholding, insurance, and other payroll-related information and enter time-related payroll information. How can all of the organization's employees interact with its payroll system quickly and easily?

One way to address the problem is to build client/server application software that uses a private network to connect remote clients with servers at the organization's administrative offices. The client portion of the application is installed on employees' laptop computers, and employees connect to payroll servers using a modem and private telephone number. The disadvantages of this approach include the cost to build and maintain a private network, the cost to develop full-fledged self-contained client-side software, and the difficulties inherent in installing, configuring, and updating client software on many different computer systems.

An alternative is to build a client/server application that uses a Web browser interface. The application program executes on a Web server that can be accessed from any computer with an Internet connection. Employees can access the payroll server with their own laptop computers or computers in a client office or hotel business suite.

Figure 13-15 shows the architecture of a typical Web-based three-layer payroll system. The client interacts with the Web server to download HTML pages that can include embedded scripts or Java applets. The bulk of the application code resides on the server, and program functions are called by the Web server in response to client requests encoded within URLs. The server-side application code interacts with a back-end database server using lower-level Internet standards such as sockets, named pipes, or RPC. All parts of the application communicate using standard Internet and Web protocols. The application can be secured via layered security protocols such as HTTPS and Kerberos.

Figure 13-15 ▶

A distributed Web-based application

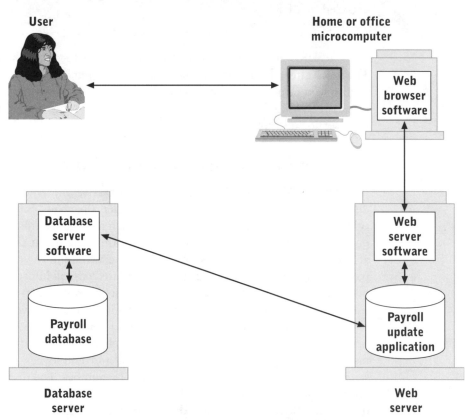

Implementing the application via the Internet expands its accessibility and eliminates the need to install custom client software on employees' laptop computers. The application can be updated by updating software only on the Web server. The application is cheaper to develop and deploy because it is built around existing Web standards and relies on Web browser software already installed on clients.

Each resource that participates in the payroll update application also can participate in other applications. For example, the database server hardware and software can host marketing and production databases. The Web server can store product manuals and host an order entry application for customers. The Web browser on an employee's microcomputer can access other Web applications or information resources anywhere on the Internet.

The primary disadvantages of implementing applications via the Internet are security, performance, and reliability. If an employee can access the system via the Web, then others might also be able to gain access. System access can be restricted by various means including user accounts and passwords, but the risk of a security breach will always be present. Performance and reliability are limited by the employee's Internet connection point and the available Internet capacity between that connection point and the application server. Unreliable or over-loaded local Internet connections can render the application unusable.

Although it's possible to implement distributed applications using Internet standards such as sockets, named pipes, HTTP, and HTML, it's not optimal because:

► With lower-level protocols, server addresses are stored in client configuration files or source code. If server resources are moved, the clients must be reconfigured or recompiled.

► Breaking up server-side processes into small manageable pieces is difficult. Each new distributed piece requires a new set of hard-coded connections.

► Developers usually create large complex server processes to avoid the complexity of large numbers of connections between many smaller server processes. But doing so reduces the chances that server processes can be incorporated into multiple distributed applications.

Newer techniques and standards for deploying and supporting distributed applications address these problems in several ways:

► They make it easier to break up large server processes into small reusable pieces by providing standards for communication among server processes.

► They improve flexibility by providing directory services so processes can locate one another using location-independent names.

► They make it easier to build systems from small reusable processes by providing the infrastructure to manage large numbers of interprocess connections.

COMPONENTS AND DISTRIBUTED OBJECTS

A **component** is a standardized and interchangeable software module that:

► Is executable
► Has a unique identifier
► Has a well-known interface

Components are ready-to-use software. They are compiled and linked for the operating system and CPU on which they'll execute, or they're Java programs that can execute in an installed Java Virtual Machine (JVM).

Every component has a unique identifier (ID), which is a number or symbolic name. Like Internet names and sockets, component names must be unique. Unique IDs enable one component to find and request services from another component. Component IDs must be registered with a directory service that can be queried by other components.

A component's interface is the set of services it provides or tasks it performs. Each service or task is similar to a function or subroutine within a program—it has a name and parameters. When one component asks another to perform a task, it sends it a message containing the task or service name and the required input parameters. The component that receives the message performs the task and, if necessary, returns results.

Component-Based Software

Components are important in modern software development because complex programs and applications can be constructed from smaller previously developed parts. Other complex products have been built from components for decades. For example, a modern automobile contains tens or hundreds of thousands of parts, relatively few of which are manufactured by the companies that assemble and sell cars. Automobile manufacturers specify standards for components such as engines, bearings, and tires that can often be used in several models. Other companies manufacture parts based on those specifications. Without component-based construction, modern automobiles would be much more expensive, much less reliable, and much more difficult to repair and modify.

Component-based design and construction provides similar benefits to complex software products. For example, consider the grammar-checking function in most word processing programs. Grammar checking can be implemented as a function or subroutine that is called by other parts of the word processing program. The grammar-checking function source code is integrated into the rest of the word processor source code during program compilation and linking. The executable program then is delivered to users.

Now consider two possible changes to the original grammar-checking function:

► The developers of another word processing program want to incorporate the existing grammar-checking function into their product

► The developers of the grammar-checking function modify it to improve speed and accuracy

To integrate the existing function into a new word processor, the developers integrate the existing source code of the grammar-checking function into the new word processor program. They add appropriate grammar-checking function calls to their word processor source code and then compile, link, and distribute the program to users.

When the developers of the grammar checker revise their source code to implement the faster and more accurate function, they deliver the source code to the developers of both word processors. Both development teams integrate the new grammar-checking source code into their word processors, recompile and relink the programs, and deliver a revised word processor to their users.

What's wrong with the above scenario? Nothing in theory, but a great deal in practice. The grammar checker developers can provide their function to other developers only as source code, which creates potential problems concerning intellectual property rights and software piracy. Also, integrating the grammar-checking function is difficult or impossible if the word processor is written in a different programming language than the grammar checker.

When the grammar-checking function is updated, the developers of both word processing programs must recompile and relink their entire product to update the embedded grammar checker. The new executable program must then be delivered to users and installed on their computers—an expensive and time-consuming process.

A component based approach to software design and construction solves all of these problems. Component developers can deliver their product as a ready-to-use executable function. Developers of the word processing programs simply plug in the component. Updating a single component doesn't require recompiling, relinking, and redistributing the entire application. Applications already installed on user machines can be updated by installing only the new component. This is exactly the mechanism used by many companies to update installed software, for example, Microsoft Windows Update and Symantec Live Update for Norton Utilities.

Components and Objects

It is difficult to talk about components without using object-oriented (OO) terminology. Components are similar to objects within an OO program because they send and respond to messages, encapsulate internal data, and interact with other components through a well-defined interface. OO program design concepts can also be applied to component-based applications. In essence, component-based design and development scales up OO programming concepts to the level of application programs and entire information systems.

Components are usually developed with OO software development tools and programming languages, but that isn't required. A component can be implemented with any programming language as long as it interacts with other components via messages and a well-defined public interface. Nonetheless, the vast majority of components are implemented with OO tools because they are naturally suited to component development. A component behaves as a distributed object regardless of its internal implementation.

Connection Standards and Infrastructure

Interoperability among hardware or software components requires well-defined and widely adopted standards. For example, consider telephone connections and the services provided by telecommunication companies (see Figure 13-16). In the U.S., all telephones are connected to the public telephone grid by a four-conductor wire with an RJ-11 connector. Each wire in an RJ-11 connector carries a specific electrical signal with known voltage and other characteristics. This standard ensures that any telephone can be connected easily to any telephone interface.

Figure 13-16 ▶

Standard connectors and infrastructure enable communication between telephones

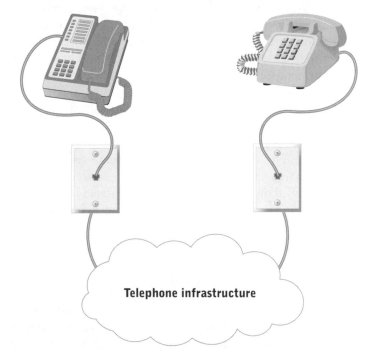

Telephone infrastructure

Internet and Distributed Application Services

Telephones rely on basic services supplied by the public telephone infrastructure. A unique telephone number identifies every telephone or set of telephones. The infrastructure provides connection services among telephones, as well as directory services to match numbers to people, organizations, and physical locations. Switching and routing are handled by the infrastructure with no specific instructions from sending and receiving telephones. Additional services such as forwarding and voice mail can also be provided. The standards that enable any telephone to access any other telephone or service have been developed over many years through industry cooperation and occasional governmental intervention.

Software components require similar standards for connections and services. They require the software equivalent of an RJ-11 plug and an infrastructure that provides directory, routing, and forwarding services. Within a single computer, operating system interprocess communication services can connect components, and configuration files can supply directory information.

Connecting components located on different machines running different operating systems requires a standard network protocol. A network protocol is desirable even when components all execute on the same machine, because it provides the flexibility needed to distribute components in the future. Standard Internet services and protocols such as TCP/IP and sockets provide part of a component connection solution. But such protocols don't address two important issues:

▶ Format and content of valid messages and responses

▶ Means of uniquely identifying each component on the Internet and routing messages to and from that component

Addressing those issues requires additional standards, protocols, and services. Many organizations have participated in developing such standards over the last two decades. Currently, four standard families are well-developed and widely implemented: CORBA, COM+, J2EE, and SOAP.

CORBA

In the 1980s, many computer hardware and software organizations joined forces to create an industry-wide component interoperability standard known today as the **Common Object Request Broker Architecture (CORBA)**. CORBA specifies the middleware used by objects to interact across networks. The two key components of CORBA are:

▶ **Object Request Broker (ORB)**, a service that maintains a component directory and routes messages among components

▶ **Internet Inter-ORB Protocol (IIOP)**, a component message-passing protocol

The ORB is a server process that can reside anywhere on a network or can be distributed across many network nodes. The ORB acts as a component registry, a message router, and a translator. Components must register themselves with the ORB before other components can connect to them. The ORB assigns a unique identifier to each registered component, so that component's methods can be invoked by other components anywhere on the Internet.

A component that wants to invoke a method within another component sends an IIOP formatted request to the nearest ORB. ORBs cooperate with one another to locate the desired component and establish a connection. Once a connection has been established, parameters can be passed in both directions. If necessary, an ORB can translate parameters from one format to another. This enables otherwise incompatible objects, such as those developed with different languages or executing on incompatible CPUs, to exchange data.

The CORBA standard is robust, inherently scalable, and independent of programming language, operating system, and CPU architecture. These features are the result of deliberate design, a long development history, and participation by a large number of computing vendors and organizations. The standard is implemented widely, with support from the Open Group and major computer and software vendors such as IBM, Hewlett-Packard, Sun Microsystems, and Oracle.

COM+

The **Component Object Model Plus (COM+)** is a Microsoft specification for component interoperability. COM+ has its roots in older Microsoft specifications including object link embedding (OLE) and the Component Object Model (COM). In the early and mid-1990s, Microsoft incorporated DCE services into COM and called the resulting specification the Distributed Component Object Model (DCOM). COM+ is the most recent upgrade of that standard.

Like CORBA, COM+ defines component registration, message routing services, and a component communication protocol. The COM+ services and protocol are similar to CORBA with the following key differences:

► Components are not assigned a permanent identifier, and their internal states cannot be stored permanently. COM+ components can't remember information from one invocation to the next. COM+ components are similar to functions or subroutines. CORBA components are objects.

► COM+ components are registered in the Windows Registry of the client machine on which they're installed. The Windows Registry stores information other than component registrations, including hardware configuration, software configuration, and user profile information. A CORBA ORB is dedicated to component services.

COM+ is widely implemented. Most developers of general-purpose application software such as word processing and spreadsheet programs use COM+.

SOAP

Both CORBA and COM+ have some significant disadvantages for building distributed component-based software. For CORBA, the primary problem is complexity. CORBA infrastructure requirements are substantial and the programming constructs needed to access those services are complex. The complexity has made companies reluctant to make significant investments in the technology and has created a shortage of skilled CORBA-trained personnel.

For COM+, the primary problem is dependence on proprietary technology and limited support outside of Microsoft products. A commitment to COM+ entails a commitment to Microsoft operating systems and other system software. Though Microsoft does dominate the desktop operating system market, most organizations have a diverse collection of server operating systems and software. Also, there are few non-Microsoft development tools that support COM+.

Simple Object Access Protocol (SOAP) is a standard for distributed object interaction that attempts to address the shortcomings of both CORBA and COM+. Unlike CORBA, SOAP has few infrastructure requirements, and its programming interface is relatively simple. SOAP is an open standard developed by the World Wide Web Consortium (W3C). Perhaps the best evidence of SOAP's long-term potential for success is that Microsoft has adopted it as the basis of its .NET distributed software platform.

A key to SOAP's simplicity and minimal infrastructure requirements are its reliance on existing Internet protocols—HTTP and XML. Messages among objects are encoded in XML and transmitted using HTTP, which enables the objects to be located anywhere on the Internet. Figure 13-17 shows a client sending a service request to a server as a SOAP message. The same transmission method supports server-to-client and peer-to-peer communication. The SOAP encoder/decoder and HTTP connection manager are standard components of a SOAP programmers' toolkit. Applications can also be embedded scripts that use a Web server to provide SOAP message passing services. SOAP messages can be transmitted using other protocols such as FTP and SMTP, but HTTP is the most common transmission protocol.

Although SOAP has been widely deployed for many types of distributed applications, it has some significant limitations. Early SOAP versions left many implementation specifics undefined, thus creating problems of interoperability. SOAP 1.2 filled many of the gaps in earlier standards, but there are still some unaddressed issues including security and message delivery guarantees. These issues will probably never be addressed by SOAP because it assumes by design that other protocols will address them. Building and deploying industrial-strength SOAP-based distributed applications isn't as simple as it might first appear because developers must employ multiple protocols. Nonetheless, SOAP's popularity will probably increase as more developers take advantage of its unique combination of minimal infrastructure requirements, simple interfaces, and vendor-independent implementation.

Figure 13-17►

Client/server
communication
with a SOAP
message

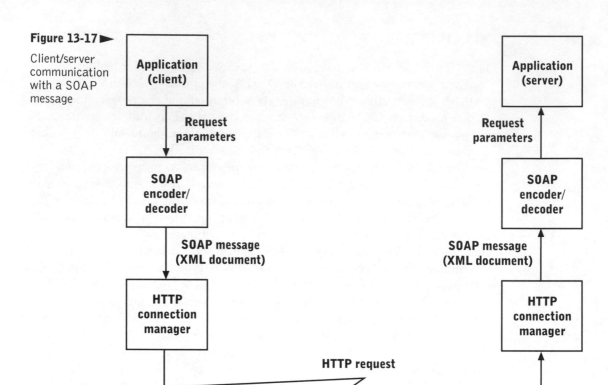

Technology
Focus

Java 2 Enterprise Edition

Java 2 Enterprise Edition (J2EE) is family of standards for developing and deploying component-based distributed applications written in Java. Figure 13-18 shows key J2EE architectural elements, which follow the three-layer architecture described earlier in this chapter. The client tier can include Java components and Web browsers displaying HTML pages with embedded scripts. Browser-based client interfaces are sometimes called thin clients since they contain little or no program code. Thick clients are collections of complete Java objects that execute under control of a Java Virtual Machine (described in Chapter 10) and communicate directly with the corresponding components in the Web/Business tier.

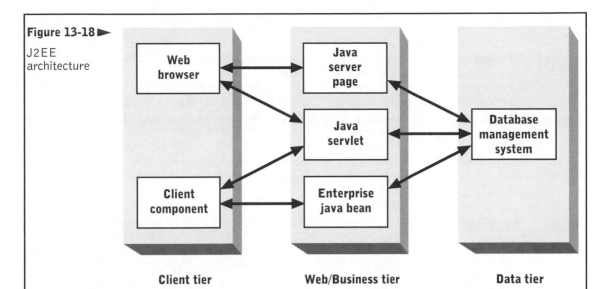

Figure 13-18 ▶

J2EE
architecture

Web browser · Java server page · Java servlet · Client component · Enterprise java bean · Database management system

Client tier · **Web/Business tier** · **Data tier**

Within the Web/Business tier, components can be **JavaServer pages** (**JSP**) and **Java servlets**, which execute under control of a Web server, and **enterprise Java beans** (**EJBs**), which execute within a system software component called a business container. JSP are components that generate formatted Web pages using embedded scripts. Java servlets are full-fledged Java programs that can perform more complex operations including computations and interactions with other EJBs and databases. JSP and servlets communicate with the client tier using HTML or XML messages.

EJBs are Java programs that perform complex "behind-the-scenes" processing. For example, an EJB might implement a distributed object with object attributes stored in the database. An EJB might also perform a complex query against multiple tables in multiple databases or manage a complex transaction such as a customer order with related financial, inventory, and shipping transactions. EJBs provide services that can be called by components executing on the client, components executing within a Web server, or other EJBs.

Component interactions are based on many standards, including:

▶ Remote Method Invocation (RMI)—enables objects executing on different computers to send messages and receive responses

▶ Java Naming and Directory Interface (JNDI)—enables objects executing on different machines to locate one another and query their available methods

▶ Java Authentication and Authorization Service (JAAS)—authenticates users and restricts access to components

▶ Java Database Connectivity (JDBC)—enables objects to interact with relational databases using SQL statements

Developing and deploying J2EE applications is a complex endeavor due to the elaborate J2EE architecture, the number of related standards, and the number of required system software elements. Many vendors, including Sun Microsystems and Oracle, offer development packages and system software suites to support J2EE applications. Developers typically require months of training to learn the intricacies of the tools and system software.

J2EE application portability is enhanced by a common programming language (Java) and well-defined standards for describing and storing components and system software configuration information in Java source code and XML files. The files have standard internal formats, standard naming conventions, and are placed within a standardized directory structure. This enables developers to move all or part of a J2EE application from one computer or system software suite to another simply by copying the corresponding files, and restarting the supporting system software services.

J2EE was developed after CORBA and COM+ but before SOAP. J2EE is incompatible with COM+ and, in its original form, was also incompatible with CORBA. However, subsequent revisions to J2EE and CORBA have brought them closer together. The current RMI standard enables CORBA components and J2EE EJBs to interact using the CORBA IIOP standard. Java components can be defined with the CORBA interface description language and registered with a CORBA ORB. J2EE also includes limited interoperability with SOAP, which is expected to expand in future J2EE standards.

J2EE has been widely deployed in enterprise-level information systems. Although its complexity rivals CORBA, it has enjoyed somewhat greater success, primarily due to significant support from Sun and Oracle. The battle for the future of distributed systems seems to be narrowing down to J2EE and Microsoft .NET. Microsoft .NET has the advantages of a complete adoption of modern Web services standards including SOAP and Microsoft's considerable resources and market presence. However, J2EE has a well-established reputation as a platform for reliable and scalable industrial-strength Web applications and a strong support base from many vendors and software developers.

DIRECTORY SERVICES

When resources are distributed across network nodes, resource users and resource providers must have some way of finding one another. **Directory services** describe middleware that:

► Stores the name and network address of distributed resources

► Responds to directory queries

► Accepts directory updates

► Synchronizes replicated or distributed directory copies

Directory services are integral components of all network operating systems. Typically, network operating system directories store information about:

► Registered users and their permissions to access directory objects

► Shared hardware resources such as printers

- Shared files, databases, and programs
- Computer systems and specialized hardware devices such as network storage appliances

Modern directory services are distributed in a manner similar to Internet name services. Directories are organized hierarchically to create a single name-space for all network resources and objects. In large networks, responsibility for maintaining directory content and answering queries is distributed throughout the network. Directory content can be replicated in multiple servers in different parts of the network to reduce response time and improve fault tolerance.

Lightweight Directory Access Protocol (LDAP)

Historically, most directory services have been tied to a specific operating system such as NDS for Novell NetWare or to a specific application type such as file sharing or e-mail distribution. Application and operating specificity is a problem in networks with multiple operating systems and many types of distributed resources and objects. Today, such networks are the norm rather than the exception, so directory services must bridge operating systems and applications.

In the 1980s, the International Telecommunications Union (ITU) developed the X.500 standard, which defines nonproprietary directory services for e-mail and network addresses. The standard was never widely implemented, though partial implementations appeared in some commercial products such as Novell NetWare.

The X.500 standard became the basis for another standard called the **Lightweight Directory Access Protocol (LDAP)**, which has been adopted by the Internet Engineering Task Force (IETF) as a formal Internet standard. The standard was updated regularly in the late 1990s and those efforts still are in progress. LDAP is implemented widely, though not all products provide all the features of the most recent standards.

An LDAP directory stores information about LDAP objects. Each object is an instance of an **objectclass**, which defines the attributes common to all member objects. For example, the objectclass Shared_Printer might define attributes named Building, Room, Manufacturer, Model, Color, Duplex, and Pages_Per_Minute. Each directory entry for a shared printer would be an object of type Shared_Printer and contain values for some or all of the defined attributes.

LDAP objects are organized into a hierarchical directory structure. Objects can be grouped into container objects that can be grouped into other container objects (see Figure 13-19). LDAP defines several standard container object types including Country (C), Organization (O), and Organizational Unit (OU). All objects in an LDAP schema have an attribute called distinguished name (DN), which uniquely identifies the object within an objectclass. A fully qualified distinguished name such as:

DN=Stephen Burd,O=Faculty,O=School of Management,OU=University of New Mexico,C=USA

specifies a complete path from a directory root node through one or more container objects to a specific object.

Figure 13-19 ▶

An LDAP hierarchy of objects and container objects

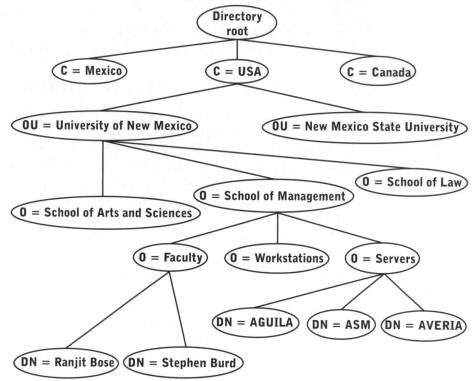

LDAP defines a standard protocol for updating and querying LDAP directories. LDAP queries contain plain text characters, though the syntax is complex and awkward. Many client programs such as e-mail clients and Web browsers provide a more user-friendly interface to LDAP services.

Strict adherence to the LDAP standard guarantees interoperability among all LDAP clients and servers. However, there is an important area of interoperability that has yet to be addressed: standard schema names and structure. Currently there are no standard objectclass or attribute names for entities and resources common to most directories such as people, shared printers, and components. As a result, there is no standard way for a directory from one vendor or organization to query the directory of another organization or vendor.

For example, one directory might define an objectclass named *Employee*, and another directory might use the name *Worker* for a similar objectclass. The two directories might use different attribute names for identical data items such as office telephone number. One directory can send a query to the other asking it to enumerate its defined objectclasses and attributes. But without a standard naming system, there is no way to determine semantic equivalence of schema classes and attributes automatically.

In the absence of standard class and attribute names, directory vendors and user organizations have developed proprietary directory schemas that usually are incompatible. Even if a set of standard schema classes and attributes is developed, the existing base of installed incompatible directory schemas will hinder true directory interoperability for years to come.

Technology
Focus

Microsoft Active Directory

Microsoft **Active Directory** is the directory service and security system built into Windows Server. Active Directory stores information about many network resources, including computers, I/O devices, and users (see Figure 13-20). Active Directory incorporates security services that allow system administrators to limit resource access to specific users or user groups. Active Directory supports large directories and distributed organizations with directory partitioning across Internet domains, directory replication across multiple servers, and automatic synchronization of replicated directories or directory partitions.

Windows XP client software queries Active Directory to locate and access network resources. For example, a user can use Windows XP Professional or Active Directory client software installed under earlier Windows versions to search for users, computers, or shared directories matching specific criteria. Programs also can interact with Active Directory by calling Active Directory-specific service functions. A word processing program might query Active Directory for all printers in a specific organizational unit, constructing and displaying the result to a user when he or she selects the Print function from the File menu.

Windows XP and Server use Active Directory to store and access security information. Every user, group, and computer object is assigned a unique security identifier. Every Active Directory resource or container object has an access control list that describes the access rights granted or denied to specific users, groups, and computers. Figure 13-21 shows the access control list for the Active Directory container object Pajaro. Members of the group Domain Admins have all permissions for Pajaro, and, by default, have all available permissions for all child objects.

Figure 13-20 ►

Active Directory
objects

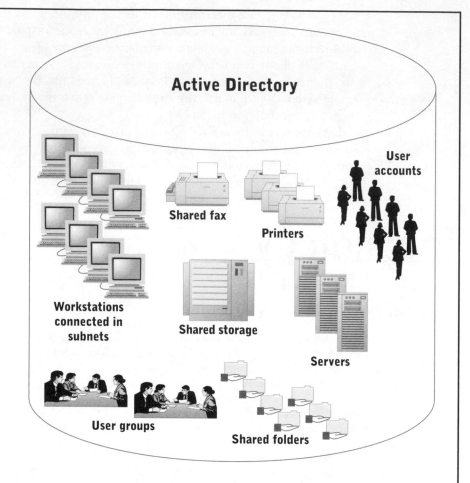

Active Directory

User accounts

Shared fax

Printers

Workstations
connected in
subnets

Shared storage

Servers

User groups

Shared folders

Active Directory access control lists and permissions inheritance can be used to distribute system administration tasks across individuals and organizations. If Active Directory objects are organized hierarchically using container objects, the directory administrator can delegate Active Directory management functions to other users by assigning rights to container objects. For example, the directory administrator can assign rights such as Create All Child Objects and Modify Permissions to other users for lower-level container objects, so those users can administer one part of the Active Directory.

Active Directory is based on LDAP and the Internet Domain Naming Service (DNS). Active Directory responds to standard LDAP information requests and uses LDAP concepts such as objectclasses and organizational units to store and hierarchically organize directory information. Active Directory clients rely on a DNS server to locate an Active Directory server. Active Directory servers register their names and services by supplying a service record (SRV record) to a DNS server. Active Directory clients query the DNS server for LDAP servers registered in their domain. The DNS server returns one or more IP addresses, which the client uses to send TCP/IP messages to an Active Directory server.

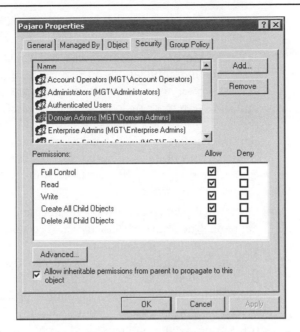

Figure 13-21▶

Access
control list

Like many other commercial products, Active Directory doesn't implement the entire LDAP protocol. Unlike other directory service software, Active Directory relies on DNS and requires DNS servers to accept dynamic updates. Although dynamic updates are an Internet DNS standard, many DNS servers do not accept them by default. DNS administrators are reluctant to enable dynamic updates, because they are a potential security and reliability threat. Microsoft's DNS server software does support dynamic updates, so most Active Directory installations use Microsoft DNS server software.

Active Directory does not support distributed or component-based software directly. Shared files, including executable files, and databases can be registered in Active Directory, which could support locating and accessing programs and data sources. However, Active Directory provides no standard way to "glue together" components and external data sources to create complex systems. Windows XP and Server rely on a parallel directory and registration system to track COM+ components and provide component-related directory and interprocess communication services.

SUMMARY

▶ Modern information systems are typically distributed across many computer systems and geographic locations. Client/server, three-layer, and n-layer architecture define frameworks for decomposing applications into multiple layers that can execute on different computers. Multi-layer architectures require standard methods and services to communicate with one another.

System software that implements communication standards and gives clients and servers the ability to interact is called middleware.

▶ A network protocol enables users and applications to interact with resources and applications on remote computer systems. The operating system service layer accepts resource access requests and routes them to appropriate remote or local service processes. Local requests to remote resources are routed through the protocol stack to distant servers, and remote requests to access local resources are received through the protocol stack. Connections to remote resources may be static or dynamic. Dynamic connections are more flexible but require a distributed registry of resource names and locations.

▶ Distributed processes must communicate with one another to exchange data and synchronize their activities. Peer-to-peer interprocess communication protocols include sockets, named pipes, remote procedure calls, and DCE. Sockets implement direct process-to-process communication via protocol stacks. Named pipes enable multiple clients to interact with a single server via a shared file system object. Remote procedure calls allow one process to execute another as a subroutine with parameter passing and format translation. DCE combines all of these approaches and adds security and minimal directory services.

▶ The Internet is a global network based on TCP/IP and many other protocols. The Web is the set of resources accessible over the Internet via standard protocols. Internet protocols define methods for sharing many types of resources including files, documents, and application programs. Multi-layer applications can be implemented over the Internet using a Web browser for the user interface and Internet and Web protocols to connect the layers.

▶ Component-based applications divide an application into many different cooperating processes or distributed objects. Each distributed object implements a public interface to the services it provides. Component-based applications require protocols and infrastructure for component registration, discovery, and communication. CORBA, COM+, J2EE, and SOAP are standards that address the full range of component infrastructure and communication services.

▶ With directory services, users, resources, and components can find one another on the Internet. Modern directory services must span the range of different resource types and Internet protocols. LDAP is a widely deployed directory service standard that can track users, distributed resources, and objects. Although LDAP defines a standard method for defining and accessing directory structure, it doesn't define standard content templates, which limits interoperability among different LDAP directories.

Key Terms

Active Directory
business logic layer
client
client/server architecture
Common Object Request Broker
    Architecture (CORBA)
component
Component Object Model Plus
    (COM+)
data layer
directory services
distributed computing
Distributed Computing
    Environment (DCE)
distributed processing
dynamic connection
enterprise Java bean (EJB)
Extensible Markup Language
    (XML)
File Transfer Protocol (FTP)
Hypertext Markup Language
    (HTML)
Hypertext Transport Protocol
    (HTTP)
Hypertext Transport Protocol
    Secure (HTTPS)

Internet
Internet Inter-ORB Protocol
    (IIOP)
Internet Message Access
    Protocol 4 (IMAP4)
intranet
JavaServer pages (JSP)
Java servlet
Java 2 Enterprise Edition
    (J2EE)
Kerberos
Lightweight Directory Access
    Protocol (LDAP)
location transparency
middleware
Multipurpose Internet Mail
    Extensions (MIME)
n-layer architecture
n-tiered architecture
named pipe
network transparency
Object Request Broker (ORB)
objectclass
peer-to-peer communication
    protocol

pipe
Post Office Protocol 3 (POP3)
protocol stack
remote procedure call (RPC)
resource registry
scripting language
Secure Shell (SSH)
server
service-oriented resource access
Simple Mail Transport
    Protocol (SMTP)
Simple Object Access Protocol
    (SOAP)
socket
static connection
Telnet
three-layer architecture
three-tier architecture
Uniform Resource Locator
    (URL)
view layer
Web
World Wide Web (WWW)

Vocabulary Exercises

1. _____ architecture divides an application into multiple processes, some of which send requests, some of which respond to requests, and others that do both.

2. System software that connects parts of a distributed application or enables users to locate and interact with remote resources is called _____.

3. A(n) _____ is a process that sends a request. A(n) _____ is a process that responds to requests.

4. A server process creates a(n) _____ that allows clients to send data or messages via a shared filename.

5. With the _____ protocol, a process on one machine can call a process on another machine as a subroutine with parameters passed in either or both directions.

6. Three-layer architecture divides an application into _____, _____, and _____ layers.

7. A(n) _____ to a remote resource must be initialized by the user or a system administrator.

8. A(n) _____ is a unique combination of an IP number and a port number.

9. _____ is the original Internet e-mail standard. More modern e-mail standards include _____ and _____.

10. The _____ is a global collection of networks that are interconnected using TCP/IP.

11. Web pages are encoded in _____ and delivered from a Web server to a Web browser via _____.

12. The _____ standard defines methods for embedding graphics and other nontextual data within e-mail messages and Web pages.

13. The _____ is a collection of resources that can be accessed over the Internet by standard protocols such as FTP and HTTP.

14. _____ is a family of component infrastructure and interoperability standards supported by Microsoft.

15. _____ defines a standard for describing and accessing directories of users and distributed resources.

16. A(n) _____ contains a protocol, Internet host, optional socket, and resource path.

17. _____ is a wide-ranging standard covering network directory services, file sharing services, remote procedure calls, remote thread execution, system security, and distributed resource management.

18. With the _____ protocol, a client can interact with a remote computer's command layer as if it were a directly connected character-based VDT.

19. A(n) _____ is a standardized and interchangeable software module that is executable, has a unique identifier, and has a well-known interface.

20. _____ is a family of component infrastructure and interoperability standards supported by a broad consortium of computer companies.

21. The set of software layers that provide low-level access to remote resources through a LAN or WAN is called a(n) _____.

22. Web/Business tier components of the J2EE distributed application architecture include _____, _____, and _____.

23. _____ is a secure version of HTTP that encrypts HTTP requests and responses.

24. _____ is a modern variant of Telnet that encrypts data flowing between client and server.

25. A(n) _____ is a shared memory region that enables multiple processes executing on the same machine to exchange messages.

Review Questions

1. Describe client/server, three-layer, and n-layer architecture. What are the differences between a client and a server? What is the function of each layer in a three-layer application? Why might more than three layers be used?

2. What is middleware?

3. What is a protocol stack? What are the components of a typical protocol in a client computer that can access many Web servers?

4. What are the differences between static and dynamic connections to remote resources? Which connection type requires a resource registry? Where should the resource registry be located?

5. A modern operating system acts as both client and server. How are software components organized to accomplish both functions at the same time?

6. List and describe three low-level peer-to-peer interprocess communication standards. What are the advantages and disadvantages of using these standards to implement distributed multi-layer applications?

7. Do the terms Internet and Web describe the same thing?

8. What are the components of a URL?

9. List and describe at least five standard Internet and Web protocols.

10. How can the Internet be used as a platform to implement distributed multi-layer applications? Which Internet and Web protocols are used and how are they used?

11. What is a component? Component-based design and development has been the norm in manufacturing durable goods for decades. Why has it only recently been adopted for designing and deploying information systems?

12. Describe the two major standard families for component infrastructure and communication. If you were forced to choose one standard to support a new large-scale information system, which would you choose? Why?

13. What are directory services? What types of information might be made available through directory services? Describe the LDAP directory services standard.

14. Describe the components pieces of the J2EE distributed application architecture. What standards govern the form of the client and business tiers? What standards govern communication among components? Is J2EE compatible with other distributed application standards?

Research Problems

1. The Kerberos security model is a part of both the DCE and CORBA standard families. Investigate the capabilities and limitations of Kerberos. What degree of security does it provide? What infrastructure components must be present to implement it? Does it address all security issues in distributed applications and resources?

2. The CORBA, J2EE, and SOAP standards are rapidly evolving. Access the Web sites www.corba.org, java.sun.com, and www.w3.org to determine what features have been added by recent standard revisions. What needs are those new features intended to satisfy? Are the newer features being implemented widely? Why or why not?

3. Distributed.net (www.distributed.net) is an organization that coordinates research on distributed computing applications. Individuals and organizations can join the organization and contribute idle computing power to ongoing research projects that require or use massively parallel approaches to solving complex computational problems. Investigate the projects currently underway and the methods and protocols that coordinate software on members' computers.

4. Investigate the middleware offered by one or two major software vendors such as Microsoft, IBM, or Oracle. List their middleware products and the primary function(s) of each. Is it possible to "mix and match" middleware products from these vendors? Do the vendors offer any products that are difficult to classify as middleware? Why are those products difficult to classify?

Chapter 14

System Administration

Chapter Goals

- ► Describe system administration responsibilities and tasks

- ► Explain the process of acquiring computer hardware and system software

- ► Describe tools and processes for evaluating application resource requirements and computer system performance

- ► Define a system security model and describe how it can be implemented

- ► Understand issues related to installing and protecting computer hardware

In previous chapters, we explored the inner workings of computer hardware and system software. In this chapter, we take a step back from bits, bytes, and instructions to examine hardware and software from the point of view of a system administrator. A system administrator's job includes many varied tasks—describing all of them in detail could easily fill an entire book. This chapter concentrates primarily on tasks that require in-depth knowledge of the topics covered in earlier chapters (see Figure 14-1).

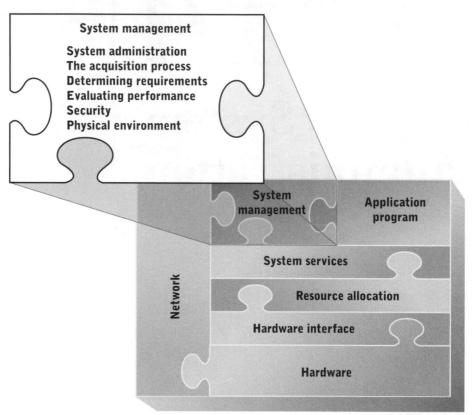

Figure 14-1 ▶

Topics covered in this chapter

SYSTEM ADMINISTRATION

Within the context of information systems, **system administration** covers a range of activities and responsibilities. The primary responsibility of a system administrator is to ensure efficient and reliable delivery of information services (IS). A typical system administrator performs many tasks to meet this responsibility, which can be grouped into the following broad categories:

- Acquiring new IS resources
- Maintaining existing IS resources
- Designing and implementing an IS security policy

The assignment of these responsibilities to specific individuals varies considerably among different organizations. In very small organizations, the user or organizational administrator might assume these responsibilities directly. In very large organizations, these responsibilities might be divided among many people. Medium-sized organizations might have a single technical specialist who assumes all system administration responsibilities.

Each activity area entails a variety of specific system administration tasks. Some of these tasks are mundane and address a relatively short time frame, such as maintaining user accounts and performing file system backups. Other tasks require many different skills and address a longer time frame, such as IS planning and resource acquisition.

Strategic Planning

IS resource acquisition and deployment should occur only in the context of a well-defined strategic plan for the organization as a whole. For the purposes of this chapter, a **strategic plan** is defined as a set of long-range goals and a plan to attain those goals. The planning horizon is typically three years and beyond. Minimally, the goals must include:

- Services to be provided
- Resources needed to provide those services

The strategic plan addresses the following issues with respect to achieving stated goals:

- Strategies for developing services and markets for them
- Strategies for acquiring sufficient resources for operations and growth
- Organizational structure and control

In all cases, strategic plans must address the basic question, "How do we get there from here?".

Information systems is only one part of the entire organization, and the IS strategic plan is only part of an organization's strategic plan. The IS strategic plan must be evaluated in concert with those of other organizational units. Information systems is normally a support service for other organizational units and functions such as customer services, accounting, and manufacturing. Thus, the strategic plan for information systems tends to follow, rather than lead, the strategic plans of other units in the organization.

Hardware and Software as Infrastructure

The resources devoted to most organizational activities can be classified into two categories—capital expenditures and operating expenditures. **Capital expenditures** purchase **capital resources**, which are assets or resources that are expected to provide benefits beyond the current fiscal year. **Operating expenditures** provide benefits only in the current fiscal year.

Typical capital expenditures include buildings, land, equipment, and research and development costs. Typical operating expenditures include salaries, rent, lease payments, supplies, and equipment maintenance. Computer and software purchases are capital expenditures, even though the expected useful lifetime of computer hardware and software has decreased in recent years due to rapid technological change.

Many capital resources provide benefits to a wide range of organizational units and functions. An office building, for example, benefits all of the units and functions that are housed within it. Such resources are referred to as **infrastructure** and have the following characteristics:

► Service to a large and diverse set of users

► Costs that are difficult to allocate to individual users

► Recurring need for new capital expenditures

► Significant maintenance costs

The computer hardware and system software that provide information systems services are infrastructure. This is obvious in organizations that rely extensively on large computer systems used by many units in the organization. It is less obvious but no less true in organizations that have highly decentralized computer hardware and system software. The IS strategic planning issues are similar to those in many other infrastructure-based service organizations.

Examples of infrastructure-based service organizations include those that provide communications, electrical power, and water. In general, the strategic issues that must be addressed by such organizations are:

► What services will be provided?

► How will service users be charged?

► What infrastructure is required to provide the services?

► How can the infrastructure be operated, maintained, and improved at minimal cost?

For example, consider the communication services provided by local telephone companies. The primary strategic question to be addressed is what types of services should be provided: basic telephone only, expanded telephone services, Internet access, mobile services, information storage and retrieval, and so on.

Answers to the primary question lead to decisions regarding the nature of the required infrastructure and its associated capital and operating expenditures.

Standards

Providing infrastructure-based services to a wide variety of users requires the adoption of **service standards**. However, standardization tends to stifle innovation and to produce solutions that are suboptimal for some users. Once again, consider the local telephone system as an example. All users agree on and abide by a number of standards for telephone service. These include acceptable user devices, basic service availability, and standards for interacting with the infrastructure (for example, line voltage and signal encoding). The nature of infrastructure requires standardization in order to provide service at a reasonable cost.

Standardization often causes problems for some users, especially for those who demand services at or near the leading edge of technology. For example, many telephones are connected to the nearest switching center with low-capacity copper wiring, which limits the availability of high-speed Internet services, which in turn limits the ability of some employees to work at home. The provider of an infrastructure-based service must constantly balance the benefits of standardization in reduced costs and simplified service against its costs in stifled innovation and failure to meet the needs of some users.

Computer hardware and system software standardization is a particularly complex issue involving many hardware and system software choices and significant compatibility issues. Standardization issues are further complicated by the diverse set of components required for information processing within even a modest-sized organization.

Competitive Advantage

Discussing computer hardware and system software as infrastructure ignores certain types of opportunities. Infrastructure management tends to concentrate on providing short-term services at minimal cost—a viewpoint that tends to preclude major technical innovations and radical redefinition of services to be provided.

The term **competitive advantage** describes a state of affairs in which one organization employs resources so as to give it a significant advantage over its competitors. This can take a number of forms, including:

► Providing services that competitors are unable to provide
► Providing services of unusually high quality
► Providing services at unusually low price
► Generating services at unusually low cost

Computer hardware and system software can be applied to achieve competitive advantage in any or all of these areas. Examples include electronic banking, bar code scanners at grocery checkouts, and coordinating mobile employees with wireless communication and computing devices.

Note that each of these examples was once novel, but is now commonplace. Such is the nature of applying technology for competitive advantage—rapid technology changes and adoption by competitors severely restrict the useful life of most technology-based competitive advantages. Furthermore, there are substantial risks in pursuing competitive advantage through new technology. The costs of technology are usually high for developers and early adopters, but then decrease rapidly after introduction, so that late adopters may incur substantially lower costs while realizing most of the benefits. Early adopters also face the inefficiency of starting at the beginning of a learning curve.

Business
Focus

A Standard Hardware Platform?

Cooper State University (CSU) is a large school with over 1000 faculty members and a full-time enrollment of 20,000 students. Academic programs span a wide range of fields including hard sciences, humanities, engineering, education, management, law, and medicine. CSU offers many graduate degree programs and is highly ranked in terms of externally funded research.

CSU has a campus-wide computing organization—Computer Information Services (CIS). CIS is responsible for supporting the computing and information processing needs of both academic and administrative users. It supports administrative users by operating a large mainframe computer, several local area networks, and many administrative applications such as payroll, accounts payable, class scheduling, class registration, and student academic records. It supports academic users with shared midrange computers, shared file servers, and a large number of microcomputers in dedicated classroom labs and general-purpose facilities. CIS also operates the campus network that links most offices, classrooms, and buildings to the Internet.

CIS has been under considerable budgetary pressure for several years. Demand for computing services has grown rapidly but funding has not kept pace. Declining hardware costs have been a budgetary bright spot, but they have been more than offset by increases in support costs for CSU's wide variety of hardware platforms, software packages, and administrative applications. Integrating this disparate collection of hardware and software with the growing campus network has been a difficult and costly task.

CIS has recently proposed standardizing the hardware platforms and operating systems used for microcomputers and midrange computers to minimize support costs. They want to contract with a single vendor for each class of machine and negotiate volume pricing on hardware, operating software, and maintenance. All CSU micro and midrange computer purchases, including those made with non-CIS funds, would be part of the contract.

Reactions to the proposal have ranged from indifference to near revolt. Most administrative departments and many academic departments are supportive, provided that CIS can assure them of a continued ability to meet their computing and information processing needs. In general, the supportive departments rely on CIS for most or all of their computing and information processing needs. Some academic departments that fund some of their own computer purchases are concerned with loss of control over the acquisition process.

The computer science, engineering, information systems, and physics departments are vehemently opposed to the proposal. They argue that their computing needs are unique and require hardware and software at the cutting edge of technology. They claim that standardized platforms would lag behind the cutting edge of technology and interfere with their teaching and research missions. They point out that most of their computing needs are met with funds derived from external sources such as research grants and contracts. They vigorously oppose restrictions on how that money is spent.

Questions

1. Are the benefits that CIS anticipates from a standardized platform likely to be realized?

2. Are the concerns of the computer science, engineering, information systems, and physics departments valid? Should any other departments share these concerns?

THE ACQUISITION PROCESS

The process of acquiring computer hardware and system software is ongoing in most organizations. New hardware and software can be acquired to:

► Support new applications

► Increase support for existing applications

► Reduce the cost of supporting existing applications

The exact nature of the acquisition process depends on which of these goals (or combination thereof) motivates the new acquisition. It also depends on a number of other factors, including:

► The mix of applications that the hardware/software will support

► Existing plans for upgrade or change in those applications

► Compatibility requirements with existing hardware and software

► Existing technical capabilities

Hardware and software exist to support present and future applications. Planning for acquisition is little more than guesswork if you don't have a thorough understanding of present and anticipated application needs.

The acquisition process consists of the following steps:

1. Determine the applications that the hardware/software will support.

2. Specify hardware and software capability and capacity requirements.

3. Draft a request for proposals and circulate it to potential vendors.

4. Evaluate the responses to the request for proposals.

5. Contract with a vendor or vendors for purchase, installation, and/or maintenance.

The acquisition process occurs whenever a new system is implemented or an existing system is upgraded. In either case, hardware and system software requirements are determined by design discipline activities of the Unified Process (UP). Activities of the requirements discipline provide detailed estimates of system activity such as transaction volume and file or database content. Decisions about what processes will be automated, type of user interfaces, operating systems and other system software, and selection of development tools are made during early design discipline activities. These decisions, combined with the activity estimates, are translated into detailed hardware requirements.

Determining and Stating Requirements

Computer system performance is measured in terms of application tasks that can be performed within a given time frame. Performance can be measured in terms of throughput, response time, or some combination of the two. The first step in stating hardware/software requirements is to state the application tasks to be performed, along with performance requirements for those tasks.

Different techniques are used to determine requirements, depending on the motivation for the proposed acquisition. For existing applications, hardware and system software requirements might be based on measurements of existing performance and resource consumption. Requirements for new applications are more difficult to derive. Techniques that might be used to determine requirements are discussed later in this chapter.

Although application requirements are the primary basis for hardware and system software requirements, other factors must also be considered, including:

► Integration with existing hardware/software

► Availability of maintenance services

- Availability of training
- Physical parameters such as size, cooling requirements, and disk space required for system software
- Availability of upgrades

These factors are integral to the overall requirements statement. Some requirements, such as physical parameters, are essential, while others are less important bases for differentiating among potential vendors.

Request for Proposals

A **request for proposal (RFP)** is a formal document sent to vendors that states requirements and solicits proposals to meet those requirements. It is often a legal document as well, particularly in governmental purchasing. Vendors rely on information and procedures specified in the RFP. They invest resources in the response process with the expectation that stated procedures will be followed consistently and completely. An RFP is generally considered a contract offer and a vendor's response is an acceptance of that offer. Problems such as erroneous or incomplete information, failure to enforce deadlines, failure to treat all respondents equitably, and failure to state all relevant requirements and procedures can lead to litigation.

The general outline of an RFP is as follows:

- Identification of requestor
- Format, content, and timing requirements for responses
- Requirements
- Evaluation criteria

The identification describes the organization that is requesting proposals. It should include the name of a person to whom questions can be addressed, as well as physical and e-mail addresses, telephone numbers, and fax numbers.

The RFP should state the procedural requirements for submitting a valid proposal and, where possible, should include an outline of a valid proposal that describes the required content of each section. In addition, the RFP should clearly state deadlines for questions, proposal delivery, and other important events.

The requirements statement comprises the majority of the RFP. Requirements should be categorized by type and listed completely. Relevant categories include:

- Hardware/software capability
- Related services such as installation and maintenance
- Warranties and guaranties

▶ Deadline dates for delivery or implementation of requested products or services

▶ Financial considerations, for example, acquisition cost, maintenance cost, lease cost, and payment options

Requirements should be separated into those that are essential and those that are optional or subject to negotiation. For example, minimum hardware capacity is generally stated as an absolute requirement, whereas some related services might be described only as desirable.

Evaluation criteria are stated as specifically as possible. A point system or weighting scheme is often used to evaluate optional or desirable requirements. Weight also might be given to factors that are not stated as part of the hardware or software requirements, such as the financial stability of the vendor and good or bad previous experiences with the vendor.

Evaluating Proposals

Proposal evaluation is a multi-step process. The usual steps are:

1. Determine acceptability of each proposal.
2. Rank acceptable proposals.
3. Validate high-ranking proposals.

Each proposal is evaluated to determine whether it meets the basic criteria, including essential requirements, financial requirements, and deadlines. Proposals that fail to satisfy minimal criteria in any of these categories are eliminated from further consideration.

The remaining proposals are ranked by evaluating the extent to which they exceed minimal requirements. This includes providing excess capability or capacity, satisfying optional requirements, and other factors. Measurements of subjective criteria such as compatibility, technical competence, and vendor stability also are considered at this stage.

A subset of the highly ranked proposals is chosen for validation. The subset should be small due to the time and expense required to evaluate proposals. To validate a proposal, the evaluator determines the correctness of vendor claims and the vendor's ability to meet commitments in the proposal. Note that the ranking process relies primarily on what the vendor asserts in the proposal. The validation stage is where the accuracy of those assertions is determined.

Various methods and sources of information can be used to validate proposals. The most reliable of these is a benchmark of the proposed system with actual applications. Benchmarks are discussed later in this chapter. For all but the largest systems, vendors usually deliver and install hardware and system software for customer evaluation. If purchased or already developed application software is available, the test system can be benchmarked with the application suite.

Alternatives to on-site benchmarking with actual application software include:

► Benchmarking at alternate sites with actual applications
► Benchmarking of test applications
► Validation through published evaluations

With systems that are difficult to install, it might be possible to test applications at an alternate site. This could be on the premises of the vendor or of another customer.

DETERMINING REQUIREMENTS AND EVALUATING PERFORMANCE

Information systems are a combination of computer hardware, systems software, and application software. One of the most difficult system administration tasks is determining hardware requirements for a specific set of application software. This task is difficult because:

► Computer systems are complex combinations of interdependent components.
► Operating and other system software configuration can significantly affect raw hardware performance.
► The hardware and system software resources required by applications cannot always be predicted precisely.

The starting point for determining hardware requirements is the application software that will run on the hardware platform. The nature and volume of application software inputs must be described to determine what application code will be executed and how often it will be executed. This determines the demands that application software will place on system software, such as number and type of service calls, and on computer hardware, such as number of CPU instructions.

If the application software has already been developed, its resource consumption in terms of hardware and system software can be measured. A number of software and hardware tools can be used to monitor resource utilization while application programs are executing, which generates precise measurements of resource requirements.

Determining resource requirements is much more complex when application software has not yet been developed. Detailed requirements and design details typically aren't available for the entire system until several UP iterations have been completed. Estimates of hardware and system software requirements are only as precise as the specifications of application software requirements. General statements of user and software requirements can generate only rough estimates of hardware requirements. More accurate hardware requirements cannot be determined until all design discipline activities are completed.

Unfortunately, application developers seldom have the luxury of waiting until all design discipline activities are completed to acquire hardware. This is especially true when implementing large systems. In order to accomplish the acquisition process described earlier in this chapter, sufficient lead time is required. The acquisition process typically is started in an early UP iteration, after the scope of the automated system has been determined.

Informal methods of estimating requirements based on requirements models can sometimes be employed. Such methods typically employ comparisons to similar application software and work volumes. For example, a developer of an online transaction processing system might attempt to identify a similar transaction processing system inside or outside the organization. The system is examined to determine the adequacy of its existing hardware and system software. If current hardware resources match current application demand, then those hardware resources can be used as a baseline for estimating the hardware requirements of the yet-to-be-developed system. Requirements can be adjusted to account for differences in transaction volume and application processing characteristics.

This basic approach to requirements estimation is used widely and successfully, particularly for smaller hardware platforms running "garden variety" application software. The risk of inaccurate estimates is minimized for smaller hardware platforms because they can be modified relatively quickly and inexpensively.

Larger hardware platforms and large-scale one-of-a-kind application software require more formal methods of estimating requirements. Lead times for hardware acquisition and configuration are longer for mainframes and large midrange computers than for smaller hardware classes. Hardware and software costs are high, which increases the economic risks of inaccurate estimates. Estimating requirements in this environment typically employs formal methods based on mathematical models.

Estimating computer requirements calls for two types of mathematical models:

► Application demand model
► Resource availability model

An application demand model represents units of work performed by application software in terms of its demand for low-level hardware services. For example, an online transaction update might be described in terms of required number of CPU instructions, disk I/O bytes, and bytes input from or output to I/O devices. A complete application demand model includes numerical representations of each type of work performed by the application.

A resource availability model describes the ability of a computer system to deliver resources to application software. The resources described in the resource availability model must match those used in the application demand model. Modeled resources may be low-level, such as CPU instructions and disk I/O bytes, or high-level, such as operating service calls or HTTP requests to a Web server.

Once both models are specified, evaluators can test them using a variety of analysis methods including integer programming and simulation. Simple models may be constructed and evaluated by information system professionals with the assistance of decision support software. However, building and analyzing complex models requires the assistance of personnel with an extensive background in mathematics and experience in developing and analyzing computer performance models.

Benchmarks

A **benchmark** is a measure of computer system performance while executing one or more specific processing tasks. There are many types of benchmarks testing many different aspects of computer performance. Benchmarks can test specific low-level aspects of hardware performance—such as CPU instruction execution speed or write throughput to a RAID storage device—as well as measure performance when executing complex tasks such as processing database queries and updates.

Two organizations that develop widely used benchmarks and perform computer system testing are:

► Standard Performance Evaluation Corporation (SPEC) – www.specbench.org
► Transaction Processing Performance Council (TPC) – www.tpc.org

Both organizations have developed many application-oriented benchmarks that simulate widely implemented information processing tasks including Web services, database processing, and online transaction processing. Major computer vendors routinely submit their computer systems for testing, and these two organizations publish the results, enabling computer purchasers to evaluate the cost/performance ratio of actual computer systems.

Using published benchmarks to evaluate new computer systems for a specific application or application suite is a complex matter because evaluators must:

► Select the benchmark test(s) most relevant to the intended application(s)
► Determine the relationship between benchmark test(s) and the actual work that the new computer system will perform

To address the first issue, evaluators examine the various benchmarks for similarities to and differences from their own application software. The testing organizations provide detailed benchmark descriptions and source code to assist evaluators in selecting appropriate benchmarks.

To determine the relationship between a benchmark test and application software, evaluators test their own application software on a system with known performance on the most relevant benchmarks(s). The purpose of the testing is to establish a mathematical relationship between the unit(s) of work in the application software and the unit(s) of work in the benchmark.

For example, evaluators testing a computer system with a known TPC-CV5 performance rating (for example, 500,000 transactions per minute) execute their own application software on the same computer system and generate a result of 300,000 transactions per minute. The ratio of the two results (300,000 ÷ 500,000 = 0.6) enables evaluators to adapt the TPC-CV5 benchmark for any computer system to their own application software. For example, if the evaluators anticipate a transaction volume of 150,000 transactions per minute, they need to acquire a computer system with TPC-CV5 benchmark of at least 150,000 ÷ 0.6 = 250,000.

It might not be possible to test computer systems with actual application software if those applications are large, difficult to install, or not yet developed. In such cases, evaluators must closely examine the performance benchmark tests and estimate the ratio computed above.

Care must be taken when using benchmarks, especially if only one benchmark is executed at a time, testing only one type of performance. Overall system performance under mixed application demand can vary substantially from individual benchmark results.

Measuring Resource Demand and Utilization

Accurate performance analysis requires accurate performance data. Such data is required as input to numerical performance models and when evaluating the performance of existing systems to determine the need for reconfiguration or hardware upgrades. Automated tools to measure resource demand and utilization include:

▶ Hardware monitors
▶ Software monitors
▶ Program profilers

The information generated by these monitors describes the behavior of specific devices, resources, or subsystems over some period of time.

A **monitor** is a program or hardware device that detects and reports processing or I/O activity. A **hardware monitor** is a device that is attached directly to the communication link between two hardware devices. It monitors the communication activity between the two devices and stores communication statistics or summaries. This data can be retrieved and printed in a report. Hardware monitors are often used to monitor the use of communication channels, disk drives, and network traffic.

A **software monitor** is a program that detects and reports processing activity or requests. Software monitors are typically embedded within operating system service routines. They are usually disabled by default, but can be enabled to monitor high-level processing requests such as file open, read, write, and close, or low-level kernel routines such as flushing file or I/O buffers, or virtual memory paging. When enabled, a software monitor generates statistics of service utilization or processing activity that can be displayed in real time or stored in a file for later analysis.

Software monitors can alter the activity being measured. For example, a software monitor that measures system-wide CPU activity generates inflated measurements because the software monitor itself requires CPU cycles to execute. Resources consumed by the software monitor must be estimated or measured and subtracted from the overall measurements to determine correct measurements for processes other than the monitor.

Monitors can operate either continuously or intermittently. A continuous monitor records all activity as it occurs. An intermittent, or sampling, monitor checks for activity on a periodic basis. The advantage of a sampling monitor is that fewer computer resources are expended executing the monitor. Another advantage is that less data is accumulated in output files, which can become very large. Continuous utilization statistics can be estimated based on the sampled activity. Continuous monitors provide complete information on activity, but their operation can consume excessive system resources.

Monitors help to identify performance bottlenecks as a precursor to configuring hardware and/or system software for maximal performance. For example, monitoring I/O activity on all secondary storage channels can indicate that some disks are used continuously while others are not used as much. Based on this information, a system administrator might decide to move frequently accessed files from heavily used disks to less used disks, or to reallocate disks among controllers or I/O channels. An entire system can be tested at full load to determine its maximum sustainable resource delivery.

A **program profiler** describes the resource or service utilization of an application program during execution. Typically, monitor subroutines are added to the program's executable image during link editing. As the program executes, these subroutines write data to a file each time the application requests a system service. They can also record other statistics such as elapsed (wall clock) time to complete each service request, CPU time consumed by service calls, and CPU time consumed by program subroutines. This information can be used to derive a resource demand model for the application. It can also be used to identify segments of the program that might be inefficiently implemented.

Technology
Focus

Windows XP and Server Performance Monitoring

Windows XP and Server provide a performance monitoring utility called System Monitor (SM). Using SM, a system administrator can monitor hardware and software resource use in real time. SM defines a number of system objects about which performance and utilization data can be captured. A partial display of these objects is shown in Figure 14-2. Monitored objects can be individual hardware devices such as the CPU, groups of hardware devices such as all physical disks, operating system services such as a network name service, or resource management data structures such as a service queue or a virtual memory paging file.

Figure 14-2 ▶

Windows XP and Server System Monitor

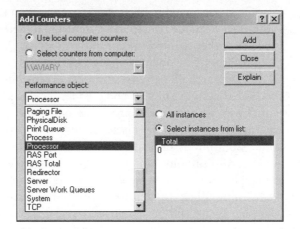

Each object has a set of data items called counters. A counter monitors a specific performance or resource utilization aspect of an object. For example, there are 19 counters for the object PhysicalDisk, including current queue length, average seconds per disk read or write operation, number of disk read or write operations per second, and number of bytes read or written per second. Using different counters, a system administrator can monitor either aggregate or highly specific aspects of performance.

Resource utilization and other system activity can be displayed in various formats. Figure 14-3 shows a line graph displaying total CPU utilization (the thin line) and virtual memory page faults per second (the thick line). Statistics such as mean, minimum, and maximum are displayed for the currently selected counter. The display can be configured for any number of counters, and each counter is assigned a specific color and type of line. A similar set of display options are available for other output formats such as bar graphs and text reports. Performance data also can be captured in a file for later analysis.

Figure 14-3 ▶

Real-time
display of
performance
data

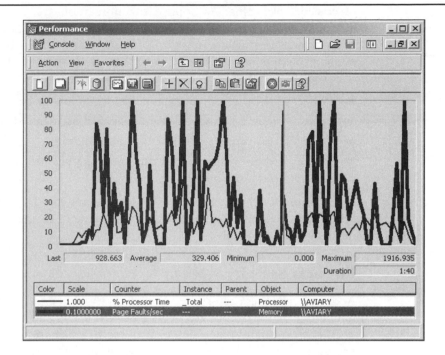

A system administrator needs accurate data to make system configuration decisions. For example, optimal decisions about changes to the size and/or location of virtual memory page files require accurate data concerning current virtual memory utilization. Optimal decisions about upgrading CPU capacity require data describing current CPU utilization and the utilization of related resources such as memory and bus I/O. SM provides a flexible and powerful means of acquiring such data.

Accurate data is not the only requirement for good configuration decisions, however. Interpreting that data requires a thorough understanding of the operating system and hardware components, which allows the system administrator to associate cause and effect and understand the trade-offs among various types of system performance. Without such an understanding, performance assessment is little more than guesswork, and configuration decisions can produce suboptimal or detrimental effects on overall system performance.

SECURITY

For most organizations, information processing resources are a sizable investment. Some of the resources, such as specific items of hardware and software, are tangible and have well-defined dollar values. Other resources, such as databases, user skill, and reliable operating procedures, are less tangible but are also of considerable value. As applied to information systems, the term security describes all measures by which the value of these investments is protected. Security measures include physical protection against loss or damage and economic protection against loss of value through unauthorized disclosure.

A well-integrated approach to system security includes measures to:

▶ Protect physical resources against accidental loss or damage
▶ Protect data and software resources against accidental loss or damage
▶ Protect all resources against malicious tampering
▶ Protect sensitive software and data resources against unauthorized access and accidental disclosure

A complete discussion of all of these measures could easily fill an entire textbook. This section presents an overview of some of the more commonly implemented security measures.

Physical Security

Access to computers and related equipment should be restricted to prevent theft, tampering, and unauthorized access. Locked doors and limited distribution of keys, key cards, and other lock control mechanisms are the most direct way to protect equipment. Additional protective measures for rooms containing servers and other dedicated equipment include architectural details such as reinforced doors, reinforced walls, and barriers above drop ceilings.

Other physical controls are required for computers located in public and semi-public areas such as offices and campus computer labs. Most computer cases can be locked to prevent removal of key system components. Computers and peripheral equipment can also be locked to desks or other furniture by cable locks.

Access Controls

All modern operating systems incorporate access control features that enable users and system administrators to restrict access to resources such as data files, programs, and hardware devices. Access control is based on two key processes:

▶ **Authentication,** which is the process of determining or verifying the identity of a user or process owner.
▶ **Authorization,** which is the process of determining whether an authenticated user or process has sufficient rights to access a particular resource.

A challenge-response dialog employing a username and password is the most common means of authentication. A user enters a name or other identifier and a password to prove his or her identity. The operating system verifies the username and password by searching a local security database or interacting with a security server such as Kerberos server.

Although password-based authentication is most common, other methods are often used as supplements or alternatives for improved security. Identification cards with bar codes or embedded ROM chips can supplement passwords. **Biometric authentication** methods are sometimes used instead of password-based authentication. Biometric authentication methods identify a specific person using physical characteristics such as:

► Fingerprint

► Facial scan

► Retinal scan

Until the last decade, authentication was a process normally handled entirely by the operating system. Each operating system maintained its own security database and enforced its own access controls. As modern organizations became more tightly integrated through information technology, authentication and access control exclusively through operating systems became problematic. If a user required access to resources on dozens of different servers, then a user account and appropriate authorizations had to be created and maintained on each server. Such an approach is too complex for both users and administrators.

Modern operating systems cooperate with one another to implement authentication and authorization. Typically, each operating system interacts with a directory or security service such as Kerberos or Microsoft Active Directory. Each operating system relies on the directory service to perform authentication and may also rely on it to perform some or all authorization functions. Because all security functions are centralized, security procedures are simpler for both users and administrators.

Once authenticated, a user or process is given a digital ID. Different operating and security systems have different names for the ID, among them user identification (UID), security identification (SID), and security ticket (we'll use the term ticket for the rest of this chapter). A ticket is the basis for determining a user's authorization to access resources managed by the operating system. Tickets can govern a user's ability to:

► Access the operating system or resources from specific locations or at specific times

► Read, write, create, and delete directories and files

► Execute programs

► Access hardware resources such as printers and communication devices

For each restricted action and resource, the operating system or security service maintains a list of users or user groups with access authority, called an authorization list or **access control list**. The list may include different authority levels for different users—for example, read authority for all users who access a file but write authority only for system administrators.

When a user or process attempts an action or requests a resource, the operating system or security service looks up the user's ticket in the access control list to determine whether the user has sufficient authority. Typically such authority checks are an integral part of the operating system service routines that application programs and administration utilities use to access resources. For example, a service routine to open a file will typically ask for a ticket and pass it to a security service for authentication before processing the request (see Figure 14-4).

Figure 14-4 ▶

Authorization within a file open service call

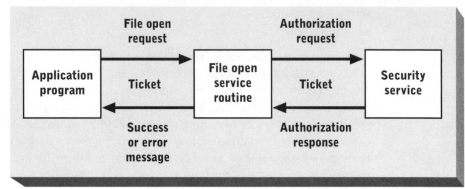

Password Controls and Security

Because password-based authentication is so common, operating systems and security services employ a number of methods to enhance it, including:

▶ Restrictions on the length and composition of valid passwords

▶ Requirements that passwords periodically be changed

▶ Analysis of password content to identify passwords that are easily guessed

▶ Encryption of passwords in files and during transmission over a network

With most operating systems, the system administrator can create and enforce password policies on a per user, per group, or per system basis. Figure 14-5 shows the password controls that can be set in Windows XP and Server, including restrictions on password length, age, complexity, and uniqueness. Operating system password policies should be supplemented with organizational policies including rules against sharing passwords, creating passwords based on personal information such as birth dates or family member names, or writing passwords down in easily accessed locations.

A related set of policies, shown in Figure 14-6, can be set to deal with failed attempts to log on. Accounts can be deactivated after a specific number of failed logons. The account can be reactivated manually by the system administrator or automatically after a specified time interval. Some of the password controls, such as password age restrictions, can be overridden for specific accounts.

Figure 14-5 ▶

Windows XP and
Server password
policies

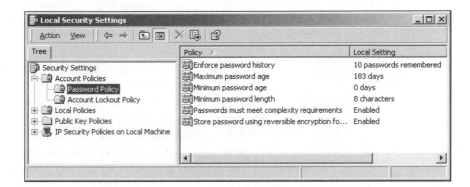

Figure 14-6 ▶

Windows XP and
Server account
lockout policies

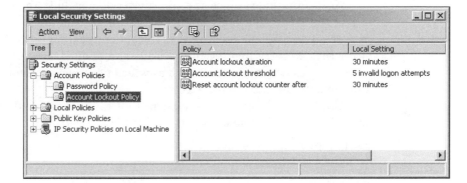

Locking out accounts after a specified number of failed logon attempts prevents unauthorized users from repeatedly attempting to guess correct passwords for valid user accounts. Software programs are sometimes used to attempt remote logon with words and/or names taken from a dictionary. Because many users choose common words or names as their passwords, it is possible for such programs to find a correct password by trying every word in their dictionary. Automatic account deactivation coupled with delayed reactivation limits the number of guesses that can be made in any time period.

Some operating systems and security services employ similar password guessing programs against local password files on a periodic or continuous basis. When passwords are guessed successfully, the operating system forces the user to change his or her password at the next logon. Similar capabilities can be embedded within the program that enables users to change their passwords. An attempt to change a password to an easily guessed value is denied.

System security might be compromised if a copy of the password file or database is somehow distributed to or stolen by an unauthorized individual. The password file is usually protected by access controls, but some degree of accessibility must be granted for normal logon processing. Most operating systems further protect the password file by encrypting part or all of its content.

Chapter 14

However, a stolen password file can be decrypted successfully if sufficient computing resources are employed to crack the encryption key.

Auditing

In accounting, auditing is the process of examining records to determine whether generally accepted accounting principles were correctly applied in preparing financial reports. With respect to computer security, the term **auditing** usually refers only to the process of creating and managing records of user activity or resource access. Those records provide data to determine whether security policy has been correctly implemented or whether resources or the system itself have been compromised.

Most operating systems and security services include an audit function that is disabled by default, but can be enabled for specific users, resources, actions, or access types. When auditing is enabled, the operating system or security service writes an entry to a log file each time an audited action is performed. The log entry includes information such as which ticket was presented to gain access and the date and time of the access.

Although auditing can be a useful tool for examining security policy and analyzing security breaches, it has several limitations, including:

► Log files can grow quickly when auditing is enabled for a large number of users, resources, actions, or access types.
► Auditing reduces system performance due to the overhead of writing log file entries.
► Auditing is backward-looking and, thus, a poor tool for preventing security breaches before they occur.
► Extracting useful information from large auditing logs requires automated search tools and a consistently implemented program of log file analysis.

Because of its limitations and effect on system performance, auditing is used sparingly, if at all. When used, it is generally only enabled for the specific aspects of a security system that an administrator wishes to evaluate.

Virus Protection

In biology, a virus is a DNA fragment that infects a living host at the cellular level and uses the host's cellular machinery to reproduce itself and to perform actions detrimental to the host. In computers, a **virus** is a program or program fragment that:

► Infects a computer by permanently installing itself, usually in a hard-to-find location
► Performs one or more malicious acts on the infected computer

► Replicates and spreads itself using services of the infected computer

Viruses come in many types including:

► Boot virus—attaches itself to code executed when the system boots, such as a BIOS or operating system startup routine

► Macro virus—embedded within a macro stored in a desktop application file such as spreadsheet or word processing document

► Worm—virus stored in a stand-alone executable program, usually sent as an e-mail attachment and executed automatically when the attachment is opened

Computer viruses are commonplace in the modern world. Viruses can perform many malicious acts including damaging or destroying important computer files, opening back doors for potential hackers, and sending sensitive information to others. No information system can be considered secure unless it includes active virus protection.

Many companies market antivirus software. Feature sets vary, but common capabilities include:

► Scanning e-mail messages and attachments for known viruses and disabling or deleting them

► Monitoring access to important system files and data structures and logging or denying such access when appropriate

► Scanning removable media such as floppy disks for known viruses whenever they are inserted

► Periodically scanning the file system and important data structures for viruses that may have escaped other scans and monitoring activities

The most important aspects of antiviral software configuration are ensuring that it is enabled and regularly updated. Antiviral software should be automatically installed with the operating system and enabled by default. Users are sometimes tempted to disable antivirus software because its continuous monitoring and scanning activities exact a performance penalty.

Modern antivirus software uses data files containing information about known viruses. Since new viruses are constantly appearing, those data files must be updated regularly. Most antiviral software includes a subscription to updated data files, which the software developer makes available as new viruses are discovered. The best approach to updated antivirus data files is one based on push technology. That is, a server continuously monitors the software developer's download site for new data files. When they are released, the server downloads the files and copies them to all of the organization's computers.

Software Updates

Modern operating systems and application software are a complex collection of interconnected components. A typical operating system or desktop application suite includes tens of millions of lines of source code. Given the size and complexity of modern software, errors, bugs, and security holes are a virtual certainty. Hackers and viruses often attempt to exploit these problems to perform malicious acts or to gain access to secure information or resources.

Software developers are in a constant race to fix bugs, errors, and security holes as they are discovered. They do so by developing new software versions or by developing software patches, sometimes called service packs, to apply to existing software installations. A key part of any security system is updating system and application software.

In the recent past, applying patches and upgrades was a manual process. Users or system administrators needed to monitor information sources about software upgrades and patches constantly and take whatever actions were necessary to apply them. Modern software includes features to streamline or automate the update process. Typically, software examines its own internal configuration whenever it executes and then sends a query to a server to determine whether its configuration is current. If the response indicates an upgrade is needed, the software can inform the user, inform the system administrator, or automatically download and install required updates. Figure 14-7 shows the configuration dialog box for Windows XP and Server automatic updates. Similar update options exist for many other operating systems and application software.

Figure 14-7 ►

Windows XP and Server automatic update configuration

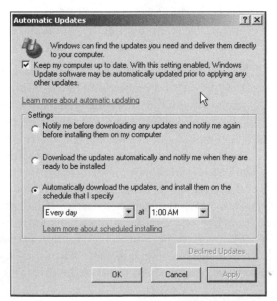

A system administrator can also control software updates via some directory services and network operating systems. Figure 14-8 illustrates the installation of an operating system service pack via a Microsoft Active Directory group policy object. In this example, the group policy object is applied to workstations running Windows 2000. Each time a workstation boots, it checks whether Service Pack 3 is installed and automatically performs the installation if necessary. Automatic software installation improves security by ensuring that the latest security patches are always installed on all of an organization's computers.

Figure 14-8 ►

Software installation policy

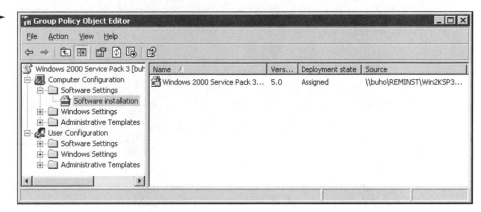

Firewalls

In architecture and machinery, a firewall is a physical barrier that prevents fire from spreading between two structures or compartments that share a common wall or divider. In computer security, a **firewall** is a hardware device, software, or a combination of hardware and software that prevents unauthorized users in one network from accessing resources on another network. Typically, a firewall is a stand-alone device with embedded software that physically separates a private network from a public network, as shown in Figure 14-9. Firewalls are widely deployed in modern information systems to protect servers and information resources from unauthorized access over the Internet.

The simplest firewall type is a **packet-filtering firewall,** which examines each packet and matches header content to a list of allowed or denied packet types. Recall from Chapter 9 that TCP/IP packets include source and target sockets where a socket is the combination of an IP address and a port number. A packet-filtering firewall consults a list of valid source and target IP addresses or address ranges and valid port numbers for those addresses. For example, if a public Web server is located within the private network, then the firewall is configured to allow incoming packets addressed to the Web server's IP address and port 80, the standard HTTP port.

Figure 14-9 ►

A firewall
between the
Internet and a
private network

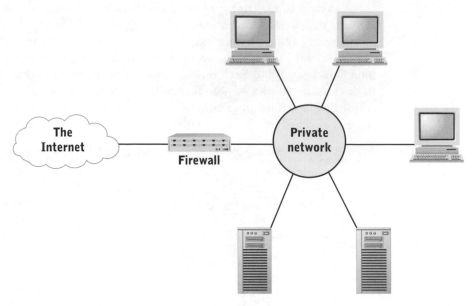

Configuring the firewall becomes more complex as the volume and variety of legitimate traffic between networks increases. For example, consider an organization that enables strategic partners such as suppliers to interface directly with its production or logistics databases. An application that supports such interactions has components running both within and outside the private network. The firewall must be configured to allow packets to be transmitted between servers and software to support the application. Specific entries must be added to the firewall database for the IP addresses used by strategic partners and the port number(s) used by the application(s).

More complex approaches to firewall use and configuration include:

► Application firewalls
► Stateful firewalls

An **application firewall**, sometimes called a **proxy server**, is a server that handles the service requests of external users of one or more applications. An application firewall accepts service requests from the Internet or other untrusted networks on behalf of servers within the internal network. It relays the request to the appropriate server or servers and relays the response back to the client (see Figure 14-10). Application firewalls improve security by shielding internal servers and resources from direct access by outside users. A hacker or other intruder must compromise two servers, usually with two different security systems, to gain access to secure resources.

Figure 14-10 ▶

An application
firewall

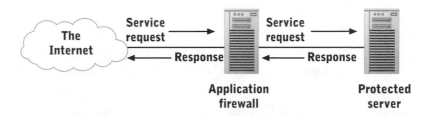

Application   **Protected**
firewall   **server**

A **stateful firewall** tracks the progress of complex client-server interactions. Stateful firewalls are most useful in client-server transaction processing scenarios where each transaction has a predictable processing pattern. The firewall is configured with the sequence of events that comprise valid client-server interactions, including packet types and their transmission sequence, port and server access order, and maximum time durations between valid requests and responses. Packets that don't match preprogrammed patterns of normal activity are automatically blocked.

Correctly installing and configuring any firewall requires a detailed understanding of network protocols, operating system security procedures, and the mechanisms by which client, server, and peer processes interact with one another. The process starts with an inventory of all existing services and the packet types, IP numbers, and ports that support client-server or peer-to-peer interaction for each service. Note that a typical information system includes many services that aren't obvious, such as remote procedure calls, authentication and authorization, and background services such as time synchronization. System administrators sometimes use a port scanner or network monitor to determine what IP numbers and ports are used by existing services.

Once the inventory is complete, the administrator determines which services need to be accessible from outside the organization's private network. Based on that information, the administrator can configure firewalls to block packets associated with unimplemented services and those from unauthorized IP addresses. Because the configuration process is complex and error-prone, any changes to firewall configuration should be thoroughly tested to ensure that unauthorized external access is blocked and that all authorized internal and external activity is enabled.

PHYSICAL ENVIRONMENT

Installing computer hardware requires special attention to many aspects of the physical environment. Issues to be considered when choosing or preparing a location for computer hardware include:

▶ Electrical power
▶ Heat dissipation

- ► Moisture
- ► Cable routing
- ► Fire protection

Each of these is discussed in the next sections.

Electrical Power

Computer hardware is very sensitive to fluctuations in power levels. Computer circuitry is designed to operate at a constant low power level. Fluctuations can cause momentary loss of operation or damage to electrical circuits. Fluctuations can be of several types:

- ► Momentary power surges
- ► Momentary power sags
- ► Long-term voltage sags
- ► Total loss of power

Power surges or spikes can be caused by a number of events. Lightning strikes in power generation or transmission facilities tend to cause the most dangerous types of power surges. Dangerous spikes can be caused by the failure of power transformers or other transmission equipment. These events lead to brief power surges of very high intensity. Because the surges are brief, they might not engage standard protection devices such as fuses or circuit breakers before significant damage has occurred. Standard fuses or breakers for a floor or entire building do not provide adequate protection for computer equipment.

Power sags normally occur when a device that requires a large amount of power is started. This can be seen in the home when air conditioners, refrigerators, and electric dryers are started, causing a momentary dimming of lights. Small power sags are almost always present when multiple devices share a single electrical circuit. Large power sags are a symptom of overloaded circuits. Longer-term power sags are often caused by the power provider. The common term for this event is a **brownout**. Brownouts occur when the demand for electricity exceeds the provider's generation and transmission capabilities during peak demand periods such as hot summer days. The power provider will reduce the voltage level temporarily on a system-wide basis to spread the available power evenly.

Most computer equipment is designed to operate reliably over a range of voltage levels. In the United States, transformers within smaller computer systems are designed for 110-volt alternating current power input. Larger computer systems in the United States can use 220- or 440-volt power inputs. Computer systems in other countries use different voltage levels. Computer system power transformers can usually deliver consistent voltage output with input voltage variations of up to 10 percent. This characteristic provides some protection against surges and sags due to the startup or shutdown of other equipment, as well as brownouts.

Auxiliary power conditioning equipment is required to deal with powerful surges and total power loss. Equipment can be protected against high power surges by a **surge protector**, which detects incoming power surges and diverts them to ground within nanoseconds. Surge protectors differ in the speed at which they react, the intensity of the surge they can suppress, and whether they can be reused after a surge. Surge protectors can be purchased as separate devices or can be integrated into other power conditioning devices.

By itself, total power loss rarely causes damage to computer hardware. However, a tripped circuit breaker or blackout is often accompanied by one or more power surges. The primary problem with power loss lies in the loss of data. Data held in RAM, including process data areas, secondary storage buffers, and communication buffers, is lost when power is interrupted.

Protection against power loss requires an auxiliary power source. This can take several forms, including a combination of redundant power circuits, auxiliary generators, and battery backup. An **uninterruptible power supply** (UPS) is a device that provides power to attached devices in the event of external power failure. UPSs vary in their power delivery capacity, switching time, and battery life. Surge protection is normally incorporated into a UPS. This is almost essential because switching between external and internal power supplies might introduce surges.

Some UPSs, particularly those with short delivery times, are designed to work in concert with a computer's operating system. They include a standard (for example, USB) communication port and cable that can be connected to the computer. When a power failure is detected, the UPS sends a signal to the computer to indicate that it has switched over to battery power. This enables the operating system to initiate protective actions prior to a total loss of power. Typically, the operating system initiates a normal or emergency shutdown procedure when such an interrupt is detected. More complex UPSs support two-way communication, which can provide the operating system with additional information such as error conditions and battery discharge rate.

Heat Dissipation

All electrical devices generate heat as a byproduct of normal operation. Excessive heat can cause intermittent or total failure of electrical circuits. Because of this, all computer equipment requires some means of **heat dissipation**. In equipment that generates little heat, vents in the equipment cabinet are normally sufficient to dissipate heat. Care must be taken to ensure that vents do not become blocked, so there is free movement of air through the cabinet.

Many computer hardware devices use fans to move air through the unit. Fans either force cool exterior air into the cabinet or draw hot interior air out.

Either method requires at least two vents and a clear pathway for air movement. Vents are normally positioned at opposite corners of the cabinet to ensure that all components receive adequate cooling. Fan-based cooling also requires some filtering to remove dust and other contaminants.

When heat is dissipated from an equipment cabinet, it collects in the room in which the cabinet is located. Heat must be dissipated from the room, especially when many hardware devices are situated in a relatively small space, such as in a server closet. Normal room or building air conditioning might be sufficient; otherwise, supplemental cooling capacity is required. Supplemental cooling more effectively counteracts heat buildup and provides additional protection if the primary cooling system fails.

In extreme cases, auxiliary cooling can be provided within an individual equipment cabinet. This can take the form of a refrigerant-based heat exchanger, a liquid cooling system, or even a liquid nitrogen system. Such measures are often used with supercomputers, where CPUs and other components operate at extremely high clock rates or access speeds.

Moisture

Excessive moisture is an enemy of electrical circuitry due to the danger of short circuits, which form when water conducts electricity between two otherwise unconnected conductors. Short circuits can cause circuit damage and fire when computer equipment is operating. Water can damage computer equipment even when power is turned off. Impurities in the water remain on computer components as the water evaporates. The impurities can corrode exposed electrical contacts or other hardware components such as power connectors, printed circuit boards, and metal cabinets. They can also form a short circuit.

Well-designed cabinets are one defense against the dangers of moisture, but they protect only against spills and leaks. Another protective measure is to mount hardware cabinets and devices above floor level, which minimizes the danger from roof leaks, broken pipes, and similar problems that can lead to standing water.

Protection must also be provided against condensation due to excessive humidity. Low humidity is also a problem. Low humidity levels increase the buildup of static electricity, which increases the likelihood of circuit damage due to inadvertent static discharges. In general, the humidity level of a room containing computer equipment should be near 50 percent.

Humidity can generally be controlled automatically by optional components of heating, ventilation, and air conditioning systems. It is possible to use a household dehumidifier in smaller rooms, but be aware that when the unit powers on, it might cause a brief dip in power level if it is on the same circuit as computer equipment. Also, household dehumidifiers collect airborne water in a pan, which must be emptied regularly. An overflowing pan can be a safety or electrical hazard.

Cable Routing

Computer facilities must provide protection for data communication lines. Because the configuration of these lines changes frequently, ease of access is also very important. Computer facilities usually deal with this problem in two ways—raised floors and dedicated cable conduits.

A raised floor is often used in a room that contains multiple hardware cabinets. The raised floor serves several purposes. The primary purpose is to provide an accessible location for cables connecting different devices. The floor is a set of load-bearing supports upon which a grid of flooring panels is laid. The flooring panels can be installed or removed easily from the grid. Cables are routed under walkway areas.

Other advantages of raised floors include protection from standing water and movement of chilled air. Several inches of water can accumulate without reaching the level of equipment cabinets. Moisture sensors placed below the floor panels can detect the buildup of standing water. The space between the actual floor and the floor panels can also be used as a conduit for chilled air. When used in this manner, equipment cabinets are vented at the bottom and top. Chilled air is forced through the floor into the bottom of the cabinets and heat is dissipated through the top.

Dedicated cabling conduits normally provide cable access between rooms or floors of a building. To prevent electromagnetic interference, these conduits should not be used to route both electrical power and electrical data communication lines. In addition, access panels should be provided at regular intervals for adding, removing, or rerouting cables. Conduits should be shielded to limit external electromagnetic interference.

Fire Protection

Fire protection is an important consideration both for safety of personnel and the protection of expensive computer hardware. As with cooling, the normal fire protection mechanisms incorporated within buildings are inadequate for rooms that house computer hardware. In fact, such measures actually increase the danger to both personnel and equipment because they usually rely on water, for example automatic sprinklers.

Carbon dioxide, fire retardant foams and powders, and various gaseous compounds are alternate methods of fire protection. Carbon dioxide is generally unacceptable because it is a suffocation hazard and because it promotes condensation within computer equipment. Fire retardant foams and powders are unacceptable due to their moisture content, and powders generally have corrosive properties.

Until recently, many large computer facilities used halon 1301 gas, which does not promote condensation and doesn't displace oxygen to the extent that carbon dioxide does. This gives personnel adequate time to evacuate a room or floor. Unfortunately, halon gas is a chlorofluorocarbon (CFC), a class of gases with ozone-depleting properties. Most industrialized countries have signed the Montreal Protocol, which has phased out CFC production and bans importation of newly produced CFCs. Various alternatives are currently being deployed including halocarbon compounds and inert gas mixtures. None has yet emerged as a clear successor to halon.

Fire detection is also a special problem within computer facilities. Electrical fires often do not generate heat or smoke as quickly as do conventional fires, and normal detection equipment might be slow to react. Fast detection is an economic necessity. A fire within one item of computer equipment can spread quickly and/or cause damage to attached equipment through power surges.

Normal building fire detection equipment is typically supplemented within a computer room. Additional smoke detectors are generally placed near large concentrations of equipment and below raised floors to detect fires in cabling.

Disaster Planning and Recovery

Because disasters such as fire, flood, and earthquakes cannot be avoided totally, plans must be made to recover from them. Disaster planning is particularly critical in online systems and systems where extended downtime will cause extreme economic impact. A number of measures normally are taken, including:

► Periodic data backup and storage of backups at alternate sites

► Backup and storage of critical software at alternate sites

► Installing duplicate or supplementary equipment at alternate sites

► Arrangements for leasing existing equipment at alternate sites, such as with another company or a service bureau

The exact measures that are appropriate for any given installation depend heavily on local characteristics.

SUMMARY

► A system administrator's primary responsibility is to ensure efficient and reliable delivery of IS services. Broad categories of system administration tasks include acquiring new IS resources, maintaining existing IS resources, and designing and implementing an IS security policy. There is considerable variation among organizations as to the assignment of specific tasks.

► IS resource acquisition and deployment should occur only in the context of a well-defined strategic plan for the entire organization. IS strategic plans tend to follow, rather than lead, the strategic plans of other organizational units. IS resources can be considered organizational infrastructure. Many IS resources are also capital assets because they are used during multiple years. Strategic issues relevant to IS include services to be provided, charging for services, infrastructure composition, and infrastructure improvement and maintenance.

► Providing infrastructure-based services to a wide variety of users requires service standards. However, standardization tends to stifle innovation and to produce solutions that are suboptimal for some users. Standardization often causes problems for users who need services at or near the leading edge of technology. Managing IS resources as infrastructure can result in ignoring opportunities to employ IS resources for competitive advantage.

► Planning for IS resource acquisition requires a thorough understanding of present and anticipated application needs. The acquisition process includes determining the applications that the hardware/software will support, specifying detailed hardware/software requirements, drafting and circulating a request for proposals (RFP), evaluating RFP responses, and negotiating a purchase and/or support contract.

► Hardware requirements depend on the hardware and system software resources required by application software. If the application software has already been developed, then its hardware and system software resource consumption can be measured. Determining resource requirements is much more complex when application software has not yet been developed. In that case, hardware and system software requirements must be determined by benchmark testing.

► A well-integrated approach to system security protects an organization's hardware, software, and data resources against accidental loss or damage, malicious tampering, unauthorized access, and accidental disclosure. Authentication and authorization enforced through the operating system or a security service are the first line of defense. Other defensive measures include password control, auditing, virus protection, regular software updates, and firewalls.

► Computer hardware installation and operation require special attention to many aspects of the computer's physical environment, with particular regard to electrical power, heat dissipation, moisture, cable routing, and fire protection.

Key Terms

access control list
application firewall
auditing
authentication
authorization
benchmark
biometric authentication
brownout
capital expenditure
capital resource
competitive advantage

firewall
hardware monitor
heat dissipation
infrastructure
monitor
operating expenditure
packet-filtering firewall
power sag
power surge
program profiler
proxy server

request for proposal (RFP)
service standard
software monitor
stateful firewall
strategic plan
surge protector
system administration
uninterruptible power supply
   (UPS)
virus

Vocabulary Exercises

1. An estimate of resource delivery by a hardware/software configuration can be obtained by a tested or published _____ .

2. _____ are expected to provide service over a period of years.

3. _____ is the process of determining or verifying the identity of a user or process owner.

4. _____ is the process of determining whether an authenticated user or process has sufficient rights to access a particular resource.

5. The term _____ usually refers only to the process of creating and managing records of user activity or resource access.

6. A(n) _____ accepts service requests from an untrusted network and relays the requests to the appropriate server or servers.

7. Because IS resources can be considered _____ , issues of service standards, cost, and cost recovery are important components of the IS strategic plan.

8. A(n) _____ is a formal (legal) document that solicits proposals from hardware and/or software vendors.

9. A(n) _____ is a program or program fragment that infects a computer by permanently installing itself, performing one or more malicious acts, and replicating and spreading itself using services of the infected computer.

10. A(n) _____ detects and reports hardware or software actions.

11. Opportunities to exploit IS resources for _____ might be missed if those resources are managed only as infrastructure.

12. A(n) _____ provides auxiliary power during blackouts and can notify the operating system when it is activated.

13. Computer hardware must be protected against _____ and _____ in electrical power.

14. A(n) _____ tracks the progress of complex client-server interactions and blocks packets that don't conform to normal activity patterns.

15. A(n) _____ is a hardware device, software, or a combination thereof that prevents unauthorized users in one network from accessing resources in another network.

16. Long-range acquisition of computer hardware and software should be made in the context of a(n) _____ for the entire organization.

17. The resource demands of an existing application program can be measured with a(n) _____ .

Review Questions

1. What is infrastructure? In what ways do computer hardware and system software qualify as infrastructure?

2. What basic strategic planning questions should be addressed with respect to infrastructure?

3. What are the advantages and disadvantages of standardization in computer hardware and systems software?

4. What is a request for proposal (RFP)? How are responses to an RFP evaluated?

5. What problems are encountered when attempting to determine hardware and system software requirements for application software that has not been designed or developed fully?

6. What is a benchmark? List two sources of widely used performance benchmarks.

7. What is a monitor? List the various types of monitors and the information that they provide.

8. Describe authentication and authorization. Which depends on the other? How and why are these processes more complex in a modern networked organization than within an organization that supports all information processing with a single mainframe?

9. What password protection measures are normally implemented by system administrators, operating systems, and security services?

10. Describe the pros and cons of enabling audits of resource accesses.

11. What is a virus? How can users and system administrators prevent virus infections?

12. Why is it important to install operating system and application software updates in a timely manner? How can users and system administrators ensure that they are installed in a timely manner?

13. Describe various firewall types and how each can improve system security. What information does a system administrator need to configure a firewall correctly?

14. Why are conventional methods of fire protection inadequate or dangerous for computer equipment?

15. What problems associated with electrical power must be considered in planning the physical environment of computer hardware?

Research Problems

1. Some manufacturers and vendors of large midrange computers and mainframes have developed documentation and software to assist engineers, account representatives, and purchasers to configure computer hardware and system software. Some of the software can help users match software demands against the capabilities of particular hardware configurations. Investigate the online offerings of several large computer companies, such as www.ibm.com, www.hp.com, and www.unisys.com.

2. Investigate firewall products from a major computer vendor such as www.ibm.com, www.cisco.com, or or www.hp.com. Which products are best suited to a small LAN with no publicly accessible resources? Which are best suited to a small LAN that contains one publicly accessible web site? Which are best suited to large organizations with e-commerce Web sites and internal resources accessed by strategic partners?

3. Locate one or more benchmarking programs for personal computers on the Web. Download and install the benchmarks and test the performance of your own computer system. What do the results tell you about your computer? Do they provide you with any guidance for improving system performance? What types of application tasks do the benchmark programs simulate? Would similar benchmark programs be useful with midrange computers or mainframes?

Appendix

Measurement Units

Many measurement units describe the capabilities of computer system components such as storage devices, I/O devices, device controllers, and computer networks. This appendix describes common measurement units, their abbreviations, and common mistakes in use and interpretation.

TIME UNITS

Time intervals, such as CPU cycle or disk access times, are expressed in fractions of a second, represented by the abbreviations and terms in Table A-1.

Table A-1 ▶

Time measurements

Abbreviation	Term	Fraction
ms	millisecond	10^{-3} (one thousandth)
ms	microsecond	10^{-6} (one millionth)
ns	nanosecond	10^{-9} (one billionth)
ps	picosecond	10^{-12} (one trillionth)

Note that both millisecond and microsecond have the same abbreviation. Microseconds are rarely used when describing computer hardware, so you can generally assume that ms refers to milliseconds.

CAPACITY UNITS

Bits and bytes are the smallest data storage and transmission units. A bit is a binary digit containing the value zero or one. A byte contains eight bits. The terms bit and byte are often quantified with a prefix to indicate larger magnitudes. Such prefixes are all based on the value 2^{10} (1024_{10}) because it approximates traditional units based on powers of ten such as thousands, millions, and billions.

Table A-2 shows the abbreviations and prefixes that are employed in capacity measurements.

Table A-2 ▶

Capacity measurements

Abbreviation	Prefix	Value
K or k	kilo	1024
M or m	mega	1024^2 (1,048,576)
G or g	giga	1024^3 (1,073,741,824)
T or t	tera	1024^4 (1,099,511,627,776)
P or p	peta	1024^5 (1,125,899,906,842,624)
E or e	exa	1024^6 (1,152,921,504,606,846,976)

Uppercase and lowercase letters are both used, though uppercase is more common. The values represented by these prefixes are slightly larger than their decimal counterparts, which can cause confusion when an abbreviation based on 1024 is used to describe a quantity based on 1000.

For example, consider the capacity of a "standard" 3.5-inch PC disk which is frequently (and incorrectly) described with a capacity of 1.44 megabytes. The actual capacity is 1,474,560 bytes (2 sides × 80 tracks/side × 18 sectors/track × 512 bytes/sector), 1440 kilobytes (1,474,560 ÷ 1024), or 1.40625 megabytes (1,474,560 ÷ 1024^2). Hard disk capacities are usually stated with similarly flawed units of measure. For example, the correct capacity of a typical "160 GB" disk is actually 149 GB (160,000,000 ÷ 1024^3).

Another source of confusion arises when the terms *bit* and *byte* are represented by the letter B or b. For example, the terms *megabit* and *megabyte* can both be abbreviated as MB. A common method of distinguishing between bits and bytes is to abbreviate the term *bit* using a lowercase letter and the term *byte* using an uppercase letter. Thus, the abbreviation GB would be interpreted as gigabyte or gigabytes and the term Gb as gigabit or gigabits. Unfortunately, this convention is not universally followed. Any abbreviation containing the letters b and B deserves careful scrutiny to determine whether bit or byte is intended.

DATA TRANSFER RATES

Data transfer rates describe the capacity of communication channels and the input speed, output speed, or throughput of I/O devices and device controllers. A data transfer rate is always expressed as a quantity of data per time interval. For example, LAN data transmission capacity might be described as 100 megabits per second. Printer output speed might be described as four pages per minute.

Communication channel capacities are usually expressed in bits or bytes per second. Magnitude prefixes such as *kilo* and *mega* can be used (for example, 500 kilobits per second and 100 megabytes per second). Bit and byte capacity measures are abbreviated as described in the previous section, and the phrase *per second*, abbreviated as ps, is appended to the measurement unit. For example, kilobits per second is abbreviated as Kbps and megabytes per second is abbreviated as MBps. As with capacity measures, any abbreviation containing the letter b or B deserves careful scrutiny to determine whether bit or byte is intended.

Printer output rates are generally expressed in pages per minute, abbreviated as ppm. The number of bytes in a page varies widely depending on the page content and data encoding method. Video output rates may be expressed in frames per second, abbreviated as fps. The number of bytes per frame depends on frame size, image content, and data encoding method.

Glossary

10 Gigabit Ethernet — Network transmission method based on the IEEE 802.11ae and 802.11ak standards.

1GL — Acronym of first generation language.

24-bit color — Coding method that represents a color as three 8-bit numbers, one for the intensity of each additive or subtractive color.

2GL — Acronym of second-generation language.

3GL — Acronym of third generation language.

4GL — Acronym of fourth-generation language.

5GL — Acronym of fifth-generation language.

802.11 — Early IEEE wireless network standard with a 2 Mbps maximum transmission speed in the 2.4 GHz band.

802.11a — IEEE wireless network standard with a 54 Mbps maximum transmission speed in the 5 GHz band using orthogonal frequency division multiplexing.

802.11b — IEEE wireless network standard with an 11 Mbps maximum transmission speed in the 2.4 GHz band using direct sequence spread spectrum transmission.

802.11g — IEEE wireless network standard with a 54 Mbps maximum transmission speed in the 2.4 GHz band using orthogonal frequency division multiplexing.

absolute addressing — Describes memory address operands that refer to actual physical memory locations.

access arm — Device within a disk drive containing one or more read/write heads mounted at one end. The other end is attached to a motor that enables the read/write head(s) to be positioned over a single track of the disk platter.

access control list — List of users or user groups with access authority to a secured object or service.

access time — Elapsed time between the receipt and completion of a read or write command by a storage device.

acknowledge (ACK) — ASCII control character sent from receiver to sender to indicate successful receipt of transmitted data.

ACK — Mnemonic of the acknowledge ASCII control character.

ACL — Acronym of access control list.

ACM — Acronym of Association for Computing Machinery.

Active Directory — Microsoft directory services software.

active matrix display — Liquid crystal or plasma display that uses one or more transistors for every pixel.

ADC — Acronym of analog-to-digital conversion.

ADD — (1) Boolean operation that computes the arithmetic sum of two numbers. (2) Central processing unit instruction that computes the arithmetic sum of two numbers and stores it in a register or memory location.

additive colors — Red, green, and blue.

address — Physical location of a data item (or the first element of a set of data items) within a storage or output device.

address bus — Portion of a bus that transmits the address of a storage location to be read from or written to.

address mapping — Synonym of address resolution.

address resolution — As performed by the control unit, converting relative or segmented address references into corresponding physical memory addresses.

addressable memory — Maximum amount of memory that can be addressed physically by the CPU.

Advanced Intelligent Tape (AIT) — Magnetic tape standard developed by Sony based on Digital Audio Tape technology.

AIT — Acronym of Advanced Intelligent Tape.

AITP — Acronym of Association for Information Technology Professionals.

algorithm — (1) Series of processing steps that describes the solution to a problem. (2) Sequence of instructions within a computer program that implements a complex processing operation.

allocation unit — Smallest unit of storage that can be allocated to an object such as a file or a program.

ALU — Acronym of arithmetic logic unit.

AM — Acronym of amplitude modulation.

American Standard Code for Information Interchange (ASCII) — Standard 7-bit coding scheme for character data and a limited set of I/O device control functions.

amplifier — Hardware device that increases the amplitude or power of a signal or waveform.

amplitude — Magnitude (highest value) of wave peaks of a waveform.

amplitude modulation (AM) — Data transmission method that encodes data values as variations in carrier wave amplitude.

amplitude shift keying (ASK) — Synonym of amplitude modulation.

analog signal — (1) Signal in which any value of an infinite set is encoded by modulating one or more carrier wave parameters in direct proportion to the encoded value. (2) Signal that varies continuously in one or more carrier wave parameters, such as frequency or amplitude.

analog-to-digital converter (ADC) — Processing device that generates a numerical representation of an analog waveform, *see* sampling.

AND — (1) Boolean operation that generates the output value false unless both input values are true. (2) Central processing unit instruction that generates the output value false unless both input values are true, and stores the result in a register or memory location.

applet — Program written in the Java programming language that is executed by another program such as a Web browser.

application development software — Software tools such as compilers and CASE tools used to develop application software.

application firewall — Server that handles the service requests of external users of one or more applications, thus shielding other servers from direct access and potential security breaches.

application layer — Application programs (software), as compared to systems software and computer hardware.

application program — Program that addresses a single specific need of a user or a specific class of information processing tasks.

application programmer — Person who creates or maintains application software.

application software — Software class consisting of all application programs.

architectural design — Selecting and describing the exact configuration of all hardware, network, systems software, and application development tools that will support system development and operations.

areal density — Surface area allocated to a bit on a storage medium, typically as measured in bits, bytes, or tracks per inch.

arithmetic logic unit (ALU) — Circuitry within a central processing unit that performs computation, comparison, and Boolean logic operations.

arithmetic SHIFT — SHIFT operation that performs division or multiplication.

array — Set of data items stored in consecutive storage areas and identified by a common name.

ASCII — Acronym of American Standard Code for Information Interchange.

ASK — Acronym of amplitude shift keying.

assembler — Program that translates assembly language source code instructions into object code or central processing unit instructions.

assembly language — Programming language that uses mnemonics to represent central processing unit instructions and storage address operands.

Association for Computing Machinery (ACM) — Professional organization for computer scientists, programmers, and engineers.

Association for Information Technology Professionals (AITP) — Professional organization for information system professionals.

asynchronous transmission — Any transmission method in which sender and receiver do not continually synchronize their clocks.

attenuation — Reduction in signal amplitude or power as the signal travels through a transmission medium.

audio response unit — Device that provides natural language instructions or responses to input from the keypad of a telephone or other device.

auditing — Process of creating and managing records of user activity or resource access.

authentication — Process of determining or verifying the identity of a user or process owner.

authorization — Process of determining whether an authenticated user or process has sufficient rights to access a particular resource.

average access time — Statistical average (or mean) elapsed time required by a storage device to respond to a read or write command.

back-end CASE tool — Software tool that creates source code instructions from a system model.

band — Portion of the message-carrying capacity of a carrier wave or transmission medium, typically expressed as a range of frequencies, such as the band between 2.4 and 2.5 gigahertz.

bandwidth — Range of carrier wave frequencies that can be transmitted over a transmission medium.

bar code — Set of dark vertical bars or a two-dimensional pattern of dark squares that encodes numerical data and can be read by a bar code scanner or other optical scanning device.

bar code scanner — Device that optically scans surfaces for numbers encoded in bar codes.

base — Multiplier that describes the difference between one position and the next in a numbering system.

baseband — (1) Transmission method that uses the entire bandwidth of a communication channel to carry one signal or data stream. (2) Local area network transmission method in which multiple digital signals are carried over a single communication channel through time-division multiplexing.

BCC — Acronym of block check character.

benchmark — Program, program fragment, or set of programs and related data inputs that test the performance of a hardware component, a group of hardware or software components, or an entire computer system or network when executing one or more specific tasks.

big endian — Describes a CPU architecture that stores a multibyte data item with the most significant byte in the storage location with the lowest-numbered address.

binary number — Number in which each digit can have only one of two possible values (0 or 1).

binary signal — Signal in which one of two values is encoded by modulating a carrier wave parameter.

biometric authentication — Authentication based on physical characteristics such as fingerprint or retinal scanning.

bit — (1) Value represented in a one position of a binary number. (2) Number that can have a value of zero or one. (3) Abbreviation of binary digit.

bit map — Image representation in which each pixel is represented by one or more bits.

bit string — Sequence of bits that describes a single data value.

bit time — Time interval during which the value of a single bit is present within a transmitted signal.

blade — Circuit board that contains most of a server computer.

block — (1) Series of logical records grouped on a storage device for efficient processing, storage, or transport. (2) Unit of data transfer to and from a storage device. (3) Portion of a program that is always executed as a unit.

block check character (BCC) — Extra 8-bit value containing error-detection information that is appended to the end of a set, or block, of transmitted data items.

block checking — Synonym of longitudinal redundancy checking.

block size — Number of bits, bytes, or records within a single storage block or allocation unit.

blocked state — State of a process or thread that is waiting for an event such as completion of another process or thread, arrival of data input, or correction of an error condition.

blocking — Storage of multiple data items or data item fragments within a single storage block or allocation unit.

blocking factor — The number of bits, bytes, or logical records grouped within a single physical record, storage block, or allocation unit.

Boolean data type — Data item class that can store only the value true or false.

Boolean logic — Formal logic system in which data inputs and outputs can have only the values of true or false.

BRANCH — (1) Processor operation that causes an instruction other than the next sequential instruction to be fetched and executed. (2) Processor instruction that implements a BRANCH operation by overwriting the current value of the instruction pointer.

branch prediction — Technique to increase the benefits of pipelining in which the CPU guesses whether a branch condition will be true or false based on past experience.

bridge — Device that copies or repeats packets from one network segment to another.

broadband — (1) Communication channel with a wide frequency range and high data-transmission capacity. (2) Transmission method in which multiple digital signals are carried over a single channel through frequency division multiplexing.

broadcast mode — Transmission method that sends a message from a sender to all devices on a communication channel.

brownout — Reduction in voltage by an electrical power provider, usually because of temporary excess of power demand over available supply.

buffer — Primary storage that temporarily stores data in transit from a sender to a receiver.

buffer overflow — Receipt of more data than can be stored in a buffer.

bus — Communication channel shared by multiple devices within a computer system or network.

bus arbitration unit — Processor that mediates competing demands for control of or access to a bus.

bus clock — Clock circuit that generates timing pulses, which are transmitted to all devices attached to a bus.

bus cycle — (1) Period of time required to perform one data transfer operation on a bus. (2) Elapsed time between two consecutive pulses of a bus clock.

bus master — (1) Any device attached to a bus that can initiate a data-transfer operation or send a command to another device. (2) Device that controls a data-transfer operation during a specific bus clock cycle.

bus protocol — Communication protocol used by all devices attached to a bus.

bus slave — Device that is incapable of acting as a bus master.

bus topology — With respect to data communication networks, a topology that uses a bus as the communication channel.

business logic layer — Software layer or component that implements the rules and procedures of business processing.

business modeling discipline — Activities of the Unified Process that develop models of an organization and the system environment.

byte — String of eight bits.

cache — Area of high-speed memory that holds portions of data also held within another storage device. A cache improves the average speed of read and write operations to or from its associated storage device.

cache controller — Processor or software that manages cache content.

cache hit — Access to data in a storage device also stored in a cache.

cache miss — Access to data in a storage device that is not also stored in cache.

cache swap — Operation that copies data from a cache to its associated storage device and then overwrites the cache copy with new data.

call — Instruction that executes a function, subroutine, or procedure.

capital expenditure — Funds expended to obtain a capital resource.

capital resource — A resource expected to provide benefits beyond the current operating period.

Carrier Sense Multiple Access/Collision Detection (CSMA/CD) — Media access control protocol that allows message collisions to occur but defines a recovery method.

carrier wave — Energy wave such as electricity or light that carries encoded messages within a communication channel.

CASE — Acronym of computer assisted software engineering.

Category 5 — Widely implemented twisted pair wiring standard that consists of four twisted pairs that transmit at speeds of up to 1 Gbps.

Category 6 — Twisted pair wiring standard with the same construction as Category 5 but higher quality wire and insulators to reduce transmission errors.

cathode ray tube (CRT) — Video display device that uses electron beams to excite phosphor(s) on the inner surface of a vacuum tube.

CD-DA — Acronym of compact disc–digital audio.

CD-R — Acronym of compact disc–recordable.

CD-ROM — Acronym of compact disc read-only memory.

CD-RW — Acronym of compact disc–read/write.

central processing unit (CPU) — Computer system hardware component that fetches and executes instructions.

channel — (1) Synonym of input/output channel. (2) Shortened form of communication channel.

character — (1) Primitive and indivisible component of a written language. (2) One byte of ASCII-encoded data or two bytes of Unicode-encoded data.

character framing — Serial data communication technique that surrounds encoded characters with specific bit patterns for purposes of clock synchronization and error detection.

chief information officer (CIO) — Manager charged with responsibility for planning, maintaining, and operating all information-processing resources within an organization.

child process — Process created and controlled by another executing process (the parent process).

chromatic depth — Synonym of chromatic resolution.

chromatic resolution — (1) Number of bits used to describe the color of each pixel in a graphic display. (2) Number of different colors that may be assigned as the value of any single pixel.

CIO — Acronym of chief information officer.

circuit switching — Technique for allocating transmission lines or circuits to senders and receivers whereby a sender and receiver are granted exclusive use of a communication channel for the duration of a communication session.

CISC — Acronym of Complex Instruction Set Computing.

class — Data structure that contains both traditional data elements and the software that manipulates the data elements.

client — Program or computer that requests and receives services from another program or computer.

client-server architecture — Organization of software and hardware components as clients and servers.

clock cycle — Time interval between two adjacent clock timing pulses.

clock rate — (1) Rate at which clock pulses are generated by a clock, stated in hertz. (2) With respect to a central processing unit, the time required to fetch and execute the simplest instruction.

cluster — A group of similar or identical computers, connected by a high-speed network, that cooperate to provide services or execute a common application.

CMY — Acronym of cyan, magenta, yellow.

CMYK — Acronym of cyan, magenta, yellow, black.

coaxial cable — Transmission medium composed of a single strand of wire surrounded by an insulator, a braided return wire, and a tough outer coating.

code — (1) Act of creating program instructions. (2) A program or set of related source or machine code instructions.

code generator — Synonym of back-end CASE tool.

coercivity — Ability of an element or compound to accept and hold a magnetic charge.

collating sequence — Order of symbols produced by sorting them according to a numeric interpretation of their coded values.

collision — Result of simultaneous transmission by two senders on a shared communication channel.

COM+ — Acronym of Component Object Model Plus.

command language — A language that directs an operating system's actions.

command layer — System software component that accepts and processes user commands to manipulate software, stored data, and hardware resources.

Common Object Request Broker Architecture (CORBA) — Nonproprietary standard that enables software components to locate and communicate with one another.

communication channel — Combination of a transmission medium and communication protocol.

communication protocol — Set of communication rules and conventions encompassing data and command representation, bit encoding and transmission, transmission media access, clock synchronization, error detection and correction, and message routing.

compact disc–digital audio (CD-DA) — Standard format for storing and distributing music on compact disc.

compact disc–read only memory (CD-ROM) — Standard 120 millimeter read only optical disc.

compact disc–read/write (CD-RW) — Standard phase-change optical storage method and medium.

compact disc–recordable (CD-R) — Compact disc containing a laser-sensitive dye that can be written with a laser switched between high and low power.

compaction — Process of reallocating storage locations assigned to one or more files or programs to eliminate empty storage locations.

competitive advantage — Use of resources to provide better or cheaper services than others can provide with those same resources.

compiler — Program that translates high-level programming language source code instructions into object code.

compiler library — Set of object-code modules intended to be linked with object-code modules produced by a compiler.

complete path — Named path from the root node of a storage volume or name space to a specific file or object through intervening layers of a hierarchical directory structure.

complex instruction — Instruction that combines primitive processing operations.

Complex Instruction Set Computing (CISC) — Processor architecture that uses complex instructions that embed many primitive processing operations, in contrast to reduced instruction set computing.

component — Executable software module or object with a well-known interface and a unique identifier.

Component Object Model Plus (COM+) — Microsoft proprietary standard that enables software components to locate and communicate with one another.

compression — Reduction in data size using a compression algorithm.

compression algorithm — Algorithm by which data inputs may be translated into equivalent outputs of smaller size.

compression ratio — Ratio of data size before and after applying a compression algorithm.

computer assisted software engineering (CASE) tool — Program or integrated set of programs used to analyze, design, and develop computer software.

computer network — Hardware and software that enables multiple users and computer systems to share information, software, and other resources.

computer operations manager — Manager responsible for the operation and maintenance of a large data-processing center or information system.

computer science — Study of the implementation, organization, and application of computer software and hardware resources.

concurrent execution — Execution of multiple processes or threads within the same immediate time frame through interleaved execution on a single processor.

condition — A comparison or other logical operation that produces a Boolean result.

conditional BRANCH — (1) BRANCH operation that is performed only if a Boolean variable is true. (2) Processor instruction that implements a conditional BRANCH operation.

conductivity — Ability of an energy wave such as electricity or light to pass through a conductor or transmission medium.

conductor — Element, compound, wire, or cable that exhibits conductivity.

connection-oriented protocol — Communication protocol that requires sender and receiver to establish a connection prior to transmitting any data.

connectionless protocol — Communication protocol that does not require sender and receiver to establish a connection prior to transmitting any data.

contiguous — Storage of a set of data elements in sequential physical storage locations with no intervening empty or unallocated locations.

control bus — Portion of a bus that transmits command and status signals.

control structure — High-level programming or job control language statement that describes the selection or repetition of other program statements, such as an If-Then or While-Do statement.

control unit — Subset of a central processing unit responsible for moving data, accessing instructions, and controlling the arithmetic logic unit.

CORBA — Acronym of Common Object Request Broker Architecture.

core memory — Antiquated form of primary storage implemented as a lattice of wires with iron rings wrapped around each wire junction point.

CPU — Acronym of central processing unit.

CRC — Acronym of cyclical redundancy checking.

crosstalk — Noise added to the signal in a wire from EMI generated by adjacent wires.

CRT — Acronym of cathode ray tube.

CSMA/CD — Acronym of Carrier Sense Multiple Access/Collision Detection.

current directory — Directory that serves as the origin for file and directory search operations.

cursor — Visual symbol on a video display, such as a colored box or underline, that indicates the current input or output position.

cycle — Single operation, function, or transfer within a series of similar operations, functions, or transfers, as performed by a processor, storage device, or communication channel.

cycle time — Duration of one cycle.

cyclical redundancy checking (CRC) — Technique for detecting errors in data transmission based on redundant data computed with a complex algorithm.

cylinder — In a disk storage device, the set of all tracks at an equivalent distance from the edge or spindle on all recording surfaces.

DAC — Acronym of digital-to-analog converter.

DAS — Acronym of direct-attached storage.

DAT — Acronym of Digital Audio Tape.

data bus — Portion of a bus that transmits data.

data declaration — Programming language statement that declares the existence, name, type, and other characteristics of a data item or structure.

data layer — Software layer or component that manages and interacts with a permanent data store such as a database.

data link layer — Layer of network system software responsible for media access and bit transmission.

data operation — Any processing operation that uses or alters data content.

data striping — Data storage method that improves read/write performance by dividing data into smaller segments and storing the segments on multiple disks.

data structure — A data item, such as an array or linked list, that contains multiple primitive data elements or other data structures.

data transfer rate — Rate at which data is transmitted through a medium or communication channel, as measured in data units per time interval.

database administrator — Manager responsible for ensuring data integrity, reliability, security, and availability.

datagram — Smallest unit of data transfer defined by the Transmission Control Protocol.

DCE — Acronym of Distributed Computing Environment.

DDS — Acronym of Digital Data Storage.

debugging tools — Type of application development software that simulates program execution and traces errors.

debugging version — Program that contains data and instructions to aid in locating and correcting errors.

decimal point — Period or comma in the decimal numbering system that separates the whole and fractional parts of a numeric value.

decoding — Process within the central processing unit of extracting the op code and operands, routing data and control signals, and signaling the ALU.

decompression algorithm — Algorithm that returns compressed data to its original state.

deployment discipline — Activities of the Unified Process that install and configure infrastructure and application software components and bring them into operation.

design discipline — Activities of the Unified Process that determine the structure of a specific information system that fulfills the system requirements.

design model — Architectural blueprint for system implementation.

detailed design — Design activities that specify system details including databases, application software, user and system interfaces, and backup and recovery mechanisms.

device controller — Processor that controls the physical actions of one or more storage devices.

Digital Audio Tape (DAT) — Early magnetic tape technology on which Digital Data Storage standards are based.

Digital Data Storage (DDS) — Family of magnetic tape standards developed by Hewlett-Packard and Sony, and based on Digital Audio Tape.

Digital Linear Tape (DLT) — Magnetic tape standard developed by Quantum.

digital signal — (1) Signal in which one of a discrete, or countable, set of values can be encoded by modulating one or more carrier wave parameters. (2) Synonym of binary signal.

digital signal processor (DSP) — Processor that manipulates continuous streams of digital data, such as audio or motion video.

digital-to-analog converter (DAC) — Processing device that generates an analog signal from a numerical representation of the signal.

digital versatile disc (DVD) — Synonym of digital video disc.

digital video disc (DVD) — Standard optical disc format for distributing movies and other audio-visual content.

digital video disc random access memory (DVD-RAM) — Standard for rewriteable digital video discs with a capacity of 2.4 GB for single-sided discs and 4.8 GB for double-sided discs.

digital video disc read-only memory (DVD-ROM) — Standard format for general-purpose read-only data storage on digital video disc.

digitizer — (1) Device that captures the position of a pointing device as input data. (2) Synonym of optical scanner.

DIMM — Acronym of double in-line memory module.

DIP — Acronym of dual in-line package.

direct access — Synonym of random access.

direct-attached storage (DAS) — Architecture in which software executing on a CPU accesses secondary storage devices within the same computer system.

direct memory access (DMA) — Method of data transfer that enables direct data movement between main memory and a storage or I/O device, bypassing the central processing unit.

directory — Data structure that contains information about files and other directories.

directory services — System software (middleware) that enables users and components to locate and access remote resources.

discipline — Related group of system development activities within the Unified Process.

discrete signal — Synonym of digital signal.

disk defragmentation — Reorganization of data on a disk drive so that all portions of an individual file are stored in contiguous sectors.

disk mirroring — Storage technique that enhances reliability by storing a redundant copy of all data on a second disk drive.

diskette — Synonym of floppy disk.

dispatching — (1) Act of giving a process or thread control of the central processing unit. (2) Act of moving a process or thread into the running state.

distortion — Undesirable alteration of one or more carrier wave characteristics resulting from interaction between signal energy and the transmission-medium or signal-propagation equipment.

distributed computing — Synonym of distributed processing.

Distributed Computing Environment (DCE) — Standard for distributed processing that covers network directory services, file sharing services, remote procedure calls, remote thread execution, system security, and distributed resource management.

distributed processing — Distribution and execution of processes or threads over multiple nodes of a computer network.

distribution version — Synonym of production version.

dithering — Simulating continuous color tones using patterns of dots containing a finite set of colors.

DLL — Acronym of dynamic-link library.

DLT — Acronym of Digital Linear Tape.

DMA — Acronym of direct memory access.

DMA controller — Dedicated processor that manages data transfers during direct memory access operations.

dot matrix printer — Impact printer that generates printed characters using a matrix of pins and an inked ribbon.

dots per inch (dpi) — Measure of print or display resolution (pixel density).

double in-line memory module (DIMM) — Small standard printed circuit board containing one or more random access memory chips with electrical contacts on both sides.

double precision — Representing a numeric value with twice the usual number of bit positions for greater accuracy or numeric range.

doubly linked list — Set of stored data items in which each element contains pointers to both the previous and next list elements.

dpi —Acronym of dots per inch.

DRAM — Acronym of dynamic random access memory.

drive array — Set of disk drives managed and accessed as if they were a single storage device.

DSP — Acronym of digital signal processor.

dual in-line package (DIP) — Early form of packaging for processor or memory circuits with two rows of electrical contact pins.

dual porting — With respect to random access memory, the ability to be read and written at the same time by two different devices.

DVD — Acronym of both digital video disc and digital versatile disc.

DVD-RAM — Acronym of digital video disc random access memory.

DVD-ROM — Acronym of digital video disc read-only memory.

DVD-RW — (1) Acronym of digital video disc read-write. (2) Rewritable DVD standard supported by the DVD Forum.

DVD+RW — (1) Acronym of digital video disc read-write. (2) Rewritable DVD standard supported by Hewlett-Packard, Philips, and Sony.

dynamic connection — Connection between a client and a server or remote resource that is not established until the client accesses the server or resource.

dynamic linking — Linking performed during program loading or execution.

dynamic random access memory (DRAM) — Type of electronic memory circuit that implements bit storage with transistors and capacitors.

dynamic-link library (DLL) — (1) A library of reusable software modules organized for dynamic linking. (2) A specific Microsoft Windows file format for storing reusable software modules.

early binding — Synonym of static linking.

EBCDIC — Acronym of Extended Binary Coded Decimal Interchange Code.

EDRAM — Acronym of enhanced dynamic random access memory.

EEPROM — Acronym of electronically erasable programmable read-only memory.

effective data transfer rate — Error-free throughput of a communication channel under normal conditions with a specific communication protocol.

EJB — Acronym of enterprise Java bean.

electromagnetic interference (EMI) — Alteration of electrical carrier wave characteristics caused by external electrical or magnetic phenomena, such as electric motors or sun spots.

electronically erasable programmable read-only memory (EEPROM) — Read-only memory device that can be erased and written by sending appropriate control signals.

enterprise Java bean (EJB) — Java program that executes within a business container on a server.

EMI — Acronym of electromagnetic interference.

enhanced dynamic random access memory (EDRAM) — Type of electronic memory circuit primarily composed of DRAM but supplemented with a small SRAM cache.

EPROM — Acronym of erasable programmable read-only memory.

erasable programmable read-only memory (EPROM) — Read-only memory device that is manufactured blank, is written (programmed) with a special EPROM writer, and erased by exposure to ultraviolet light.

Ethernet — Widely implemented local area network standard family.

even parity — Error-detection method that appends a parity bit to each transmitted or stored character. The value of the parity bit is zero if the number of one valued bits in a character is even; it is one otherwise.

excess notation — Coding method that represents integer values as bit strings such that the value zero (the midpoint of the numeric range) is represented by all zero bits except for the most significant bit (which contains a one).

exclusive OR (XOR) — (1) Boolean operation that generates the output value true if only one of the input values is true. (2) CPU instruction that generates the output value true if only one of the input values is true.

executable code — Program that consists entirely of CPU instructions that are ready to be loaded and executed.

execute — Process of carrying out an instruction within the central processing unit.

execution cycle — Process of carrying out an instruction in the central processing unit and completing the operation it specifies.

explicit priority — Process scheduling through explicitly stated priority levels and a corresponding scheduling algorithm.

Extended Binary Coded Decimal Interchange Code (EBCDIC) — IBM mainframe coding standard for representing character data in an 8-bit format.

extensible markup language (XML) — Standard Web format that encodes document structure and content.

external function call — Placeholder within object code for missing executable code.

FAT — Acronym of file allocation table.

father — Original version of a file, after updates have been applied to generate a new version (the son).

fault tolerance — Characteristic of a computer system or file management system that enables rapid recovery from the failure of a hardware component without data loss.

FCFS — Acronym of first come, first served.

FDM — Acronym of frequency division multiplexing.

ferroelectric random access memory — Type of nonvolatile random access memory that stores bit values within metallic crystals.

fetch cycle — Portion of a central processing unit cycle in which an instruction is loaded into the instruction register and decoded.

fiber-optic cable — Transmission medium for optical signal propagation, generally consisting of one or more plastic or glass fibers sheathed in a protective plastic coating.

field — Component data item of a record.

fifth-generation language (5GL) — Nonprocedural programming language used to develop software that mimics human intelligence.

file allocation table (FAT) — Data structure used in MS-DOS and early Microsoft Windows versions to record the allocation of storage device locations to files and directories.

file association — Relationship between a file type and the program or operating system utility that manipulates it.

file close operation — Operation that severs the relationship between a file and a process, flushes file I/O buffers, and deletes operating system data structures associated with the file.

file control layer — System software layer that accepts file manipulation requests from application programs and translates them into corresponding low-level processing commands to the central processing unit and secondary storage device controllers.

file management system (FMS) — Entire collection of system software that performs file management, usually part of the operating system.

file migration — Management technique for secondary storage in which older versions of a file are moved automatically to less costly storage media or devices such as magnetic tape.

file open operation — Process of associating a file with an active process by allocating buffers and updating internal tables.

File Transfer Protocol (FTP) — Older Web protocol for copying files from one Internet host to another.

firewall — Hardware device, software, or a combination of hardware and software that prevents unauthorized users in one network from accessing resources on another network.

firmware — Software that has been permanently stored in read-only memory devices.

first come, first served (FCFS) — Scheduling method that executes processes in order of their request arrival.

first-generation language (1GL) — Synonym of machine language or central processing unit instruction set.

fixed length instruction — One member of an instruction set in which all instructions contain the same number of bits.

flag — Single bit within a larger data unit or the program status word that represents one data item or condition.

flash memory — Type of electronically erasable read-only memory that requires relatively little time to update memory contents.

flash RAM — Synonym of flash memory.

flat memory model — Memory organization and access method in which memory locations are described by single unsigned integers that corresponds to linear position.

flat panel display — Display device that is thin, flat, and lightweight.

floating point notation — Method of encoding real numbers in a bit string consisting of two parts—a mantissa and exponent.

floppy disk — Small, removable magnetic disk storage medium encased in a protective cover.

FM — Acronym of frequency modulation.

FMS — Acronym of file management system.

font — Named set of display formats for printable characters and symbols with a similar appearance or style.

formula — Processing problem that can be solved by executing a fixed sequence of instructions.

fourth-generation language (4GL) — Programming language that supports nonprocedural programming, database manipulation, and advanced I/O capabilities.

fragmented — File or storage device characteristic in which noncontiguous storage locations are allocated to one or more files.

frequency — Number of complete waveform transitions (change from positive to negative to positive energy peak) that occur in one second.

frequency division multiplexing (FDM) — Communication channel-sharing technique in which a single broadband channel is partitioned into multiple baseband subchannels (frequency bands), each of which can carry a separate data stream.

frequency modulation (FM) — Data transmission method that encodes data values as variations in carrier wave frequency.

frequency shift keying (FSK) — Synonym of frequency modulation.

front-end CASE tool — Program that supports creating and modifying system models.

front-end processor — Synonym of I/O channel, particularly as applied to minicomputers and mainframes not manufactured by IBM.

FSK — Acronym of frequency shift keying.

FTP — Acronym of File Transfer Protocol.

full backup — Data store containing copies of all data in a directory or storage device.

full-duplex — Type of communication channel composed of two transmission media that enables simultaneous communication in both directions.

fully qualified reference — Synonym of complete path with respect to a specific file or object.

function — Named segment of a high-level language program that is always executed as a unit.

gallium arsenide — Compound with both electrical and optical properties.

gate — Processing device that implements a primitive Boolean operation or processing function by physically transforming one or more input signals.

gateway — Computer system or hardware device that is connected to two or more network segments and forwards packets among the segments.

general-purpose processor — Processor that can be directed to perform a wide variety of specific tasks.

GHz — Abbreviation of gigahertz.

GIF — Acronym of Graphics Interchange Format.

Gigabit Ethernet — One gigabit per second Ethernet standard based on the 802.3z and 802.3ab standards.

gigahertz (GHz) — Unit of wave or clock frequency, one billion cycles per second.

---

grandfather — Version of a file prior to the father version.

graph directory structure — (1) Directory structure in which allowable relationships between directories can be represented as a graph. (2) Directory structure in which a file or subdirectory can be contained within two or more directories.

graphics accelerator — High-performance video controller with an embedded processor and software that converts incoming image description language commands into appropriate input to a video display.

Graphics Interchange Format (GIF) — Standard format for storing compressed images.

grayscale — Describes a display device or image encoding method that can display or represent, black, white, and shades of gray, but no other colors.

grid — Group of dissimilar computer systems, connected by a high-speed network, that cooperate to provide services or execute a common application.

Grosch's Law — Outdated mathematical relationship between computer size and cost per unit of instruction execution that states that cost per executed instruction decreases as computer system size increases.

half-duplex — Type of communication channel that uses only a single transmission medium where sender and receiver take turns transmitting data to one another.

half-toning — Simulating shades of gray with dithering using only two colors—black and white.

HALT — Instruction that suspends the execution of any further instructions by the central processing unit.

hard disk — (1) Storage medium consisting of a rigid platter coated with a metallic oxide on which data are recorded as patterns of magnetic charge. (2) Synonym of hard disk drive.

hardware independence — Independence of a program or processing method from the physical details of computer system hardware.

hardware monitor — Program or device that records and reports processing or communication activity within or between hardware devices.

HCA — Acronym of host channel adapter.

head-to-head switching time — Elapsed time required to switch shared read/write circuitry between two adjacent read/write heads.

heat dissipation — Conducting heat away from a device, thus reducing its temperature.

heat sink — Thermal mass placed in direct contact with a processor or other device to increase heat dissipation.

helical scanning — Tape recording method in which data is read and written by rotating the read/write head at an angle, moving from tape edge to tape edge.

hertz (Hz) — Unit of measure for signal frequency defined as one cycle per second.

hexadecimal notation — Numbering system with a base value of 16, which uses digit values ranging from 0 to 9 and from A to F (corresponding to decimal values 0 to 15).

hierarchical directory structure — Multilevel system of directories in which directories and files may be related to one another in a hierarchy or inverted tree.

high order bit — Synonym of most significant digit

hit ratio — Ratio of cache hits to read accesses.

home directory — Primary directory associated with and owned by a single user.

host channel adapter (HCA) — Interface or controller that connects a device that can initiate or respond to data transfer commands to an Infiniband or Fibre Channel network.

HTML — Acronym of Hypertext Markup Language.

HTTP — Acronym of Hypertext Transport Protocol.

HTTPS — Acronym of Hypertext Transport Protocol Secure.

hub — Network interface device that serves as the network connection point for multiple computer systems.

Hypertext Markup Language (HTML) — Device-independent formatting language that describes Web documents and in which links to other documents can be embedded.

Hypertext Transport Protocol (HTTP) — Protocol by which clients request HTML or XML documents and how servers respond to those requests.

Hypertext Transport Protocol Secure (HTTPS) — Encrypted version of the HTTP protocol.

Hz — Abbreviation of hertz.

H.323 — Oldest and most widely deployed VoIP protocol suite.

IA5 — Acronym of International Alphabet Number 5.

IC — Acronym of integrated circuit.

ICMP — Acronym of Internet Control Message Protocol.

IDL — Acronym of image description language.

IEEE — Acronym of Institute for Electrical and Electronics Engineers.

IEEE 802 standards — Standard family for voice, data, and computer networks.

IIOP — Acronym of Internet Inter-ORB Protocol.

image description language — Language that symbolically describes the content of a printed or displayed image.

IMAP — Acronym of Internet Message Access Protocol.

implementation discipline — Activities of the Unified Process that build, acquire, and integrate application software components.

inclusive OR — Synonym of OR operation.

incremental backup — Data store containing copies of data sets (files) that have been altered since the most recent incremental or full backup.

index — Stored set of paired data items. Each pair contains a key value and a pointer to the location of the data item possessing (or corresponding to) that key value.

indirect addressing — Any addressing method where a program's memory address operands do not necessarily correspond to physical memory storage locations.

Infiniband — Standard for high-speed interconnect of computer networks, servers, and storage devices.

information architecture — Set of requirements and constraints that define important characteristics of information processing resources and how those resources interact with one another.

infrastructure — Facilities or resources that provide services that are pervasively available and commonly used by a large numbers of users.

ink-jet printer — Printer that produces printed images by placing small drops of liquid ink onto paper.

input output (I/O) unit — Device that enables communication between a user and software or among computer systems or their components.

input pad — Input device that converts pressure into usable computer input, such as for capturing written signatures or drawings.

Institute for Electrical and Electronics Engineers (IEEE) Computer Society — Subgroup of the IEEE that specializes in computer and data communication technology.

instruction — Bit string containing an operation code and zero or more operands that cause a processor to perform a processing operation.

instruction cycle — Synonym of fetch cycle.

instruction explosion — Correspondence of a single high-level programming language statement to multiple central processing unit instructions.

instruction format — Length and order of the operation code and operands within a machine language instruction.

instruction pointer — Register that stores the address of the next instruction to be fetched from primary storage.

instruction register — Register that holds an instruction prior to decoding.

instruction set — Set of all machine language instructions that can be executed by a central processing unit.

integer — Whole number, or a value that does not have a fractional part.

integrated circuit (IC) — Semiconductor device, manufactured as a single unit, that incorporates multiple gates.

interleaved execution — Scheduling technique by which a processor allocates CPU cycles to multiple active processes in very small time slices.

International Alphabet Number 5 (IA5) — International equivalent of ASCII.

International Organization for Standardization (ISO) — International body with functions similar to those of the American National Standards Institute.

Internet — Global collection of networks interconnected using TCP/IP.

Internet Control Message Protocol (ICMP) — Internet protocol for message exchange among gateways and routers.

Internet Inter-ORB Protocol (IIOP) — Component message passing protocol in CORBA.

Internet Message Access Protocol version 4 (IMAP4) — Protocol that extends Post Office Protocol 3 to enable management and storage of e-mail messages on a server.

Internet Protocol (IP) — Internet protocol for packet switching and routing.

interpretation — Process for translating and executing source code instructions that interleaves instruction translation and execution.

interpreter — Program that performs interpretation.

interrupt — Signal to the CPU that some event requires its attention.

interrupt code — Numerically coded value of an interrupt, indicating the type of event that has occurred.

interrupt handler — Program or subroutine that is executed in response to an interrupt.

interrupt register — Register in the control unit that stores an interrupt code received over the bus or generated by the central processing unit.

intranet — Internal, private network that uses Internet protocols but is only accessible to a limited set of internal users.

I/O — Acronym of input/output.

I/O channel — (1) Device controller dedicated to a mainframe bus port that enables many devices to share access to (and the capacity of) the port. (2) Specific hardware component of an IBM mainframe computer system.

I/O port — (1) Communication pathway from the CPU to a peripheral device. (2) Memory address, or a set of contiguous memory addresses, that can be read or written by the CPU and a single peripheral device.

I/O wait state — Idle processor cycle consumed while waiting for data transmission from a secondary storage or input/output (I/O) device.

IP — Acronym of Internet Protocol.

IPv6 — Internet Protocol standard version 6 that uses 128-bit addresses.

ISO — Acronym of International Standards Organization.

iteration — Four to six week period of a system development project that produces a specific set of deliverables including a prototype or the final system.

J2EE — Acronym of Java 2 Enterprise Edition.

Java 2 Enterprise Edition (J2EE) — Family of standards for developing and deploying component-based distributed applications written in Java.

Java — Object-oriented programming language used to construct programs that can be executed on different hardware and system software platforms.

Java server page (JSP) — Server-side components that generate formatted Web pages using embedded scripts.

Java servlet — Server-side Java programs that can perform complex operations and interact with EJBs and databases.

Java Virtual Machine (JVM) — Interpreter that translates and executes Java programs.

JCL — Acronym of job control language.

job control language (JCL) — Control language used to direct the batch execution of programs or groups of programs.

Joint Photographic Experts Group (JPEG) — (1) Organization that promulgates image representation standards. (2) Standard method of compressed image representation for single images.

journaling — Synonym of transaction logging.

JPEG — Acronym of Joint Photographic Experts Group.

JSP — Acronym of Java server page.

JUMP — Synonym of BRANCH operation.

JVM — Acronym of Java Virtual Machine.

keyboard controller — Microprocessor, integrated within a keyboard, that converts keystrokes to coded computer inputs or scan codes.

Kerberos — Security model that defines standard interactions between clients, services, and a security service.

kernel — Portion of the operating system that manages resources and directly interacts with computer hardware.

L1 — Acronym of level one.

L2 — Acronym of level two.

L3 — Acronym of level three.

label — Mnemonic that represents the memory address of a program instruction.

LAN — Acronym of local area network.

large format printer — A printer, typically using inkjet technology, that prints large images such a posters and blueprints.

laser printer — Printer that uses a laser to charge areas of a photoconductive drum and paper.

late binding — Synonym of dynamic linking.

Latin-1 — Standard character coding table containing the ASCII character set in the first 128 table entries and most of the additional characters used by Western European languages in the second 128 table entries.

LCD — Acronym of liquid crystal display.

LDAP — Acronym of Lightweight Directory Access Protocol.

least significant byte — Byte within a multiple byte data item that contains the digits of least, or smallest, magnitude.

least significant digit — Bit position within a bit string that represents the least, or smallest, magnitude.

level one (L1) cache — Within-CPU cache when all three cache types are used.

level two (L2) cache — Primary storage cache implemented within the same chip as a processor.

level three (L3) cache — When three levels of primary storage cache are used, the cache implemented outside the microprocessor.

Lightweight Directory Access Protocol (LDAP) — Internet standard formed from the X.500 standard and adopted by the Internet Engineering Task Force.

line turnaround — Signal sent between a sender and receiver that causes them to reverse roles (the sender becomes the receiver and vice versa).

linear address space — Logical view of a secondary storage or I/O device as a sequentially numbered set of storage locations.

linear recording — Recording of data onto a tape in which bits are placed along parallel tracks that run along the entire length of the tape.

Linear Tape Open (LTO) — Proprietary magnetic tape standard developed by Hewlett-Packard, IBM, and Seagate.

link — (1) A type of connection between a directory and another directory or file used to implement a graph directory structure. (2) A pointer that connects two data items within a data structure. (3) An external function call within an object code file.

link editor — Program that combines multiple object code modules into an integrated set of executable code with a consistent scheme of memory addresses and references.

link map — Synonym of memory map.

linked list — Data structure in which each data item contains a pointer to the previous or next data item.

liquid crystal display (LCD) — Display device that uses liquid crystals that can be changed from transparent to opaque.

little endian — Describes a CPU architecture that stores a multibyte data item with the least significant byte in the storage location with the lowest-numbered address.

load — Copying a word from primary storage to a register.

local area network (LAN) — Network that spans a limited area such as a single building or office floor.

location transparency — Characteristic of software such that resource access is implemented identically whether the resource is part of the local computer system or located within or controlled by a remote computer system.

logic instruction — Instruction that implements a boolean operation such as ADD, AND, OR, NOR, or XOR.

logical access — Access to a storage location within a linear address space.

logical record — Record formatted and described in terms of a logical (rather than physical) file structure.

logical SHIFT — Shift operation used to extract a single bit from a bit string.

logical topology — Path that messages traverse as they travel among network nodes.

long integer — Double precision representation of an integer.

longitudinal redundancy checking (LRC) — Parity checking method in which parity bits are determined based on equivalent bit positions in a group of bytes or characters.

lossless compression — Compression algorithm in which data content is unchanged when compressed and then decompressed.

lossy compression — Compression algorithm in which data content is altered or lost when compressed and then decompressed.

low order bit — Synonym of least significant digit.

LRC — Acronym of longitudinal redundancy check.

LTO — Acronym of Linear Tape Open.

MAC — Acronym of media access control.

machine data type — Synonym of primitive data type.

machine independence — Synonym of hardware independence.

machine language — Language consisting solely of instructions from a specific central processing unit instruction set.

machine state — Current processing state of a program or the central processing unit, as represented by the values currently held in registers.

magnetic decay — Loss in strength of a stored magnetic charge over time.

magnetic leakage — Reduction in strength of a stored magnetic charge because of interference from one or more adjacent magnetic charges of opposite polarity.

magnetic tape — Polymer ribbon coated with a metallic compound used to store data.

magneto-optical — Secondary storage device that reads and writes data bits using a combination of magnetic and optical methods.

main memory — (1) Synonym of primary storage. (2) Set of devices that implement primary storage excluding cache.

mainframe — High-capacity computer system designed to support hundreds of interactive users and processes simultaneously.

Mammoth — Magnetic tape standard developed by Exabyte based on the Digital Audio Tape standard.

mark sensor — Input device that recognizes printed marks at predetermined locations on an input document.

media access control (MAC) — Set of rules that regulate access to a transmission medium.

megahertz (MHz) — Measurement of wave or clock frequency, one million cycles per second.

memory — (1) Synonym of primary storage. (2) Random access memory devices used anywhere within a computer system, peripheral device, or device controller.

memory allocation — Allocation of primary storage resources to active processes.

memory map — List of module and data names and memory addresses produced by a link editor.

message — (1) With respect to communication networks, a command, request, or response sent from one network node to another. (2) With respect to components and object-oriented programs, a request sent from one object or component to another.

method — Program within a class that manipulates data.

MFLOPS — Acronym of millions of floating point operations per second.

MHz — Abbreviation of megahertz.

microchip — Semiconductor device that implements integrated electronic components in a single unit.

microcomputer — Computer system that can meet low-intensity processing needs of a single user.

microprocessor — Microchip containing all of the components of a central processing unit.

middleware — System software that enables clients and servers or distributed components to locate and communicate with one another.

midrange computer — Computer system that meets the processing needs of a small- to medium-sized group of interactive users.

MIDI — Acronym of Musical Instrument Digital Interface.

millions of floating point operations per second (MFLOPS) — Measure of processor or computer system speed in terms of the number of floating point computation operations executed per second.

millions of instructions per second (MIPS) — Measure of processor or computer system speed in terms of the number of central processing unit instructions executed per second.

MIME — Acronym of Multipurpose Internet Mail Extensions.

minicomputer — Synonym of midrange computer.

MIPS — Acronym of millions of instructions per second.

modem — Acronym of MOdulator-DEModulator.

modulation — Varying the amplitude, frequency, or phase of a carrier wave.

modulator-demodulator (modem) — Device that translates analog signals into digital signals (and vice versa), enabling computer hardware to communicate over voice-grade telephone lines.

monitor — (1) Video display device. (2) Hardware or software element that monitors and reports processing or communication activity.

monochrome — Describes a display device or encoding method that can display or represent only two colors, typically, black and white.

monophonic — Capable of generating only one audible frequency (note) or timbre (voice) at a time.

Moore's Law — Statement that the transistor density and power of microprocessors doubles every 18 to 24 months with no per unit cost increase.

most significant byte — Byte within a multiple byte data item that contains the digits of most, or largest, magnitude.

most significant digit — Bit position within a bit string that represents the most (largest) magnitude.

mouse — Interactive pointing device used to control cursor position and to select objects on a video display.

MOVE — Instruction that can copy data bits among any combination of registers and primary storage locations.

Moving Pictures Experts Group (MPEG) — (1) Organization that promulgates motion video image representation standards. (2) Method of compressed motion video-image representation.

MP3 — (1) Audio encoding and compression standard. (2) Layer 3 of the MPEG-1 standard.

MPEG — Acronym of Moving Pictures Experts Group.

multicore architecture — Microprocessor architecture that embeds multiple CPUs and cache memory on a single chip.

multilevel coding — Describes any method that encodes multiple bits within each instance of a modulated carrier wave or signal.

multimedia controller — Device controller for multiple audio or visual I/O devices.

multimode cable — Fiber-optic cable type that surrounds plastic or glass fibers with a reflective coating to contain light.

multinational character — Character, such as ñ or é, similar to an English language character but used by Western European languages other than English.

multiprocessing — Any CPU architecture in which duplicate CPUs or processor stages can execute in parallel.

Multipurpose Internet Mail Extensions (MIME) — Protocol that enables non-text content to be included in e-mail messages and other Internet data transmissions.

multitasking — Ability of an operating system to support multiple active processes.

multithreaded — Process divided into two or more threads, each of which may be scheduled and executed independently.

multi-CPU architecture — Computer system architecture that employs two or more CPUs on a single motherboard or set of interconnected motherboards.

Musical Instrument Digital Interface (MIDI) — Standard method of encoding control information for musical instruments and synthesizers.

n-layer architecture — Client-server architecture that uses more than three layers.

n-tiered architecture — Synonym of n-layered architecture.

NAK — Mnemonic of the negative acknowledge ASCII control character.

named pipe — Pipe with a name permanently placed in a file system directory that enables communication among processes executing on different computers.

NAS — Acronym of network-attached storage.

native application — Application compiled and linked for a particular central processing unit and operating system.

negative acknowledge (NAK) — ASCII control character sent by a receiver to indicate unsuccessful receipt of transmitted data.

network administrator — Manager charged with operating and maintaining a local or wide area network.

network computer — Personal computer or workstation with no locally stored operating software, application software, or configuration information.

network interface card (NIC) — Synonym of network interface unit (NIU), particularly with respect to a microcomputer, network computer, or workstation.

network interface unit (NIU) — Physical network interface device used by a single computer system.

network layer — Layer of network system software that routes packets to their destination.

network topology — Structure defined by the organization of network devices, routing of network cabling, and flow of messages through network nodes.

network transparency — Synonym of location transparency.

network-attached storage (NAS) — Architecture in which a dedicated storage server is attached to a general-purpose network to service storage access requests from other servers.

NIC — Acronym of network interface card.

NIU — Acronym of network interface unit.

noise — Combination of electromagnetic interference and distortion that modifies a carrier wave or signal during transmission.

noncontiguous — Any storage allocation scheme under which multiple storage locations allocated to a file or program need not be sequential nor physically adjacent to one another.

nonprocedural language — Programming language that describes a processing requirement without describing a specific procedure for satisfying the requirement.

nonvolatile — Term describing storage devices that retain their contents over long periods of time.

nonvolatile memory (NVM) — Memory with long-term or permanent data retention.

NOT — (1) Boolean operation that generates the value false if its input is true, and true if its input is false. (2) Central processing unit instruction that implements the boolean NOT operation and stores the result in a register or memory.

numeric range — Set of all data values that can be represented by a specific data-encoding method.

NVM — Acronym of nonvolatile memory.

object — One instance, or variable, of a class.

object code — Output of an assembler or compiler that contains machine instructions and unresolved references to external library routines.

object-oriented programming (OOP) — Software construction method or paradigm that configures software as cooperating objects.

Object Request Broker (ORB) — CORBA service that maintains a directory of components and routes messages among them.

objectclass — A Lightweight Directory Access Protocol (LDAP) concept that defines the attributes common to all members of a class.

OCR — Acronym of optical character recognition.

octal notation — Base 8 numbering system that uses digit values ranging from 0 to 7.

odd parity — Error-detection method that appends a parity bit to each transmitted or stored character. The value of the parity bit is zero if the number of one-valued bits in a character is odd, and one otherwise.

offset register — Register that holds a memory address offset, which is added to each explicit memory reference.

on-off keying (OOK) — Synonym of pulse code modulation.

OOK — Acronym of on-off keying.

OOP — Acronym of object-oriented programming.

op code — Synonym of operation code.

Open Systems Interconnection (OSI) model — Standardized seven layer architecture for computer network software and hardware.

operand — Component field of an instruction containing a data value or address used as input to or output from a specific processing operation.

operating expenditure — Funds expended during the current operating period to support normal operations.

operating system — Set of software programs that manage and control access to computer resources.

operation code — Coded value or bit string representing the function or operation to be performed by a processor.

optical character recognition (OCR) — Methods, programs, and devices by which printed characters are recognized as computer system input or data.

optical scanner — Device that can scan printed graphic inputs and convert them to computer system input or data.

OR — (1) Boolean operation that generates the value true if either or both of its inputs are true. (2) Central processing unit instruction that generates the value true if either or both of its inputs are true.

ORB — Acronym of Object Request Broker.

OSI — Acronym of Open Systems Interconnection.

overflow — Error condition that occurs when the output bit string of a processing operation is too large to fit in the designated register.

packet — Fundamental unit of data communication in a computer network.

packet switching — Form of time division multiplexing in which multiple message streams are broken into packets and individually transmitted and routed to their destination.

packet-filtering firewall — Firewall that examines each packet and matches header content to a list of allowed or denied packet types.

page — Small fixed-size portion of a program, swapped between primary and secondary storage under virtual memory management.

page fault — Condition that occurs under virtual memory management when a program references a memory location in a page not currently held in primary storage.

page frame — Portion of primary storage designated to hold a page under virtual memory management.

page file — Synonym of swap space or swap file.

page hit — Reference to a page held in memory under virtual memory management.

page table — Table of pages and information about them maintained by an operating system that implements virtual memory management.

palette — Set or table of colors used by a video display device or controller.

parallel access — Simultaneous access to multiple portions of a data item through multiple communication channels or transmission media.

parallel transmission — Simultaneous transmission of multiple parts of a single message over multiple transmission media or channels.

parent process — Process that initiates and controls the execution of another (child) process.

parity bit — Bit appended to a small data unit, such as an ASCII character, that stores redundant information used for error checking. *See* even parity and odd parity.

parity checking — Act of validating data by recomputing the value of a parity bit.

passive matrix display — Liquid crystal display that shares transistors among rows and columns of pixels.

PC — Acronym of personal computer.

PCB — Acronym of process control block.

PCM — Acronym of pulse code modulation.

PDF – Acronym of Portable Document Format.

peer-to-peer bus — Bus in which more than one attached device may control access to the bus.

peer-to-peer communication protocol — Protocol that enables processes to communicate synchronously across a network.

peripheral device — Device on a system bus other than the central processing unit and primary storage.

peripheral processing unit — Synonym of I/O channel, particularly as applied to minicomputers and mainframes not manufactured by IBM.

personal computer (PC) — Synonym of microcomputer.

phase — Characteristic of an analog wave that indicates its current cycle position (in degrees) with respect to the cycle origin.

phase shift keying (PSK) — Synonym of phase shift modulation.

phase shift modulation — Data transmission method that encodes data values as variations in carrier wave phase.

phoneme — Individual vocal sound comprising a primitive component of human speech.

photosensor — Device that generates an electrical signal when light is shone upon it.

physical layer — Layer of network system software responsible for physical transmission of signals representing data.

physical memory — Physical primary storage capacity of a computer system, in contrast to virtual memory capacity and addressable memory.

physical record — Unit of physical data transfer to or from a storage device.

physical topology — Physical placement of cables and connections in a network topology.

pipe — Region of shared memory through which multiple processes executing on the same machine can exchange data.

pipelining — Method of organizing CPU circuitry to enable multiple instructions to be in different stages of execution at the same time.

pixel — (1) Abbreviation of the term "picture element." (2) Single unit of data in a graphic image. (3) Single point on a display surface.

plasma display — Display that generates light by applying an electrical charge to neon gas.

platter — One disk within a disk storage device, the surface or surfaces of which store data.

plotter — (1) Device that generates printed output through the movement of paper and one or more pens. (2) An ink jet printer that can print wide sheets or rolls of paper. (3) Synonym of large format printer.

point — (1) One seventy-second of an inch. (2) Measurement unit for font size.

pointer — (1) Data element that contains the address (location in a storage device) of another data element. (2) Device used to input positional data or control the location of a cursor.

polymer memory — Type of nonvolatile memory that stores bit values in plastic with electrical resistance that can be increased or decreased by an electrical field.

polyphonic — Capable of generating many audible frequencies (notes) or timbres (voices) simultaneously.

pop — Process of removing an item from the top of a stack and copying its content to processor registers.

POP3 — Acronym of Post Office Protocol version 3.

port — Physical connection point on a bus, communication channel, or input/output device identified by a unique integer number.

Portable Document Format (PDF) — An image description language (a superset of PostScript) developed by Adobe Systems Incorporated.

Post Office Protocol version 3 (POP3) — Protocol that standardizes the interaction between e-mail clients and servers.

PostScript — Language for representing and transmitting complex images and display contents.

power sag — Momentary reduction in the voltage or amperage of electrical power.

power surge — Momentary increase in the voltage or amperage of electrical power.

preemptive scheduling — Scheduling method that enables a higher-priority process to interrupt and suspend a lower-priority process.

presentation layer — Layer of network system software responsible for input and output to network nodes.

primary storage — High-speed storage within a computer system, accessed directly by the central processing unit, used to hold currently active programs and data immediately needed by those programs.

primitive data type — Integer, real number, character, Boolean, memory address, and double precision data types supported by a central processing unit.

priority-based scheduling — Scheduling method that determines which ready thread or process should be dispatched to the central processing unit based on user or process priority.

procedure — Group of instructions that are always executed as a unit (similar to a function or subroutine).

process — (1) Program or program fragment that is separately managed and scheduled by the operating system. (2) To transform input data by the application of processing operations. (3) Representation of data or information processing on a data flow diagram.

process control block (PCB) — Data structure that contains information about a currently active process.

process family — Parent process and all its descendants.

process list — Synonym of process queue.

process queue — Data structure containing process control blocks for all current processes.

processor — Any device capable of performing data transformation operations.

production version — Program that contains no debugging information and has been optimized for minimal consumption of hardware resources.

program — Sequence of processing instructions.

program counter — Synonym of instruction pointer.

program editor — Software that assists programmers in creating valid source code.

program profiler — Software utility that monitors and reports the activities and resource utilization of another program during execution.

program status word (PSW) — Bit string held in a control unit register that stores status information (flags) in each bit.

program translator — Program that translates instructions in one programming language or instruction set into equivalent instructions in another programming language or instruction set. *See* assembler, compiler, and interpreter.

programmer — Person who creates or maintains programs.

programmer's workbench — Integrated set of programs that support program creation, editing, translation, and debugging.

programming language — Any language in which computer-processing functions or instructions can be expressed.

protocol stack — Set of software programs, services, or device drivers, each of which implements one protocol in a layered set of protocols.

PSK — Acronym of phase shift keying.

PSW — Acronym of program status word.

pulse code modulation (PCM) — Signal coding method under which bit values are represented as bursts of light or electrical voltage.

push — Process of copying register values to the top of a stack.

quality of service — Guaranteed minimum data transfer throughput between a sender and receiver.

quantum measurement problem — Behavior where the multiple states of a qubit are disturbed by any attempt to measure their value.

QIC — Acronym of Quarter Inch Committee.

Quarter Inch Committee (QIC) — Committee that promulgates standards for cartridge magnetic tapes.

qubit — Data stored in a single quantum particle.

radio frequency (RF) — Electromagnetic radiation propagated through space.

radix — Base of the numbering system, such as 2 for the binary numbering system and 10 for the decimal numbering system.

radix point — Period or comma that separates the whole and fractional parts of a numeric value.

RAID — Acronym of Redundant Array of Inexpensive Disks.

RAM — Acronym of random access memory.

random access — Ability of a storage device to access storage locations directly, or in any desired order.

random access memory (RAM) — (1) Generic description of semiconductor devices used to implement primary storage. (2) Device used to implement primary storage that provides direct access to stored data.

raw data transfer rate — Number of bits per time interval that can be sent through a communication channel.

read-only memory (ROM) — Primary storage device that can be read but not written.

read/write head — Mechanism within a storage device that reads and writes data to/from the storage medium.

read/write mechanism — Synonym of read/write head.

ready state — State of an active process or thread that is waiting only for access to the central processing unit.

real number — Number that can contain both whole and fractional components.

real resource — Hardware or software resource that physically exists.

real-time scheduling — Any scheduling method that guarantees the complete execution of a program or program cycle within a stated time interval.

record — (1) Data structure composed of data items relating to a single entity such as a person or transaction. (2) Unit of data transfer. (3) Primary component data structure of a file.

recording density — Synonym of areal density.

Reduced Instruction Set Computing (RISC) — (1) Use of a small set of simple central processing unit instructions that cannot be decomposed into other processor instructions. (2) Processor architecture that emphasizes short, primitive instructions that reference operands contained in registers, in contrast to complex instruction set computing.

Redundant Array of Inexpensive Disks (RAID) — Set of techniques for improving storage device performance or ensuring data integrity using multiple disk storage devices.

refresh cycle — (1) Period during a dynamic random access memory refresh operation when the storage device is unable to respond to a read or write request. (2) Transfer of one full screen of data to a video monitor.

refresh rate — Number of times per second a refresh operation occurs on a video display device.

register — High-speed storage location within a central processing unit that can hold a single word.

relative addressing — Synonym of indirect addressing.

relative path — Path that begins at the level of the current directory.

remote procedure call (RPC) — Protocol that enables a process on one computer to call a process on another computer.

request for proposal (RFP) — Formal document stating hardware or software requirements and requesting proposals from vendors to meet those requirements.

requirements discipline — Activities of the Unified Process that develop models of system requirements.

resistance — Phenomenon in energy wave transmission where energy is converted from one form to another, for example, electrical energy to heat.

resolution — Number of pixels displayed per linear measurement unit.

resource availability model — A model of the computing resources that can be provided by a given combination of computer hardware and system software.

resource registry — Database of services and resources available within a local or wide area network.

return — Instruction executed at the end of a function, subroutine, or procedure to return control to the calling function, subroutine, or procedure.

return wire — Transmission line that completes an electrical circuit between the sending and the receiving device.

RF — Acronym of radio frequency.

RFP — Acronym of request for proposal.

RGB — Acronym of red, green, and blue.

ring topology — Network configuration in which each network node is connected to two other network nodes, with the entire set of connections and nodes forming a closed ring.

RIP — Acronym of Routing Information Protocol.

RISC — Acronym of Reduced Instruction Set Computing.

Rock's Law — Observation that states that the cost of fabrication facilities for the latest generation of semiconductor devices doubles every four years.

ROM — Acronym of read-only memory.

rotational delay — Waiting time for the desired sector of a disk to rotate beneath a read/write head.

router — Device that examines packet destination addresses and forwards them to another network segment, if necessary.

Routing Information Protocol (RIP) — Internet protocol for the exchange of routing information among network nodes.

routing table — Table of network node addresses and transmission lines or connection ports.

RPC — Acronym of remote procedure call.

run queue — Synonym of thread list.

running state — State of a process or thread that currently is executing within the central processing unit.

sampling — (1) Process of measuring and digitally encoding one or more parameters of an analog signal at regular time intervals. (2) Process of digitally encoding analog sound waves.

SAN — Acronym of storage area network.

sandbox — Protected area of a program within which a Java applet is executed.

SAS — Acronym of serial-attached SCSI.

SATA — Acronym of serial ATA.

scaling up — Increasing available processing and other computer system power by using ever larger and more powerful computers.

scaling out — Increasing available processing and other computer system power by partitioning processing and other tasks among multiple computer systems.

scan code — Coded output generated by a keyboard for interpretation by a processor or keyboard controller.

scanning laser — Laser that is automatically swept back and forth over a predefined viewing area, as used in devices such as bar code readers.

scheduler — Program within the operating system that controls process states and access to the CPU.

scheduling — Process of determining and implementing process priorities for access to hardware resources.

scripting language — (1) Simple programming language that can be embedded within HTML pages. (2) Stored set of operating system commands or a job control language program.

SCSI — Acronym of Small Computer System Interface.

SDLC — Acronym of systems development life cycle.

SDRAM — Acronym of synchronous dynamic random access memory.

second-generation language (2GL) — Synonym of assembly language.

secondary storage — Set of computer system devices that provide large-capacity and long-term data storage.

sector — (1) Smallest accessible unit of a disk drive. (2) For most disk drives, 512 bytes.

secure shell (SSH) — Modern variant of Telnet that encrypts data flowing between client and server.

segment register — Register that holds the base address of a memory segment, as used in a central processing unit that uses a segmented memory model.

segmented memory model — Memory allocation and partitioning based on equal sized segments and segmented memory addresses.

semiconductor — Material with resistance properties that can be tailored between those of a conductor and an insulator by adding chemical impurities.

sequential access time — Time required to access the second of a stored sequential pair of data items.

serial access — Access technique where data items are read and written in an order

corresponding to their position within allocated storage locations.

serial ATA (SATA) — Storage device and cabling standard, compatible with older parallel ATA standards, but using serial transmission.

serial transmission — Transmission method in which individual bits are transmitted sequentially over a single communication channel.

serial-attached SCSI (SAS) — Storage device and cabling standard, compatible with older parallel SCSI standards, but using serial transmission.

server — Computer system or process that manages hardware or software resources and makes those resources available to clients.

service call — Request for an operating system service by an application program.

service layer — Portion of the operating system that accepts service calls and translates them into low-level requests to the kernel.

service-oriented resource access — Operating system architectural or behavioral feature in which all resources are managed and accessed through server processes.

service standard — Standard for interacting with a shared resource.

session layer — Layer of network system software responsible for establishing and managing communication sessions.

shell — User interface (or command layer) of an operating system.

SHIFT — Operation whereby bits of a bit string are moved left or right by a stated number of positions, empty positions are filled with zeros, and bit values that shift beyond the bounds of the bit string are discarded.

shortcut — Synonym of link, as implemented within Microsoft Windows for graph directory structures.

shortest time remaining — Scheduling method that assigns highest priority to processes or requests with shortest time remaining until completion.

sibling process — Another child process created, or spawned, by the same parent.

---

signal — (1) Specific data transmission event or group of events that represents a bit or group of bits. (2) Message sent from one active process to another.

signal-to-noise (S/N) ratio — Mathematical relationship between the power of a carrier signal and the power of the noise in the communication channel, measured in decibels.

signal wire — Component of an electrical communication channel (circuit) used to transmit a carrier wave or signal.

SIMM — Acronym of single in-line memory module.

Simple Mail Transport Protocol (SMTP) — Earliest e-mail protocol that defines how text messages are routed through the Internet.

Simple Object Access Protocol — Standard for distributed object interaction.

simplex — Type of communication channel that enables transmission in only one direction.

sine wave — Waveform that varies continually between positive and negative states.

single in-line memory module (SIMM) — Small printed circuit board with memory chips on one or both sides and electrical contacts on one edge.

single-mode cable — Highest quality optical cable constructed so that light waves travel down the center of the fiber.

singly linked list — Data structure in which each data item contains a pointer to the next data item.

skew — Timing differences in the arrival of signals transmitted simultaneously on parallel communication channels.

Small Computer System Interface (SCSI) — Family of standard buses designed primarily for connecting secondary storage devices.

SMTP — Acronym of Simple Mail Transport Protocol.

S/N — Acronym of signal-to-noise.

SOAP — Acronym of Simple Object Access Protocol.

socket — Internet standard for service identification that combines an Internet address with a service port number, such as 129.24.8.4:53.

software monitor — Program that monitors and reports the utilization of a software resource.

son — File containing the results of updating a master file (the father) based on the contents of a transaction file.

sound card — Hardware device containing components for sound input and output.

source code — Instructions or statements in a high-level programming language.

spawn — Creation of a child process by a parent process.

speaker dependent — Speech recognition system that recognizes input from only one user and that must be trained to accurately recognize that user's specific speech patterns.

special-purpose processor — (1) Processor with a limited set of processing functions. (2) Processor capable of executing only a single program.

speculative execution — Executing instructions after branch prediction but before the final branch condition value is known with certainty.

speech recognition — Process of recognizing human speech as computer system input.

speech synthesis — Process of generating human speech based on character or textual input.

SRAM — Acronym of static random access memory.

SSH — Acronym of secure shell.

stack — (1) List of data items maintained in last-in, first-out order. (2) Set of registers or memory locations used to store the register values of temporarily suspended processes.

stack overflow — Condition that occurs when an attempt is made to add data to a stack that is already at its maximum capacity.

stack pointer — Register containing the primary storage address of the last (most recently added) stack element.

star topology — Physical network connection pattern in which all nodes are attached to a central node.

start bit — Zero bits that are added to the beginning of each character in asynchronous serial data transmission.

stateful firewall — Firewall that tracks the progress of complex client-server interactions.

static connection — Mapping between a local resource name and a remote resource that does not change once initialized and that must be explicitly initialized prior to use.

static linking — Linking process in which library and other subroutines cannot be changed once they are inserted into the executable code.

static random access memory (SRAM) — Type of random access memory that implements bit storage with a flip-flop circuit.

storage allocation table — Table that correlates storage device allocation units with specific files, directories, or unallocated space.

storage area network (SAN) — High speed interconnection among general-purpose servers and a separate storage server.

storage input/output control layer — System software layer that controls input and output to and from storage devices.

storage medium — Device or substance within a storage device that physically stores data.

store — Operation in which register content is copied to a memory location.

store and forward — Interconnected system of end nodes and transfer points used to route data among end nodes.

strategic plan — Long-range plan stating the services to be provided by an organization and the means for obtaining and using needed resources.

string — Ordered set of related data elements, usually stored as a list or an array.

subroutine — Named segment of a high-level language program that is always executed as a unit.

subtractive colors — Cyan (absence of red), magenta (absence of green), and yellow (absence of blue).

supercomputer — Computer designed for very fast processing of real numbers, typically implemented with some form of parallel processing.

supervisor — Operating system program that serves as the master interrupt handler.

surge protector — Hardware device that controls electrical power surges.

sustained data transfer rate — Maximum data transfer rate that can be sustained by a device or a communication channel during lengthy data-transfer operations.

swap file — Synonym of swap space or page file.

swap space — Area of secondary storage used to hold virtual memory pages that cannot fit into primary storage.

switch — (1) Device that exists in one of two states and can be instructed to alternate between them. (2) Network node that quickly routes packets among network segments with switching technology.

symbol table — List of program modules and data names and their allocated primary storage locations.

symbolic debugger — Debugger that can report program execution progress and errors in terms of symbolic module and data names used within program source code.

synchronous dynamic random access memory (SDRAM) — Read-ahead random access memory that breaks read and write operations into a series of simple steps that can be completed in one bus clock cycle.

synchronous idle message — Control message sent continually between sender and receiver to keep their internal clocks synchronized.

synchronous transmission — Reliable high-speed method of data transmission in which data are sent in continuous data streams.

system administration — Managerial actions required to ensure efficient, reliable, and secure computer system operation.

system bus — Bus shared by most or all devices in a computer system.

system development tools — (1) Software tool designed to aid in creating groups of programs constituting an entire system. (2) Approximate synonym of computer assisted software engineering tool.

system overhead — Resource(s) consumed for resource allocation and translation functions.

system requirements model — Model that provides the detail necessary to develop a system that meets the needs the user requires.

system software — Software programs that perform hardware interface, resource management, or application support functions.

systems analyst — Person who performs business modeling and requirements discipline activities.

systems architecture — Structure, interaction, and technology of computer system components.

systems designer — Person who performs design discipline activities.

systems development life cycle (SDLC) — Process for developing an information system following a methodology or series of steps or activities such as the Unified Process.

systems programmer — Person who creates or maintains systems software.

systems survey — Process of initially ascertaining user processing requirements and determining the feasibility of meeting those requirements with computer-based solutions.

tape drive — Device that contains motors that wind and unwind tapes and read/write heads to access their content.

target channel adapter — Interface or controller that connects a device that can only respond to data transfer commands to an Infiniband or Fibre Channel network.

TCA — Acronym of target channel adapter.

TCB — Acronym of thread control block.

TCP — Acronym of Transmission Control Protocol.

TCP/IP — Acronym of Transmission Control Protocol/Internet Protocol.

TDM — Acronym of time division multiplexing.

Telnet — Protocol in which users on one Internet host can interact with another host's operating system command layer.

terminal — Synonym of video display terminal (VDT).

testing discipline — Activities of the Unified Process that verify correct functioning of infrastructure and application software components and ensure that they satisfy system requirements.

TFT – Acronym of thin-film transistor.

thin client — A synonym of network computer.

thin-film transistor — Transistor embedded within a device such as a display surface and constructed using semiconductor fabrication technology.

thread — Subcomponent of a process that can be independently scheduled and executed.

thread control block (TCB) — Data structure in which an operating system keeps track of thread-specific information.

thread cycle — Set of program instructions that can be executed repetitively.

thread list — Set of threads awaiting execution represented by a list of thread control blocks.

three-layer architecture — Client-server architecture composed of three layers: the view layer, the business logic layer, and the data layer.

three-tier architecture — Synonym of three-layer architecture.

third-generation language (3GL) — High-level programming language that does not possess advanced capabilities for interactive input/output, database processing, or nonprocedural programming.

time division multiplexing (TDM) — Technique that subdivides the capacity of a communication channel into discrete time slices, each of which may be allocated to a separate sender and receiver.

timer interrupt — Interrupt automatically generated by the central processing unit in response to the passage of a specific interval.

token — Control packet that regulates access to a communication channel.

token passing — Media access protocol that uses tokens passed among network nodes.

trace — Connection within a microchip or printed circuit board that enables electrons to travel from one place to another.

track — Set of sectors on one side of a disk platter that form a concentric circle.

track-to-track seek time — Time required to move a disk read/write head between two adjacent tracks.

transaction — (1) Event or system input that initiates the execution of one or more program segments. (2) Real-world event from which data input is derived, such as a sale. (3) A general term describing any change to stored data, such as the addition of a record to a file.

transaction logging — Method of file system update in which all changes to file and directory content also are written immediately to a separate log, or journal, which is used to recover lost data in the event of a system crash.

transistor — Solid state electrical switch that forms the basic component of most computer-processing circuitry.

Transmission Control Protocol (TCP) — Internet protocol for translating messages into packets and guaranteeing their delivery.

transmission medium — Physical communication path through which a carrier wave is propagated.

transport layer — Layer of network system software responsible for converting network messages into packets or other data structures suitable for transmission.

tree directory structure — Synonym of hierarchical directory structure.

truncation — Act of deleting bits that will not fit within a storage location.

twin-axial cable — Transmission medium composed of two wire strands surrounded by an insulator, a braided return wire, and a tough outer coating.

twisted pair wire — Transmission medium consisting of two electrical conductors (wires) twisted around one another.

two's complement — Notation system that represents positive integers as an ordinary bit string, and negative integers by adding one to the bit string that represents the absolute value.

Type I error — (1) Percentage of data items containing errors that will be incorrectly identified as error free. (2) Percentage of a sample or population that will be incorrectly identified as satisfying the null hypothesis.

Type II error — (1) Percentage of data items without errors that will be incorrectly identified as containing errors. (2) Percentage of a sample or population that will be incorrectly identified as not satisfying the null hypothesis.

unblocked — Storage of a single logical record within a physical record (block).

unconditional BRANCH — Synonym for BRANCH instruction.

undelete operation — Act of restoring a record or file by re-creating its index information (directory entry) and recovering its previously allocated storage locations.

underflow — (1) Condition that occurs when a value is too small to represent in floating point notation. (2) Overflow of a negative exponent in floating point notation.

Unicode — Standard 16-bit character coding method.

Unified Process (UP) — Iterative system development life cycle based on object-oriented techniques.

Uniform Resource Locator (URL) — Unique name that identifies a Web resource and is composed of a protocol, host, port, and resource.

uninterruptible power supply (UPS) — Device that provides electrical power when normal power inputs are interrupted.

unresolved reference — Synonym of external function call.

unsigned integer — Data type which stores positive integer values as ordinary binary numbers.

UP — Acronym of Unified Process.

UPS — Acronym of uninterruptible power supply.

URL — Acronym of Uniform Resource Locator.

utility program — Program that performs a commonly used function, such as printing a text file or copying a file, typically provided by the operating system either as separate tool or as part of the service layer.

variable — (1) Mnemonic that represents a data item memory address in assembly language. (2) Name that represents a data item memory address in a high-level programming language.

variable length instruction — One instruction from a set of central processing unit instruction formats that vary in length.

VDT — Acronym of video display terminal.

vector — (1) Line segment that has direction and length. (2) One-dimensional array.

vector list — List of line descriptions used for the input or output of graphic images.

versioning — Process that archives original file versions as they are modified.

vertical redundancy checking — Synonym of parity checking.

victim — Memory page chosen to be swapped to secondary storage under virtual memory management.

video controller — Device connected to the system bus and a video monitor that accepts input from the CPU or main memory and generates output appropriate to the video monitor.

video display terminal (VDT) — Input/output device comprising a keyboard, video display surface, and one or more communication ports.

video random access memory — Specific random access memory type used within a video controller (usually dual ported).

view layer — Division of the client-server architecture that accepts user input and formats and displays processing results.

virtual local area network — Two shorter local area networks joined by a bridge, also called a bridged local area network.

virtual memory management — Mode of operating system memory management in which secondary storage is used to extend the capacity of primary storage.

virtual resource — Resource visible to a user or program, but not necessarily available at all times or physically extant.

virus — Program or program fragment that infects a computer by permanently installing itself, performs one or more malicious acts on the infected computer, and replicates and spreads itself using services of the infected computer.

volatile — Storage devices that cannot retain their contents indefinitely.

voice over Internet Protocol (VoIP) — Family of technologies and standards that enables voice messages and data to be carried over a single packet-switched network.

VoIP – Acronym of voice over Internet Protocol.

volatility — (1) Rate of change in a set of data items. (2) Lack of ability to permanently store data content.

VRAM — Acronym of video random access memory.

wait state — Idle processor cycle consumed while waiting for a response from another device.

WAN — Acronym of wide area network.

wavelength-division multiplexing (WDM) — Form of frequency division multiplexing that simultaneously transmits multiple messages over a fiber-optic cable.

WDM — Acronym of wavelength-division multiplexing.

Web — Synonym of World Wide Web (WWW).

wide area network (WAN) — Network that spans large physical distances, such as multiple buildings, cities, regions, or continents.

wire — Connection that enables electrons to travel from one place to another.

word — Unit of data processed by a single central processing unit (CPU) instruction.

working directory — Synonym of current directory.

workload model — Synonym of application demand model.

workstation — (1) Powerful microcomputer designed to support demanding numerical or graphical processing tasks. (2) A synonym of personal computer or microcomputer.

World Wide Web (WWW) — Collection of resources accessed over the Internet by standard protocols such as FTP and HTTP.

WWW — Acronym of World Wide Web.

XML — Acronym of eXtensible Markup Language.

XOR — Acronym of eXclusive OR.

INDEX

quantum processing, 158–159
quantum states, 158
qubits, 25–26, 158

R

RAB (RAID Advisory Board), 503
radio frequencies, 304, 316
radix, 69
radix point, 69, 83–84
RAID (Redundant Array of Inexpensive Disks), 503–506
RAID 0, 503
RAID 1, 503
RAID 1+0, 505
RAID 10, 505
raised floors, 585
RAM (random access memory), 34, 170–172, 176–178
random access storage devices, 172
range of electromagnetic frequencies, 316
raw data transfer rate, 316
read access, 494
Read right, 483
read threshold, 186
read-ahead memory, 178
read/write circuitry, 178
read/write heads, 184, 193
read/write mechanism, 168
ready state, 442–443
real data type, 145
real numbers, 82–87
real resources, 437–438
Real-time priority, 450
real-time scheduling, 449–450
receivers, 324, 336
recordable DVDs, 203
recording density, 187
records, 96, 103–104, 479
recovering files, 501–502
recovery utility, 501
red, 263
redundant bit transmission, 344
redundant transmission, 339–340
refresh cycles, 177, 273
refresh rate, 273
registers, 33, 114–115, 135–137, 146
relative addressing, 183
relative path, 484
reliability, 437, 535
remote resources, accessing, 520–524
removable devices, 172
removable magnetic media, 188, 199
removable MO (magneto-optical) disks, 206
removable storage devices, 30
removable storage media, 172–173
repeaters, 323
Repeat-Until control structures, 405
Request signal, 226
requirements
    estimation, 566
    hardware and software, 562–563
requirements discipline, 4–5
resident attribute, 496–497
resistance, 149, 155
resolution, 261
resource allocation, 434–437, 444
resource availability model, 567
resource locator, 522–523

resource registry, 523
resources, 12
    accessing remote, 520–524
    allocating, 56
    defining characteristics of, 6
    detailed records of available, 437
    external networks, 55–56
    locations of, 523
    managing, 8–9, 434
    measuring demand and utilization, 568–569
    names of, 523
    network access, 519–524
    periodical literature, 13–14
    professional societies, 17
    real, 437–438
    restricting access to, 572–574
    scheduling, 437
    shifting among programs, 438
    static connections, 520–521
    substituting, 438
    technology-oriented Web sites, 14–16
    unconcerned with availability of, 438
    vendor and manufacturer Web sites, 16–17
    virtual, 437–438
    Web, 529–530
return instruction, 407
return wire, 325
REV disks, 199
RF (radio frequency), 314, 324, 343
RFPs (request for proposals), 563–564
RGB (red, green, blue), 263
ring networks, 366, 368
ring topology, 360
RIP (Routing Information Protocol), 373
RISC (Reduced Instruction Set Computing), 130–133
RISC processors, 130–133
RJ-45 jacks, 322
RMI (Remote Method Invocation), 543
RO (read-only) segment, 466
Rock, Arthur, 153
Rock's Law, 153
roles of personnel, 9–12
ROM (read only memory), 176, 179
root directory, 483
rotational delay, 195–196
routers, 369–370
routing, 363–365, 372, 453
routing table, 363
RPCs (remote procedure calls), 527–528, 534
running state, 443
run-time errors, 416
RW (read-write) segment, 466

S

sampling, 289–290
sampling rates, 293
Samsung, 205
SAN (storage area network), 507–509
sandbox, 414
SAS (serial-attached SCSI), 330
SATA (serial ATA) standard, 330
satellite transmissions, 314
scalability, 523
scaling out, 243–244
scaling up, 243–244
scanning lasers, 286

scheduling, 445–450
sharing CPU, 441
sharing resources allocated to parent process, 441
states, 442–443
STR (shortest time remaining) scheduling, 448–449
terminating execution, 443
timer interrupts, 447–448
updating state, 445–446
waiting for access to CPU, 442–443
waiting longest, 448
three-layer architecture, 517–518
three-tier architecture, 517
threshold, 312
throughput, acceptable levels of, 437
tickets, 528
Time Critical priority class, 450
time intervals, 593
time stamps, 483
timer interrupts, 447–448
timing fluctuations, 336
Tn3270, 531
token, 366
token passing MAC protocol, 366
tools for application development, 391–392
Torvalds, Linus, 453
Toshiba, 205
touch screens, 284
TPC (Transaction Processing Performance Council), 567
traces, 149, 154–155, 157
trackballs, 283
tracks, 192, 194
track-to-track seek time, 195–196
transaction logging, 500–501
transaction processing, 449
transaction processing programs, 498
transactions, 500
transistors, 151, 154–155, 158, 177
translating instructions into CPU instructions, 49
transmission lines, 327
transmission media
    atmosphere, 314
    attenuation, 319
    bandwidth, 316–319
    capacity, 315–316
    characteristics, 315
    coaxial cable, 322
    copper wire, 314
    distortion, 319
    electrical and optical cabling, 321–323
    fiber optic cable, 322–323
    frequency, 316–319
    length of media, 315
    media bandwidth, 317–318
    multiple media segments interconnection, 315
    optical fiber, 314
    rate which bits are encoded, 315
    signal-to-noise ratio, 319–320
    space, 314
    speed, 315–316
    twin-axial cable, 322
    twisted pair wire, 321–322
    wireless data transmission, 324
transmitters, 344
transport layer, 372
tree directory structure, 483
trinary signals, 310
Tru64 UNIX, 452

"true" data, 465
Tuxedo, 519
twin-axial cable, 322
twisted pair wire, 321–322
two-dimensional bar codes, 287
two's complement notation, 80, 84
    8-bit, 125
    arithmetic SHIFT instructions, 122
    less processing complexity, 86
    memory addresses, 93
    negative value, 124
    numeric range, 81–82
    subtraction, 124
TXT (text) file, 480
Type I errors, 340
Type II errors, 340
typewriter, 278

U

UID (user identification), 573
unblocked file read and write operations, 489
unconditional BRANCH instruction, 123, 136, 405
undelete operation, 494
underflow, 85
Unicode, 87, 91–92, 279
Unicode Consortium Web site, 91
Universal Serial Bus, 271
UNIX, 48–49, 452–454, 527
    access controls, 494
    Bourne Shell, 531
    links, 485
unresolved reference, 409
unsigned binary memory addresses, 93
unsigned integer data type, 145
unsigned integers, 78
UP (Unified Process), 562
    business modeling discipline, 4–5
    deployment discipline, 7–8, 391
    design discipline, 5–7, 390–391
    disciplines, 3
    implementation activities, 391
    implementation discipline, 7, 418
    iterations, 3
    maintenance, 8
    requirements discipline, 4–5, 390
    software development project, 390
    systems evaluation, 8
    testing discipline, 7
Update-First-Name method, 105
upgrading
    network capacity, 345–346, 382
    storage, 345–346
UPS (uninterruptible power supplies), 583
URL (Uniform Resource Locator), 529–530
U.S. DOD (Department of Defense), 373
U.S. Federal Communications Commission, 380
USB (Universal Serial Bus), 199, 287
USB drive, 199
USB flash memory, 172
USB port, 199
user interfaces
    accessing local and remote resources, 521
    to operating system, 433
    structure of, 5
user-defined directories, 497
user-oriented management programs, 46